A LIFE OF INVERTEBRATES

W. D. RUSSELL-HUNTER

Syracuse University and
The Marine Biological Laboratory, Woods Hole

A Life of

INVERTEBRATES

Macmillan Publishing Co., Inc.

NEW YORK

Collier Macmillan Publishers

LONDON

Macmillan Publishing Co., Inc.
866 Third Avenue, New York, New York 10022

Collier Macmillan Canada, Ltd.

Library of Congress Cataloging in Publication Data

Russell-Hunter, W. D., (date)
   A life of invertebrates.

   Bibliography: p.
   Includes index.
   1. Invertebrates.   I. Title.   [DNLM: 1. Inverte-
brates,   QL362 R968L]
QL362.R82   592   77-27065
ISBN 0-02-404620-5

Printing: 1 2 3 4 5 6 7 8   Year: 9 0 1 2 3 4 5

PERCONTARI AUDE

For **P. D. R-H.** and **M. R-H.**,
my different drummers for aye.

PREFACE

Nothing in nature stands alone . . . .   John Hunter, 1786

. . . dans cette dépendance mutuelle des fonctions et dans ce secours qu'elles se prêtent réciproquement . . . il est évident que l'harmonie convenable entre les organes qui agissent les uns sur les autres, est une condition necessaire de l'existence de l'être auquel ils appartiennent . . . .   Georges J-L. N-F. Cuvier, 1800

THIS book is intended for the beginning student of the invertebrates and for the numerous specialist biologists—from biochemists to ecologists and from palaeontologists to molecular geneticists—whose professional subdisciplines allow them but little time for the study of these myriads of "animals without backbones." There are probably nearly two million living species of animals, approximately half of which have been formally described—and some 96% of *these* are "invertebrates."

The book does not assume more than background knowledge of some aspects of basic biology. Its title is deliberate: it is *A Life of Invertebrates,* providing only one of the many possible surveys of this diversity of animals. Much has had to be omitted or given an overly narrow synoptic treatment, while a few topics are perhaps overemphasized. The title also echoes in homage J. Z. Young's fine text on the vertebrates which first appeared about a quarter of a century ago. At that time, I had recently returned to teaching in a university department of zoology. Older colleagues, and even my contemporaries, were at first dubious of Young's great book because, in dealing with the *life* of vertebrates, it seemed to us—trained, to a man, in the classical schools of vertebrate morphology which had lasted from the beginning of the nineteenth century into the first half of the twentieth—to omit much more than was decent of the extensive and detailed comparative anatomy of the vertebrates that all previous texts and teachers had told us was important. We simply noted the omission of extensive evolutionary studies of the anatomy of each organ system and did not immediately recognize that by combining only the highlights of anatomy with a partial survey of comparative physiology and of ecology in vertebrates, this new and seemingly eclectic book moved us toward a better appreciation of the *evolution* of vertebrates. Treating the *life* of these vertebrate animals as whole organisms made it somewhat easier to conceive of their evolution as a series of changes in their total organization to fit certain environments. The present book attempts a similar compass for the invertebrates.

Although this book is obviously derived from the two paperbacks which have preceded it by nearly a decade, it is intended to be a book for today. The last three decades have seen great changes in the science of biology and in the emphases of biological teaching. From the pedagogical point of view, the necessarily increased emphases on biochemistry and biophysics at one end of the biological spectrum, and on ecology and ethology at the other, have had some paradoxical results. One of these is that many well-trained biologists of today know less about the life of animals as such, than did preclinical students of medicine a half-century ago. There is so little time available during the training of most professional biologists for the comparative anatomy which constituted the major part of the training of any zoologist or botanist of two or three decades ago that it can only be presented in the most limited and arid of digests. However, comparative anatomy must be part of any more integrated, or at least more rounded, contemporary introduction to the biology of either vertebrates or invertebrates. To this comparative anatomy must be added the comparative study of function and, to these combined as functional morphology, such further modern insights from neurophysiology and behavior and from physiological ecology and population biology, as are mandated for any current survey of the "unities in diversity" shown by major animal groups.

In such syntheses, the biology of vertebrates—as a relatively compact system of closely interrelated groups—will always seem "easier" to students. The richer diversity of the invertebrates makes any attempt at a contemporary *biological* account of them appear more "difficult," as it did any account of their comparative anatomy in the 1920's. Modern biology, at the level of the organism, has made some of its greatest advances in such fields as whole-animal integration (including control physiology and behavior) and in the huge field of population biology (including population dynamics, population genetics, and actuarial bioenergetics). Thus, at the very time when classical morphological understanding of particular animal groups is being supplemented and complemented not only by physiological studies but also by behavioral and ecological ones, the time available to acquire any basic understanding of whole-animal biology is steadily decreasing.

This loss of time for zoöcentric training in zoology in turn has bred a further paradox in formal biological education. It is that many students who are first attracted to biology by a confessed interest in microbes, or in plants, or in animals often find the truncated courses they receive on the structural organization of vertebrates, invertebrates, plants, or microorganisms the least satisfying parts of their formal training in biology. Thus, it behoves us, who wish to communicate our interest in living animals, not to present some dry digest of the admittedly encyclopaedic knowledge that *does* exist of the diverse anatomy of vertebrate or invertebrate animals but to attempt an account of their diverse "life-styles."

Obviously, this book will not suffice for anyone preparing for a professional career in invertebrate biology. In fact, most invertebrate biologists today are specialists—crustacean population ecologists or molluscan neurophysiologists, for example—rather than being concerned with the whole field of invertebrates. (It will be obvious that this author has more direct research experience with the bivalves than with the corals, and—even in a smaller compass—of the basommatophoran gastropods than of the higher cephalopods.) Where choices of content have had to be made, it has been decided in most cases to sacrifice comprehensiveness for a readier comprehensibility. The book is intended for use as an introductory or intermediate text rather than as a reference volume. Systomatics as such has therefore been played down, particularly in some large and relatively uniform classes of animals. This is perhaps most obvious in the treatments of the protozoans and the insects, where the synoptic—and, in some cases, eclectic—surveys certainly do not provide the kind of systematic coverages necessary to satisfy protozoologists or entomologists as adequate introductions to their subdisciplines. But, this book is not intended to provide those.

The fact that we have to deal with roughly one million living species of invertebrates is itself another reason against attempting an inclusive listing with notes on anatomical diversity. Even a summary catalogue or directory listing of invertebrate systematics and anatomy could readily fill a book of this size, and the attempt must be made to incorporate some more contemporary biology. The way out of this *impasse* adopted here results from two natural characteristics of invertebrate diversity and two arbitrary decisions personal to the author. The first significant characteristic is that underlying the fantastic structural and functional diversity of the invertebrates are certain *functional homologies* of considerable extent. In other words, if comparative physiology is studied at the level of the integration of functions in the whole animal, certain unifying features become apparent in the ways in which each of the 32 animal phyla carries out and regulates its major biological processes (see Chapter 2). A second feature of the diversity is also important. Without becoming too involved in the difficult question of what constitutes "evolutionary success," we can be justified in regarding certain phyla as of major importance and others (the majority) as of minor importance. As discussed in Chapter 2, in addition to such criteria as numbers of species in each phylum, there is a quantifiable ecological sense in which designation of "minor phyla" can be vindicated. This book emphasizes the functional homologies of major phyla at the expense of the rest.

The other limitations of the treatment are more arbitrary, but still defensible. As has been discussed many times, biological organization can be studied at several grades of complexity (such as ecosystem, community, population, individual organism, tissue, cell, or molecule) and by asking questions at about four conceptual levels. These latter range from the structural-descriptive (encompassing anatomy, histology, and even some part of molecular biology on the one hand and ecological zonation on the other) through the mechanistic-physiological ("how does it work?") and the adaptive-functional ("what is the biological value of this structural organization or group of processes to the individual animal and/or to a population of that species?") to the evolutionary-historical ("what is the history of this organ or process in time—how has it evolved?"). The treatment of the biology of the invertebrates here is intermediate in regard to both grade of organization and level of concept: it is concerned with whole animals considered mainly at the mechanistic-physiological and adaptive-functional levels of explanation. Such "zoological" biology (involving the study of the whole animal as an integrated functional system) may be considerably older than the presently more fashionable "molecular" biology, but *both* must be, equally, parts of the biology of the future. Even on the pragmatic level of what it "profits" man to understand, we shall need zoöcentric biologists. In rather different terms, the book involves my attempt to steer between pure comparative anatomy with classification on the one hand and physiology at the biochemical level on the other. Obviously, it would be possible to devote several books of this size to a detailed account of comparative anatomy in one major phylum—say, the Mollusca. Alternatively, a single book could discuss the occurrence through the invertebrate groups of a few chosen enzyme systems or blood pigments (incidentally, the data are not yet available for many enzyme systems). Neither kind of book is attempted here. What *is* discussed is largely functional morphology within each of the various groups, and this discussion uses the adaptive level of consideration of form and function in an attempt to explain the way in which invertebrates of diverse groups fit and function in their particular habitats. There is, perhaps, some overemphasis on the mechanics of locomotion and of nutrition.

This functional morphology is arranged and connected by use of those parts of systematics which are felt to be phyletically significant, and by discussions of the evolution of certain characteristic functions. Other links are provided by emphasizing the existence of a few successful plans of functional organization (i.e., the existence of functional homologies within each major phylum). The student should realize that both strands of continuity—the phylogenetic classification and the recognition of homologous features—are matters involving hypotheses. On the other hand, the principal content of this book—the functional morphology—is matter involving only simple observations and comparative experiments which can be verified, or expanded, or corrected by any student, *without* elaborate apparatus but *with* access to living specimens of the invertebrate animals concerned. The use of "representative types" is deliberately avoided. Our understanding of the biology of invertebrates has suffered in the past through a pedagogical devotion to *Hydra* as "the representative coelenterate," "the earthworm" as "type annelid," *Helix* as "type mollusc," and "the crayfish" as "type crustacean." Apart from certain basic biological absurdities in this usage, the actual animal forms most employed as "types" were often (as in the above four examples) rather unrepresentative of form and function for the phylum concerned. In instructional writing—as in teaching—the author prefers all the dangers involved in setting up a "hypothetical primitive form" as a model and then discussing the facts of diversity in a group in relation to that model to using as a starting point "the type"—the detailed study of a specialized form. The general disadvantages resulting from the use of models should be well-known to every scientist, and the peculiar dangers of "archetypes" (with their values in comprehensibility) are discussed in Chapters 2, 26, 35, and elsewhere.

Thus, one obvious difference between this book and most other textbooks of invertebrate biology results from the deliberate omission of complete series of descriptions of external appearance and gross anatomy of "type" or representative animals for every group. All good courses, and even independent students, can have opportunities provided to observe such animal anatomy personally in laboratory or field. This book treats invertebrate biology largely phylum by phylum. If the significance of functional homology within each phylum (see Chapter 2) is acceptable to the reader, then it will be realized that this treatment is followed not because it happens to have been the classical pattern for texts on invertebrate anatomy but because the author believes it to be conceptually in line with current thinking about comparative physiology and thus appropriate to a contemporary survey entitled *A Life of Invertebrates*.

I am grateful to all the individuals and organizations who helped with illustrations, and due acknowledgments are made in the appropriate captions. As with three previous books, two associates deserve my special thanks in this regard. Dr. Douglas P. Wilson, with whom my scientific exchanges extend back over thirty years, has contributed more of his superb photographs of marine animals. Mr. H. Peter Loewer not only has continued to create a valuable and integrated series of line diagrams out of my colored cartoons, but also has produced for the Who's Who of Chapter 1 some more representational—and more elegant—invertebrates.

As regards the text, it is a pleasure to thank once again those who read the original paperbacks in manuscript: Drs. John H. Welsh, Martyn L. Apley, Albert J. Burky, Robert J. Avolizi, and Jack S. Mattice. After publication of the paperbacks, helpful suggestions were contributed by Drs. Robert D. Beeman, Albert R. Mead, James G. Morin, Richard P. Nickerson, Robert F. McMahon, R. Douglas Hunter, Arnold E. Eversole, Christopher H. Price, Robert A. Browne, and George Hechtel, while basic reediting was done by Sandra E. Belanger, and I am grateful to them all. Translations of the earlier paperbacks into Spanish and into Malaysian have been most encouraging, but I am particularly grateful to my three translators into Portuguese (Drs. D. D. Corrêa, C. G. Froehlich, and E. Schlenz of the University of São Paulo) not only for kind comments in their preface to the second volume translated by them, but also for remaining completely faithful to the original text in regard to certain phyletic concepts with which they do not agree. I am also most grateful to those who read the entire final manuscript of this book and helped improve it, including my research associates Barry S. Payne and Jay Shiro Tashiro, and another valued collaborator from an earlier book, Mrs. Elisabeth H. Belfer of Macmillan. Particular thanks are due to my cherished colleague, Dr. John M. Anderson of Cornell, whose constructive criticisms of the final manuscript were as valuable as his expert help with the original paperbacks had been a decade earlier. Finally, the book could not have come into existence without the support and active help of my wife, Myra Russell-Hunter, and of my son, Perry Russell-Hunter, at all stages of its preparation.

None of those I have thanked share my responsibility for the errors that remain, for a culpable level of eclecticism in some topics, or for the uneven emphasis on certain invertebrate groups.   W. D. R-H.

CONTENTS

1 INTRODUCTION I: WHO'S WHO AMONG INVERTEBRATES

FOR the purposes of this book, the term "the invertebrates" encompasses all groups of living animals which do not have backbones. Although over one million species of animals have been described, the great bulk of them have body designs based on only nine general structural plans. These nine distinctive styles of animal architecture have each certain essential groupings of structural units put together on a common ground plan. This common architecture is associated with considerable functional unity. Within each major structural design, all member animals carry out and regulate their major living processes in a similar fashion. In other words, each major kind of animal machine operates in one particular pattern, exhibiting not only certain structural units in common but also a communality of physiological functions. To some extent, both the possibilities of and the limits to each of these basic patterns of organs and functions determine the possible ecology of that animal group. Thus the distribution of different kinds of animals in space and time has been broadly determined by the advantages and disadvantages of each particular pattern of anatomy and function.

In the second part of the Introduction (Chapter 2), we will consider both theoretical and pragmatic aspects of animal classification. In the meantime, each major kind of body design is ranked as a phylum of animals. A total of 32 phyla have to be at least briefly considered in this book. Of these, nine are much more important in terms not only of species numbers but also of their ecological importance and their evolutionary significance. Estimates of the numbers of living species in each phylum are set out in Table 1-1 along with common names for the group where they exist. The following "biographical" paragraphs are not intended to be complete diagnostic statements or encyclopaedic surveys; rather, they represent an unscientific "cast list" to introduce the invertebrate "players" to those who have not viewed this part of the evolutionary "theater" before.

TABLE 1-1
Approximate Numbers of Living Species of Many-Celled Animals[a]

PHYLUM	SPECIES	
Cnidaria	11,000	jellyfish, sea-anemones, corals, hydroids, etc.
Ctenophora	80	comb-jellies, sea-gooseberries
Porifera	4,200	sponges
Mesozoa	50	—
Platyhelminthes	15,000	flatworms: planarians, flukes, tapeworms
Rhynchocoela (= Nemertea)	600	ribbon-worms
Gnathostomulida	90	—
Entoprocta	60	moss-animals[b]
Rotifera	1,500	wheel-animalcules
Gastrotricha	150	—
Echinorhyncha	100	—
Nematomorpha	250	horsehair-worms
Acanthocephala	300	spineheaded-worms
Nemathelminthes	80,000[c]	roundworms (free-living and parasitic)
Annelida	8,800	segmented worms
Arthropoda	>800,000	insects, crustaceans, spiders
Onychophora	80	walking-worms
Tardigrada	170	water-bears
Linguatulida	60	tongue-worms
Echiuroidea	80	—
Mollusca	110,000	snails, clams, octopus, etc.
Bryozoa (= Ectoprocta)	4,000	moss-animals[b]
Priapulida	5	—
Phoronida	15	—
Brachiopoda	310[d]	lampshells
Sipunculoidea	275	—
Chaetognatha	60	arrow-worms
Echinodermata	6,000	starfish, sea-urchins, sea-cucumbers, etc.
Hemichordata	100	acorn-worms
Pogonophora	100	—
Chordata	45,000	vertebrates, etc.
includes "invertebrate Chordata"	2,100	lancelets, sea-squirts

Notes:
[a] Of the 32 phyla considered in this book, only one (the phylum or subkingdom Protozoa, numbering possibly 50,000 living species) does not appear in this table. The other 31 animal phyla are all metazoan or "many-celled."
[b] The common name moss-animals is applied to two distinct phyla.
[c] This is a compromise figure between widely differing estimates; see text.
[d] Plus at least 12,000 described fossil species.

Nine Important "Kinds" and Relations

Not only are the first nine phyla to be considered the more important ones, but they are also more likely to be familiar to the nonzoologist. Since all readers of this book have backbones, and therefore all belong to the major vertebrate subdivision of the phylum Chordata, it may be best to proceed from the more familiar to the less familiar and start this survey of the nine important phyla with the chordates. Thus we will proceed from the structurally more complicated toward apparently more simple forms of animal organization. In terms of Table 1-1, we will be proceeding with the *larger* phyla from bottom to top of the table, and then returning from top to bottom with the more minor groups.

Phylum CHORDATA

Within the phylum Chordata, vertebrates outnumber invertebrates by about twenty to one. We are vertebrates; *Homo sapiens* is a species of mammal, that is, a live-bearing vertebrate with hair and milk-glands. The vertebrate mammals, birds, and reptiles are the more highly developed types of chordates that dominate terrestrial and aerial environments. In addition, vertebrates include the successful amphibians and the more extensive groups of fishes which are the dominant members of many animal communities in the sea and in fresh waters. The vertebrates number approximately 43,000 species, but the phylum Chordata also includes a little over 2000 species which do not have backbones. These invertebrate chordates are our particular concern here; they fall into two groups: one rather rare and the other moderately successful.

The rare ones are lancelets of the genus *Amphioxus* (Figure 1-1), and they are marine animals found in shallow water over cleaner sand substrata in somewhat restricted localities, but in all the world's oceans. They have been called "headless fish," and they are about 4 centimeters long, laterally flattened, spindle-shaped, and nearly translucent when they are alive. Although they can swim quite efficiently by lateral flexures of the whole body, as do most fish, they spend most of their life half-buried with the anterior end protruding above the surface of the sand. They feed by filtering from a water current passed through their numerous gill clefts.

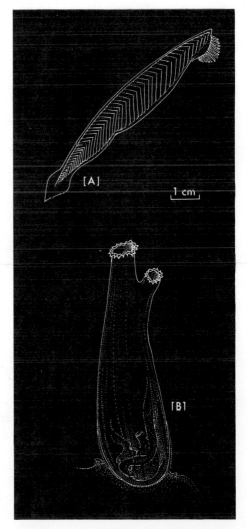

FIGURE 1-1. Invertebrate representatives of the phylum Chordata. **A:** The lancelet or "headless fish", *Amphioxus*. **B:** A typical solitary sea-squirt of the genus *Ciona*.

Although technically an invertebrate, *Amphioxus* is almost universally studied in courses on vertebrate morphology because it displays in a simplified form the general structural plan exhibited with considerable uniformity, particularly in the developmental and young stages, by all vertebrates: fishes, amphibians, reptiles, birds, and mammals. It is probable that in the very earliest chordates, only a larval stage was fish-like (or, more correctly, like *Amphioxus*), and the life-history almost certainly involved a sessile adult stage. The more successful subgroups of invertebrate chordates show this adult form today. They are the sea-squirts or tunicates or ascidians, soft bag-like marine organisms which live attached to rocks or dock pilings or ships' bottoms and extract a living from the sea by continuously passing a stream of water through a perforated pharynx which acts as an internal filter (Figure 1-1). Their common name comes from the fact that the two openings to the bag-like body squirt a stream of water when the tunicate is squeezed or, in many cases, even when it is touched. Sea-squirts either are individual animals like small plastic or leathery bags attached to the rocks, or else they form colonies with great numbers of smaller bag-units imbedded in a common gelatinous mass. The great majority of both solitary and colonial ascidian species are completely sessile as adults, cemented down and lacking all means of locomotion. However, a few forms, including the salps, live in the plankton (the drifting life of the sea). The larval stages of most of these sea-squirts have a considerable resemblance to *Amphioxus*, and thus a more distant structural resemblance to fishes and to the embryonic stages of other vertebrates. If the vertebrates had not evolved, today's sea-squirts would be the most successful animals built on the chordate ground plan, since there are over two thousand species and since on suitable rocky seashores they can make up a considerable part of the animal biomass. In an evolutionary sense, the vertebrates are aberrant members of the phylum Chordata whose ancestors eliminated the bottom-dwelling sessile adult and turned their larval stages into "adults." Thus, although it may seem difficult from our vertebracentric position as humans, the sea-squirts are not degenerate members of our phylum, but rather more conservative chordates which have kept more strictly to the life-style of our common ancestors. However, the vertebrates undoubtedly show more complicated mechanisms of self-maintenance, and thus can occupy a greater variety of habitats, than any other group of animal organisms. If one defines the higher forms of life as those whose independence of the environment involves more physicochemical improbabilities, then it is on the homeostatic vertebrates that the claims of the phylum Chordata to be regarded as a dominant group must be based.

Phylum ECHINODERMATA

The starfish and their allies, which form the phylum Echinodermata, are one of the most readily recognized groups of animals. They are exclusively marine, though they are widely distributed through the seas, and their six thousand or so species form a successful group by any measure. The ground plan of their anatomy is radial, usually pentaradiate (i.e., based on five or ten axes radiating from a central mouth). They are all very slow-moving or sessile animals and the calcareous ossicles embedded in the skin give the group its name. Associated with this peculiar skin-skeleton is an equally peculiar hydraulic system of in-

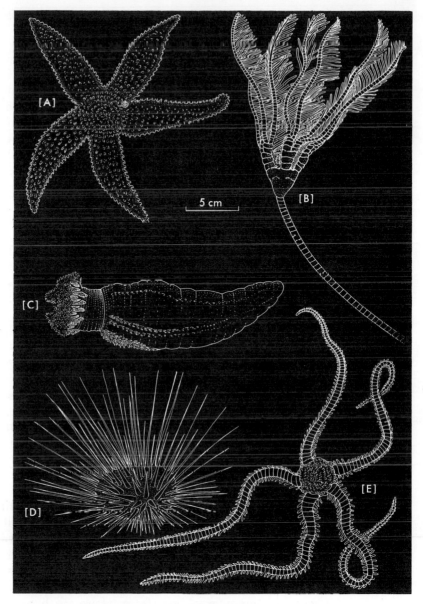

FIGURE 1-2. Representatives of the major phylum Echinodermata. A: A typical starfish or seastar of the genus *Asterias*. B: A crinoid sea-lily. C: A soft-bodied sea-cucumber. D: A long-spined sea-urchin of the tropical genus, *Diadema*. E: A typical ophiuroid or brittle-star.

ternal water-tubes, linked to very large numbers of tiny "tube-feet" which can be used both for slow locomotion and for the capture of prey. The tube-feet, along with the skeleton of dermal ossicles supporting them and the internal system of canals forming a water-vascular system, are unique to the echinoderms, as is the basically pentaradiate symmetry.

The presently living echinoderms (Figure 1-2) fall into five distinct groups. Perhaps the most familiar are the true starfish, with usually five arms which are not clearly marked off from the central disc and contain lobes of the alimentary canal, gonads, and other internal structures. Starfish have tough but flexible skins since their ossicles are separately embedded in the dermis. Not unlike them are the brittle-stars where the arms are always sharply demarcated from the disc and do not contain lobes of the internal organs. The dermal ossicles of brittle-stars form an almost continuous articulated armor, particularly along the arms.

Somewhat less familiar animals of the sea bottom are the sausage-shaped sea-cucumbers or trepangs, in which the central body is elongated and the arms greatly reduced. However, sea-cucumbers retain the pentaradiate symmetry of the group, along with the water-vascular system and tube-feet, although their ossicles are relatively small and scattered in a leathery skin. In the fourth group, the sea-urchins, the body is spherical, heart-shaped, or disc-shaped, and the ossicles form a continuous, fused armor or test. On the outside of this firm shell-like test are attached long, movable spines which in many cases assist the tube-feet in locomotion.

The last group of living echinoderms, comprising both the sea-lilies and the feather-stars, are rarer forms which are stalked, or at least sessile, for the whole or part of their lives. They resemble the brittle-stars in having their ossicles almost continuous but articulated, and they have a greatly reduced central disc with the arms showing a great development of branching. Although present-day forms are rare, there is a relatively enormous fossil record of sea-lilies. In general, there are huge numbers of echinoderm fossils (possibly 40,000 species have already been described), and several groups of them were established as successful marine animals over long periods of geological time.

Phylum MOLLUSCA

One of the most successful patterns of animal construction is the molluscan plan. There are more than twice as many species of molluscs as there are of vertebrates, and only the arthropods are clearly a more numerous and more successful group. There is usually a soft body within a hard calcareous shell, and most molluscs are readily recognizable as such. Another characteristic is the extensive use of ciliary and mucous mechanisms in feeding, locomotion, and reproduction. A characteristic form of ciliated gill, the ctenidium, is found throughout most of the phylum Mollusca; and a semiinternal respiratory chamber, the mantle-cavity, is an even more truly universal feature of their anatomy. Among the few animals, other than vertebrates and insects, which would almost certainly be recognized by a majority of nonscientists would be *Octopus* and several species of snails and clams. The basic anatomy of these different molluscs conforms to a remarkably uniform pattern, but they show extreme diversity in their external body form. There are three major groups of living molluscs (Figure 1-3), and a number of smaller ones.

The snails or gastropods form a large, diverse group with the molluscan calcareous shell usually in one piece. It may be coiled as in typical snails, conical as in limpets, or secondarily reduced or absent as in the "slugs." Most gastropods are marine, but many are found in fresh waters, and gastropods are the only molluscs living as air-breathing, terrestrial animals. In fact, next to the land arthropods (insects, spiders, and their allies) and the higher vertebrates, pulmonate land snails are, somewhat surprisingly, the most successful land animals. The marine gastropods are also the most diverse: living between tide-marks on the seashore, on the bottom in all depths of the ocean, and even (with reduced shells) drifting near the sea surface.

The second great division of the molluscs is the bi-valves, with the shell in the form of two calcareous valves united by an elastic hinge ligament. They constitute a considerably more uniform group, and mussels, clams, and oysters are familiar bivalves even to those whose only invertebrate interests are gastronomic. The shell valves can be closed by pow-erful adductor muscles, and bivalves are mostly rather sedentary creatures, not moving from place to place but creating water currents to bring them their food. The group is mainly marine with a number of species in estuaries and in fresh waters. There can be no land bivalves since their basic functional organi-zation is as filter-feeding machines.

The third major group of the molluscs, the cepha-lopods, includes the most active and most special-ized molluscs—the octopods, cuttlefish, and squids —along with a few forms with chambered, coiled shells. They not only are the most highly organized molluscs but can be claimed as the most highly orga-nized invertebrate animals. Although they are all marine, some of them show as complicated patterns of innate behavior as do some birds, and several have a capacity to learn or be trained which is exceeded only by certain birds and mammals. The majority can swim by jet-propulsion at speeds in excess of those attained by any other invertebrate and compa-rable to those of the faster fishes. Certain giant squids are the largest living nonvertebrate animals. Al-though the cephalopods are clearly constructed on the basic molluscan pattern of structure and func-tion, they include large, fast-moving, "brainy" ani-mals. Although the 350 living species of cephalo-pods represent less than 0.5% of living molluscs, they are of disproportionate importance in the ecol-ogy of the sea and, as predaceous carnivores, are dominant organisms in many marine food-chains. Further, there are at least seven thousand known fos-sil species of cephalopods, some of which were among the commonest marine animals of their times.

In addition to the three main groups of living mol-luscs—snails, bivalves, and cephalopods—there are four more minor kinds. The minor molluscan types exhibit further diversity of external body form, but they all conform in their basic anatomical layout to the standard molluscan pattern.

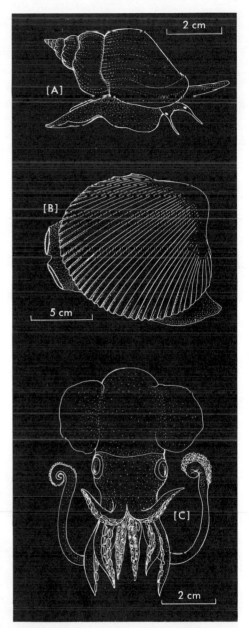

FIGURE 1-3. Representatives of the major phylum Mollusca. **A:** A marine "whelk" or carnivorous snail of the genus *Buccinum.* **B:** A typical filter-feeding bi-valve, a "cockle" of the superfamily Cardia-cea. **C:** A small cuttlefish of the genus *Ros-sia,* representing the zenith of invertebrate evolution, the class Cephalopoda (squids, cuttlefish and octopuses).

Phylum ARTHROPODA

This enormous phylum of the most successful invertebrate stocks comprises the insects, crustaceans, spiders, and allied groups, all with jointed limbs and segmented bodies covered with exoskeleton (Figure 1-4). The jointed chitinous procuticle, which serves as both a protective armor and an exoskeleton for muscle attachments, is characteristic. All arthropods are superbly efficient locomotory machines, constructed as interacting systems of levers and based on limbs which are tubes of hardened procuticle linked by flexible joints. These jointed limbs, which give the phylum its name, are attached in pairs to many, or to some, of the body segments.

By any measure, this pattern of animal machine is enormously successful, with species numbers (see Table 1-1) in excess of those for all the other phyla added together. Ecologically, the arthropods make up a disproportionately large share of the animal biomass alive in the world today. The forms included in the phylum Arthropoda in the strict sense fall into ten groups, living in the seas and in fresh waters, and forming the commonest of land animals. Of these ten classes, three are enormous: the insects, the crustaceans, and the spiders and their allies.

Insects are primarily land animals, with their whole concert of structures and functions adapted for life in air. As such, they are the most numerous land animals and the most numerous class of living animals, numbering 700,000 species. They are remarkably stereotyped in structure and function, with the body segments arranged in three groups of head, thorax, and abdomen. The three segments of the thorax can bear two pairs of wings and bear the only three pairs of adult walking legs (providing the alternative name Hexapoda for the insects). The largest subdivisions of the insects are the beetles and the Diptera or

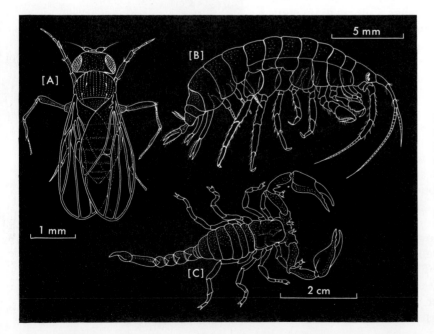

FIGURE 1-4. Representatives of the largest phylum, the Arthropoda. **A:** Representing the insects, the largest class in the animal kingdom, a female fruitfly of the genus *Drosophila*. **B:** Representing the crustaceans, a marine amphipod of the genus *Gammarus*. **C:** A typical scorpion, representing the Arachnida, the third largest class of arthropods which also includes spiders, ticks and and mites.

true flies. It is probable that some species of house-fly (possibly of the genus *Musca*) are the most numerous land animals, in terms of individual numbers.

The success of insects obviously depends upon the jointed exoskeleton and limbs. Such physiological processes as aerial respiration, control of water losses, and methods of terrestrial locomotion in insects show efficiencies (in energetic terms) unsurpassed by those of any other organisms.

The crustaceans form the second great class of arthropods. In contrast to the insects, they are primarily a marine group although there are many representatives in fresh waters and a few on land. Although basically all gill-breathing, they show much greater diversity of anatomy and physiology than do the other arthropod groups. Body form, especially limb pattern, and life-style vary greatly within the Crustacea. At one extreme would be the more familiar "higher" crustaceans, including the many kinds of crabs, lobsters, shrimps, prawns, and crayfish. In contrast to these relatively large and complex forms are the tiny copepods, almost certainly the most numerous animals in the world, which live in the ocean plankton and feed as herbivores directly on the world's largest crop of green plants—the marine phytoplankton. Between these extremes are many other kinds of crustaceans, for example, the barnacles, specialized as adults for an entirely sessile mode of life, cemented to rocks and to the bottoms of ships. One major difference between the simpler (and probably more primitive) forms of crustaceans and the more complex ones, such as crabs, is in the degree of specialization of the paired, jointed limbs. In primitive crustaceans, and almost certainly in the most primitive arthropods, the paired limbs are very much alike from one end of the body to the other; while in the higher forms there can be as many as seven types of body limbs, each one specialized for particular functions.

The third enormous division of the arthropods comprises the arachnids, consisting of spiders, ticks, and mites, along with some rarer allied forms such as scorpions and harvestmen. Like the insects, the arachnids are basically a terrestrial group. The most abundant forms, spiders and mites, have rather globular bodies of largely fused segments and mostly have four pairs of walking legs (never wings) and another two pairs of appendages associated with the mouth. Apart from the scorpions, the arachnids are nearly as stereotyped in their body form as the insects are in theirs, and show nothing of the diversity of limb form and function which we encounter in the crustaceans.

Apart from these three enormous groups of arthropods, there are seven other subdivisions in the phylum. Of these, four are tiny groups of minor ecological importance, and one relatively large group, the trilobites, consists entirely of fossils. Although extinct, trilobites have considerable significance, being a likely stem group for many, if not all, of the other arthropod classes. The two remaining groups are of considerable extent and number several thousand species each: the centipedes and the millipedes. Centipedes, with fewer and longer limbs, are fast-moving animals which live as predaceous carnivores on other small arthropods. On the other hand, millipedes, with many more but shorter limbs, are slower-moving, omnivorous animals which shove their way through soil and leaf-litter, in and on which they feed.

Despite all this diversity, there is clearly a unique structural and functional basic plan common to all arthropods: such features as the chemical and mechanical properties of the exoskeleton and the hormonal control of molting are common to insects, to crustaceans, to arachnids, and to the myriapodous groups. The great bulk of available evidence suggests that the arthropods evolved from the soft-bodied worms of the other great group of metamerically segmented animals, the annelids.

Phylum ANNELIDA

Several kinds of animals are called "worms." The most highly organized ones are those of the phylum Annelida, the metamerically segmented worms of the sea, of fresh waters, and of terrestrial soils (Figure 1-5). Their metamerism involves a longitudinal division of the body wall and of the body cavity into a series of segments, each containing a nearly identical set of organs. Mechanically, the existence of a fluid-filled body cavity is of paramount importance to annelids, since it provides for the hydrostatic or hydraulic transmission of forces produced by the longitudinal and circular muscles of each segment of the body wall, and hence for the efficient propulsion of the worm through its environment. Locomotion in these soft-bodied worms involves either waves of alternate elongations and bulgings or lateral undulatory movements of the entire body.

   Apart from two very minor groups, the phylum has three subdivisions: polychaetes, oligochaetes, and leeches. The most numerous and the most diverse group is formed by the polychaetes or bristle-worms, an almost entirely marine group. Their segments bear pairs of flap-like parapodia with numerous bristles or setae which are used in swimming, walking, and burrowing. There are more

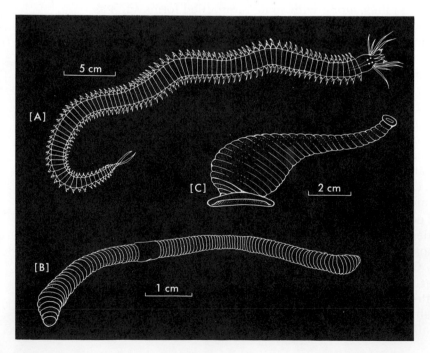

FIGURE 1-5. Representatives of the major phylum Annelida. **A:** An unspecialized marine polychaete, the ragworm, *Nereis*. **B:** A typical earthworm of the genus *Lumbricus*. **C:** The leech is typical of the specialized class of segmented worms, the Hirudinea.

than five thousand species—some very active, some sedentary or sessile. They include predaceous carnivores, carrion-feeders, herbivores, detritus-feeders, and even some highly modified filter-feeders.

There are more than three thousand species of oligochaetes, mainly earthworms and a few smaller freshwater allies. Both the land and freshwater worms in this group have fewer and smaller setae on the segments, never have flaplike parapodia, and usually have somewhat poorly developed heads compared to those of polychaetes.

The third group of annelid worms, the leeches, numbers only a few hundred species. They have a much more stereotyped pattern of body with attachment suckers and a modified mouth and gut. There are leeches in the sea, in fresh waters, and on land (the terrestrial ones being principally found in the tropics). Many leeches are predaceous carnivores rather than blood-suckers and prey upon other invertebrates, such as insect larvae, slugs, and earthworms.

Despite the major differences in body wall and locomotion between the arthropods and the annelid worms, these two major phyla are undoubtedly linked by their metamerism. Not only their segments, possessing unit subdivisions of the several organ systems, but also their embryonic and larval development, including the topography of the "budding zone" for new segments, are features common to all arthropods and to all segmented worms.

Body Cavities and Basic Plans

Before passing from these five relatively well-known phyla to less familiar and less highly organized patterns of animal machines, we should introduce some concepts and definitions regarding basic animal architecture. The fundamental body plan which is termed triploblastic coelomate is common to chordates, echinoderms, molluscs, arthropods, and annelids. The term triploblastic implies that developmentally there are three major cell layers: ectoderm on the outside, endoderm as a lining to the gut and a few other internal organs, and between them a distinctly derived epithelial mass of cells termed the mesoderm. The term coelomate implies that they have a body cavity or coelom enclosed by this mesoderm. As shown in Figure 1-6, the internal organs are slung on mesodermal mesenteries, and there is a mesodermal covering to the gut-tube endoderm as well as a mesoderm lining within the ectoderm of the outer body wall. If not in their adult anatomy, at least at some stage in their embryonic development, all members of the chordate, echinoderm, mollusc, arthropod, and annelid phyla would show in cross section the basic plan of tissue layers and cavities which we term triploblastic coelomate (Figure 1-6).

The only other type of body cavity is that termed a pseudocoel, which, as shown in Figure 1-6, occupies a space between an inner mesoderm layer of the body wall and the endoderm of the gut. In this plan there are no mesenteries suspending the internal organs and, since developmentally there is no mesoderm surrounding the endoderm of the gut, in the adult anatomy there are no muscle layers around the endoderm of the alimentary canal. Triploblastic pseudocoelomates include one major successful phylum, the nematode worms (roundworms), and six more minor phyla.

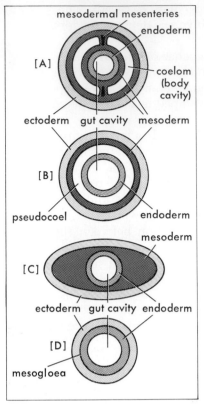

FIGURE 1-6.   Basic plans for many-celled animal phyla in representative cross sections. (These are discussed more extensively in Chapter 8.) **A:** The triploblastic coelomate condition as it occurs in the "higher" major phyla, Chordata, Mollusca, Arthropoda, and Annelida. **B:** The triploblastic pseudocoelomate condition as it is found in the major phylum Nemathelminthes and several minor phyla. **C:** The triploblastic acoelomate condition, occurring in the phyla Rhynchocoela, Platyhelminthes and certain other minor groups. **D:** The simplest plan, or diploblastic acoelomate condition, as it occurs in the major phylum Cnidaria and the minor phylum Ctenophora.

Other less highly organized phyla of animals are acoelomate, with no internal cavities in the body other than the lumen of the gut. The flatworms of the phylum Platyhelminthes, and certain other minor phyla, can be termed triploblastic acoelomate (Figure 1-6). A final simplification of basic body plan is found in the phylum Cnidaria (jellyfish, sea-anemones, corals, and their allies) where there are only two basic cell layers in development, the ectoderm and the endoderm. Although there are certain semantic difficulties which will be discussed later, these forms can be termed diploblastic acoelomates (Figure 1-6).

Another feature which is common to the chordate, molluscan, arthropodan, and annelid phyla (though not to adults of the phylum Echinodermata) is bilateral symmetry. Animals which have a definite front end and rear end, with different upper and under surfaces, and therefore left and right sides, are said to possess true bilateral symmetry. Conceptually, in such animals there is a plane surface in the midline (i.e., running anterioposteriorly and dorsoventrally) on either side of which are found "mirror images" of the tissues and organs. Apart from the four "higher" phyla named, there are many other groups of animals showing bilateral symmetry. Even in the phylum Echinodermata, the majority of the larval forms, though very few of the adults, show bilateral symmetry.

In brief diagnoses, we can now describe our own phylum, the Chordata, as "triploblastic coelomate bilateria"; starfish and their allies in the Echinodermata as "triploblastic coelomate bilateria, secondarily radiata as adults"; and the jellyfish and corals of the phylum Cnidaria as "diploblastic acoelomate radiata." It is worth noting that we are not implying degrees of closeness of relationship by these terms; all "radiata" are not necessarily closely allied.

Phylum NEMATHELMINTHES

The one phylum constructed on the pseudocoelomate plan that is at all successful has been enormously successful. The nematode "roundworms" are completely unsegmented and have a simple tubular gut which runs through the pseudocool from the mouth at one end to the anus at the other (Figure 1-7). They have also a thick, continuous cuticle (which is molted in growth) and never have any ciliated cells. Although perhaps best known as parasites of both animals and plants, they are also found free-living in all environments: in the sea, in fresh waters, and on land. They probably number about 80,000 species which would place them third (behind the arthropods and the molluscs, but well ahead of the chordates) in such lists as Table 1-1, however, any estimate of species numbers for this group must be arbitrary. Individual numbers are also very high in the free-living forms, which are mostly microscopic or a few millimeters long. All nematodes—both free-living forms and parasites—are structurally very alike, so much so that there can be no higher levels of systematics in the phylum. All the vast arrays of nematode worm species are placed in one class, the Nematoda, which is thus the only class in the phylum Nemathelminthes. All the terrestrial soils and the majority of marine and freshwater bottom deposits contain hundreds of species, and tens of millions of individuals, of adult nematodes, active or in a resting state, and of their eggs. All available evidence supports the literal truth of the statement that one or more species of nematodes are found as parasites in every species of higher plants and animals. Certainly, most vertebrate animals are hosts to about ten species of roundworms each. About fifty species of nematodes are known to occur in man, and probably more than 70% of the readers of this book are presently hosts to a nematode infection. Paradoxically, this emphasizes one aspect of the success of nematodes as parasites: that they are mostly nonlethal to their hosts and often unobtrusive and ignored. Such parasites can be claimed to show much greater efficiency in their chosen way of life than those which cause serious diseases.

Many of the features possessed by all nematodes, including the free-living ones from soils and bottom muds, seem to "preadapt" them for the parasitic mode of life. In reproduction, they produce numer-

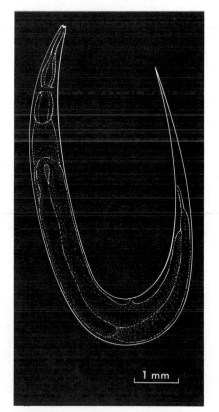

1 mm

FIGURE 1-7 A representative roundworm of the major phylum Nemathelminthes.

ous eggs, and both eggs and the resting stages of adults can remain inactive in a diapause for long periods of time. Secondly, the muscles of the body wall can do little more than flex the body from side to side, and consequently, while most nematodes move readily through soils or tissue cells, directional locomotion over surfaces or through the open sea is impractical. Thirdly, as already noted, the pseudocoelomate condition implies that there are *no* muscles at all round the gut, so that ingestion and the peristaltic movement of solid food through the gut-tube are also impracticable. Despite this, there are nematodes everywhere, from the polar regions to the tropics, and from the tops of mountains to the greatest depths of the oceans, as well as *within* all other higher animals of suitable size and the majority of macroscopic plants. In the days of non-disposable felt mats to be placed under mugs or *Bierseidel* in *Bräuhausen* and *Biergartens*, there were even special species of nematodes that lived in them, nourished only by beer slop.

Phylum PLATYHELMINTHES

The flatworms of the phylum Platyhelminthes are a relatively successful group (about 15,000 species) of rather simply organized animals. They are triploblastic but acoelomate, lacking any body cavity, and have the gut in the form of a blind sac with a single opening to the exterior or, in the case of some parasites, no gut at all (Figure 1-8). Although they are bilaterally symmetrical, they are much flattened dorsoventrally and are therefore physiologically rather "thin" organisms. Their total lack of any circulatory system or additional body cavities is clearly related to this functional aspect. All processes of excretion and respiration can be carried out over the general body surface, since no internal tissue is more than a few cells away from the outside of the flatworm. Their nervous systems and muscular organization are also rather simple, but their reproductive structures are hermaphrodite and form a complex system of organs.

There are three main subdivisions of the phylum, one free-living and two parasitic. The free-living flatworms or turbellarians are mostly small (never more than a few centimeters) and are found in many marine and freshwater habitats, and even exceptionally on land in certain highly humid habitats. Most of them have a mouth placed centrally on the underside of the body, and somewhat surprisingly, they are mostly carnivores or carrion-feeders. Living turbellarians are so thin that they conform very closely to any plants or rocks over which they are moving, and they seem to glide like a film or flexible animated leaf slowly over the surfaces. In addition to sexual reproduction, the majority of flatworms can also reproduce asexually by binary or multiple fission; and, as is perhaps fitting in such simply organized creatures, most have great powers of regeneration if they are injured naturally or mutilated experimentally. Turbellarians have long been popular laboratory animals for studies both of differentiation in regeneration and of the levels of behavior which can be based on such neural and sensory simplicity.

The second major group consists of the parasitic flatworms called flukes, whose general body form is not unlike that of turbellarians. Most have ex-

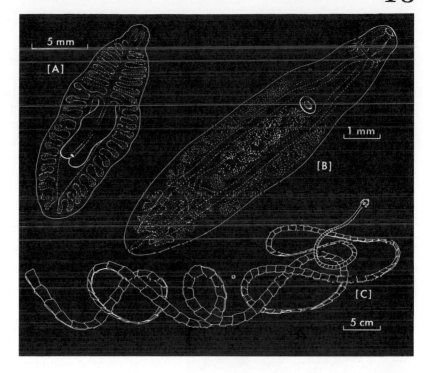

5 mm

[A]

1 mm

[B]

[C]

5 cm

FIGURE 1-8. Representatives of the phylum Platyhelminthes. **A:** A typical free-living freshwater flatworm or triclad turbellarian. **B:** A flatworm parasitic in man, the Chinese liverfluke. **C:** A typical tapeworm or cestode.

tremely complicated life-histories, the adults living in the gut, in the bloodstream, or in the actual tissues of their final hosts. Many typical life-histories involve stages in three kinds of animals as hosts; some of the best-studied forms live as adults in the liver and bile-ducts of man and of certain farm animals. Typically, their larval stages live in intermediate hosts, first in aquatic snails and then in fishes. One of the most widespread and deadliest of human tropical diseases, bilharziasis, is caused by three species of flukes which live in larger vessels of the bloodstream and cause either rectal or bladder haemorrhaging when they lay large batches of spined eggs.

The third group of flatworms are also endoparasites, the tapeworms. A budding series of hermaphrodite adult units makes up the "tape" which lives in the gut of suitable hosts, usually vertebrates. The larval stages, or "bladderworms," generally live in the tissues of an animal which is fed upon by the primary host. Man is the primary host to several kinds of tapeworms whose larval or transmission stages occur in the flesh of pigs, cattle, and fishes. Thus physicians refer to human tapeworms as pork tapeworms, beef tapeworms, or fish tapeworms.

Both flukes and tapeworms almost certainly have been evolved from the free-living flatworms or turbellarians. Further, although there are many theories regarding the origins of such higher phyla as the Annelida, Mollusca, and Chordata, the free-living flatworms must at least resemble a central basic stock which gave rise to all of these more complex patterns of animal organization. In turn, it is relatively easy to see resemblances between the simplest flatworms and certain larval forms among the cnidarians.

Phylum CNIDARIA

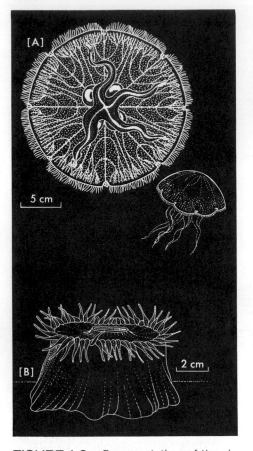

[A]

5 cm

[B]

2 cm

FIGURE 1-9. Representatives of the phylum Cnidaria. **A:** A free-swimming jellyfish of the genus *Aurelia*. **B:** An attached sea-anemone of the genus *Calliactis*.

Somewhat more familiar than flatworms, animals such as jellyfish, sea-anemones, corals, and hydroids make up the phylum Cnidaria and are often called the coelenterates (Figure 1-9). They include the most simply organized of many-celled animals and can be referred to as diploblastic and acoelomate. Although they certainly lack a true mesoderm in the sense of a major epithelial mass in development stages, some biologists would prefer to avoid the term diploblastic. However, it is always correct to refer to the coelenterates as having a "tissue grade" of organization, since they do not have their tissues arranged into "organs," such as are found in the flatworms and all "higher" metazoans. The general body form is of two cell layers forming the wall of a basic tubular body, the cavity of which is the blind gut. There are no specialized tissues for respiration or excretion; even contractility is a shared function (there are no specialized muscle cells); and the scattered nerve cells form essentially nondirectional networks. All of them possess stinging capsules or nematocysts which are used both in defense and as a means of obtaining food. It is the possession of these stinging capsules that allows jellyfish, corals, and the other coelenterates to live as predaceous carnivores, despite their lowly "tissue grade" of anatomical organization. Nematocysts are formed and used only by cnidarians.

The phylum Cnidaria is not an unsuccessful phylum, numbering about 11,000 species. Its members are abundant in a wide variety of habitats in the sea, but there are only a few freshwater forms. Of course, animals like jellyfish and sea-anemones, with thin, exposed layers of living tissues, could never be land animals. Ecologically, cnidarians are only of major importance in certain regions of tropical seas, where coral animals build calcareous reefs and form a significant fraction of the biomass. What has long made them an important group to zoologists is the near certainty that cnidarians are at least closely related to the stock from which all other metazoan groups have been derived. Obviously, it will always remain in the realm of untestable hypotheses both that the coelenterate-like forms were the most primitive of many-celled animals and that the first more complex forms to evolve from them were like certain turbellarians alive today, but the circumstantial evidence is very strong for both of these phyletic conceptions.

The coelenterates are clearly metazoans, or many-celled animals. Since it is somewhat ambiguous to refer to the Protozoa as single-celled, we have to embark upon a somewhat more complicated definition of the animal protistans. It is that the protozoan body never has any specialized parts of the cytoplasm under the sole control of a nucleus. (There can, in fact, be a few nuclei, or many nuclei, rather than one, but no single nucleus ever has separate control of any part of the protozoan specialized for a particular function.) In contrast, in the metazoans, there are always many nuclei, each in charge of cells specialized for particular purposes, as muscles, as gland-cells, or as nerves. If this, at first sight rather elaborate, distinction is used then the division between the Protozoa and the Metazoa is clearly definable. As we will discuss later, the distinction between the Protozoa and the lower plants is far less clearcut than the even more significant distinction between eucaryotes and procaryotes. This again serves to emphasize the artificiality of the "Invertebrata" as a primary division of living organisms.

Phylum (or Subkingdom) PROTOZOA

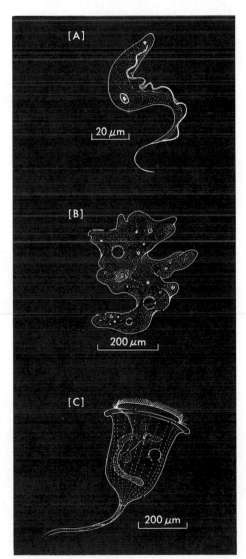

Almost all protozoans are minute, ranging downward from about 1 millimeter in length to about one hundredth of that. They are essentially noncellular, with all living functions being carried out within a single cell membrane. Thus all the more complex functions of locomotion, of food capture, and of mating, which in the metazoans involve special tissues or organs, are here combined with the normal intracellular functions, involving the endoplasmic reticulum, the mitochondria, and one or more nuclei. However, certain cytoplasmic parts can be differentiated for particular functions and are termed "organelles." In terms of numbers of species, the Protozoa are clearly a major group; about 25,000 have been identified. They show a variety of forms of asexual as well as sexual reproduction, and there are often periods of encystment (with reduced metabolism) as parts of their life-cycles. Locomotion is by pseudopodia, or flagella, or cilia. In this survey, we need consider only the four main body forms and life-styles in the phylum Protozoa (Figure 1-10).

The first kind comprises the flagellates, most of which have a single nucleus and one or a few flagella. The group includes both plant-like forms with chlorophyll and others more animal-like in their nu-

FIGURE 1-10. Representative single-celled animals (phylum Protozoa). **A:** A flagellate of the parasitic genus *Trypanosoma*. **B:** A free-living sarcodine, *Amoeba*. **C:** An attached, highly complex ciliate, *Vorticella*.

trition. A second group closely allied to the first includes the sarcodines like *Amoeba* which (at least as adult forms) move by one or more pseudopodia. They usually have a single nucleus, and a large number of them secrete either calcareous or siliceous shell capsules or skeletal spicules. There are some parasitic forms of both flagellates and sarcodines, but such are not much modified from their free-living allies.

In the third major group, the sporozoans, the adult stages usually lack locomotory organelles and have complex nuclear processes associated with complicated life-cycles and reproduction. They are internal parasites of other animals, with intracellular infection in the majority of cases. Several species of the genus *Plasmodium*, the causative organisms of various malarias, are typical sporozoans.

The last group comprises the most complex and most highly organized protozoans, the ciliates. They typically have a stouter pellicle augmenting the cytoplasmic membrane, and in it, with elaborate infrastructure, are embedded the locomotory organs, the "regimented" rows of cilia. Most of them have a cell-mouth or cytostome and two types of nuclei. Of these, the micronuclei contain the reserves of DNA used only in reproductive activities, while one or more meganuclei contain the genetic material responsible for running the "somatic" life of the animal. Ciliates are very varied in external shape, and include free-swimming forms, surface-crawling forms, and parasitic forms occurring in the guts and similar cavities of higher animals.

The nine phyla introduced above comprise all but 1% of living animal species. We have moved, anthropocentrically, from the relatively familiar animals related to man to the markedly less familiar and minute protozoans. The remaining animal species fall into 23 phyla, and we shall survey these even more briefly and with reference to the differences from, and resemblances to, the nine important "kinds," proceeding in this case (see Table 1-1) from the more simply organized back toward higher levels of organization.

Twenty-three Minor "Kinds"

Of these minor phyla, the first five to be considered lack body cavities and can be termed acoelomate like coelenterates and flatworms. The next six minor groups are constructed on a ground plan similar to that of the nematodes and can be termed pseudocoelomates. The remaining twelve minor phyla are all coelomate in general body pattern, having a true coelom lying between mesodermally derived tissues covering the gut wall and lining the body wall. Of these twelve, the first four share a number of common features with the enormous phylum Arthropoda, while the last two have some rather more distant concordances with our own phylum Chordata.

Phylum (or Subkingdom) PORIFERA

To some extent the sponges of the phylum Porifera are intermediate between the protozoans and more integrated many-celled organisms such as cnidarians

(Figure 1-11). They are a medium-sized group of more than four thousand species and are almost entirely marine but with a few freshwater species. Sponges are without organs and, for most purposes, can be regarded as loose aggregations of cells (some flagellated) without definite form or symmetry. They live by creating a water current and feeding on filtered particulate food. Many of their physiological processes are essentially intracellular like those of protozoans.

Phylum CTENOPHORA

Among the most beautiful of marine animals, ctenophores are pelagic and structurally and functionally close to the cnidarians (Figure 1-11). Their most characteristic features are ctenes or comb-plates of fused cilia. They lack nematocysts but are predaceous carnivores living on other animals of the plankton. There are only about eighty species of ctenophores.

Phylum MESOZOA

There are only about fifty known species of mesozoans and all are minute parasites in higher invertebrates. They are similar to larvae of certain cnidarians or else to much simplified flatworms (Figure 1-12). They have a solid two-layered construction with the inner cells as reproductive cells. Their life-cycles have some of the reproductive complexities of parasitic protozoans.

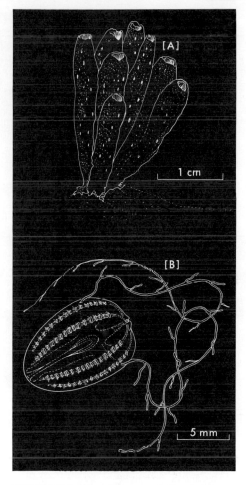

FIGURE 1-11. Representatives of two minor phyla that exemplify evolutionary byways. **A:** A relatively simple "colony" of a calcareous sponge (phylum Porifera). **B:** A sea-gooseberry or comb-jelly of the Phylum Ctenophora.

Phylum RHYNCHOCOELA (= NEMERTEA)

The minor phylum of ribbon-worms, sometimes called nemertines, is obviously closely allied to the true flatworms of the phylum Platyhelminthes. They are relatively complex and massive triploblastic acoelomates. Although some are extremely long (the marine "bootlace-worms" and ribbon-worms can reach 2 meters) they show no trace of segmentation (Figure 1-12). There are about six hundred species with four different basic patterns of muscle layers. Apart from their size, one consistent difference from the flatworms lies in their possession of an eversible proboscis which, when retracted, is enclosed in a special cavity lying just above the anterior part of the gut. Secondly, the gut is a

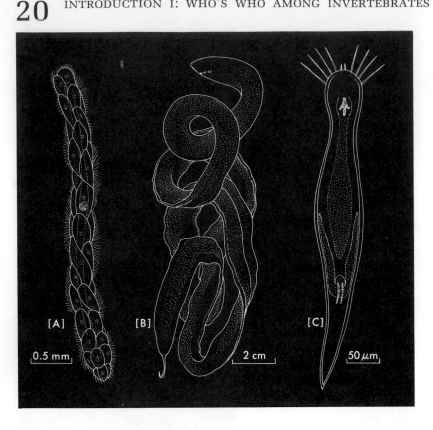

FIGURE 1-12. Representatives of three minor phyla of primitive acoelomate "worms." **A:** An infusiform stage of a dicyemid parasite of cephalopod kidneys (phylum Mesozoa). **B:** A typical marine ribbon-worm or nemertean of the phylum Rhynchocoela. **C:** A representative of the minute interstitial worms of the most recently designated phylum Gnathostomulida.

through-tube ending in an anus; and thirdly, they possess a primitive sort of blood vessel. A few species have invaded fresh waters and certain tropical land habitats, although the majority of nemertines are marine.

Phylum GNATHOSTOMULIDA

There may be more than ninety species of gnathostomulid worms already described. All are minute forms living interstitially in marine deposits. They are triploblastic acoelomates like tiny flatworms but possess toothed oral plates and jaws (Figure 1-12). This most recently designated animal phylum was set up in 1956.

Phylum ENTOPROCTA

There are about sixty marine species of entoprocts which are tiny ciliary filter-feeders (Figure 1-13), living a sedentary life and looking superficially like colonial hydroid cnidarians which have developed some mesodermal structures. They are pseudocoelomate and have some histological resemblances to the nematode worms. They were once associated with the phylum Ectoprocta as "moss-animals," but that phylum consists of coelomates of totally distinct relationships.

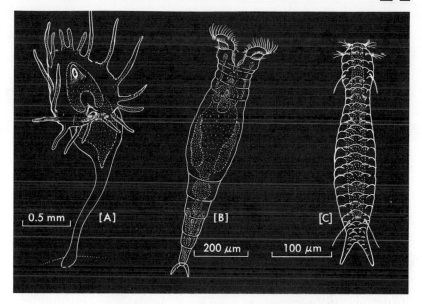

FIGURE 1-13. Representatives of three more minor phyla, all pseudocoelomate. **A:** A solitary entoproct of the genus *Loxosoma* (phylum Entoprocta). **B:** A representative of the microscopic pseudocoelomate rotifers occasionally numerous in fresh waters (phylum Rotifera). **C:** Another minute pseudocoelomate form, a gastrotrich of the genus *Lepidoderma* (phylum Gastrotricha).

Phylum ROTIFERA

The "wheel-animalcules" are rotifers and were favorite objects of early microscopists (Figure 1-13). There are about 1500 species living mainly in fresh water, with a few marine species. Although metazoan and pseudocoelomate, they are mostly microscopic, living with and competing with the ciliates. Most of them have a corona or wheel organ of ciliated cells which is used both in locomotion and in feeding. In many genera there is marked sexual dimorphism, and females capable of parthenogenesis produce true genetic "clones" to stock fresh waters seasonally. Both adult rotifers and their eggs resemble nematodes in their capacities for diapause and encystment. The resulting ability to resist desiccation and freezing is responsible for their remarkable passive dispersal: to mountain tops, to oceanic islands, and to polar icecaps.

Phylum GASTROTRICHA

There are about 150 species of gastrotrichs living in fresh waters and in the sea. They are all minute and not unlike small flatworms, except for bearing thin cuticle with regular rows of scales or spines (Figure 1-13). Structurally, they show resemblances both to nematodes and to rotifers, and they are clearly pseudocoelomate with no trace of metameric segmentation.

Phylum ECHINORHYNCHA

There are about one hundred marine species of kinorhynchs or echinoderes. They are microscopic and have a regularly segmented covering of cuticle (Figure 1-14). Growth in later stages involves molting as in arthropods and nematodes, but kinorhynchs have no internally segmented structures, and they are clearly pseudocoelomate.

Phylum NEMATOMORPHA

There are about 250 species of horse-hair-worms or nematomorphs. Most are exceptionally long, thin worm-like animals which can be up to 1 meter in length (Figure 1-14). In most species, the adults are free-living, though the larval stages are parasites in arthropods. Almost all the adults live in fresh waters, although one genus is marine. They have a thick fibrous cuticle, nematode-like muscles, and a pseudocoel partly filled with mesenchyme cells.

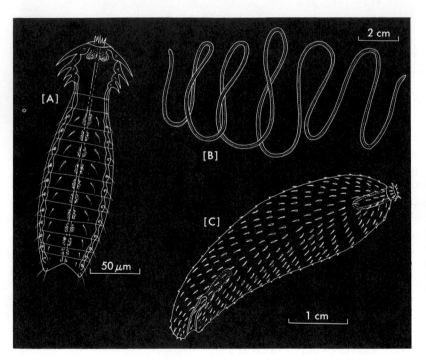

FIGURE 1-14. Representatives of three further pseudocoelomate minor phyla. **A:** A typical kinorhynch, another minute interstitial form from marine silts (phylum Echinorhyncha). **B:** An elongate "hairworm" of the phylum Nematomorpha. **C:** A typical endoparasitic worm of the minor phylum Acanthocephala.

Phylum ACANTHOCEPHALA

The three hundred species of acanthocephalons are endoparasites in various vertebrates from fish to mammals (Figure 1-14). They all have complex lifecycles with one or two intermediate hosts which are usually arthropods. They are mostly 1 or 2 centimeters in length, though a few reach 50 centimeters and some are only 2 millimeters as adults. The worm-like gutless adult body has a proboscis with recurved hooks which are attached to the gut wall of the host. They are pseudocoelomate with some resemblances to nematodes and rotifers.

This phylum is the sixth and last minor pseudocoelomate phylum, and the following minor phyla are all triploblastic coelomates. Of the twelve, the next four to be considered are allied to the annelid-arthropod stem.

Phylum ONYCHOPHORA

The walking-worms or onychophorans number about eighty species of terrestrial animals living in the tropics and south temperate regions. They are mostly 4–10 centimeters long, and they have numerous paired, stumpy legs (Figure 1-15). The soft body is anatomically annelid-like, and the limbs and respiratory organs (tracheae) are like those of primitive arthropods. For this reason, the onychophorans are phyletically one of the most interesting of the minor phyla.

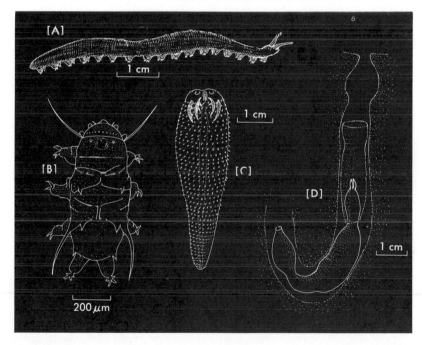

FIGURE 1-15. Representatives of four minor metameric phyla. **A:** *Peripatus* (phylum Onychophora), sometimes regarded as an intermediate form between annelid worms and arthropods. **B:** A young specimen of the tiny freshwater tardigrades (phylum Tardigrada). **C:** A typical young parasitic pentastomid-worm of the phylum Linguatulida. **D:** A filter-feeding echiuroid worm, genus *Urechis,* of the phylum Echiuroidea, in its burrow.

Phylum TARDIGRADA

Tardigrades are microscopic segmented animals with a nonchitinous exoskeleton (Figure 1-15). There are about 170 species, mostly living in fresh waters on mosses, lichens, and liverworts, but a few species are marine. Their worldwide distribution results in part from their capacity for diapause. They have a molt-cycle and most of their internal organs show cell-number constancy. In their anatomy and embryology, they show features associated with a wide variety of phyla from arthropods to chordates.

Phylum LINGUATULIDA

The pentastomid-worms (tongue-worms) or linguatulids are a minor phylum of about sixty species of parasites (Figure 1-15). They have a thick, chitinous cuticle which is molted in a regular cycle, but metameric segmentation is somewhat blurred. Typically, the adult linguatulids live in carnivorous vertebrates, and their larval stages are found in herbivores fed upon by the host of the adult.

Phylum ECHIUROIDEA

There are about eighty species of echiurids, which make up another group of marine unsegmented worms. Except for a prominent, nonretractable, and grooved proboscis, most of their anatomy is like that of annelid worms but without metameric segmentation (Figure 1-15). Their embryonic and larval development is closely similar to that in annelids, and they even have a few setae. They are sedentary animals, living in semipermanent burrows and feeding by filtration or by mucous collection of detritus.

Phylum BRYOZOA

The ectoprocts or true bryozoans are the coelomate moss-animals, with a superficial resemblance to the entoprocts (Figure 1-16). Although the phylum is of minor ecological significance, there are probably about four thousand living species, and three to four times that number are known as fossils. Most are

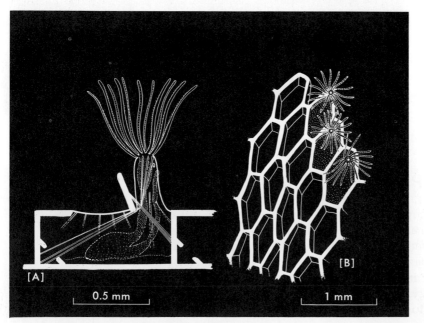

0.5 mm

1 mm

[A]

[B]

FIGURE 1-16. Representative individual with protruded lophophore (**A**) and colony skeleton or "houses" (**B**) of a coelomate moss-animal, or ectoproct, of the phylum Bryozoa.

marine, but there are some freshwater and estuarine species. Individual bryozoans are microscopic, but they form extensive and polymorphic colonies, in
which each individual feeds by protrusion of a ciliated lophophore for the collection of particulate food. They have a basically tripartite coelom which is
used hydrostatically in the protrusion of the lophophore.

Phylum PRIAPULIDA

There are only five described species of priapulids, another group of unsegmented marine worms (Figure 1-17). Their outer cuticle is like that found in
rotifers; there is a molt-cycle, and they have a large eversible proboscis-like presoma covered with spines. They have an open body cavity used as a hydrostatic
skeleton, and evidence suggests that this is a true coelom.

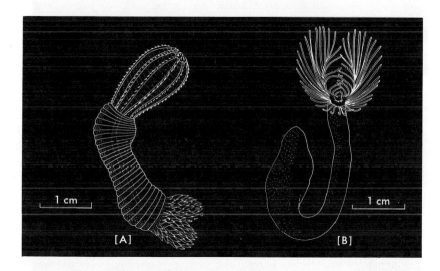

FIGURE 1-17. Repre-
sentatives of two minor
"worm" phyla. **A:** A
marine priapulid worm
(phylum Priapulida). **B:** A
lophophore-bearing worm
of the phylum Phoronida.

Phylum PHORONIDA

There are only about fifteen species of phoronids, marine tubiculous worms
with a lophophore (Figure 1-17). They use this organ for ciliary filter-feeding,
and it closely resembles similar structures both in ectoprocts and in the lampshells. Although the body is worm-like, the gut is U-shaped with an anus near
the mouth. There is a bipartite coelom.

Phylum BRACHIOPODA

Once a dominant phylum in the marine fauna with at least 12,000 described
fossil species, the lampshells, or brachiopods are now reduced to about three
hundred living species. Although completely unrelated to the bivalve mol-

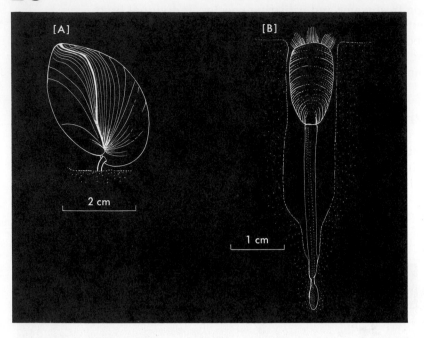

2 cm

1 cm

FIGURE 1-18. Representatives of the phylum Brachiopoda, a minor living remnant of a dominant faunal group in the fossil record. **A:** A typical lampshell with its dorsal and ventral valves. **B:** An ecardinate brachiopod of the genus *Lingula* in its burrow.

luscs, they have two unequal shell valves (actually dorsal and ventral) hinged together and closed by adductor muscles (Figure 1-18). They have a complex lophophore which is used in the collection of suspended particles from the water current which it circulates. In one major group of lampshells, the lophophore has an elaborate internal skeleton. In the details of their anatomy and embryology, the brachiopods show a variety of features not assembled together in any other single phylum. Once a major phylum, they are still a most distinctive one with no clear relationship to any other.

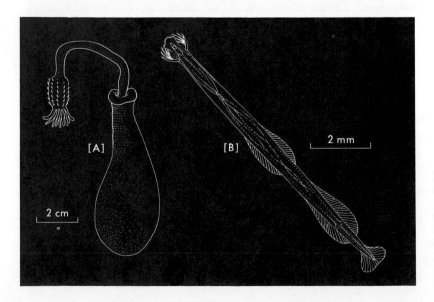

2 mm

2 cm

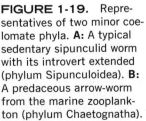

FIGURE 1-19. Representatives of two minor coelomate phyla. **A:** A typical sedentary sipunculid worm with its introvert extended (phylum Sipunculoidea). **B:** A predaceous arrow-worm from the marine zooplankton (phylum Chaetognatha).

Phylum SIPUNCULOIDEA

There are nearly three hundred species of sipunculids, another group of unsegmented worm-like marine animals (Figure 1-19). The anterior part of the body
can be retracted into the trunk or protruded for feeding. They burrow actively
in muddy sandflats, and seem to be detritus-feeders. The gut is U-shaped but
they have sense-organs and some nervous elements like annelids, and a very
well-developed coelom which is used hydrostatically.

Phylum CHAETOGNATHA

Living as predaceous carnivores in the marine
plankton are the sixty species of arrow-worms or
chaetognaths. They are coelomate and unsegmented and their transparent, torpedo-shaped
bodies and heads with grasping spines all show
perfect bilateral symmetry (Figure 1-19). Their
muscles are unusually undifferentiated, resembling those of nematodes, while some aspects of
their development have parallels in the chordates.
Some arrow-worms are important biological indicators used to track movements of bodies of oceanic
water.

Phylum POGONOPHORA

There may be more than the one hundred species of
pogonophorans thus far described. All are marine
worm-like tube-dwellers from soft sediments in the
deeper parts of the oceans (Figure 1-20). Because of
this habitat and their relatively recent description,
there could be more species and their distribution
could be worldwide. The long thin body (like a
"living thread") is divided into three regions, but
lacks any mouth, alimentary canal, or digestive
organs. They have a tentacular organ, presumably
for food collection, which in some species is elaborated into a complex structure like a lophophore. It
is assumed that nutrients are taken up through the
general body wall. They show some annelid and
some chordate characteristics.

Phylum HEMICHORDATA

The acorn-worms or hemichordates number about
one hundred species of marine worm-like animals
(Figure 1-20). They have pharyngeal gill slits like
those of chordates and a short section of hollow

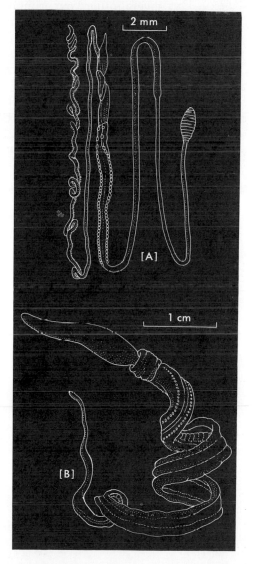

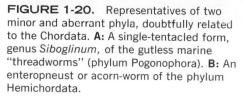

FIGURE 1-20. Representatives of two
minor and aberrant phyla, doubtfully related
to the Chordata. **A:** A single-tentacled form,
genus *Siboglinum,* of the gutless marine
"threadworms" (phylum Pogonophora). **B:** An
enteropneust or acorn-worm of the phylum
Hemichordata.

nerve cord, but they do not have a notochord at any stage of the life-cycle, and are therefore not placed in the phylum Chordata. Apart from two genera of minute, sessile colonial forms, hemichordates live in semipermanent burrows on sandy shores between tidemarks. They show both worm-like feeding (engulfing the surface sand and passing it through the gut) and a ciliary filter-feeding mechanism similar to that of the other protochordates. Apart from such living forms as acorn-worms, some fossil evidence suggests that there were several other "experimental stocks" of protochordates in the Palaeozoic, from which survive only the minor hemichordate phylum and the major chordate phylum.

In this survey of the "cast" of invertebrates, we moved through the 9 important patterns of animal body design and then considered 23 more minor groups. These 32 kinds of basic body pattern are ranked as phyla. As discussed in the following chapter, this level of classification—the phylum—is important both in pragmatic terms, by making possible some comprehension of invertebrate diversity, and in evolutionary terms as involving sufficient evidence of structural and functional homologies to allow deduction of closer common ancestry.

2 INTRODUCTION II: FUNCTIONAL HOMOLOGIES AND PHYLA

THERE are probably nearly two million living species of animals, approximately half of which have been formally described. Of these, approximately 8% are protozoan and approximately 3.5% are vertebrate, leaving slightly less than 90% of described animal species as "many-celled invertebrates." A synoptic view of the 31 phyla of many-celled animals is provided in Table 1-1 (Chapter 1, page 2). It should be noted that the numbers of species listed are approximate and in some cases compromise figures between widely differing estimates. Further, estimates of the total number of extinct species of animals are mostly around seven times the number presently alive, and the proportion of these which could be termed "many-celled invertebrates" is probably about the same. In other words, there may have been just over 13 million invertebrate species in the course of our planet's history.

In this chapter, first, the taxonomic categories used in systematics to classify this animal diversity will be summarized. Secondly, some of the evolutionary implications associated with structural and functional homologies will be discussed. Thirdly, after a brief discussion of the bases on which "major phyla" can be distinguished from "minor phyla," the nonconformist but fundamentally phyletic arrangement of the rest of the book will be outlined.

Animal Phyla and Taxonomy

As is well known, species can be delimited on the basis of a "biological species definition" which is based on the natural (nonarbitrary) criterion of reproductive isolation. The practical application of this criterion in taxonomic discrimination is usually difficult, and almost all described species involve arbitrary separations based on subjective inferences about genetic groupings, which are themselves derived from "hard" data on morphological groupings and upon geographical and ecological distribution. All taxa above the species level are admittedly ideal, arbitrary categories of this kind rather than natural categories. However, it has often been recognized that, among the higher taxa of systematics, a phylum—particularly a phylum of metazoans—is normally less subjective than such a taxon as class or order. Although a logically defined grouping like these intermediate taxa, a phylum may approach the objective reality of a species more closely.

The aspect of this most often stressed is that the 31 animal phyla are characterized by a unity of basic structural pattern in each. In other words, though the members of a phylum may vary in external features, comparative anatomical studies reveal that they form an assemblage, all constructed on the same ground plan with certain essential groupings of structural units. Much less emphasis has been placed on the considerable functional unity which exists within each phylum. If comparative physiology is studied at the level of the organization of the whole individual animal, certain unifying features become apparent in the ways in which each animal phylum carries out and regulates its major biological processes. In other words, underlying the fantastic structural and functional diversity of the invertebrates are certain functional homologies of considerable extent. Throughout this book and its predecessor paperbacks, an attempt is made to stress these homologies. The concept of homology—whether applied to structures or functions—involves the hypothesis of (theoretically) traceable derivation from, and origin in, some *common* anatomical or physiological precursor in a *common* ancestral animal. Thus, like all other concepts involving phylogeny (the circumstantially deduced history of genealogical descent), including "archetypes as models of ancestors," homologies are hypothetical. Obviously, we intend such concepts, when used in systematics, to reflect the realities of differing degrees of genetic relationship or, in the historical-evolutionary sense, of lesser or greater distance in time (in generations) from a common ancestor. Some biological implications underlying the "more objective" usage of the phylum as a taxonomic concept will be discussed in the next section of this chapter. Meanwhile, it is safest to emphasize the pragmatic value of functional homologies as they can be postulated for each phylum, in that they can allow human comprehension of the diversity of form and function in the invertebrates.

As already noted, the taxonomic groups used *between* the phylum and the species levels are ideal arbitrary categories rather than relationships which can be "proved" experimentally. In contrast, the modern biological definition of a species is based on genetic properties which, at least in theory, could be confirmed or nullified by appropriate breeding experiments. A *species* is defined as being made up of a series of populations of organisms which share a common gene pool and are actually or potentially capable of freely interbreeding and producing fertile offspring. The importance to this species definition of the fertility of the offspring of an experimental crossing has implicit in it the significance of genetic compatibility. The "null" part of the species hypothesis is provided by such hybrids as mules, which demonstrate the genetic incompatibility of their hybridization by their infertility. Put another way, the specific separation of the horse species from the donkey species is established by the inability of their hybrid, the mule, to have offspring in turn. Despite this, the majority of described species have involved arbitrary separations since the practical application of this experimental criterion of fertile interbreeding is usually difficult. Thus the majority of species, and all taxa above the species level, including genus, family, order, class (and phylum), are based on inferences about genetic groupings rather than upon provable genetic relationships. The hard data on which these inferences are made involve mainly features of comparative anatomy, supplemented where possible with data on physiological group characteristics, and on patterns of geographical and ecological distribution.

It is important to emphasize the pragmatic value of systems of taxonomy. No matter how much the intermediate categories are human artifacts imposed on natural diversity, the convenience of them as a filing or index system to allow some comprehension of the diversity cannot be overemphasized. Most beginning biologists have learned that the binomial Latin combination of generic and specific name has a value in international scientific communication that no common name for an organism can ever approach. Snail or *Schnecke* or *escargot* can each mean a great variety of different species, but *Helix aspersa* names only one species (in this case defined by breeding experiments as well as by morphological and enzymatic characteristics). Further, it is easy to see that, using only the five categories from species to phylum (Table 2-1), one can define a very large number of organisms efficiently with a set of six names. (For example, at the levels of ten species in each genus, ten genera in each family, ten families in each order, and so on, we can describe and classify in six words each a total of 100,000 species.) Given this power relationship, even with three units in each category, we can still cover 243 organisms with our six names. Of

TABLE 2-1
Hierarchy of Categories Employed
in Animal Classification

Phylum	(phyla)
Class	(classes)
Order	(orders)
Family	(families)
Genus	(genera)
Species	(species)

Where needed, additional steps may be interpolated, for example:
    class—subclass—series—superorder—order
or:
    superfamily—family—subfamily—tribe—genus

course, as set out in Table 1-1, the actual numbers of species in our 31 many-celled animal phyla vary very greatly. Particularly in the larger phyla, additional steps have been introduced into the systematic series by the use of the prefixes "super-," "sub-," and, rather more rarely, "infra-." A few other levels are used in certain groups of animals. Somewhat confusingly, division is sometimes used as a category below phylum, and in other cases between subclass and superorder (at which level cohort or series may also be used). The level "tribe" is sometimes inserted between family and genus.

The greatest elaboration of systematic levels is found in large groups of considerable diversity where investigations have been intensive enough to reveal an extensive hierarchy of levels of similarities and hence of probable levels of relationships, such as the class Crustacea (see Table 14-1). In other cases, a group consisting of a large number of species may nevertheless be relatively stereotyped in form and function. An inferred close relationship may involve some systematic levels being essentially unused in classification; for example, the phylum Nemathelminthes (though consisting of an enormous number of species) is generally considered to be made up of a single class Nematoda, the phylum having uniformly all the features we regard as diagnostic of the class. Above the level of phylum it has been customary to set up kingdoms, subkingdoms, and superphyla. For reasons that will be discussed later (see Chapters 6, 27, and 35), such categories have not been standardized and are of doubtful significance.

It is important to realize that the value of a classification system as a reflection of supposed phylogeny varies greatly from group to group. When we speak of a particular group classification as being "artificial," we imply that ignorance of possible relationships has made us adopt a classification, useful as a filing system, but largely arbitrary, and liable to be changed if future research allows us to deduce more regarding the relationships of the animals concerned. In contrast, a classification is referred to as "natural" if it reflects reasonably well-established hypotheses regarding the phylogeny of a group—for example, those which are consistent with many different kinds of circumstantial evidence of relationships. In the case of a very well-studied group of animals, it may be impractical to try to express all the inferred interrelationships in a classification simple enough to be pragmatically useful. What is done in such cases is to set up a classification which is as consistent as is possible with the best established parts of the phylogeny. Some of the consistencies between classifications in different textbooks result, not from their validity as evolutionary relationships, but from a conservatism in systematics which holds that a well-established system of classification should not be changed unless new studies show that it is clearly inconsistent with deduced phylogeny. No man-made classification of a large group of animals like a phylum can ever be "correct," but some classifications are likely to be more nearly correct (in reflecting relationships) than others. It is important to stress this artificiality, and that such taxa as represented by the words order and family correspond only to hypotheses set up by systematic zoologists and not to any particular divisions of nature in space or in time. It is best to regard them as essentially the key words of an index system that allows the zoologist to comprehend the diversity of species in the group (despite their continuity in space and time), and is at best a

shorthand notation which, if correctly interpreted, forms a statement of that zoologist's ideas on their probable interrelationships, and thus on their evolutionary history.

Structural and Functional Homologies

In discussing the bases of systematic zoology, we have just stressed the practical value of discerning homology, but now some assessment of the implicit biological meaning of structural and functional homologies is in order. Much of
biology as a science has always depended upon the perception of unities of pattern in the prodigious versatility of structure and function within the fantastic
diversity of higher organisms. A half-century before Charles Darwin's great
reorientation of evolutionary theory in 1859, Georges Cuvier had adumbrated
and Richard Owen had elucidated the basic concept of homology. In their classical studies of comparative anatomy, they distinguished between two patterns
of resemblance in organ structures, which they termed homology and analogy.
It is useful to note that this distinction can still be set out in pre-Darwinian
terms. Without introduction of any evolutionary explanation, an organ in one
form of animal can be termed *homologous* with a corresponding organ in another form, regardless of the functions they perform, if they are similar in their
fundamental structural plan, in their anatomical associations with other
organs, and in their embryonic development. In contrast, the organs in the two
distinct animal forms can be termed *analogous* if they perform the same function but are fundamentally different in structural plan, anatomical connections,
and development. Thus the bird's wing, the whale's flipper, and the human
forelimb (arm and hand) are homologous structures, whereas the bird's wing
and the wing of an insect are analogous. When the distinction is placed in an
evolutionary framework, homologous organs occurring in two or more distinct
animal forms are theoretically claimed as being derived from a common precursor organ in a common ancestral animal. From the extensive studies of comparative anatomy in both invertebrates and vertebrates carried out in the nineteenth and early twentieth centuries, all of the more important (and more
convincing) structural homologies had been worked out for the major phyla by
about 1930. In the middle years of this century, at the same time as the work of
functional morphologists and comparative physiologists had begun to establish additional patterns of *functional homologies*, the advent of modern genetics required of us a greater sophistication in our concept of "derivation from a
common precursor organ in a common ancestral form." Of course, what animals actually derive from their ancestral forms are solely replicated DNA molecules, encoding the genetic information which can be used in morphogenesis
by cell differentiation of particular structures or organs. This could imply that,
under certain circumstances, structures considered as homologous in two
forms need not correspond to any structures necessarily present as a patterned
set in their common ancestor. Their common inheritance could be merely the
genetic basis for the potentiality to evolve the structures in question under *similar parallel* selection pressures. The tendency for the concept of homology to
become somewhat fuzzy in this more modern framework has led a few of

today's biologists to despair of providing evolutionary interpretations for data from either comparative anatomy or comparative physiology. I believe such a pessimistic view of hypotheses of homology and phylogeny is unjustified, even in these current decades of the flowering of molecular biology. The powerful increase in our perception of fundamental unities of pattern in the bricks and mortar need not lead us to undervalue study of similar unities at the organismic level of architecture. At present it may need an author of admittedly zoöcentric bias to remind some biologists that processes of evolution have been concerned with *whole organisms*. The invertebrates have evolved by processes of natural selection acting upon whole animals and upon their efficiencies as machines to make more of their kind in immediately succeeding generations, while retaining sufficient flexibility of genetic response (or, in a sense, machine ineffi-ciency) to ensure persistence of the species through many generations. Thus, rather than single alleles, patterns of functional interdependence (encoded by largish bundles of integrated genetic material) are what I think to be important in making hypotheses about all evolutionary matters, including phylogeny. Continuity of certain patterns of functional integration has been an essential part of animal evolution.

Within each phylum of animals, the common anatomical ground plan im-plies a unique network of relationships between the groups of structural units which compose it. Some of this pattern of relationships is "stamped in" early in development by the need for certain early sequences of differentiation to pre-cede the more elaborate processes of morphogenesis of organs which must fol-low. The significance of this necessity for stereotyped pattern was outlined by O. Hertwig and R. Hertwig at the end of the nineteenth century, and since has intrigued (and sometimes misled) many developmental biologists (see discus-sion in Chapter 27). Another significant feature of the pattern of the interrela-tionships is functional. All animals must be efficient machines, in which the whole concert of organs and functions operates in an integrated fashion. Fur-ther, all ancestors of animals must have been working, efficient machines with similar functional integration. The need for physiological interaction in the way each major biological process is carried out and regulated has imposed pattern. The 32 "patterns" that we call phyla *do* work as efficient machines. Natural selection has not only tended to maintain each stereotyped network of anatomical relationships (and eliminated almost all random variants), but it has promoted increasing functional interdependence in a pattern of whole-animal function characteristic for each phylum.

Perhaps this somewhat theoretical discussion of the nature of functional ho-mologies can be elucidated by a few specific negative examples. Characteristic hollow skin structures termed tube-feet or podia are found throughout the phy-lum Echinodermata. Although employed in a variety of ways—for the manipu-lation of food or protective materials, for respiration, for sensory purposes, and for locomotion—podia are recognized, on the basis of extensive morphological evidence, as structurally and functionally homologous in about six thousand species of living echinoderms (and probably in at least another forty thousand extinct forms). They appear as numerous delicate external projections from the surface of the animal, each having: intrinsic muscles for withdrawal and for postural "pointing," a hydraulic mechanism for protraction depending on the so-called water-vascular system within the animal, a characteristic pattern of internal ciliation for fluid circulation, and individual innervation which allows either local reflex responses or movements integrated throughout the whole animal. Podia or tube-feet are efficient structures for certain kinds of slow locomotion and manipulation and could seemingly be used by many aquatic animals belonging to other phyla. *Only echinoderms have them.* Within the echinoderm body, the podia are part of an integrated functional system. Mechanically, they could not possibly function unless they were supported by the unique skeleton of dermal ossicles which give the phylum its name. Although the details of protraction vary in different echinoderms, all podia are extended hydraulically by pressure of fluids within the water-vascular system, a subdivision of the coelom. Thus, they depend on the peculiar interrelationships of development and adult function of the coelomic spaces which are unique to the phylum Echinodermata.

Many animals, other than coelenterates, could use the characteristic stinging-cells (nematocysts) for defense or offense. *No others make and use them.*

A characteristic form of gill, the ctenidium, is found throughout most of the phylum Mollusca and is inferred (on evidence that leaves little room for any doubt) to be structurally and functionally homologous throughout those animals where it occurs (i.e., in perhaps 75,000 molluscan species—see Chapters 19, 20, and 26). It is functionally a most competent organ, with patterns of ciliated epithelia and blood vessels arranged to create a highly efficient counter-flow system for oxygen exchange between blood and water, characteristic cleansing mechanisms (both ciliary and muscular), and typical mechanical arrangements of supporting skeletal elements. There are many aquatic animals belonging to other phyla which seemingly could make good use of a ctenidium. *No nonmolluscan animal has one.* Within the molluscs, the ctenidia are part of an integrated functional system: the heart and other blood vessels, certain glands and sense-organs, the external openings of genital and renal systems, and the posterior part of the alimentary canal are all structurally and functionally stereotyped in their relationships to the ctenidia. Many more examples of this sort could be set out.

A related topic is the use of "hypothetical primitive forms" or "archetypes" as models. Biologists will never know with certainty the characteristics of the ancestors of *any* stocks of animals. All modern scientists should be well aware of the general pitfalls—as well as the pragmatic values—of the use of models. The peculiar dangers in evolutionary discussions of setting up an "archetype" seem to result from assembling together in the unfortunate hypothetical animal a group of incompatible structures, all thought to be "primitive" within the stock. I feel that many of these can be avoided if, when a hypothetical ancestral type is constructed, an attempt is made to create a working archetype—one in which the concert of organs and functions could operate as a whole, in an integrated functional plan, as in all living organisms. It is perhaps significant that discerning functional homologies and setting up archetypes is much more difficult in the phyla of *less complex* animals. Significant functional unity is more apparent in the Mollusca, the Arthropoda, or the Echinodermata than it is in phyla with less complexity of structure and function such as the Cnidaria or the Platyhelminthes. It seems that the network of functional relationships is less closely integrated in these "lower" groups, and thus the anatomical and physiological pattern is not so stereotyped. "Experimental variants" of pattern can be thought of as having a better chance of survival in a group with less complexity of interdependence in its basic archetype.

Arrangement of Major and Minor Phyla

In Chapter 1, we introduced the 32 kinds of animal body designs which can each be ranked as a phylum, and noted that, of these, only eight patterns of many-celled design (along with the Protozoa) can be regarded as being important, or as constituting *major* phyla. Once we accept that the logically defined groupings of many-celled animals into the 31 phyla of Table 1-1 can reflect some objective reality of common descent and interrelationship, then it becomes immediately obvious from the listed species numbers alone that some patterns of animal design are much more successful than others. Of course, among these species, some can be regarded as "rare," because they each consist of small numbers of individuals, whereas other species are each abundant as individuals showing ubiquitous dominance within many faunal communities. For most authorities, the distinction between minor and major phyla is based on the number of individuals, the number of species, or the number of "successful, dominant" species encompassed in the groups or on (and this is almost impossible to quantify) the relative "phyletic importance" of the groups. On all such counts, four phylum groups at least are clearly major: arthropods, chordates, molluscs, and nematodes. More than twelve phyla of invertebrates are equally clearly minor.

Less often recognized is that such designation as major and minor can also be justified ecologically, on the basis of a relatively simple aspect of community ecology which has become more and more obvious to me over the last twenty years. Briefly, it is that the bulk of the animal biomass in the great majority of natural communities is made up of individuals belonging to a few major phyla of animals, and that the minor phyla make up only a tiny fraction of the animal tissue alive in the world today. To put it another way, of the solar energy first incorporated into green plants, a disproportionately large share flows through (or is utilized by) representatives of such major phyla as the Arthropoda, whereas the energy-flow through the representatives of a phylum such as the Entoprocta is normally several orders of magnitude smaller. In terms of quantitative ecology, these two assessments are measures of standing-crop biomass or production [expressed as mass of organic carbon, or ash-free tissue dry weight, per area—for example, in grams per square meter (g/m^2)] and of productivity rate [expressed in similar biomass values or, better, in energy units, such as calories or joules, per area *and* time—for example, as joules/$(m^2 \cdot year)$], respectively. Within any ecosystem, both bioenergetic assessments are highest for herbivore species and lower for each successive level of carnivores and for parasites. It follows that no animal group consisting exclusively of carnivorous species, or of parasitic ones, could be regarded as a major one in this ecological sense. There are many abundant herbivorous species in such phyla as the Arthropoda and the Mollusca. If we rank the animal phyla in terms either of biomass or of rate productivity, it seems justifiable to refer to a phylum like the Entoprocta as being "relatively unsuccessful" in the competition for the *finite* amount of organic energy originating in the green plants of the world, and to refer to it as a minor phylum. Applying a combination of the above criteria to the 31 phyla of many-celled animals, we can regard 8 as major invertebrate phyla, perhaps 3 others as important, and the rest as minor phyla of little phyletic or ecological significance.

The minor phyla are not treated extensively here. As in the "Who's Who" of Chapter 1, emphasis is given to the successful patterns of animal machine (the major phyla), with some precedence to the more familiar rather than the obscure. The arrangement of the book is further dictated by the importance of structural and functional homology *within* each phylum. Thus the biology of invertebrates is presented almost entirely *phylum by phylum,* and the order in which both the major phyla and the minor phyla are treated generally corresponds to increasing complexity of structure and function. We begin in the next two chapters (Chapters 3 and 4) with the relatively well-known phylum Cnidaria, which includes some of the most simply organized of many-celled animals. Three minor phyla—the Ctenophora, the sponges (Porifera), and the Mesozoa—are dealt with next. The complexities of the single-celled protozoans are surveyed in Chapter 6, and we then move to the triploblastic level of organization with the flatworms and their allies in Chapters 7 and 8. Some minor phyla, both acoelomate and pseudocoelomate, are included here, and with Chapter 9 we move to the only major phylum of pseudocoelomates, the phylum Nemathelminthes, or nematode worms. The remaining 22 chapters all are concerned with animals built on the triploblastic coelomate plan, including the three most important animal phyla. Chapter 10 introduces the metamerically segmented worms and arthropods; Chapters 11 and 12 deal with the annelid worms, and Chapters 13–17 with the arthropods. In order to deal with the enormous phylum Arthropoda, a somewhat arbitrary treatment has been evolved. First, the biology of one major class, the Crustacea, is treated more extensively, largely because the occurrence of relatively primitive marine forms within it makes crustacean functional morphology a somewhat more rewarding study. Then, in Chapters 16 and 17, systematics of arachnid and of insect diversity are surveyed only synoptically, but certain aspects of the comparative physiology of terrestrial arthropods—including respiration, excretion, molt-cycle, control, land locomotion, and flight—are discussed more extensively. The minor metameric phyla are briefly considered in Chapter 18. The next group of chapters is concerned with another huge phylum, the Mollusca. Chapters 19–23 explore the diversity of structure and function in the major molluscan groups, Gastropoda and Bivalvia, and their basis in a remarkably uniform ground plan. Chapters 24 and 25 deal with the most highly organized invertebrate animals, the molluscan class Cephalopoda, including *Octopus* and its allies, and briefly discuss the importance of these animals in research on brain function and learning. Chapter 26 deals with the minor molluscan groups and with molluscan phylogeny. In Chapters 27 and 28, we are concerned with further minor phyla (all coelomate), with larval types, and with some aspects of the suspect interrelationships between phyla. Chapters 29 and 30 deal with the major phylum Echinodermata; and Chapters 31 and 32 present the invertebrate members of the phylum Chordata, to which we, as mammalian vertebrates, belong. The last three chapters (Chapters 33–35) provide brief recapitulations regarding the invertebrates of the marine plankton, the fossil record of invertebrates, and some hypotheses concerning the phylogeny of invertebrates.

3 LOWER METAZOA: THE COELENTERATES

THE relatively familiar animals which make up the phylum Cnidaria, including such forms as jellyfish, sea-anemones, corals, and hydroids, are commonly called coelenterates. Strictly speaking, the term also includes the "comb-jellies," which are treated here separately as the phylum Ctenophora. These two groups were formerly characterized as the diploblastic metazoans. Modern biologists have tended to abandon the term diploblastic, being embarrassed by the existence in many coelenterates of cellular elements in an intermediate layer between the epidermis and the lining of the gut cavity. If, however, one considers the development of the cells, there is still a clear distinction between these elements and mesoderm as it is found in the truly triploblastic phyla. The cells making up intermediate layers in the coelenterates have always been derived as wandering amoeboid cells which have been detached individually from the outer epidermis, and never originate (as does the great bulk of the mesoderm in the "higher" phyla) as an initially epithelial mass of tissue derived from the endoderm of the developing animal. However, without becoming further involved in the semantics of development which have provided metabiological controversy for over a century, there is a good functional basis for the separation of these lower phyla from the rest of the metazoans. It is that they, with certain minor exceptions, lack organs, having their cells organized into tissues which are specialized for various functions, but not going beyond this "tissue grade" of inter-dependence of their parts to the development of the organs of particular func-tion so characteristic of the higher metazoans.

Although the coelenterates are the most simply constructed of many-celled animals, the Cnidaria are a relatively successful phylum. They are abundant in a wide variety of habitats in the sea, though there are only a few freshwater forms and their general functional organization makes them totally unable to exist as terrestrial animals. Under a strict application of our ecological assess-ment, the Cnidaria do not really constitute a major phylum, although in some regions of the tropical seas they are ecologically important, making up a signif-icant part of the biomass. On the other hand, they constitute an important group because they have in the past received a disproportionate degree of at-tention from zoologists. This amount of study has been based on two concep-tions, both involving uncertainties which have often been attacked, but both still largely defensible. The first is simply that the coelenterates are at least closely related to the stock from which the other metazoan groups eventually were derived. Obviously, it is impossible to prove that they are the most primi-tive of many-celled animals; without a time machine this will always be un-provable. (With cnidarians, we must also face the unlikelihood of important fossil evidence ever being discovered.) The second conceptual basis for interest in the coelenterates is that their organization involves a degree of structural and functional simplicity which is not found in any other living metazoans. This particular hypothesis has attracted two groups of biologists to them: de-velopmental biologists interested in problems of regeneration and differentia-tion and neurophysiologists. Both groups of investigators have uncovered greater and greater degrees of complexity in recent years than were first thought to exist in these forms.

The lack of organs, and the consequent occurrence of functional interdepen-dence only at a lower level of complexity, is partially responsible for some dif-ficulty in applying the concepts of functional homologies—as discussed in Chapter 2—to these "lower" phyla. As several of my colleagues, but notably Robert K. Josephson, have pointed out, however applicable my idea of signifi-cant functional unity is to phyla such as the molluscs or echinoderms, it is more open to criticism when applied to the phylum Cnidaria. For example, two things which show clear functional homologies within the higher phyla, the nervous system and the skeletal mechanisms, vary widely throughout the coe-lenterates. As regards the first, each of the three main groups of coelenterates seems to have evolved its own method of coordinating its activities through conducting—or nervous—cells. One result is that it is almost impossible, on the present evidence, to decide which type of nervous organization was arche-typic for the group. Similarly, the skeletal mechanisms, or the ways in which the animal maintains its shape in reaction to the environmental forces acting upon it, vary and form a poor subject for generalization in the coelenterates. It appears, then, that the stereotyped network of interrelationships of functions so typical of each of the higher phyla does not need to exist at the simpler level of organization found in coelenterates. To the contrary, the case for functional homology can be defended by the fact that it is possible to give a general diag-nosis for the group which is on the one hand simpler, and on the other more universal, than can be applied to any other phylum of animals. It is that the coelenterates are predaceous carnivores of the tissue level of organization, utilizing nematocysts as both a means of defense and of obtaining their prey. Nematocysts are *formed and used* only by cnidarians and by one species of ctenophore.

Also
d in of lagellates
!

Nematocysts

Each nematocyst or stinging capsule is contained within a parent cell, the cnidoblast. Figure 3-1 shows a generalized diagram of a cnidoblast with contained nematocyst before and after discharge. It is somewhat difficult to say anything about nematocysts that would be true for all cnidarians. Their size varies from 5 microns to over 1 millimeter, and 27 basic types have been described. There is an elaborate classification of nematocysts based on the number of barbed spines and like structures which form an armature around the hollow thread, but with few exceptions the classification of nematocyst forms does not correspond to the presumed major relationships of the various coelenterates which bear them. More than six of these forms can be found in one species of coelenterate. As shown, the capsule of this larger type of nematocyst has a wide invagination, known as the butt, which bears large barbs folded within it and then narrows into a coiled tubular thread that may or may not bear smaller barbs or hooks. Appropriate histochemical tests should eventually show whether the toxin of the stinging cell is found in the undischarged condition within the butt or is in the separate fluid in the remainder of the capsule. In normal circumstances, extrusion of the tube is accomplished in a matter of less than a second after appropriate stimulation. Many years ago, careful observations of partially extruded nematocysts with the light microscope had revealed that on discharge not only the butt with its prominent barbs was everted, but also the whole length of the hollow thread was discharged through itself and turned inside out as it went. It had also been observed that the final length of thread in a wholly discharged nematocyst was greater than the coiled length within the undischarged stinging cell. Recent electron microscopy of sections of discharging nematocysts has revealed the reasons for this apparent length increase (Figures 3-2 and 3-3).

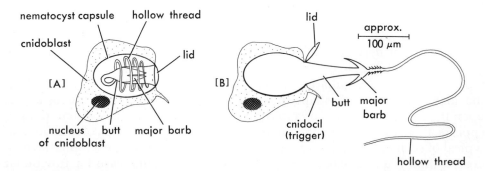

FIGURE 3-1. A typical coelenterate stinging cell: stylized diagrams of a stenotele nematocyst before (**A**) and after (**B**) discharge.

[A]   [B]

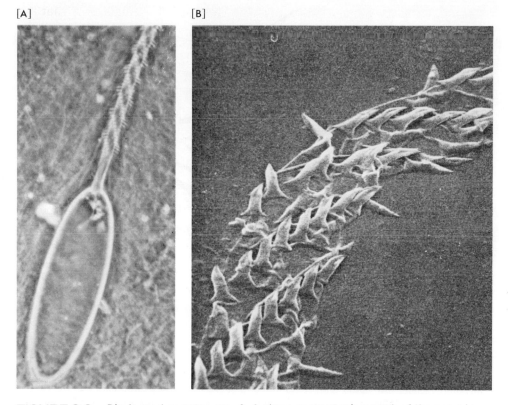

FIGURE 3-2. Discharged nematocysts. **A:** A phase contrast micrograph of the capsule and part of the hollow thread of a discharged *holotrichous isorhiza* from the anemone *Corynactis californica,* showing the barbs arranged in three spiral rows around the hollow thread. The capsule is about 38 microns in length. **B:** A scanning electron micrograph of part of the hollow thread of a similar nematocyst (compare with the diagrammatic section in Figure 3-3). The flattened thread is about 2.5 microns across. [Micrographs courtesy of Dr. Richard N. Mariscal, Florida State University, Tallahassee.]

It seems that the undischarged thread has the appearance of a screw because of three left-handed helical spiral folds in its wall. After discharge the thread is a smooth, slightly tapered tube, the process of discharge apparently involving unfolding of these pleats. In a highly enlarged longitudinal section of a partially discharged thread, where evenly spaced equal-sized barbs are present, the appearance is that diagrammed in Figure 3-3. This new evidence of the microanatomy of nematocyst threads has made it easier to understand how these threads are able to penetrate the exoskeleton of the more highly organized animals such as crustaceans which provide a large part of the diet of small Cnidaria. From the point of view of the prey, one has not only to imagine a double tube everting toward one at a fantastically high velocity, but the advancing edge of that double tube (i.e., the fold between the tube already discharged and that undischarged contained within it) revolving at enormous speed as the spiral pleats unfold, and at each turn a new ring of sharp barbs springing forward and then flicking backward like the blades of penknives.

The poisons released by the discharge of nematocysts appear to be of several

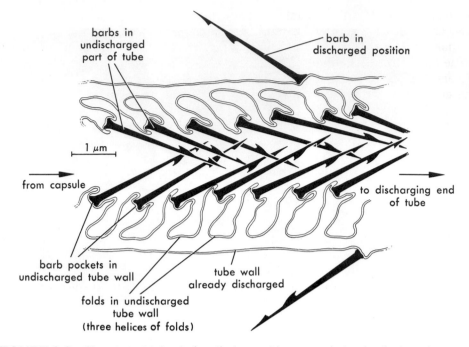

barbs in
undischarged
part of tube

barb in
discharged position

1 μm

from capsule

to discharging end
of tube

barb pockets in
undischarged tube wall

tube wall
already discharged

folds in undischarged
tube wall
(three helices of folds)

FIGURE 3-3. Nematocyst tube during discharge. Diagrammatic longitudinal section through the partially discharged tube of a *holotrichous isorhiza*. Within the smooth-walled tube already discharged with its widely spaced and backwardly directed barbs is the as yet undischarged tube with spiral folds in its wall and packed forwardly directed barbs. [Based on electron micrographs from L. E. R. Picken and R. J. Skaer in *Symposia Zool. Soc. London*, **16**:19, 1966.]

sorts, though they are probably all proteins, and mostly paralyzing neurotoxins. Certain enzymes, such as cholinesterase and phosphatases, are also present in some nematocysts. Though many theories have been advanced, we still have no complete explanation of the mechanism of discharge in nematocysts. Perhaps the most significant observations are those which support the hypothesis that the initiation of discharge depends on a rise in pressure within the capsule fluid. It is obvious even in a crude diagram, such as Figure 3-1, that the volume of fluid within the capsule, butt, and thread must increase rapidly during eversion if a high positive pressure is to be maintained and discharge to be successful. Freezing-point determinations of osmotic pressure on the capsular fluid both before and after discharge have now been made, and these values suggest that the fluid within the capsules (of one moderately large nematocyst type) has an extremely high osmotic pressure in the undischarged condition (corresponding to a notional pressure of about 140 atmospheres). Other observations, as yet difficult to correlate with the foregoing, suggest that the velocity of the tip of the everting thread remains constant throughout the discharge.

Basically, cnidoblasts have been regarded as independent effectors. In other words the cnidoblast-nematocyst unit fires as a direct response to stimulus; it is not fired by the parent organism. Many nematocysts require a dual stimulus for their effective discharge. Various experiments can be performed by the student to demonstrate this. For example, many do not discharge on stimulation of their cnidoblasts with a chemically clean and sterile glass rod, but if various animal extracts are added, even in low concentration, the nematocysts respond to the mechanical stimulation. It can be demonstrated in Hydra that two types of nematocysts (desmonemes and stenoteles) are more easily provoked to discharge in the presence of a crustacean extract, whereas the discharge of a third type (atrichous isorhizas) is almost completely inhibited. Observations suggest that this third type of nematocyst is employed by Hydra in adhesion during locomotion, rather than in the capture of prey like the other two kinds.

It seems that there are several other circumstances during apparently suitable stimulation when nematocysts are not discharged. Commensal fishes and protozoa are not stung, though they cause mechanical stimulation at least as intensive as many sorts of prey organisms. A certain amount of evidence has been accumulated which suggests that the threshold for effective excitation of nematocysts varies with the physiological condition of the coelenterate as a whole, and in some cases appears to be under nervous control. However, cnidoblasts and their contained nematocysts may still be considered as independent effectors in most of their actions. The toxins are effective against higher organisms, and suitable doses prove fatal to many small vertebrates. The tentacles of the "Portuguese man-of-war," Physalia (see Figures 4-8 and 4-9), can raise inflamed welts on the unwary swimmer in tropical waters, and certain small cubomedusae in the coastal waters of Australia have been responsible for at least fifty human deaths in recent years. It is important to realize that in most cases nematocysts are effective only because they are discharged in large numbers at a time. Discharged nematocysts, if not pulled away by struggling prey, are shed at varying times after their use. This implies that both the nematocyst and the cnidoblast surrounding it are expendable and replaceable, and there are now many well-authenticated instances of migration of cnidoblasts bearing nematocysts into sites where they can be used. These cnidoblasts have always arisen from the interstitial cells, which we will discuss below.

Polymorphism and General Physiology

Another general character of the coelenterates is the occurrence of polymorphism within many species. This may be temporal, with the species passing through a succession of different body forms in the course of its life-cycle, or several different morphs of the species may occur simultaneously in a colony. In colonies of polymorphic Cnidaria, one finds considerable interdependence and division of labor among individuals (rather than between organs as in the higher Metazoa). There are two basic body forms or morphs: the medusoid and the polyp or hydranth. These are illustrated in Figure 3-4, and although the medusoid is generally motile and the polyp typically sessile, the basic features of the two can be homologized.

There is only one internal cavity, the gastrovascular cavity, and the mouth is the only opening from the exterior into it. The gastrovascular cavity is lined with endodermally derived gastrodermis, and the outside of the coelenterate is covered with the ectodermal epithelium or epidermis. Between these two

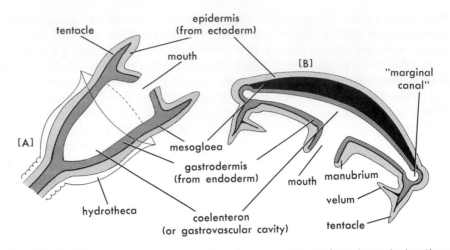

FIGURE 3-4. The two basic body forms in cnidarians—the sessile polyp or hydranth and the motile medusa.

layers of living cells lies the mesogloea, usually but incorrectly designated as nonliving and consisting in the simplest coelenterates of a structureless gel. This noncellular mesogloea is made up of more than 90% water with less than 1% organic materials, and an average of about 5% inorganic salts. Both in the outer epidermis and in the lining tissue of the cavity, the commonest cell type is the myoepithelial cell. Each has contractile elements formed on its mesogloeal face and also constitutes an epithelium with its neighbor cells. In the Hydrozoa the contractile tails or "fibers" are arranged longitudinally in the epidermis and circularly in the gastrodermis. In both layers there are also comparatively undifferentiated cells lying close to the mesogloea. These are the interstitial cells, which are characterized by their lack of a free margin, either to the exterior or to the cavity.

As already noted, the processes of feeding and digestion are relatively uniform throughout the coelenterates. Captured food is passed into the mouth by the tentacles or by tracts of cilia. The coelenteron or gastrovascular cavity is usually capable of great distention, and many gland cells which secrete enzymes upon the stimulus of feeding open onto it. The enzymes secreted into the cavity are mainly proteases, and the extracellular digestion which goes on in the cavity, aided by flagellar stirring, results in a disintegration of the food organisms rather than a complete digestion. Amoebic ingestion of food follows this and digestion is completed intracellularly in food vacuoles like those of protozoans (see Chapter 6). It seems likely that carbohydrates and fats can only be digested intracellularly. There is little functional difference in the more highly organized and more complex coelenterates, although in them one may find specialized areas of gastrodermis concerned *only with* secretion of enzymes *or with* absorption of partially digested food material. Some of the functional simplicity of coelenterates is possible only because they are *thin* organisms. In other words, most of their living cells lie at, or very close to, exposed surfaces. Both epidermis and gastrodermis absorb oxygen for respiration directly from the environment or gastrovascular cavity. There is simple direct diffusion to any underlying tissues, and diffusion outward of certain products of metabolism including carbon dioxide and nitrogenous wastes. Most nitrogenous excretion is accomplished by direct diffusion of soluble ammonia. Amoeboid cells carry away larger particles of waste material which join the undigested "faeces" to be expelled from the gastrovascular cavity through the mouth.

In the sexual reproduction of coelenterates, the cells that take part are, like those that give rise to nematocysts, the interstitial cells. Interstitial cells from both epidermal and gastrodermal layers wander into the sites of the gonads. These are hardly organs, but merely accumulations of sex cells, which eventually will open to the exterior by a simple rupture of the overlying tissues. Except for the specialized nonmarine *Hydra* and the recently discovered minor group (the order Actinulida), the gonads of the Hydrozoa and Scyphozoa are always developed within medusae or on sessile gonophores (which represent reduction of a medusoid body form). The gonads produce eggs and sperms which are essentially similar to those produced by other many celled animals. The fertilized egg, whether retained or floating freely in the sea, undergoes holoblastic division to form a hollow blastula. Then migration of endoderm cells to the center forms a solid internal mass, and the embryo becomes a typical ciliated planula larva. The mass of endodermal tissue splits to give rise to the coelenteron, and a mouth forms at what is functionally the posterior end. After a free life of a few hours, or in some cases a few days, the planula settles on its anterior end (or on its side) and develops into a polyp (or the first hydranth individual of a new colony).

Asexual reproduction is common throughout coelenterates. Medusae are nearly always formed by budding asexually from the hydroid, and further hydranths or polyps may also be budded off. These may drop away to form new separate individuals or remain linked by tissues to form colonial forms. In fact, colonial groups of polyp-individuals are perhaps the most characteristic coelenterates. As noted previously, cooperation between different polymorphic individuals within a colony of coelenterates is often analogous to the interdependence of different organ systems in a more complex animal. It is conceptually possible to regard the coelenterate colony as a *superorganism* of interdependent individual zooids. Along with their capacity for asexual reproduction, all coelenterates show great powers of regeneration. Many of them also have capacity for dedifferentiation, so that they can "return" to a simpler structural plan or to a more embryonic level of organization. (On the other hand, starved sea-anemones grow smaller but retain all their adult structures.)

Neural Physiology

In coelenterates, nerve cells form nets immediately below the epithelia. The cells are usually multipolar—though in a few places in more highly organized cnidarians they may be bipolar—and are arranged in discontinuous two-dimensional nets. It is worth stressing discontinuity: these cells have synaptic contacts only (and no cytoplasmic continuity) with their neighbors, just as in the neural organization of higher many-celled animals. However, coelenterate synapses show a major functional difference and appear to lack any polarization in transmitting action potentials. Further, any branch or "fiber" of a cnidarian nerve cell can conduct in either direction, and any normal multipolar neuron can conduct from any branch in any direction to any other branch. There are only rarely concentrations of nerve cells in cnidarians and never anything approaching a central nervous system.

Nerve cells are grouped together in the "ring-nerves" of hydrozoan medusae, at the marginal organs (compound sense-organs) of Scyphozoa, and in a few tracts in siphonophores. Even in these "active" forms the great bulk of nerve cells are scattered in the nets and not grouped together. Recent electron microscopy has revealed two characteristic differences in ultrastructure. First, at each synapse between two cells, there are secretion vesicles on both sides of the synaptic gap. This could be regarded as the cytoanatomical basis for the lack of polarity. Secondly, there appears to be a complete absence of any sheathing cells around the "axons"; that is, there is no glial material whatsoever. If this is correct the cnidarian "axons" are the only truly naked nerve fibers in animals. A similar uniqueness has been established by recent pharmacological research on the Cnidaria. The response pattern of scyphozoan nerve preparations to neurotransmitter substances and neurohormones common in other metazoan animals, to tetrodotoxin, and to a wide variety of other drugs, indicates that, in cnidarian nerve cells and neuromuscular functions, the basic biochemistry of neural membranes is different in kind from that pertaining in vertebrates and in all "higher" invertebrates.

As was first elegantly demonstrated by the late C. F. A. Pantin and his associates, the degree to which a stimulus affects a cnidarian such as a sea-anenome depends on the extent of its spread through the nerve net from the site of stimulus. This in turn depends on decremental conduction and on interneural facilitation. The latter occurs when a subthreshold stimulus (not producing a response) followed closely by a second similar stimulus results in a response, indicating that further spread in the nerve net had been made possible by the first stimulus altering the properties of an interneural gap or gaps. In the last fifteen years, the supposed "simpler" neurophysiology of coelenterates has been greatly complicated by the discovery (using modern microelectrodes and suction electrodes) of several conducting systems, not all neural, coexisting in certain coelenterates and all contributing to the complexity of behavior which can occur. In fact, the sea-anemones seem to have the simplest system, with one comparatively simple sheet of nerve net. Even octocorallians may have greater complexity, and in the Scyphozoa there are at least two systems—one fast-conducting, controlling the swimming muscles, and the other slower, controlling tentacular movements. In some hydrozoans there are greater degrees of complexity. R. K. Josephson has found four conducting systems—each with its own conduction speeds and effector connections—in the gymnoblastic hydroid *Tubularia*.

Some distinguished workers on coelenterates have claimed that the simplicity of nervous system in sea-anemones is evidence in favor of their being a primitive "stem-group" within the Cnidaria. It seems equally likely that coelenterates with structurally simpler but functionally more efficient nervous organization could have evolved from forms with several diffuse conducting systems. If conduction was a general property of the first cells to be arranged in tissues, then the evolution of conducting systems could follow a similar course to that known for the evolution of epithelial sensillae into complex sense-organs (by increasing specificity of response). In this way, too, an essentially random but extensive arrangement of functional connections could give rise by progressive structural simplification and functional polarization to a system capable of much greater *specificity* in its conduction.

4 THE MAIN KINDS OF CNIDARIA

THE Cnidaria can be divided into three well-defined classes: the Hydrozoa, the Scyphozoa or true jellyfish, and the Anthozoa, which include the corals and sea-anemones. The gastrovascular cavity, or coelenteron, is always a blind sac with only one opening (functionally both mouth and anus); this sac carries out the functions of both gut and circulatory system in higher animals. It is usually a simple bag in the Hydrozoa, has more complicated branches forming canals and pockets in the Scyphozoa, and is partially divided by ingrowth of a series of septa in the Anthozoa. The functional significance of both types of complication lies in the increased surface area of gastrodermis provided in both jellyfish and sea-anemones.

Hydroida

The class Hydrozoa encompasses coelenterates with a simple gastrovascular cavity without septa, with a noncellular mesogloea, and without organs other than groups of nematocysts. They are mostly polymorphic and many of them have both hydroid and medusoid forms developed at different stages in their life-cycles. The accepted subdivisions of the class are outlined in Table 4-1. The first order, the Hydroida or Hydromedusae, is one of the biggest orders of coelenterates, encompassing more than two thousand known species. Two of its four subdivisions

TABLE 4-1
Outline Classification of the Cnidaria

Class A HYDROZOA
    Order 1 Hydroida (Hydromedusae)
      Suborder Gymnoblastea
      Suborder Calyptoblastea
      Suborder Hydrida
      Suborder Hydrocorallinae
    Order 2 Trachylina
      Suborder Narcomedusae
      Suborder Trachymedusae
    Order 3 Actinulida
    Order 4 Siphonophora
      Suborder Calycophora
      Suborder Physophorida
Class B SCYPHOZOA
    Order Discomedusae
    Order Rhizostomeae
    Order Stauromedusae
    Order Coronatae
    Order Cubomedusae
Class C ANTHOZOA
    Subclass I Octocorallia (Alcyonaria)
    Order Stolonifera
    Order Telestacea
    Order Alcyonacea
    Order Coenothecalia
    Order Gorgonacea
    Order Pennatulacea
    Subclass II Hexacorallia
    Order Actinaria
    Order Madreporaria
    Order Antipatharia
    Order Zoantharia
    Order Ceriantharia

are much larger, and one of these, the Calyptoblastea, includes the typical colonial hydroid *Obelia*, the life-cycle of which is outlined in Figure 4-1. Colonies of such hydroids (*Obelia* and such other genera as *Campanularia*, *Sertularia*, and *Plumularia*) vary somewhat in form as a result of their patterns of growth and branching, but all are surrounded with an external tubular skeleton of chitin. This tube—the so-called perisarc—is prolonged to surround each individual in a cup-like theca (see Figure 3-4). The majority of individuals in any hydroid colony are normal feeding polyps or hydranths (such as illustrated in Figure 4-1). There are usually also nonfeeding individuals called blastostyles, which are responsible for the asexual budding off of the free-swimming medusae. In a few calyptoblast colonies a third type of individual is found: for example, in *Plumularia*, we find that there are nonfeeding individuals lacking mouths which are studded with many nematocyst-batteries and are functionally defensive individuals or dactylozooids. In such colonies, the feeding individuals are usually known as gastrozooids, and blastostyles as gonozooids.

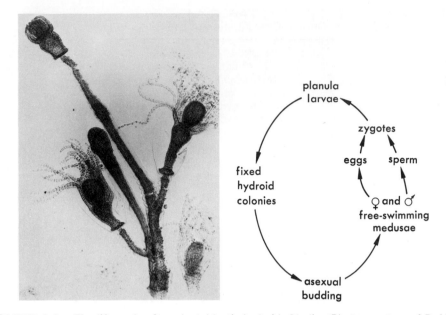

FIGURE 4-1. The life-cycle of a calyptoblastic hydroid, *Obelia*. [Photo courtesy of B. Michael Hollick.]

The protective cup surrounding a feeding polyp is called a hydrotheca, and the more vase-shaped protection of the asexually reproducing blastostyle is termed a gonotheca. The medusae of calyptoblastic hydroids are typically flattened, saucerlike "leptomedusae," with a proportionately small velum and four clear radial canals traversing the four gonads. Calyptoblastic hydroids are typically shallow-water marine colonies living attached to hard substrata or to seaweeds in temperate latitudes. They often show much seasonal variation: with extensive late spring and early summer asexual proliferation and dying down in the autumn.

In the only other large group of the Hydrozoa, the Gymnoblastea, the perisarc does not usually extend around polyps or blastostyles. A few are solitary, but one of the most common patterns of colony formation is for a creeping stolon, ramifying over a hard substratum, to support many separate upright individuals (Figure 4-2). Whereas the life-cycle in the more primitive forms is similar to that of *Obelia*, various gymnoblasts have modified this, both by the development of a wide range of different ways of producing medusae and by progressive reduction of the importance to the life-cycle of this free-living stage. In some gymnoblasts, single complete medusae are budded off near the base stalks of polyps. In other cases, there are naked blastostyles producing many medusae in a similar fashion to the gonozooids of the calyptoblastic hydroids. In some others, there are so-called sporosacs whose development reveals them to be medusa-buds which remain permanently attached to the parent colony. In all cases where medusae are released, and in several cases where they remain attached to the colony, the medusae are structurally characteristic. These "antho-medusae" are bell-shaped, with a very long manubrium which bears the gonads, a very large velum, and often only four sets of marginal tentacles (Fig-

FIGURE 4-2. Hydrozoan polyps. An expanded colony of the gymnoblastic hydroid *Tubularia*. [Photo © Douglas P. Wilson.]

ure 4-3) A few cases are known where these anthomedusae can reproduce asexually by budding off new medusae from the tentacle bases or from the manubrium. A typical gymnoblastic colony is represented by *Tubularia* (Figure 4-2). Many species are epizoic; some few are ectocommensal; and even ectoparasitic forms have been evolved. The genus *Hydrichthys* lives on fish skin, with the colony stolon growing over, and occasionally down through, the skin tissues. The polyps lack tentacles, and bend over to feed by sucking blood and tissue cells from the parts injured by the stolon growth.

Several genera of gymnoblastic hydroids show the threefold polymorphy already mentioned; that is, they have gastrozooids, dactylozooids, and gonozooids dispersed through the colony. A somewhat greater degree of polymorphy is found in the encrusting form, *Hydractinia*, which normally grows on the snail shells occupied by certain hermit-crabs. Five types of polyps arise individually from an encrusting base, and each of these takes on special functions for the colony as a whole. There are feeding individuals or gastrozooids (Figure 4-4A) which are "normal-looking" polyps, each with many tentacles around the mouth and a relatively large gastrovascular cavity. There are two types of dactylozooids (Figure 4-4B,C) or protective individuals, which have no tentacles and no mouth, but masses of nematocysts. The commoner (C) is called a spiral-zooid (from its method of retraction by coiling back on itself) and bears vestiges of the tentacles as knobs. The larger type of dactylozooid (B) looks like a long, single tentacle, and occurs only near the edge of the hermit-crab's shell. The reproductive individuals are blastostyles or gonozooids (Figure 4-4D) bear-

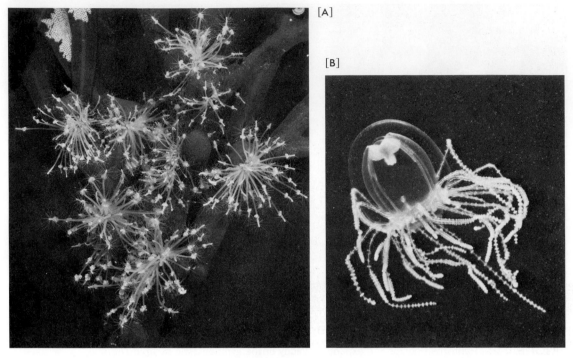

FIGURE 4-3. Hydrozoan polyps and medusa. **A:** Colonies of the gymnoblastic hydroid *Clava* photographed under water growing on seaweed. **B:** A hydrozoan medusa, *Gonionemus*. Although generally classified as a trachyline medusa, *Gonionemus* shows most of the features of the bell-shaped anthomedusae produced by many gymnoblastic hydroids. [Photos © Douglas P. Wilson.]

ing medusa-buds which are never released. The gonads develop in the bud (or gonophore) and eggs or sperm are eventually shed into the sea. It should be noted that colonies of these (as of most hydroid coelenterates) are dioecious, that is to say, the colony bears either male or female medusa-buds, but not both together. Finally, the skeletozooids (Figure 4-4E) are just a thin covering of tissue over spiny projections of chitin arising from the chitinous mass of the encrusting base. Regeneration is possible in *Hydractinia*, but with extremely interesting limitations. Taken from adult colonies, fragments of each zooid type will regenerate only that type. However, since each colony arose from a zygote, which gave rise to a planula which settled and formed the first gastrozooid of the colony, there is obviously a stage in the development of any colony when individuals first become special morphs and their fragments "determinate" in a morphogenetic sense. For these reasons, developmental biologists are working on *Hydractinia* and on other polymorphic hydroids at the present time.

Spawning in *Hydractinia* is apparently synchronized by a photo-sensitive mechanism so that adjacent male and female colonies will release gametes together when exposed to certain light intensities after a period of dark adaptation. There is some recent evidence that different wavelengths of light are important, and it is significant that different physiological races occur which are characterized by their different colors. However, there is no clearcut evidence

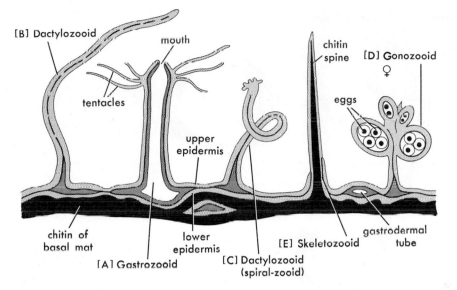

FIGURE 4-4. Polymorphy in *Hydractinia*. Five types of polyps occur in the colony. For further details, see text.

that different pigments result in different screening of the light-sensitive cells and therefore in different spawning reactions.

In many gymnoblastic hydroids, and in a few other Hydrozoa, the medusoid generation is reduced so that the asexually derived buds are retained as gonophores on the hydroid colony and there is no free-living jellyfish stage. The gonads mature within the gonophore, and may liberate sperms and eggs to give rise to a free-living planula, or may retain the eggs after fertilization. Even the free life of the planula can be suppressed. Before considering aspects of this, it is worth reviewing the normal process of asexual budding which gives rise to normal free-living medusae in a gymnoblastic hydroid. This is illustrated in Figure 4-5, where it can be seen that the bell rudiment and the spadix (or rudiment of the manubrial portion of the future medusa's gastrovascular cavity) are the first-formed elements of the bud. The characteristic four radial canal rudiments are also notable. Along with the rupture of the inside of the bell, which gives rise to the characteristic velum of the group, the fact that the rupture which separates the medusae from the parent colony takes place on the surface which will become the upper surface of the little jellyfish is characteristic. Many of these features are recognizable in the gonophores of several hydroids; however, in others—for example, the sporosacs (Figure 4-6)—little trace of medusoid structure is retained. As noted earlier, the gonophores (like the free medusae) can only be male or female. The male gonophores invariably liberate their sperms into the sea. The female gonophores can retain their eggs for varying periods of time: in some cases until after fertilization, in others until the planula is formed (see *Clava*, Figure 4-6), and in others still later until the planula has become transformed into a miniature hydroid polyp. This is then liberated ready for settlement. This last case is exemplified by *Tubularia* (see Figures 4-2 and 4-6) where the development stage eventually liberated is called an actinula. It swims, then wanders upside-down over the substratum for a

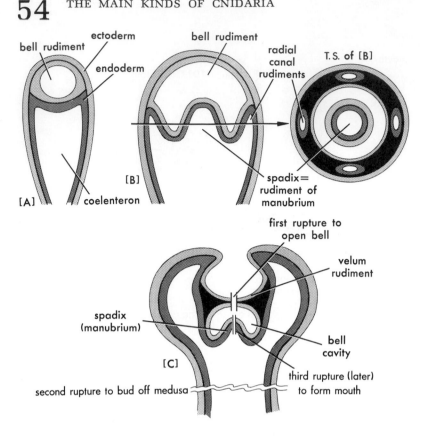

FIGURE 4-5. The process of asexual budding of a medusa in a gymnoblastic hydroid. For further details, see text.

while before settling and attaching to become the first polyp individual of a new colony.

There remain two relatively minor suborders of the order Hydroida (Hydromedusae). The first of these is the suborder Hydrida, a small group probably close to the gymnoblastic hydroids, which was set up for *Hydra* and its allies. In this group of freshwater coelenterates, the polyps are solitary: because budded individuals separate before attaining full size, no colonies are formed. There is no secreted skeleton and the individual animals are capable of considerable locomotion. There is no medusa stage at all, and no differentiation of gonad-bearing regions in the polyp. If these animals originated from a gymnoblastic stock with the usual life-cycle involving medusae, then they show a loss of medusoid structure unusual within the Hydrozoa, where gonophores usually retain some of the developmental features of the budding of free medusae and some structural vestiges of medusoid organization. There are only a few dozen species scattered in the fresh waters of the world, and most authorities assemble them into three genera: *Hydra, Pelmatohydra,* and *Chlorohydra.* These differ from each other only in body proportions and in the presence or absence of symbiotic algal cells in the gastrodermis. All their morphological simplifications, and the peculiarities of their reproduction, can be regarded as specializations for life in fresh waters. Typically, they reproduce by asexual budding throughout the spring and summer and then, toward fall, turn to sexual reproduction and produce fertilized eggs surrounded by protective coats.

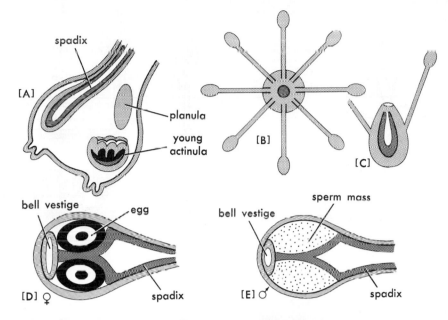

spadix

[A]

planula

young
actinula

[B]

[C]

bell vestige

egg

[D] ♀

spadix

bell vestige

sperm mass

[E] ♂

spadix

FIGURE 4-6. Reduction of life-cycle in gymnoblastic hydroids. **A:** Gonophore of Tubularia, in which the fertilized egg is retained through the planula stage and liberated as a miniature hydroid polyp, or actinula. **B** and **C:** Free actinulae, in top view and vertical section. **D** and **E:** Female and male "sporosacs" or gonophores of *Clava*.

These so-called winter eggs can survive freezing and desiccation. Thus, the only group of coelenterates to be successful in fresh waters have modified their life-cycle to the sort of annual pattern achieved by more highly organized animals living in the same environment. The buildup of population numbers in spring and early summer, and the provision of resistant stages to over-winter, uses a different mechanism but almost exactly parallels the result achieved in certain freshwater Cladocera (or the "water-fleas" of the freshwater plankton) by their process of switching from production of diploid eggs parthenogenetically through spring and early summer to normal haploid eggs in the fall. Animals as diverse as freshwater sponges and coelenterates, oligochaete worms, rotifers, ectoprocts, and planktonic crustacea all show similar patterns of life-cycle (though achieved in different ways). All reflect adaptation to the peculiar needs and stresses of life in fresh waters in the temperate latitudes of the world.

The fourth group, suborder Hydrocorallinae, is again a relatively small group of a few genera, which can be moderately common organisms in tropical seas. The genera *Millepora*, *Stylaster*, and *Distichopora* are all colonial, secreting massive calcareous skeletons. In *Millepora*, only the outermost layer of skeleton contains living tissue, while in the stylasterines the entire skeleton is permeated with living tissue. This results from the fact that the epidermis in *Stylaster* (and presumably also in *Distichopora*) secretes the skeleton *inwards* as a sort of "epidermal endoskeleton." Another difference is that species of *Millepora* have free medusae, while the stylasterines have only vestigial medusabuds (or gonophores). It is possible that they are not all closely related; that is, the Hydrocorallinae may be a polyphyletic group. Characteristically, they are

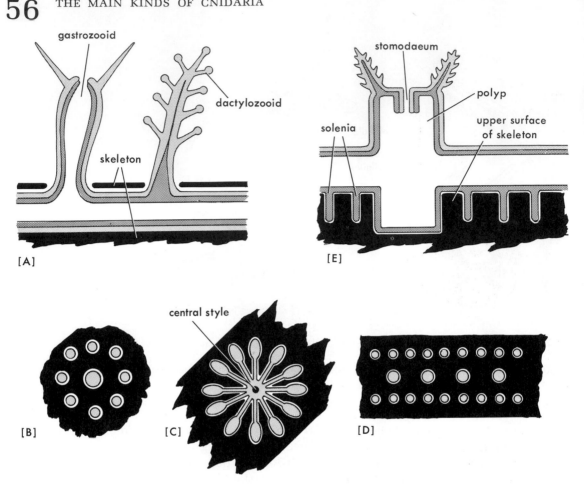

FIGURE 4-7. **A–D:** Hydrocorallinae or hydrozoan "corals." **A:** Vertical section through part of a colony of *Millepora*. Surface views show the characteristic patterns in *Millepora* (**B**), *Stylaster* (**C**) and *Distichopora* (**D**). **E:** Vertical section through an octocorallian, *Heliopora* (the blue coral), which secretes a massive endoskeleton.

always dimorphic (with two polyp types: gastrozooids and branched dactylo-zooids) and these polyps can always contract *within* the stony calcareous se-creted skeleton (Figure 4-7). The dactylozooids are always arranged in a spe-cific fashion around the feeding polyps or gastrozooids, and Figure 4-7 illustrates the patterns found in the three genera. The hydrocorallines mostly live with the true reef corals which will be discussed later, though they are sometimes found at much greater depths than the true corals. The batteries of nematocysts found in the knobs of the branched dactylozooids of *Millepora* constitute one of the few coelenterate dangers to man, the nematocysts being particularly virulent. It should be noted that some species of *Millepora* produce free medusae from special pits in the skeleton, and although these minute jelly-fish are liberated without mouth or canal system they do have separate sexes and bear gonads and have some structural features resembling anthomedusae. Some zoologists take this as evidence for a close relation between the hydro-corallines and the gymnoblastic hydroids discussed earlier.

Minor Hydrozoa

Another relatively minor subdivision of the class Hydrozoa is the second order, the Trachylina, a pelagic group of somewhat aberrant medusae which lack a hydroid generation. In our ecological sense "minor," these could be regarded as the unsuccessful jellyfish—the successful pattern of pelagic medusae being placed in a separate class, the Scyphozoa. The trachyline medusae are craspedote (i.e., with a velum as an inwardly developed shelf of the bell). Thus they resemble the anthomedusae and leptomedusae of more typical hydrozoan groups. On the other hand, the true jellyfish or Scyphozoa are velum-less (acraspedote). These less successful trachyline jellyfish are placed in two suborders, the Narcomedusae and the Trachymedusae, each numbering about twenty genera mainly found in the open waters of tropical oceans. Apart from certain peculiar anatomical features such as tentacles derived from the upper surface and greatly dilated radial canals in the Narcomedusae, and centripetal canals and an enormous manubrium in the Trachymedusae, biological interest in these jellyfish lies in two facts.

First, they are clearly Hydrozoa, more closely related to forms like the Hydromedusae than to the true jellyfish; secondly, they have no trace of a *normal* hydroid in the course of their life-cycle. Unfortunately, we know practically nothing about the life-cycle of the more primitive Narcomedusae, but the pattern of development which has been found in such genera as *Liriope* and *Aglaura* (both trachymedusans) is significant. In these, the fertilized egg develops directly into an actinula larva, resembling that of gymnoblastic hydroids like *Tubularia*. The actinula changes directly into a medusa which becomes adult as gonads develop. In one particular narcomedusan, *Cunina,* actinulae are known to multiply asexually before they all turn into adult medusae. Further, it has been claimed that in another narcomedusan, *Pegantha*, similar budding of the actinula stage involves a primary individual, which proliferates other individuals by budding. These become medusae while the primary actinula remains a "polyp." If this is so (and the actinula larva of *Pegantha* is epizoic on other medusae, and therefore rather rarely seen), then within the Trachylina we have representatives which show the more usual life-cycle involving alternation of polyp and medusae found in the other Hydrozoa. It is worth mentioning here that two very unusual medusae found in tropical or warmish fresh waters, *Limnocnida* from Africa and *Craspedacusta* from South America, may belong in the order Trachylina. Not only the phylogeny of these two medusae but also the physiology of their osmoregulation and ionic regulation remain obscure. Incidentally, both were first discovered after being introduced, along with giant water lilies from the tropics, to botanical ponds in Europe and North America.

The third order of the Hydrozoa, the Actinulida, was set up in 1959 to comprise two families each with a single genus (*Halammohydra* and *Otohydra*). In turn these genera each includes a few species of small hydrozoans found in the interstitial fauna of marine sands. The Actinulida are minute (about 1 millimeter in length), free-living, and solitary with the actinula type of organization as the adult form. They develop directly from the egg and at no stage show medusoid organization or the typical organization of a hydrozoan polyp. Their size and the simplicity of their organization undoubtedly reflect the peculiar needs of their rather specialized habitat. Some authorities claim, however, that the Actinulida originated from an ancestral actinula type which could have been closely related to the ancestor of Hydrozoa as we know them. Other authors have regarded them as evolved by neoteny from the trachyline medusae which have an actinula in their development. This theory implies their origin to be trachyline larvae which acquired gonads (and capacity to reproduce) before the assumption of the adult trachyline medusan form. These alternative theories are significant to questions of the evolution of the coelenterates as a whole, which will be discussed later.

Oceanic Hydrozoa

The fourth and final subdivision of the class Hydrozoa is the order Siphonophora. These pelagic, swimming or floating colonies of hydrozoan individuals have strong claims to be numbered among the most peculiar animals which have ever been described. They are characterized by a greater extent of polymorphism than is found in any other group of coelenterates, with both modified hydroid and medusoid individuals being associated together in colonies. The so-called Oceanic Hydrozoa have attracted a series of famous zoologists since the time of T. H. Huxley. Essentially, the polymorphic individuals are attached to a central column or string or disc, and individuals may be specialized for feeding, protection, reproduction, buoyancy, or swimming, and may result from modification of hydroid or medusoid organization.

The following sorts of hydroid morphs are found:

1. Gastrozooids, or feeding individuals, often consisting only of a mouth without tentacles, though some have one long contractile tentacle arising from their base;
2. Dactylozooids, or protective individuals, resembling the gastrozooids but usually lacking a mouth and most often having the long basal tentacle;
3. Gonozooids, or reproductive individuals, which correspond to the blastostyles of colonial hydroids and which produce medusa-buds or gonophores;

and the following medusoid morphs are found:

4. Nectophores, or swimming individuals, each consisting of merely the bell of a medusa; that is, they lack tentacles or manubrium but are well muscularized for the propulsion of the colony as a whole;
5. Pneumatophores, or gas-sac floats, each a modifed medusa bell enclosed as a bladder into which gas is secreted;
6. Oleocysts, or oil-floats, of similar structure to the last;
7. Phyllozooids, or bracts, which are other defensive structures, usually leaf-like and sometimes relatively elaborate, but always studded with nematocysts and serving for the protection of other morphs, and always obviously derived from modified medusae; and
8. Gonophores or gonad-bearing individuals, which are usually borne on gonozooids, though they rarely develop to free medusae and almost always are retained as attached buds, which in turn bear the gonads producing the gametes.

There are two distinct suborders of the oceanic siphonophores. The first, Calycophora, includes colonies which are linear like a long string or cord, the apical end of which always consists of one or more nectophores or swimming bells. Behind these nectophores the individuals are grouped as in Figure 4-8, in sets, each set being repeated many times over on the long stem. A typical set (or

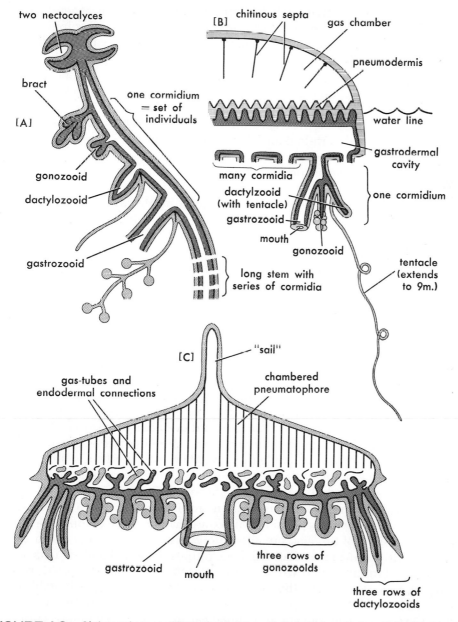

FIGURE 4-8. Siphonophora or "Oceanic Hydrozoa." **A:** Apical part of a calycophoran colony. **B:** Partial section through *Physalia* ("Portuguese man-of-war"), showing only one of the many sets of individuals. Compare this with Figure 4-9. **C:** Vertical section through a colony of *Velella*.

cormidium) would consist of a bract, plus a gonozooid, plus a dactylozooid with its tentacle, plus a gastrozooid and its tentacle. It should be noted that in these linear colonies, although there are relatively few individuals with open mouths, the gastrovascular system into which the mouths open is continuous throughout the whole colony. Systematics and taxonomy are relatively concise for these cord-like colonies, most genera being defined as having x nectophores plus y cormidia.

The other major subdivision is the suborder Physophorida, where there are always floats of one sort or another present at the apex of the colony. This is a much more diverse group which is usually split into three superfamilies. The first of these is the Physonectae, which have linear colonies like the last except that there is an apical float followed by a number of nectophores and then by a number of cormidia. Some of the typical genera in this group, such as *Agalma* and *Halistemma* from the Mediterranean, show a budding zone between the swimming bells and the typical cormidia. This zone produces both more swimming bells and more cormidia and the colony continually elongates.

Much better known than these are the Rhizophysaliae, where there is a large float above the water level and the colony is not usually linear. These are exemplified by *Physalia*, the "Portuguese man-of-war," illustrated in Figure 4-9. The bright blue floats of *Physalia* are blown along by the wind on warmer seas throughout the world. The gas in the float chamber varies in composition from specimen to specimen and can differ significantly from air (up to 90% nitrogen, 5–20% carbon monoxide remarkably secreted by *living* ectodermal tissues, and 1.5% argon). An oval disc on the underside of the float contains not only the gas-secreting tissue but also a common gastrodermal cavity for the entire colony (Figure 4-8B). From the underside of this disc hang groups of cormidia —each including one gastrozooid, a branched blastostyle which bears both male and female gonophores, and a large dactylozooid with an enormous nematocyst-bearing tentacle. These tentacles, seen clearly in Figure 4-9, can extend 9 meters from a 12-centimeter-long *Physalia* and, after catching prey which can include relatively large fish, can contract to about 10 centimeters in length. Obviously, the neuromuscular arrangements of tentacles capable of hundredfold contraction must be peculiar. The nematocysts of the tentacles are particularly virulent to small vertebrates, and even dangerous to man. When a fish is caught by these tentacles, more seem to close in on it and discharge nematocysts; then they are all hauled up together until the fish is brought into contact with the mouths of the gastrozooids which wrap around it. They pull away only when little but the skeleton of the fish remains.

Even casual observation of *Physalia* in tropical waters will reveal another movement which involves considerable coordination of the colony as a whole. It is that the sides of the float are alternately pulled down under water to wet the tissue and secreted material which surrounds the gas chamber. This occurs at intervals of minutes, and whether it proves to be a reflex promoted by local drying, or a spontaneous rhythmic occurrence, it does imply considerable coordination of the contractile myoepithelial cells associated with different members of the colony. The concept mentioned earlier, of the coelenterate colony as a "superorganism" of interdependent individual zooids, is never so obvious as when contemplating a working healthy colony of *Physalia*.

However, perhaps the highest degree of interdependence is found in the

[A]

FIGURE 4-9. *Physalia physalis*, the "Portuguese man-of-war." **A:** The whole colony, with its pneumatophore or gas float above and tentacles trailing from the groups (cormidia) of individual zooids below. **B:** Closeup view of a captured fish being held by the feeding polyps (gastrozoids), whose trumpet-shaped mouths almost cover the fish. [Photos © Douglas P. Wilson.]

[B] **61**

third superfamily, the Chondrophorae, which are the most highly organized colonies in the animal kingdom. Such genera as *Velella* (Figure 4-8C) and *Porpita* have a single large central gastrozooid, with the only mouth in the colony, surrounded by concentric rings of other zooids. These small colonies—*Velella* is only about 2 centimeters across—are supported by a chambered pneumatophore, which in the case of *Velella* bears a diagonal sail. Obviously, they catch prey considerably smaller than that of *Physalia*, and fish only immediately below the surface. Even the developmental stages of the life-cycle are completely oceanic; the fertilized egg is released from the colony and sinks slowly into the open waters of the sea. It hatches out into a so-called "medusa" larva which swims up again, gradually growing into the colonial form of a *Velella* as it reaches the surface. The most impressive feature of these small colonies is the fact that the individual zooids are not arranged in cormidia but instead are definitely placed in a regular pattern around the single gastrozooid, so that even under close examination this hydrozoan colony looks like a single, highly organized, individual animal. Some authorities claim that such genera as *Velella* and its allies are not related to the other siphonophores, but can be interpreted as compound heads of gymnoblastic polyps, which have evolved for detached pelagic life.

It is most likely that the siphonophores arose from those anthomedusae which are capable of budding (see page 51).

True Jellyfish

Class Scyphozoa encompasses the exclusively marine, true jellyfish. In these, the medusa is the dominant part of the life-cycle and, while all jellyfish are more or less alike (see *Aurelia*, Figure 4-10), individuals can be far larger than

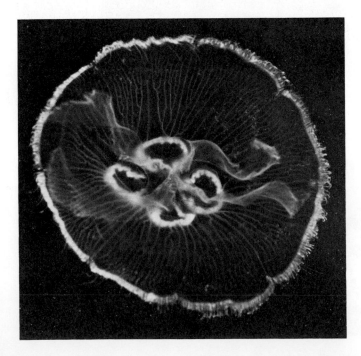

FIGURE 4-10. A typical scyphozoan medusa. An adult specimen of *Aurelia* is viewed from below. [Photo © Douglas P. Wilson.]

the medusae of the hydrozoan groups—some species are nearly 2 meters across. They are all acraspedote and have a complex gastrovascular cavity and complicated marginal sense-organs. Apart from this, the general form is like that of the medusae discussed earlier, but although the thickness of tissues has not increased greatly the mesogloea makes up an even greater proportion of the bulk of a larger true jellyfish. The class is divided into five orders, but the great majority of species belong to the first, the Discomedusae, and bear a close resemblance to *Aurelia*. Polyp stages are always small, and peculiarly "square" in cross section (Figure 4-11).

The gastric cavity in the Scyphozoa differs from that in the Hydrozoa, being divided into a central stomach and four gastric pouches. These are formed in development in a totally different fashion from the four radial canals of the other groups. The septal edges between the gastric pouches, and certain other surfaces within the cavity, bear characteristic gastric filaments carrying batteries of nematocysts which kill any food organism still alive when taken into

[A]

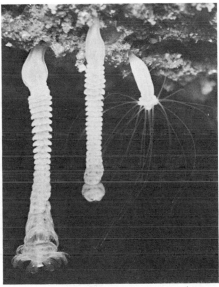

[B]

[C]

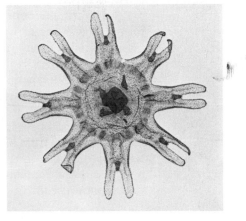

FIGURE 4-11. The life-cycle of a scyphozoan medusa. **A:** An underwater rock-cave supports several scyphistoma (polyp) stages of *Aurelia*, most of which are strobilating. An ephyra larva has just been released (bottom right). **B:** Closeup view of two strobilating scyphostoma stages and one earlier *Hydra*-like polyp with long tentacles. **C:** An ephyra larva after release. The spaces between the eight arms will fill up as it grows into a subadult medusa. [Photos **A** and **B** © Douglas P. Wilson; photo **C** courtesy of Martyn L. Apley from a preparation by the author.]

the gut. These filaments are also the sites of concentrations of enzyme-secreting cells and the main tracts of absorptive cells for digestion. Further complexity of the gastrovascular system results from the massive mesogloea, necessitating efficient circulation between the widely separated living epithelia. The gastrovascular system is lined with cilia for continuous circulation. Water is drawn in orally to the stomach; thence it passes into the gastric pouches and then by adradial canals into the circular canal at the periphery of the umbrella. It then passes back by the branched interradial and perradial canals to exhalant grooves on the oral arms. This circulation of water by ciliary currents serves to take oxygen, and partially digested food, from the open stomach to all parts of the tissues. It should be noted that all the canals are lined with endodermal cells so that presumably the final intracellular digestion can take place anywhere in the system, to meet local needs of growth or metabolism. Thus in the Schyphozoa there is better justification for the use of the term gastrovascular (implying combined digestive and circulatory functions) than in any of the other coelenterates. There is no inturning of the ectoderm as a stomodaeum in the Scyphozoa, but the angles of the mouth are continued into four oral arms which are grooved and often frilled and always heavily ciliated.

While all large jellyfish of the class Scyphozoa—like all cnidarians—are predaceous carnivores, the size ranges of their preferred prey vary greatly. Some, including *Aurelia,* are essentially suspension-feeders on the marine plankton, trapping it in mucus on the subumbrellar surface while slowly sinking between pulses of the swimming musculature of the bell. Mucous-bound plankton is moved by cilia to the bell margin and then passes by the grooves of the oral arms to the mouth. Other jellyfish of similar size including species of *Chrysaora* and *Cyanea* (with longer oral arms and many long trailing marginal tentacles, Figure 4-12) compete with *Physalia* and its allies in feeding on small and medium-sized fishes, larger crustaceans, and a variety of other invertebrates. These fish-eating forms rely to a greater extent on the batteries of nematocysts on their tentacles and oral arms for prey capture. A piece of natural history advice for human swimmers in the temperate waters of the North Atlantic (applicable both to the mid-Atlantic coast of North America and to the western coasts of Europe) is to ignore jellyfish which have bluish or milky translucency with a flattened discoid umbrella and no trailing tentacles (*Aurelia,* Figure 4-10), while avoiding brownish-purple-pink and golden-brownish globose jellyfish with numerous long trailing tentacles (*Cyanea* and *Chrysaora,* Figure 4-12). An armament of nematocyst batteries adapted to kill fish, as in the latter forms, rather than to assist in the capture of zoöplankton (*Aurelia*) is more likely to cause injury or discomfort on human contact. The fatal "Lion's Mane" of one of Sherlock Holmes' adventures was *Cyanea capillata.*

In the Scyphozoa the muscular system of contractile epithelial cells is essentially limited to the surface under the umbrella. This forms the bell musculature which is used in slow continuous swimming and also in the righting reflexes discussed earlier. There are normally eight compound marginal sense-organs, which each include two sensory pits (assumed to be concerned with chemoreception), one ocellus or eye-spot, one or two tactile sensory "lappets," and one tentaculocyst. It is the last which seems to be most concerned with postural control of the jellyfish. These compound marginal sense-organs

FIGURE 4-12. Another living scypho-zoan jellyfish. A slowly swimming specimen of the compass jellyfish, *Chrysaora hyso-scolla.* Such forms are best avoided by swimmers since their nematocysts are much more virulent to vertebrates than those of *Aurelia* (see Figure 4-10). [Photo © Douglas P. Wilson.]

have long been known to be the main sources of stimulation of the bell muscu-lature in swimming. However, even if all but one of them are destroyed, under certain conditions there can still be coordination of swimming activity. Usually when the last compound sense-organ is experimentally destroyed, all muscular contraction ceases, but, in some cases, trapped contraction waves occur and have been known to continue for eleven days.

The gonads lie on the floor of the gastrovascular cavity, and gametes pass from each of the four of them into the gastric pouches, thus being spawned to the exterior via the mouth. In most cases the ripe sperms are liberated directly, but ripe eggs are not. They are fertilized in the female stomach and only pass out through the mouth after a few cleavages—though they are usually extruded well before the planula stage. The coincidence of digestive enzymes and de-veloping embryos in one cavity of an animal is remarkable and might well repay further investigation.

Most discomedusans like *Aurelia* have a life-cycle pattern which involves a settled polyp of a peculiar form. The released planula invaginates and settles, becoming a miniature polyp attached to seaweed. This first hydroid stage was known to earlier naturalists as *"Hydra tuba"* (Figure 4-11). It is totally unlike the polyps of other groups, and has four gastric ridges as septa dividing the cavity, and four ectodermal pits sunk into the gastrovascular cavity from the oral surface. A functional aspect of this peculiar anatomy is that the longitudinal muscles of the polyp are provided on the faces of these ectodermal pits—in other words, the subumbrellar musculature on the oral surface of the polyp is carried down into these pits to allow shortening of the organism. This polyp, which is usually called a scyphistoma, begins to divide transversely after a certain period of growth (Figure 4-11). This transverse division gives rise to a strobila like a column of little discs. Each of these discs then becomes detached as an ephyra larva (Figure 4-11), which swims freely in the plankton. The ephyrae grow by filling up the spaces between the eight arms and, by increasing elaboration of the gastric cavity and increase in the bulk of the mesogloea, become adult jellyfish. The scyphistoma lives on, growing from its base, regrowing tentacles, and (if surviving to the next year) strobilating again. Neuromuscular changes are involved in the transition of a sessile polyp into an actively swimming medusa. Near the end of Chapter 3, we noted that the scyphozoan jellyfish had two nerve nets with distinctly different conduction rates. Recent electrophysiological work has revealed that the rapidly conducting system, the so-called "giant fiber nerve net," controlling the swimming muscles and coordinating the active life of the adult medusa only develops and becomes functional in the ephyra larva. The other slow-conducting system, or "diffuse nerve net," controlling tentacle movements in the adult jellyfish is present in the sessile polyp stage, where it controls both protective and feeding reflexes, and persists through the ephyra stage and into the medusa, though with reduced functional responsibility.

Most free-living Scyphozoa have this life-cycle (planula → scyphistoma polyp → strobila → ephyra larva → adult medusa), although a few deep-sea genera, such as *Pelagia,* cut out the polyp generation and larval stages so that each medusa gives rise to miniature medusae. Another large jellyfish of the deep sea, *Stygiomedusa,* has solved the problem of this scyphistoma stage in another way. In it the polyp stage or scyphistoma is retained in the subgenital pits of the adult medusa and is nourished there by special maternal tissues (rather like the foetus in a viviparous vertebrate).

The second order of the Scyphozoa is the Rhizostomeae. This is a relatively minor group of medusae in which the oral arms are greatly enlarged, and the mouth is not open in the adult. The oral arms are much perforated by a system of pores, and the animal feeds on very minute plankton sucked in through them. *Rhizostoma* and several other genera are found in tropical waters, particularly of the Indo-Pacific. There are also a few aberrant forms which live, mainly in mangrove swamps, in shallow water in the tropics. Some of these are semisessile, living upside down and, by pulsations of their fixed bell, bringing in a water current with plankton. *Cassiopeia* is an example: it even has a sucker of columnar cells formed on the "upper" surface of the umbrella away from the mouth and is fixed upside down by this.

The third group is the order Stauromedusae, in which there is no normal free-living medusa but only a "pseudohydroid." The organism has the typical form of a scyphistoma slightly elaborated, and a strong case can be made for

FIGURE 4-13. Living stauromedusans or "attached jellyfish." Attached to branches of the red seaweed *Ceramium* are specimens of *Haliclystus auricula* (left) and *Craterolophus convolvulus* (right). [Photo © Douglas P. Wilson.]

this apparent hydroid actually being a modified medusa. They are smallish organisms only a few centimeters in height, placed in genera such as *Haliclystis* and *Lucernaria* (Figure 4-13). They are typically square in section with eight lobes bearing small tentacles. There are four gastric ridges in the coelenteron and four ectodermal pits invaginated from the oral end. A simplified diagram of their complexity is shown in Figure 4-14. Stauromedusae occur throughout the colder waters of the world attached to seaweeds. Some species are permanently cemented; others detach and reattach. Some species absorb pigments from the plant to which they are attached and become closely similar in color to their seaweed hosts. In most, the fertilized eggs give rise to planulae which crawl for a few days and then give rise to this "adult polyp" with gonads. Perhaps the easiest way to regard the stauromedusan is as a combination of polyp and medusa. It seems most likely that the Stauromedusae are not in any way ancestral, but were derived from forms which already had both polyp and medusa as stages in their life-cycle. Put more formally, they probably result from a secondary phylogenetic fusion of two ontogenetic stages.

There are two remaining small orders of the Scyphozoa. The Coronatae are typically jellyfish of the deeper waters of the ocean. They have the upper surface of the umbrella subdivided by a horizontal circular groove, the coronal groove, and the gastrovascular system is peculiar, with the stomach in the form

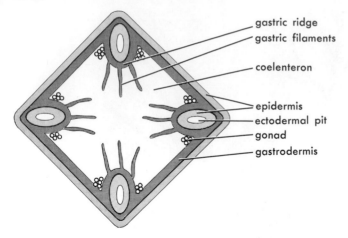

gastric ridge
gastric filaments

coelenteron

epidermis
ectodermal pit
gonad
gastrodermis

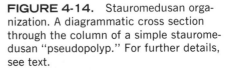

FIGURE 4-14.   Stauromedusan organization. A diagrammatic cross section
through the column of a simple stauromedusan "pseudopolyp." For further details,
see text.

of a wide ring-sinus. Little is known of their development. The Cubomedusae
have cuboidal bell-shaped umbrellae, making them distinctly square in transverse section. From each lower corner of the cube springs a group of tentacles
which gives them a characteristic appearance. They are found in warm shallow
waters in the tropics, and are strong swimmers and voracious feeders, the bell
pulsating at rates of two to three contractions per second. They have been
called "sea wasps" because of this, and because of the already-mentioned viru-
lence of their nematocysts. Some authorities regard the Cubomedusae as being
near to the stem-group of the Scyphozoa. In general terms, the Stauromedusae
and the Cubomedusae seem to be closer than any other scyphozoans to the
basic hydrozoan stock of the coelenterates, and an obvious feature of this is
their tetraradiate symmetry.

Life-cycles

Before passing to the last class of the coelenterates, the Anthozoa, it is worth
discussing the various forms of life-cycle found in the Hydrozoa and Scyphozoa and their possible evolutionary significance. Table 4-2 sets out the life-
cycles in a range of forms which we have already mentioned. The classical interpretation of these facts is that the hydroid or polyp is a larval stage. This
implies that the sexual adult is the medusa which bears gonads. It also implies
that the larval stage may be omitted and there can be direct development from
adult to adult, or that the adult stage may be omitted by neoteny occurring (in
other words, by the larva forming mature gonads). An alternative view would
suggest that metagenesis had occurred. This would imply that an original sessile adult form of polyp secondarily gave rise to detached medusae bearing
gonads. Obviously, the two theories of life-cycle evolution could be used to
support alternative versions of evolution within the coelenterates, one implying evolution from primitive Hydrozoa to more advanced, although simplified,

TABLE 4-2
Hydroid and Medusoid Stages in Cnidaria (A Partial Summary)

"PRIMITIVE LIFE-CYCLE"

1. *Obelia* and many other hydroids: fertilized ovum → planula → polyp (asexual) → medusa (sexual) → sperm or ova

MODIFICATIONS

2. *Tubularia:* fertilized ovum → polyp with gonophores (sexual) → sperm or ova
3. *Hydra:* fertilized ovum → polyp (sexual) → sperm or ova (sexual generation pushed back)
4. *Gonionemus* (special case): fertilized ovum → planula → reduced polyp → medusa (sexual) → sperm or ova
5. Most Trachylina: fertilized ovum → planula → medusa (sexual) → sperm or ova (complete reduction of hydroid)
6. *Aurelia* and most Scyphozoa: fertilized ovum → planula → scyphistoma (polyp, asexual) → dominant medusa (sexual) → sperm or ova
7. *Pelagia:* fertilized ovum → medusa (sexual) → sperm or ova (complete reduction of hydroid)
8. *Haliclystis:* fertilized ovum → planula → polyp-like attached medusa (sexual) → sperm or ova
9. All Anthozoa: fertilized ovum → planula → polyp (sexual) → sperm or ova (complete loss of medusa)

Anthozoa, and the other implying evolution from Anthozoa to Hydrozoa. These evolutionary theories will be discussed again later. Meanwhile, one kind of evidence, often ignored by protagonists of the anthozoan origin, is important. It is that in cases where gonophores are formed attached to polyps in Gymnoblastea, these clearly are reduced medusae, often showing such medusan structures as the velum and gastrovascular canals. Indeed, all gonophores, wherever attached on the various Hydrozoa in which they occur, clearly are reduced medusae.

Functional and ecological aspects of the case which may be very important are also often ignored. It seems to me that sessile organisms like coelenterates face some of the problems that animal parasites do. The autecology of each population involves two major phases: buildup of species-biomass and dispersal. In the coelenterates, dispersal forms may be medusae, or actinulae, or planulae, or other larvae. The buildup of biomass may be achieved to a great extent by asexual budding, and this may occur at any stage. We have noted the multiplication of polyps in many instances by budding, and the occasional multiplication of medusae in the same way. It is also notable that some planulae can bud and multiply. Further, the scyphistoma in Scyphozoa may bud in more than one fashion, producing more polyps or producing ephyrae by strobilation. Thus, not merely one but many asexual generations may intervene between each sexual one. N. J. Berrill has pointed out the unbelievable plasticity of pathways of development in scyphozoan life-cycles. It is therefore almost impossible to say which type of life-cycle is most primitive in the more complex coelenterates, though it is perhaps reasonable to postulate that, within the Hydrozoa alone, alternation of a medusoid stage with a polyp stage is the most primitive condition.

Anthozoa

The remaining class of the coelenterates is the Anthozoa, comprising corals and sea-anemones among other forms. No anthozoan has any medusoid stage in its life-cycle; the polyp gives rise to fertilized eggs, which develop into planulae, which give rise to adult sexual polyps again. They may be solitary or colonial, but typically have a cylindrical column rising from a basal disc with a ring of tentacles surrounding an oral disc at the other end. The gastrovascular cavity is characteristically divided by radial mesenteries, like the gastric ridges

[A]

[B]

FIGURE 4-15. Living anthozoans. Typical sea-anemones, expanded and "fishing": *Tealia felina* (**A**) and *Anemonia sulcata* (**B**). [Photos © Douglas P. Wilson.]

of the Scyphozoa, but more numerous and more complicated. The ectoderm is always tucked into a characteristic stomodaeum, and the mouth is typically slit-like or oval. In general, the Anthozoa tend to exhibit bilateral (or at least biradial) symmetry, never tetraradial, as in the preceding groups. Anthozoan internal organization is alway bilateral—even in the sea-anemone, which externally appears to show beautiful radial symmetry like that of a flower (Figure 4-15)—and this primary bilateral organization arises in embryonic development and persist into the adult. The gonads are endodermal and usually borne on the mesenteries. Figure 4-16 shows something of the characteristics of organization of these mesenteries, in that they are in-pushings of the endoderm up to the level of the stomodaeum but above this they fuse with it. The free edges of the mesenteries below the stomodaeum are usually thickened and form mesenterial filaments (which may be much-folded to become considerably longer than the mesentery itself). At the ends of the mouth slit there are (either one or two) ciliated gutters or siphonoglyphs, and the mesenteries associated with them differ from the others in their arrangement of muscle bands and are called directive mesenteries. The tentacles are hollow in all Anthozoa and contain extensions of the gastrovascular cavity. Myoepithelial cells are organized into well-developed muscles, including the longitudinal bands on the mesenteries and circular sphincters around the disc and stomodaeum. The elaborate pattern of these well-developed muscles allows for quite complex behavior, especially in sea-anemones. The mesogloea is always relatively thick and often supported with fibrous material. The nematocysts are usually relatively powerful and often localized into batteries. Besides the simple sexual life-cycle outlined above, there is extensive asexual reproduction by budding and by fragmentation.

The class Anthozoa is conventionally divided into two subclasses, the Octocorallia (or Alcyonaria) and the Hexacorallia (or Zoantharia). This primary di-

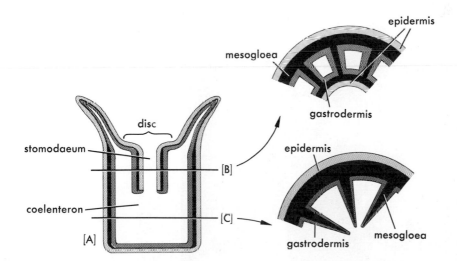

FIGURE 4-16. Anthozoan organization. Longitudinal section (**A**) and partial cross sections (**B** and **C**) through a typical anthozoan polyp. Contrast with hydrozoan organization as shown in Figure 4-7.

vision appears to reflect a natural difference in organization and in relationships. The Octocorallia have their tentacles and mesenteries arranged in an
eightfold symmetry and are always colonial. Their tentacles are pinnate, and
there is a single siphonoglyph. The supporting skeleton can be of various sorts
but is always an endoskeleton, often relatively soft and occasionally built up of
separately secreted spicules. Below the stomodaeum (see Figure 4-18), two
mesenteries differ from the other six. These, the dorsal or directive mesenteries,
are highly ciliated, while the rest are lobed and bear not only digestive cells but
also the gonads. The ciliated dorsals are longer, go farther into the coelenteron,
and are responsible for circulation of water. In several groups of Octocorallia,
there is dimorphism of individuals within the colonies. Autozooids have all
the structures described above, feed, and bear gonads, while siphonozooids do
not feed or bear gonads, but have a mouth with a well-developed siphonoglyph

FIGURE 4-17. Living colony of
the octocorallian *Alcyonium digita-
tum,* with polyps expanded. Compare
with Figure 4-18A. [Photo © Douglas
P. Wilson.]

and only the two ciliated mesenteries. These siphonozooids are concerned with water circulation through the colony as a whole.

As shown in Table 4-1, the Octocorallia comprise six orders of very unequal size.

Soft-corals

One of the largest groups of Octocorallia is the order Alcyonacea, or soft-corals, representatives of which (including the type genus *Alcyonium*, Figure 4-17) are not uncommon in moderately deep waters along our Atlantic coast. However, they are commonest in warmer waters, particularly of the Indo-Pacific. The polyps have the form illustrated in Figure 4-18, with a skeleton of separate calcareous spicules embedded in a massive mesogloea. This endoskeleton is com-

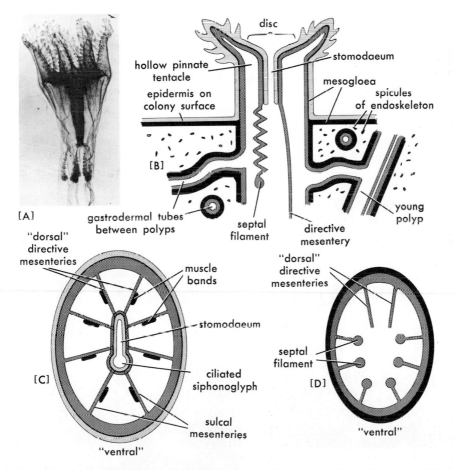

FIGURE 4-18. Octocorallian organization as shown in polyps of *Alcyonium*. **A:** Individual polyp removed from endoskeleton. **B:** Longitudinal section through a polyp embedded in the endoskeleton. **C** and **D:** Cross sections of a polyp. See also whole colony of *Alcyonium* in Figure 4-17 and text for further details and discussion. [Photo courtesy of Martyn L. Apley, from a preparation by the author.]

mon to the whole colony and canals of endodermal origin run through it connecting the gastrovascular cavities of individual polyps. The soft mass of the common skeleton (termed the coenenchyme) has also given the popular name "dead-man's fingers" to castup pieces of colony. Except for the fact that the colonies are not dimorphic, but consist entirely of normal feeding polyps. *Alcyomium* has all the characteristics listed above as diagnostic of the subclass Octocorallia. This is undoubtedly one of the simplest body forms to be found in the Anthozoa as a whole and has some claims to being an ideal model for an ancestral anthozoan polyp. Many of the more complex Hexacorallia, such as sea-anemones and true corals, go through a stage in their development in which they closely resemble a young alcyonarian polyp. Further, there are very close structural and functional resemblances between this type of polyp and the four-cornered polyp stage in the life-cycle of scyphozoan jellyfish.

Three orders are of relatively minor importance, both ecologically and otherwise. The order Stolonifera includes *Sarcodictyon*, which has the simplest sort of anthozoan colony: a creeping stolon with separate polyps arising from it, and no true common skeletal mass or coenenchyme. The group also includes the more complex colonies of *Tubipora*, the so-called organ-pipe coral, in which the skeleton (made up of calcareous spicules stained red with iron salts) is built into a series of tubes strengthened by lateral platforms and also by smaller internal partitions cutting across each tube (Figure 4-19).

In another minor group, the order Telestacea, the colonies are again simply branched stems, but each main branch consists of a much elongated single polyp, budding lateral polyps alternately on either side. The third small group is the order Coenothecalia. This includes only one genus, *Heliopora*—the blue

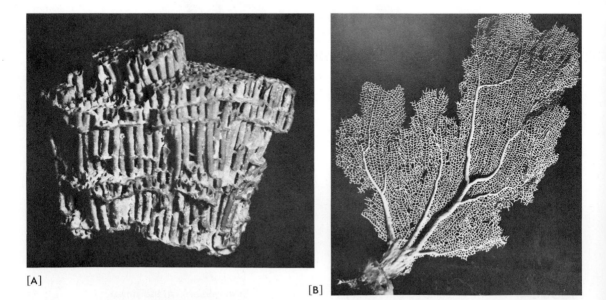

[A]
[B]

FIGURE 4-19.   Skeletons of two octocorallian Anthozoa. **A:** *Tubipora*, the "organ-pipe coral" (Octocorallia: Stolonifera). **B:** *Gorgonia*, a "sea-fan" (Octocorallia: Gorganacea). [Photos by the author, **B** collected and prepared by Robert J. Avolizi.]

coral which lives along with the true corals on the reefs of the Indo-Pacific. It is the only member of the whole group of Octocorallia in which the secreted spicules of the skeleton are not separate but form a massive basal block. The surface of this block is perforated by blind pits of two sizes, larger ones into which the polyps can contract and smaller ones termed solenial tubes (Figure 4-7). The living tissues of the colony are limited to the upper surface of the skeletal block and the colony slowly rises as more skeletal material is laid down.

The remaining two orders of the Octocorallia include many more species and are of some ecological significance. The sea-fans or horny-corals form the order Gorgonacea, in which the colony usually branches in one plane and has an axial skeleton of horny proteinaceous material as well as masses of calcareous spicules arranged in various ways around the polyps. They are widespread throughout the world, but normally require a hard substratum of rock or other coral on which to grow. They are found at all depths but are perhaps most abundant on the Atlantic coral reefs of Florida and the Caribbean, where they make up some of the most distinctive and attractive elements of submarine "coral gardens" and "coral grottoes." Their colors range from yellows through orange to reds, pinks, and purples. Like trees on land, their growth forms are very often orientated to the direction of the prevailing currents (Figure 4-19). The precious coral of commerce, the bright pink *Corallium rubrum*, belongs to this group and is dredged mainly off Japan and in the Mediterranean.

The sea-pens, or order Pennatulacea, represent the octocorallian form specialized for life on soft, muddy sea bottoms. The fleshy colonies consist of a primary axial polyp modified to support lateral branches of secondary polyps.

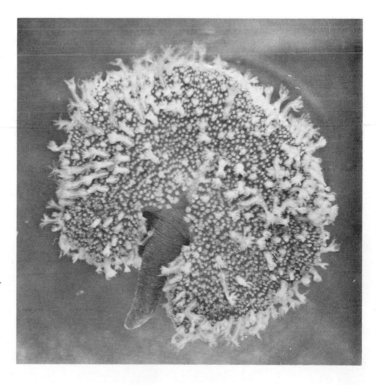

FIGURE 4-20. A colony of the sea-pansy *Renilla*. Note some expanded polyps and the colony peduncle (or "root") used as an anchor in the soft sea-bottom, which is derived from the primary or axial polyp of the colony. [Photo by the author.]

The whole structure is erectile, being inflated by water drawn in through the siphonoglyphs of the individual polyps. The lower end, or peduncle, is embedded in the mud and lacks branches or open polyps. There is usually a central axial rod strengthening the rachis. All colonies are dimorphic with autozooids and siphonozooids, and many forms show quite elaborate bilateral symmetry of the colony as a whole. The central polyp is capable of considerable coordinated movement, since the gastrodermal musculature is well-developed down the stalk and capable of bending the colony as a whole. Further, peristaltic movements of the peduncle are possible, and serve to dig the colony deeper into the substratum. A few forms are capable of actual burrowing. Colonies range in size from a few centimeters to over 1 meter in length, and there is some variation in the colony form. Certain deep-water species like *Umbellula* have a very long axial stem supporting only a terminal tuft of polyps. *Pennatula* itself is more typical of the forms with polyps borne on side arms, while *Renilla*, the so-called sea-pansy, has the polyps grouped on the upper surface of a flat disc like a flower (Figure 4-20).

Sea-anemones

The subclass Hexacorallia is ecologically the most important group of coelenterates because it includes the true corals. The tentacles and mesenteries are often in multiples of six (but never in eights). The tentacles are always simple (not divided or pinnate), and the skeleton may be calcareous or horny but is never made up of spicules and is an external mass secreted by cells of ectodermal origin. About half the known species in the group are colonial; the other half are solitary. Five orders are shown in Table 4-1, but of these only two are of major importance: the Actinaria or sea-anemones and the Madreporaria or true corals.

Most typical sea-anemones (order Actinaria) are solitary polyps living in the littoral zone or in shallow water of the sea, mostly attached to rock or other hard substrata, although a few forms can burrow into soft substrata and a few other forms live regularly as epizoic animals. As is characteristic of the Anthozoa as a whole, the longitudinal musculature is restricted to the faces of the mesenteries. The large number of mesenteries found in an adult sea-anemone have a definite arrangement in spite of the superficial appearances. There are twelve, or six pairs, of primary mesenteries fused with the stomodaeum near the oral end of the polyp, and in many there are two siphonoglyphs and two pairs of directive mesenteries. Figure 4-21 illustrates the condition and shows where the secondary, tertiary, and quaternary mesenteries can develop. Essentially the spaces between mesenteries are of two kinds, termed entocoelic and exocoelic, and additional mesenteries can develop only in the exocoels. The arrangement of the longitudinal muscle bands (shown here in cross section) is also very characteristic. No obvious functional significance can be attached to the specific order in which facing bands of muscle are developed, but the pattern follows an unvaried rule throughout the sea-anemones and corals. In the development of many sea-anemones there is a stage where only four pairs of mesenteries are laid down (Figure 4-21). This so-called *"Edwardsia"* stage may

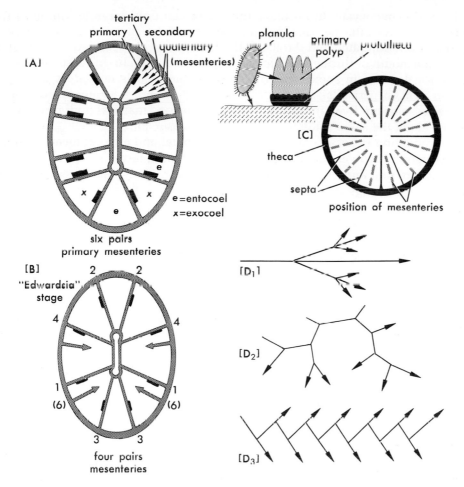

FIGURE 4-21. Sea-anemones and corals. **A:** Development of mesenteries in a typical sea-anemone. **B:** The earlier *"Edwardsia"* stage of sea-anemone development, which has a significant resemblance to octocorallian organization. Compare with Figure 4-18. **C:** The initial settlement of a coral, formation of a prototheca, and arrangement of protothecal septa. **D:** Patterns of branching in the growth of corals.

be regarded as having a significant resemblance to octocorallian structure, despite the difference in one pair (the "ventral" or sulcal pair) of muscle bands, which relates to the two siphonoglyphs of the Hexacorallia.

The mesenteries are involved in what amounts to considerable specialization of different areas of the gastrodermis lining the coelenteron. Below the stomodaeum, the free edges of the mesenteries or mesenterial filaments are each three-lobed in cross section. Two of these lobes are ciliated, and the central one is glandular. Lower down, the ciliated lobes disappear while the glandular lobe remains and bears the principal absorptive part of the gastrodermis. The same region is concerned with excretion of particulate matter into the coelenteron. In many common sea-anemones, long thread-like structures called acontia are developed from the lower ends of the mesenteries. They contain further gland cells and nematocysts. In ordinary life, these can only serve to kill any food remaining alive after passage into the gastrovascular cavity. If an anemone is grossly overstimulated and retracts fully, the acontia are occasionally extruded out of the mouth or through special pores in the column. In this case they could serve for defense. In some longer anemones, the food is moved by muscular action akin to peristalsis, and it is possible that here a sequential digestive process can occur. Even in these animals, all the food material undigested must return along the same route to the mouth for expulsion.

Reproduction in sea-anemones is by both sexual and asexual means. Sexual reproduction involves the development of a planula larva, which settles on a hard substratum and grows up to be the polyp. In many anemones, however, the developing eggs are retained within the gastrovascular cavity of the parent in a form of viviparity, and the planulae only emerge when they are effectively ready for settlement. Asexual methods of reproduction involve not only budding similar to that of hydrozoans, but also a peculiar process of fragmentation. In many common genera, as the adult slowly crawls along, little lobes are torn off from the attachment disk and grow up to be a series of little anemone-pups along the path taken by the parent. Such little pup-anemones often show some irregularities in the organization of their mesenteries and siphonoglyphs. However, in *Metridium* there in one morphogenetic observation on the growth of these pups which would repay further investigation. It is that a new siphonoglyph will appear only when a couple of the mesenteries left in the pup reverse their faces, thus assuring the proper organization of the directive mesenteries as the anemone grows up.

Behavior

The behavior, and the underlying neuromuscular organization, of anemones such as *Calliactis* and *Metridium* has been much studied. Both nerves and muscles display rather more obvious specialization in these anemones than they do in hydrozoan polyps. The more clearly defined muscle strands consist of folded masses which are, however, still made up of myoepithelial cells as in other coelenterates. The most massive muscles are the longitudinally arranged parietal and retractor muscles on each mesentery and the circularly arranged muscles which form a sphincter around the disc (Figure 4-22). Similarly, although the nerve cells still form a network, there are certain through conduction lines. In the Hydrozoa most nerve cells are multipolar, but it seems that in anemones, particularly on the mesenteries, more bipolar cells with a vertical orientation occur. Thus there are seemingly more directional paths of conduction, though there is no concentration of nerve cells, nor real polarity of conduction, as in the higher Metazoa.

Out of these seemingly simple structural arrangements, anemones achieve quite complex patterns of behavior, both in terms of specific reflex responses and slower spontaneous activity. Any student with access to a live, healthy sea-anemone, and without any elaborate apparatus, can make useful observations on the reflex responses. Any gross stimulation (either mechanical or electrical) of any part of the surface of an anemone results in contraction of the whole

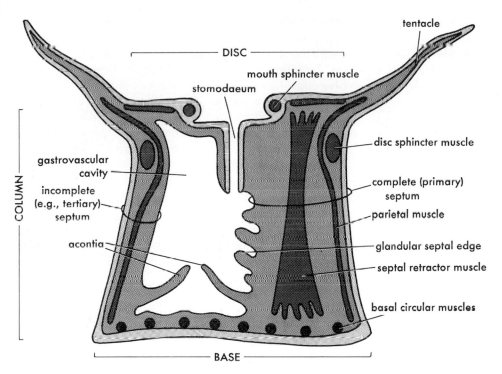

FIGURE 4-22. A diagrammatic longitudinal section through a sea-anemone like *Calliactis* or *Metridium* to show the principal muscles. The section passes through a complete primary mesentery or septum, on the right, and an incomplete (secondary, tertiary, or quaternary) mesentery on the left. The most massive longitudinal muscles are septal retractors and parietals; circular muscles include disc and oral sphincters and basal circular muscles.

animal in a fast reaction. This involves simultaneous contraction of the retractor musculature and of the sphincters, water being expelled through the mouth. Actually, a rather important (but often missed) aspect of this fast retraction is that the radial muscles of the disc contract early in the process, and thus hold the mouth and the valve of the stomodaeum open during the total contraction which is the rest of the response. An obvious observation is that a sea-anemone contracts fast, but reexpands very slowly. Since muscles can only contract (and must, before they can contract again, be stretched by a force outside themselves to their precontracted length), we must look elsewhere for the forces which reexpand the sea-anemone. They are of two sorts. Part of the slow recovery of the precontracted shape after relaxation of the muscles results from the elasticity of the mesogloea. In *Calliactis* and other anemones, fibers have been found in the mesogloea in a type of geodesic or cross-lattice structure, particularly immediately under the epidermis and gastrodermis. In other words, under each surface of living cells we have fibers arranged like the strands of a fence of diagonal wire netting. These lattices of fibers are deformed during contraction of the anemone and aid its extension when the muscles relax. In addition, there is a slow buildup of water pressure in all the spaces of the gastrovascular system with water introduced by the siphonoglyphs, the cilia of which beat continuously inward. Thus in the fully extended anemone there is a slight positive pressure of water maintained (about 0.5 centimeter of water). In addition, the stomodaeum is so arranged that it acts as a flap-valve to maintain this slight positive internal pressure. Incidentally, when an anemone is making movements which do not involve complete retraction with the mouth open, it makes them around this contained mass of water as a hydrostatic skeleton. In other words, when a sea-anemone sways from side to side the longitudinal muscles of one side are acting as the antagonists of the longitudinal muscles of the other around a hydrostatic bag of unchanging volume. Contraction of the longitudinal muscles on one side of the anemone means that the longitudinal muscles of the other side are being stretched, if the stomodaeal valve is closed, and thus the volume of the bag between the muscles is unchanged (Figure 4-23). Similarly, elongation of an anemone by contraction of circularly arranged muscles inevitably means that all its longitudinal muscles are being stretched. Lastly, contraction of the longitudinal muscles with the mouth closed will necessarily result in a shorter but fatter anemone with stretched circular muscles (Figure 4-23).

If a graded series of stimuli are applied to an anemone (and this is best done with electrical stimulation, varying both the frequency of stimulus and its intensity), it can be demonstrated that excitation of the nerve net and thence of the muscles spreads at a definite rate and is not instantaneous. Further, below certain thresholds there is no response to stimulus. In other words, taken unit by unit, there is an all-or-nothing response similar to that known in the neuro-muscular arrangements of higher metazoans. A whole series of specific reflex responses occur when the tentacles and oral disc are stimulated at a series of increasing intensities and frequencies. We see differing degrees of involvement of the whole anemone as responses to increased stimulation. Slight stimulation will cause movement of one tentacle. Increased stimulation may involve several tentacles and a small sector of the disc in the response. Still greater stimulation will cause the disc and tentacles to wrap themselves around the stimulating electrode, and a slightly greater degree will cause the anemone to attempt the ingestion of the electrode through its mouth. Of course, still greater stimulation evokes contraction of the whole animal in the manner already discussed, with the radial muscles of the disc holding the stomodaeum open so that the fluid of the hydrostatic skeleton is expelled to the exterior. There appears to be inhibition of contraction of one system by another. For example, when the circular musculature of any region is already contracted, it requires much greater stimulation to make the parietals contract.

Besides these reflex responses, there is a surprising amount of spontaneous activity which is mostly much slower. It is therefore only revealed to devoted anemone-watchers, and most usually studied by means of time-lapse photography (in many cases a sixtyfold speeding of the motion picture film is required to show the movements of the anemone). When this is done, it can be seen that the column elongates (by contraction of the circular muscles), and sways from side to side (the parietal and retractor musculature working on either side in antagonism). Peristaltic movements occur and these are involved in defecation and in spawning. Perhaps most surprising, in time-lapse films the anemone appears to walk slowly from place to place. This, the slowest spontaneous move-

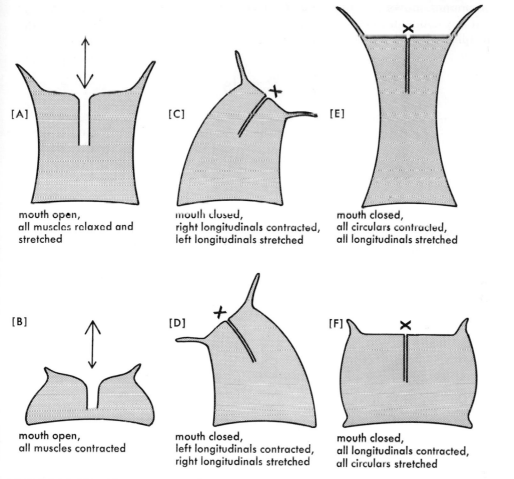

[A] mouth open,
all muscles relaxed and
stretched

[C] mouth closed,
right longitudinals contracted,
left longitudinals stretched

[E] mouth closed,
all circulars contracted,
all longitudinals stretched

[B] mouth open,
all muscles contracted

[D] mouth closed,
left longitudinals contracted,
right longitudinals stretched

[F] mouth closed,
all longitudinals contracted,
all circulars stretched

FIGURE 4-23. The movements of sea-anemones. Note that the "escape" withdrawal reaction is accomplished with the mouth open (**A** and **B**). Note that in **C–F** the mouth is closed (by the flap-valve reaction of the stomodaeum to increased internal pressure), and thus the volume of seawater in the gastrovascular cavity is unchanged between **C** and **D**, and between **E** and **F**. This constant enclosed volume acts as a hydrostatic skeleton, making possible the antagonistic action of circulars *versus* longitudinals in **E** and **F**, and of opposite longitudinals in **C** and **D**. See Figure 4-22 for further details and text for further explanation.

ment, is somewhat complicated to explain mechanically. It involves a graded contraction of sectors of the concentrically arranged circular muscles of the pedal disc, which allows directional movement. The rate can reach about 1 centimeter per hour, but is usually considerably slower. The ability to swim has arisen independently in at least three groups of anemones. In each case it involves sustained periods of rhythmical contractions. Pacemaker systems within the nervous system must be involved, but it is beyond the scope of this book to consider the problems of their physiological integration. They are currently being investigated.

In general, study of the reflex responses and the spontaneous activity of sea-anemones shows the considerable variety of behavior which is possible without any concentration of neurons in a central nervous system such as is found in the higher many-celled animals.

Corals

The other large group of the Hexacorallia is the order Madreporaria or true corals. In fundamental anatomy of the mesenteries and gastrovascular cavity, they closely resemble sea-anemones. The major differences involve the colonial pattern in some corals and the secretion of a massive skeleton. Essentially, the polyp develops from a planula which settles down and begins to secrete a skeleton rudiment almost at once, the polyp being exactly similar to an actinarian except for the absence of a differentiated pedal disc. The initial formation of a prototheca is illustrated in Figure 4-21; it is secreted by the ectoderm first as a basal plate. Almost at once the undersurface of the juvenile polyp develops radial folds which secrete septa, and at the same time a rim is formed and built up as a thecal wall around the polyp. Actually the septa of the skeleton are built up from the bottom, and the animal is lying on top of these as they are secreted by the ectoderm. Then further skeletal material is secreted into the gaps between the septa. Thus, while the skeleton grows, the septa of the skeleton usually alternate with the mesenteries of the living coelenterate (Figure 4-21). In life, when a coral expands, it is mostly clear of all skeletal material; but when it contracts, it comes down hard on the skeleton and the coelenteron or gastrovascular cavity is pushed out of existence. Like all fixed animals, corals are in constant danger of being buried by sediment, and an obvious adaptation is that they have cilia all over their exposed bodies. The cleansing cilia around the mouth are particularly prominent—beating centrifugally (that is, outward from the mouth).

In a few forms there are solitary individuals—the so-called cup-corals—but the majority of species are colonial. The many forms taken up by corals result from there being several degrees of association of the asexually budded polyps and two major patterns of branching. These patterns are illustrated in Figure 4-21, and it can be seen that in the first there is always a primary polyp at the top of the colony with lateral branches on either side (for example, the stagshorn-coral, *Ctenopora*). In the second pattern, every budded individual is potentially equal to the others. With persistent growth this gives rise to forms with solid colonies which may be encrusting or foliaceous. A subtype, again involving equal division but in this case alternate growth, gives rise to cymose branching. The degree to which the daughter polyps remain together after asexual division also determines the form of the coral. They may retain a common skeletal base but have widely spaced cups as in *Oculina*, or the thecae may be placed closer and share common walls as in *Favia*, or the polyps may be confluent into rows not separated by skeletal material as in the meandrine group of corals. In these, the winding grooves or valleys on the surface of the calcareous mass are occupied in life by fused lines of polyps, with each row of mouths opening into a common gastrovascular cavity. These meandrines are the characteristic "brain-corals" which can grow into massive coral boulders or amalgamate into reefs (Figure 4-24A).

[A]

[B]

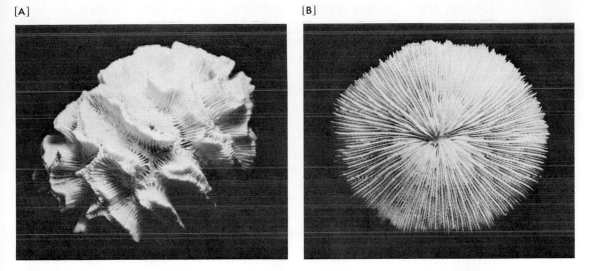

FIGURE 4-24. Skeletons of other Anthozoa. **A:** A young colony of a meandrine or "brain" coral (Hexacorallia: Madreporaria). **B:** A motile "mushroom-coral," *Fungia* (Hexacorallia: Madreporaria). [Photos by the author of personal specimens.]

A few solitary corals are not sessile in adult life. In *Fungia*, the "mushroom-coral" of the Indo-Pacific, the planula gives rise to a single stalked individual like a flattened cup-coral. After growth this breaks away from the stalk and the free coral disc continues to grow, and it is carried by water and sand movements as an unattached organism (Figure 4-24B). It can uncover itself if buried in sand (by action of its cilia and by inflating the coelenteron with water). There are several well-authenticated accounts of specimens of *Fungia* being able to turn over if accidentally inverted on the sand, and individuals have been reported to move several centimeters across sand in the absence of all water currents, in the course of one night. In some species of *Fungia*, the stalk remaining when the disc breaks off is able to go on to form further disc-individuals. Other fungiid corals are capable of several other kinds of asexual multiplication.

Biology of Reefs

Any claim that the phylum Cnidaria constitutes a major phylum in our ecological sense is based on the enormous success of the reef-building corals in those specific regions of the tropical seas where they make up a significant part of the biomass and provide a variety of habitats for a diversity of other organisms. Particularly over the last 25 years, there has been considerable controversy about the nutrition and general bioenergetics of these reef-building (or hermatypic) corals. It had been long known—particularly as a result of the detailed work on the physiology of coral polyps conducted by C. M. Yonge on the Great Barrier Reef in Australia in 1929—that, like all nematocyst-bearing coelenterates, corals are almost entirely carnivorous. Individual coral polyps can be observed to feed on planktonic organisms, either capturing them by the tentacles, as do sea-anemones and other Anthozoa, or collecting them on mucus by temporary reversal of the patterns of cleansing cilia arranged radially on the disc around the mouth. Many of them expand only at night to feed on the zoöplankton.

In many corals and other anthozoans—including not only some of the most important reef-builders but also several kinds of octocoralline "soft-corals"—there are minute algal cells, or zooxanthellae, living in the tissues of the coral polyps. There has been considerable controversy about the exact relationship between corals and their contained algae. The work of coelenterate physiologists such as C. M. Yonge had suggested that the reef-building coral polyps were essentially heterotrophs, ingesting animal prey in a typically carnivorous nutrition. This did not deny the status of true symbionts to the contained zooxanthellae (which are modified single-celled dinoflagellates—see Chapter 6), though their benefits to the coral polyps were assumed to be largely the automatic removal of waste products in the course of nutrition of the plant cells. In this way, not only carbon dioxide but residues involving nitrogen, phosphorus, and sulfur were "mopped up" by the algal cells in the coral tissues, thus saving the coral polyp's "animal" economy any energetic expense for excretion to the exterior.

In 1954 a totally new view of the economy of coral reefs was propounded as a result of the work of Eugene P. Odum and Howard T. Odum on Eniwetok Atoll and elsewhere. Employing a variety of ecological methods which included monitoring oxygen changes in the water flow over the intact reef community and using radioactive tracers, the Odums demonstrated that sections of coral reef were, on balance, autotrophic in their economy. In other words, the reef as a whole must, on energetic balance, be made up more of photosynthetic plant material than of heterotrophic animal tissues. Further, the observed high organic productivity rates of reefs as communities could be claimed to depend upon efficiencies of energy exchange and of nutrient recycling resulting from the symbiotic links between plant and animal components. The Odums and other workers thus came to the conclusion that the zooxanthellae provided a major nutritional source for coral polyps, and that any feeding as carnivores on planktonic organisms provided purely a proteinaceous food supplement. This deceptively simple question, whether reef corals are heterotrophic or autotrophic in their nutritional economy, became obscured in the controversial interpretation of different experiments which occurred in the next few years, but the seeming controversy itself stimulated much valuable modern research.

Largely as a result of recent physiological experiments conducted in situ on coral organisms by the late Thomas F. Goreau and others, it is now possible to say rather more about the significance of zooxanthellae to the reef-building corals and to certain other coelenterates which harbor them. It is quite clear that most of the reef-building corals can live and grow without algae, but there is some evidence that corals without algae do not grow so fast as those with the contained symbionts. Some studies using ^{14}C-labeled algae reveal that plant-synthesized organic materials do pass into the polyp tissues. If coelenterates which harbor zooxanthellae are maintained for periods in the dark, the zooxanthellae do not grow nor reproduce, and after a period of days or weeks the polyps actually expel their zooxanthellae. Many reef-building corals can survive such loss of their zooxanthellae without major change. However, a few other forms of reef-building corals show some atrophy of tissues and consequent loss of contained algae, after exposure to darkness. Some workers would claim that this is evidence for starvation of the coral polyps, while others would assume that the ill-health of the coral polyps arises merely from the accumulation of waste products of metabolism normally removed by plant cells, the coral polyps having lost the capacity for a complete and efficient removal of these wastes by their own anabolic efforts.

Other features of coral biology elucidated by Goreau, Yonge, Muscatine, Johannes, and other recent investigators provide further evidence for the continuing heterotrophic condition of their mixed nutrition. It is important to distinguish between hermatypic (or reef-building) corals, with their contained zooxanthellae, and ahermatypic (non-reef-building) corals, which do not normally contain zooxanthellae and do not build as massive calcareous skeletons. It is also important to distinguish between both these kinds of "true" or "stony" corals and certain octocoralline "soft-corals" (see pages 72–76), some species of which not merely contain zooxanthellae but also "farm" these algal cells as their principal nutritional source. The hermatypic corals require warm shallow waters (normally above 21°C). Coral reefs are therefore limited to the Indo-Pacific, the central western Pacific, and the Caribbean regions north to Bermuda. In contrast, ahermatypic corals live at moderate depths throughout the world, including colder latitudes, and a few of them grow at great depths in water of around 4°C. Both hermatypic and ahermatypic corals retain all the anatomical apparatus of predaceous carnivores including nematocysts (discharging only in the presence of suitable animal prey) and mesenterial filaments and gland cells specialized for the digestion of such prey. In contrast, several of the zooxanthellae-farming soft-corals have lost their capacity to catch zooplankton and similar prey, and a few of them have even lost their nematocysts. Earlier investigations had demonstrated that a few important species of reef-building corals show a net daily uptake of oxygen from the environment, this being evidence for a heterotrophic nutrition (if they were largely plant-like in their economy, the processes of photosynthesis would involve a net production of oxygen over each 24-hour period). On the other hand, Goreau, working with radioactive carbon and calcium as tracers, has clearly related calcification in corals to the activities of their contained zooxanthellae.

After developing methods for measuring the rate of calcium deposition by corals under different conditions in situ on Jamaican coral reefs, Goreau produced conclusive evidence that the zooxanthellae are responsible for the exceptionally high rates of calcification which occur. In this, there was a clear distinction between hermatypic (or reef-building) corals and ahermatypic corals which live elsewhere. The method involved sealing glass jars round corals in position on the reefs and inoculating them with radioactive calcium as a tracer. Where zooxanthellae were present, he was able to demonstrate that growth of the calcareous skeletal material was, on average, ten times faster in sunlight than in darkness. This was true for fourteen species of reef-building corals studied, and in most cases where zooxanthellae had been removed by sustained periods of darkness the lower rates of calcification persisted even in sunlight. Goreau also studied mechanical aspects of the shape of coral formations by different species and found that some important reef-builders which create massive blocks in the well-lit waters near the surface produce a skeleton more flattened and plate-like at greater depths where zooxanthellae contribute less to processes of calcification. It is important to note that at those greater depths the coral polyps and their tissues are still healthy, although they are laying down proportionately less calcium carbonate. There is an intriguing set of observations that suggest that the plate-like colonies of deeper water are somewhat more resistant to boring organisms, including sponges, than the apparently more massive blocks near the reef's surface and its growing-edge.

In general terms, the contained zooxanthellae of reef-building corals assist in the removal of animal metabolites and are of major importance in relation to processes of skeletal secretion, but the coral polyps remain carnivorous organisms. Another feature is that the majority of reef-building corals show typical behavioral responses in response to certain amino acids in their environment. Goreau and Yonge also observed similar feeding responses produced in corals by seawater which had previously contained living zoöplankton. They also observed under natural conditions in the field that certain reef-building corals will expand in apparent anticipation of plankton drifting toward them and deduced that this was due to each polyp's ability to sense (taste) the diffusing halo of metabolites, including amino acids, which usually surrounds swarms of zoöplankton organisms. Another important recent discovery was that of direct utilization by coral polyps of organic material present in the dissolved or colloidal state in the seawater bathing them. There is considerable ultrastructural, histochemical, and physiological evidence for the uptake of amino acids by the epidermis along its free cell border. Goreau's studies have also emphasized the importance of yet another supplementary food source in hermatypic corals, that of suspended organic detritus, collected largely by ciliary reversal of the "cleansing systems." In certain corals with small polyps, which show reduced tentacles but hypertrophied mesenterial filaments, this detrital food source could be of particular significance. Thus the polyps of reef-building corals can have four nutritional sources: first, from "normal" carnivorous feeding on zoöplankton using nematocysts; secondly, from unspecialized detritus feeding; thirdly, from direct uptake of dissolved or colloidal organic matter from the seawater; and fourthly, from transfer of organic materials synthesized by their contained zooxanthellae.

At the same time, it has become clear that certain coelenterates (other than true corals) with contained zooxanthellae are almost totally dependent upon *their* contained algal cells for survival. These are all relatively rare forms of no major ecological significance. They include the octocorallian soft-coral *Xenia*, probably closely related to *Alcyonium* (see page 72), and certain forms in the aberrant hexacorallian order Zoantharia, including *Zoanthus sociatus* itself. These forms show significantly greater uptake of carbon dioxide than do reef-builders with or without zooxanthellae, and they lack all normal feeding responses to animal prey. Most significantly, *Zoanthus* has only a few rather disordered nematocysts, and *Xenia* totally lacks these stinging cells which are so typical of the carnivorous way of life in all other coelenterates. It should again be emphasized that the majority of reef-building corals show a full complement of nematocysts.

On the ecological side, on the apparent autotrophy of coral reefs as whole systems, it has become increasingly clear that a typical block of reef or an isolated coral head may contain as much calcareous algal tissue as it does animal tissue of the founding reef-building polyps. These reef-building calcareous algae are, however, living outside the coral organisms, and they build up a cement in the interstices of larger coral masses as well as creating their own calcareous masses in the reef community. It is often forgotten how very slight is the layer of living polyp tissue which covers the massive calcareous blocks which make up a coral reef. The spectacular success of the reef-building corals perhaps makes us exaggerate the amount of food required for the nutrition of this attenuated superficial layer of living coral tissue.

Other recent studies have emphasized that not only the thin layer of living coral tissue but also the superficial layers of the secreted calcareous skeleton may be relatively translucent. This means that many algal cells can live and grow in the interstices of older skeletal material. Another important discovery has been that algal products may be used by corals for enhanced calcification at sites some distance from the illuminated algae; in other words, translocated materials produced synthetically by the zooxanthellae in one part of the coral may increase the rate at which "coral skeleton" is laid down in others.

To summarize the physiological aspects, reef-building corals, like all nematocyst-bearing coelenterates, are carnivorous. Ahermatypic (non-reef-building) corals do not normally contain zooxanthellae and do not build as massive skeletons as do the hermatypic (reef-building) corals. Ahermatypic corals are widely distributed in the colder waters of higher latitudes and to greater depths in the sea; in contrast, hermatypic corals are limited to the tropics and to relatively shallow waters. Obviously, the less widely distributed hermatypes are far more numerous and of vastly greater ecological significance. The other contrasting set of conditions—between the hermatypic corals retaining carnivorous feeding as well as zooxanthellae, and those octocoralline "soft-corals" which have numerous zooxanthellae, but which have lost their capacity to feed actively and, in some cases, even their nematocysts—should also be emphasized. In a summary of the ecological aspects, it is important to emphasize that it is the autotrophic capacity of the reef community as a whole that creates and sustains the significance of coral reefs as areas of high productivity in the otherwise relatively sterile tropical oceans. Since the productivity around coral atolls in the central Pacific is many times that of the open ocean, Eugene P. Odum has compared these atolls to "oases in a desert." Apart from a local "black body" effect resulting from the photosynthetic organisms of the reef, both outside and inside the coral polyps, being close to the tropically illuminated surface, much of the high productivity appears to depend upon efficient, and sometimes very local, recirculation of nutrients. In other words, on and around the reef, little of each year's primary production of organic matter drops into deep water and is lost, as happens elsewhere in the tropical oceans. On and around the reef, invertebrates of higher trophic levels and fishes are much more abundant than elsewhere, and the great bulk of mechanically degrading organisms (boring sponges, bivalves, and the rest) along with the true decomposers are localized within the shallow waters created by the coral growth. The spatial autonomy of a coral reef depends on this local recycling, as well as upon the reef being a locally photosynthetic "energy trap." True symbioses, and community interdependence at all levels, contribute to the ecological efficiency of reef communities. Eugene P. Odum has used the tropical coral reef as a sociopolitical object lesson for man: the key to maintaining human prosperity in a world of limited resources is efficient recycling of materials and use of energy within the mutualism between producers and consumers.

There are three minor groups of Hexacorallia. The sea-trees or black-corals form the order Antipatharia, typically found in the deeper parts of the oceans as slender, branching, plant-like colonies with horny stems. Their polyps have ten mesenteries and two siphonoglyphs but no mesenterial muscle bands. Thus the polyps cannot contract as rapidly as those of the other Anthozoa. Species of the order Zoantharia are usually colonial without a skeleton, and live as epizoic growths on other animals. They have more numerous mesenteries (fourteen pairs in some forms), and are more like some fossil corals than most living Anthozoa. The last order, the Ceriantharia, consists of a few genera of solitary elongate sea-anemones living embedded in sand in a tube of mucus. Again, they have a large indefinite number of mesenteries, a single siphonoglyph, ectodermal bands of vertical muscles (not on the mesenteries), and a terminal pore opening from the gastrovascular cavity to the base. This pore is not known to function as an anus. Cerianthids have planktonic polyp-larvae, somewhat like stout inflated actinulae, but details of their complete life-cycle and of settlement behavior are unknown.

Without discussing the relationship of the cnidarians to other many-celled animals for the present, we can speculate on phylogeny within the group. It seems that a stem form for the anthozoans could readily be a simplified octocorallian polyp (see page 73). The stem form of the true jellyfish or Scyphozoa is most likely to have been a polyp like the present scyphistoma or like a simplified stauromedusan (see page 67). Both of these archetypes could readily be related to a stem form of the class Hydrozoa such as a simplified Trachyline. The basic cnidarian, therefore, would have a fourfold symmetry, possibly with four hollow tentacles, and would combine some of the features we associate with polyps with some of those we ascribe to medusae. Allowing the very important distinction between archetype and ancestor (see pages 30 and 613–614), it is possible that the first coelenterate was a ciliated, nematocyst-bearing organism resembling a simplified version of the actinula described earlier (page 53). Since nematocysts have now been discovered in a single species of ctenophore, this description might also serve for a remote ancestor of that, also tetraradiate, group.

5 THREE EVOLUTIONARY BYWAYS: CTENOPHORA, PARAZOA, MESOZOA

WHEN alive, ctenophores are among the most beautiful of marine organisms. They are pelagic, transparent animals, with radial symmetry based on an underlying bilateral symmetry. They are never colonial, have no fixed stage, and, with one exception, lack nematocysts. Their most characteristic features are eight vertical rows of plates of fused cilia—the ctenes or comb-plates. Each row has the ctenes placed across it and they beat one after the other from the aboral to the oral pole. These meridionally arranged rows are termed costae. The effective beat of the ctenes in the costa is toward the aboral end, so that ctenophores swim mouth first in the opposite fashion to medusae. The beating is metachronal and, when the sense-organ at the aboral pole is functioning properly, all eight costae beat in unison. Since the ctenes are iridescent, a flock of ctenophores swimming together in still sunlit water is an unforgettable sight.

Comb-jellies

The basic form of ctenophores is perhaps best illustrated in one of the sea-gooseberries, *Pleurobrachia*, which occurs in colder waters of both the Atlantic and the Pacific oceans. Apart from the diagnostic character of eight costal rows of ctenes, the globular body has two lateral pits into which long, branched tentacles can be completely withdrawn. At the aboral pole is a complicated sense-organ, and from the oral pole a long stomodaeum opens into a complex gastrovascular system. This system is basically biradial or tetraradial in its symmetry and is best understood with reference to Figure 5-1. There is a small central gastric cavity or stomach from which arise a series of canals. Two run on either side of the stomodaeum; the infundibular canal, runs in the midline to just under the aboral sense-organ and then divides into two small "anal" canals which open to the surface through tiny pores. In spite of doubts expressed by many textbook writers, the expulsion of undigested material from the coelenteron does occur here—thus these are functional anuses. In the equatorial plane; two perradial canals leave the sides of the gastric cavity. Each of these divides into three and thence into five. The threefold division gives rise to one larger central tentacular canal, which terminates blindly at the inner surface of the tentacle sheath, and two interradial canals, each of which branches into two adradial canals which run toward the underside of costae. The eight adradial canals end by opening into the meridional canals which run from pole to pole, one under each costal row of ctenes.

The apical sense-organ lies in a depression of the aboral pole covered by a transparent dome. It consists of four sets of bristle-like structures called "springs" or "balancers" which support a single calcareous statolith. Ciliated tracts connect each row of ctenes to the bases of these springs. Both the transparent dome and the springs seem to be made up of modified fused cilia, and thus resemble the organelles of the more complex ciliate protozoa. These ciliary tracts that connect the apical sense-organ to the eight costal rows are func-

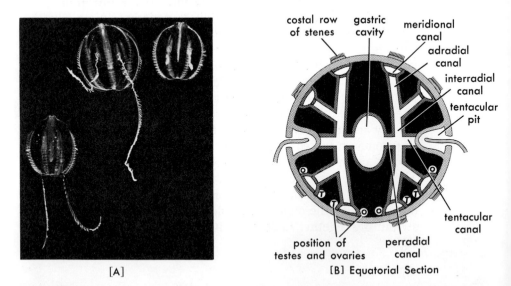

[A]

[B] Equatorial Section

costal row of stenes — gastric cavity — meridional canal — adradial canal — interradial canal — tentacular pit — tentacular canal — perradial canal — position of testes and ovaries

FIGURE 5-1. Ctenophora or "comb-jellies." **A:** Living specimens of *Pleurobrachia pileus* from plankton in temperate waters of the Atlantic. **B:** Ctenophore organization as shown in an equatorial section through *Pleurobrachia*. The peculiar organization of the sixteen strands of gonad on the meridional canals is indicated by the letters **O** (ovary) and **T** (testis). [Photo © Douglas P. Wilson.]

tionally analogous to axons of motor nerves. Stimuli from the apical sense-organ are undoubtedly responsible for modifying the pattern of beating in the comb rows and thus maintaining the posture of the animal as a whole. These ciliated "axons" probably involve cell-to-cell electrical transmission throughout. They make up only one of four conducting systems which have been recently demonstrated to be involved in the coordination of ctenophores. Secondly, there is an ectodermal nerve net, which appears to inhibit the ciliated system, but also acts on certain muscle fibers. Thirdly, there is apparently a slow cell-to-cell transmission between muscle cells, and, fourthly, there is a conducting system in many ctenophores for waves of luminescence. This last system is possibly endodermal.

Like the cnidarians, all ctenophores are carnivorous and many use tentacles to capture their food. All ctenophores use colloblast cells (sometimes miscalled lasso-cells), each of which consists of a hemispherical sticky head fastened to the muscular core of the tentacle by two strands: a straight fiber and a spiral contractile fiber. Normally the prey is captured by the attachment of very many colloblast cells whose filaments cannot all be broken by its struggles. In tentaculate ctenophores, the prey is then wiped into the stomodaeum.

Ctenophores are hermaphrodite and gonads are developed from endodermal cells on the walls of the meridional canals, that is, under the rows of ctenes. Although the wall of each meridional canal bears both a testis and an ovary, there is a peculiar symmetry in their organization which is best explained with reference to Figure 5-1. Eggs and sperms are spawned out through the mouth and the zygote shows determinate cleavage and mosaic development. Many of the more aberrant ctenophores pass through a cydippid stage resembling *Pleurobrachia* in their development before taking up their adult form. One genus has a planula larva resembling that of cnidarians.

The ctenophores are usually divided into five clearly defined orders, the first of which, Cydippida, has been described in the above account of the typical genus, *Pleurobrachia*. A second order is the Lobata, where the body is laterally compressed with two large oral lobes and the costal rows of ctenes occur as four long and four short. A typical genus is *Mnemiopsis*, which is abundant in late summer in the waters of Cape Cod. It is one of the many ctenophore genera capable of luminescing. Since individual *Mnemiopsis* grow to about 3 centimeters across, and since they often occur in immense swarms, night sailing through a luminescent swarm can be a memorable experience. Most lobate ctenophores pass through a cydippid stage. A third order, Cestoidea, is made up of only two genera, each of several species. *Cestus veneris*, or "Venus's girdle," is a typical Mediterranean species, an elongate ribbon form up to 1 meter in length which moves by undulating muscular movements. Its costae are spectacularly iridescent. Ctenophores of the order Beröidea have an elongate thimble shape due to the development of an enormous stomodaeum. They may be up to 20 centimeters in length, and are often pinkish in color. The typical genus *Beröe* is found throughout the oceans of the world and is often extremely abundant for short periods. The last and most aberrant order of ctenophores is the Platyctenea, consisting of a few genera of flattened ctenophores which have taken to crawling on the bottom. The ciliated surface on which they crawl is modified from the stomodaeum.

Apart from the recent discovery (1957) of true nematocysts in one ctenophore species, *Euchlora rubra,* many features link the ctenophores with the cnidarian coelenterates. These include: the properties and cell types of meso-gloea, the nature and organization of the gastrovascular system and of the nerve nets, the general tetraradial symmetry, and the general lack of organs (apart from statocysts and compound sense-organs in general) reflecting the tissue grade of organization. On the other hand, it is impossible to derive the cteno-phores from any of the present stocks of cnidarian coelenterates. The once much-discussed theory of the origin of turbellarian flatworms from platyctenid ctenophores was largely based on indefensible homologies and can be disre-garded.

Parazoa

Although they constitute less highly integrated animal organisms than coelen-terates, treatment of the sponges (phylum Porifera) has been postponed until now, largely because they appear to involve an evolutionary dead-end. This is often emphasized by classifying them in a separate subkingdom, the Parazoa, distinct on the one hand from the rest of many-celled animals (Metazoa), and on the other from the acellular Protista. In terms of numbers of living forms, the sponges (of which there are probably more than four thousand species) do not constitute a minor group. Ecologically, however, they only make up a measur-able part of the biomass in relatively shallow waters in some areas of the seas. A few species have colonized certain freshwater habitats.

Most of the diagnostic characteristics of the phylum Porifera are negative. Sponges are without organs and with little in the way of definite tissue layers. For most physiological purposes, they can be regarded as loose aggregations of cells without definite form or symmetry. In fact, a few of the smaller simple sponges do show a loosely organized form of radial symmetry. They are with-out a true gut—if we define a gut as a digestive tube lined by endodermal epi-thelium opening from a mouth. The only internal space is the so-called para-gaster, often in the form of a series of cavities, pores, canals, and chambers through which water is caused to flow. This water current, which subserves all functions of feeding, respiration, excretion, and reproduction of the sponge cells, is created by the peculiar cells which line the paragaster. These cells, which are diagnostic of the group as a whole, are the choanocytes or collar-cells. The flagella of the choanocytes create the water current which allows sponges to feed on fine suspended food particles. When filtered out of the water current these particles are ingested by the choanocytes and then undergo intra-cellular digestion such as is found in protozoans (see Chapter 6).

The essential features of the functioning of sponges arise from the fact that the cells act more or less independently and show little cooperation or coordi-nation or interdependence. Both the basic architecture of sponges, and their archetypical functioning, can be best understood by study of a young growing sponge at the *Olynthus*-stage (in which simplicty of form is retained in a few species as adults—the so-called *Ascon* sponges). As illustrated in Figure 5-2 the little sponge vase has one larger opening, the osculum, out of which water

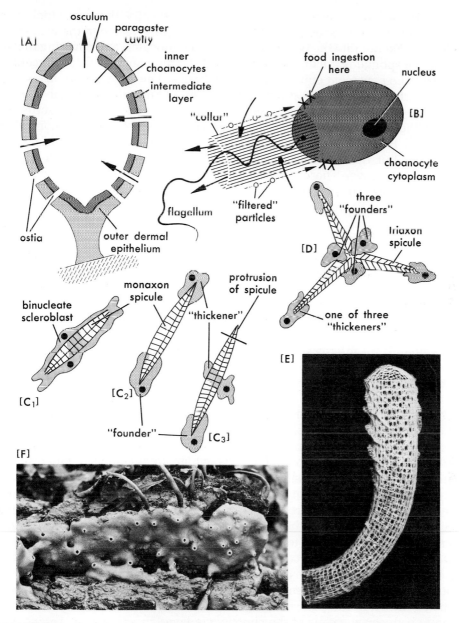

FIGURE 5-2. Organization of sponges. **A:** An *Olynthus*-type simple sponge in vertical section. **B:** Reconstruction of the feeding structures in an individual choanoctye. **C1–C3:** Stages in the secretion of a monaxon calcareous spicule. **D:** A late stage in the secretion of a triaxon spicule. **E:** Skeleton of the hexactinellid sponge *Euplectella*. The six-rayed siliceous spicules are arranged to form rectangular units (often cuboidal) in the skeleton. **F:** A living specimen of an encrusting sponge, *Halichondria,* exposed on a rock surface near low tide level. [**C1–C3** and **D** adapted from W. Woodland, in *Quart. J. Micros. Sci.,* **49:**231–325, 1905, photo **E** by the author; photo **F** © Douglas P. Wilson.]

passes, and a series of many smaller pores, termed ostia, through which water passes into the paragaster. There are structurally three layers in the wall of this little vase. Facing into the paragaster are the choanocytes, and covering the outside of the sponge is the dermal epithelium, a single layer of flat, thin cells resembling the pavement epithelium of vertebrates. These dermal "pinacocytes" have a certain degree of contractility and can change the shape of a simple sponge to some extent. In some cases they can form specialized contractile cells around the pores of the ostia—the porocytes, which can act as controlling sphincters. Between these two layers of cells we find not only spicules forming a skeleton but other nonliving gelatinous material through which discrete amoebocytes move. These amoebocytes of the so-called mesenchyme have many functions. They can carry food material from the choanocytes to the rest of the sponge cells, or store food material, or carry and secrete excretory materials or pigments. They also secrete the spicules or spongin fibers if they occur, and are also the precursors of the cells involved in both sexual and asexual reproduction. Their ability to be transformed into dermal epithelial cells (pinacocytes), where and when required, reveals that the "three" structural layers of the sponge are constructed on only two basic cell-types: choanocytes and common ectomesenchyme cells.

The continuing healthy life of any sponge depends on the continuing activity of its choanocytes. If we ignore for the moment the structures which appear under good light microscopy as transparent contractile collars, each choanocyte closely resembles a protozoan with a single flagellum. The flagellum has a normal active and recovery stroke, and its ultrastructure as revealed by electron microscopy is exactly similar to that of protistans. Its effective stroke is away from the cell body, and there is no coordination between adjacent cells. They all simply beat their flagella in this direction thus causing an increased pressure in the paragaster and thence a flow out of the osculum. This is made up by a flow of water in through the ostia, which are openings of much smaller diameter (Figure 5-3). All sponges, no matter how complex their paragaster system becomes, function in this way. Statements that the flagella of choanocytes beat in a particular direction, thus directly creating a specific directional current of water through the canal and chamber systems of more complex sponges, are made in many textbooks and are totally erroneous. In all cases, an inwardly facing surface of choanocytes causes a local increase of pressure in the center of a cavity and the resultant water flow is always out through a single or a few relatively large channels (oscula) and water is replaced in the system by being drawn in through relatively tiny pores (ostia) which must open functionally between the cells of the choanocyte layer.

Many years ago, careful light microscopy revealed that the choanocytes were also involved in the ingestion of food particles, the sites of ingestion being invariably around the outside of the collar as shown in Figure 5-2. Until recently it was difficult to understand how these food particles got there. Electron microscopy has now elucidated the ultrastructure of the collar, which proves to resemble nothing so much as a rather open palisade or picket fence. As reconstructed in Figure 5-2, the water which is impelled by the choanocyte toward the center of the paragaster must first pass through the tiny gaps in this microscopic fence. In living healthy choanocytes, the "collar" or "fence" is with-

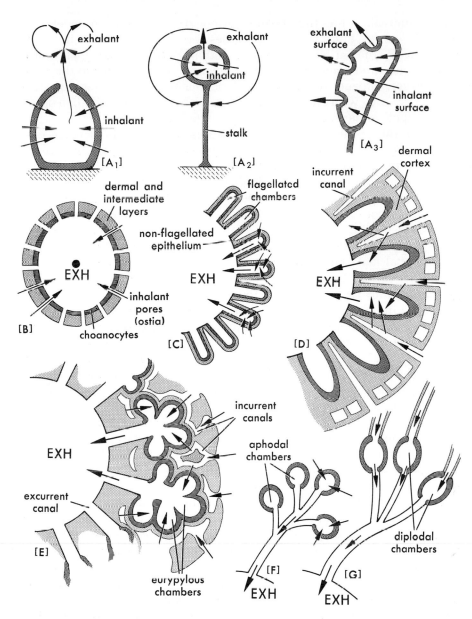

FIGURE 5-3. Adaptations of more complex sponges. **A1–A3:** Shapes of upright sponges in which repeated circulation of water is prevented. **B–G:** Horizontal sections through sponges of increasing structural complexity. **B:** *Ascon* type. **C:** Simpler *Sycon* type. **D:** *Sycon* type with dermal cortex. **E:** *Leucon* type with eurypylous chambers. **F:** Aphodal chambers (second *Leucon* type). **G:** Diplodal chambers (third *Leucon* type).

95

drawn at intervals. Thus the choanocyte not only creates the water current using its flagellum, but also filters particles from the water with its "collar," and then ingests the particles by amoebic action. Food vacuoles are formed just as in protistans, and some or all of the digestion is carried out in the choano-cyte. However, choanocytes may pass food material for further digestion to amoebocytes, along with the products of digestion, and these may then be stored or carried to other parts of the sponge organization.

Of course, the creation of positive pressures and currents of water on any scale requires that the cell layers of more complex sponges be supported against deformation by skeletal elements. These may be spicules of varying form and mineral composition or fibers of secreted protein termed spongin. Specialized amoebocytes of the mesenchyme secrete both: scleroblasts the for-mer, and spongocytes the latter. The processes of spicule secretion were worked out by careful microscopy (more than sixty years ago) for those sponges with calcareous spicules and the processes are still better understood with this type of skeletal material. Each original scleroblast is apparently a binucleate cell. Calcium carbonate is first laid down around an organic thread; then the cell separates into two, termed the founder and the thickener. The simplest case is found in the secretion of monaxon spicules where the founder remains at the inner end as the size of the spicule is increased (Figure 5-2). Finally, when the spicule is complete both cells move off through the mesenchyme. More complicated spicules involve more cells: the triaxon being secreted by three binucleate scleroblasts which later separate to three founders and three thickeners (Figure 5-2).

The processes as regards siliceous spicules are less well studied, but it seems that single cells are usually involved. The elaborate framework of a hexactinel-lid sponge, however, is produced by a mass which is syncytial and multinu-cleate. Again, the silica is laid down in concentric cylinders around an initial organic strand. Spongin fibers are apparently secreted by spongocytes which arrange themselves in rows in the mesenchyme and then lay down a fiber of protein, usually as part of a network or branching system. It is obvious that the sponges which produce siliceous spicules must utilize markedly different bio-chemical processes from those in sponges which secrete calcareous spicules. Further, since in both cases the inorganic salts must be concentrated from the surrounding seawater, and since the concentration of silica is several orders of magnitude less than that of calcium, the siliceous sponges must expend much more energy in obtaining their skeletal material. A simple experiment is possi-ble with calcareous sponges. If these are placed in calcium-free seawater, not only are they unable to form new spicules, but those already present seem to dissolve out and the sponge as a whole collapses and degenerates.

The more complicated types of sponge structure which occur have often been described and are illustrated in Figure 5-3. These developments are best understood if two functional aspects are emphasized. First, any folding of the wall of the paragaster will increase the surface of choanocytes per unit volume of sponge. Secondly, if the above account of the functioning of a simple sponge is accepted, then it will be obvious that a large number of small flagellated chambers will be more efficient than fewer larger chambers. The size of the out-let provided will determine the direction of flow and also its rate. Bearing these functional implications in mind, it is possible to survey the increasing complexity as reflecting potentially increased efficiency in creating the multipurpose water flow. In very few adult sponges does the simplest form—the ascon type—persist (Figure 5-3B). The simpler sycon involves a folding of the walls and a restriction of the choanocytes to the invaginations of the folds (Figure 5-3C), while a more complex sycon arises from the growth of a cortex over the outside creating a series of subdermal spaces and inhalant canals. A greater level of complexity is reached with the leucon types where the choanocytes are limited to flagellated chambers: these may be eurypylous, aphodal, or diplodal. These increasing degrees of "efficiency" are best understood with reference to Figure 5-3E, F, G. At the leucon level of complexity, some larger food particles may be taken up by cells lining the incurrent canals, although the minute organic particles and microorganisms forming the principal food of the sponge are still ingested by the choanocytes lining the small flagellated chambers. Very many sponges are simply encrusting masses, but a few more upright types show some organization of the sponge as a whole (Figure 5-3A). Again the significance of most shapes is understandable as a series of adaptations tending to prevent repeated circulation of the water current, by isolation of the inhalant ostia from the exhalant oscula. An ecological fact obviously related to this is that most encrusting forms of sponges will not live in completely still water.

Sponges are usually divided into three major and one minor classes. The Calcarea include those whose skeleton is made up of separate calcareous spicules which may be one-, three-, or four-rayed. The second class is the Hexactinellida, or glass sponges, whose skeleton is made up of six-rayed siliceous spicules formed into a basically rectangular pattern. The third class is the Demospongea where the skeleton is either siliceous or made up of the protein spongin or both. One subgroup within this, the Keratosa, have entirely proteinaceous skeletons which, with all the living cells rotted away, provide the bath sponges of commerce. A fourth class, Sclerospongiae, was set up in 1970 to include a small number of species associated with coral reefs—the so-called coralline sponges, which have a calcareous outer "shell" associated with both spongin fibers and siliceous spicules internally.

Sponges reproduce both sexually and asexually. It is first important to note that sponges are capable of considerable "regeneration." The term has here a somewhat different meaning from that which we use for other many-celled animals which have a capacity to regenerate lost parts. Any piece of any sponge containing both amoebocytes and choanocytes can, if kept in good conditions, grow into a full-sized specimen of the appropriate species. To put it another way, what we are concerned with is a plant-like capacity for regrowth, involving replication of tissues and structures without obvious symmetry, rather than the redifferentiation of a replacement part meant by the term in most many-celled animals. An experimental method which demonstrates the fantastic capacity of sponge cells to reorganize themselves was devised by H. V. Wilson in 1907 and has been used by workers in this field extensively since. If living sponges are squeezed through fine silk cloth, they are broken up into clumps of cells. Apparently random movements of the amoebocytes lead to aggregation into masses of amoebocytes and choanocytes (which at this stage lack their collars). The latter become enclosed within the mass and form flagellated chambers, and the amoebocytes give rise to the layered arrangement of a small sponge. It has been postulated that gradients of oxygen tension, or of electrical charge, are important in this reorganization, but the experimental work does not allow of clearcut generalization. It seems that choanocytes alone cannot re-form a sponge. One other point is emphasized by these studies. It is that there are no distinctive epithelial cells: the so-called pinacocytes arise in these re-united masses from modified amoebocytes.

Related to this experimental capacity is the fact that many sponges are capable of forming balls of cells to act as resting buds. This may occur in adverse circumstances when the main mass of sponge tissue collapses and disintegrates leaving only these buds, each of which contains a sufficient mass of amoebocyte cells to regrow in appropriate conditions into a sponge. This can be adopted as a regular method of asexual reproduction, and is so in the life-cycle of freshwater (and a few marine) sponges where gemmules are formed. Each gemmule is a mass of amoebocytes surrounded by two membranes with spicules embedded in them. There is a small pore through this double shell called the micropile. Under favorable conditions, usually involving a rise in external temperature, hatching occurs: cells stream out from the small opening and reorganize into a tiny sponge. It has been known for some time that the responsiveness of gemmules of freshwater sponges to improved conditions varies with their age, and more recently it has been discovered that they can be subject to processes akin to "vernalization" in plants. In other words, if they have been subjected to a period of low temperature, the gemmules then hatch more rapidly when placed in suitable conditions. In nature, gemmules withstand freezing and a considerably greater degree of desiccation than do adult sponges. Regular seasonal production of gemmules (in the fall) can be seen as adaptation to the peculiarities of the temperate freshwater environment.

In the sexual reproduction of sponges, differentiated amoebocytes form the gamete mother cells and give rise to eggs or sperms. Most sponges are hermaphrodite but a few have separate sexes. The egg is usually retained within the parent until fertilization and holoblastic cleavage have occurred, giving rise to a flagellated motile blastula which swims out through an exhalant osculum and, after further development, settles down as a new sponge. In some simpler sponges such as *Sycon*, there is a typical stage called an amphiblastula, with a hemisphere of flagellated cells and a hemisphere of granular cells, and after settling the flagellated part becomes invaginated into the nonflagellated part and differentiation leads to the *Olynthus*-stage which we discussed earlier. There is considerable difference from one sponge to another in the mode of invagination which gives rise to the first chambered sponge. However, it seems clear that this is always totally different from the processes of gastrulation in other many-celled animals. In fact, when there is a simple invagination, it is the cells derived from the animal pole of the egg which invaginate within those derived from the vegetal pole (not the opposite). Thus the lining cells cannot be homologized with the developmental endoderm of other metazoans.

As emphasized, all aspects of the physiology of sponges depend on the continuity of the water current through the sponge. It can be calculated that 1 cubic centimeter of sponge tissue from a form such as *Leucandra* can propel more than 20 liters of water per day. There is no evidence of extensive coordination though most sponges have some cells capable of slowish contraction. Although there is no good evidence of nerve connections, many sponges can close their ostia and oscula, and some do regularly with a tidal rhythm. Others respond slowly to mechanical stimuli or to light.

If it is conceded that many-celled animals most probably arose from flagellate protistans, it is most reasonable to suppose that either the Porifera evolved from a different group of flagellates than the rest of the Metazoa, or else they diverged very early from the metazoan stem. It is impossible to homologize the development of sponges with the embryology of other Metazoa, and sponges lack both true polarity and division of labor among specialized tissues. It can be assumed that the Porifera represent an evolutionary dead-end. For these and other reasons, classifying them in a totally separate subkingdom—the Parazoa —is probably justified.

Mesozoa

The Mesozoa constitute a minor but enigmatic phylum of many-celled animals. There are about fifty known species: all are minute parasites found in the body cavities of various higher invertebrates. While they could be regarded as the simplest organized of many-celled animals, they are thought by some to be extremely degenerate derivatives of flatworms. They are not unlike the planula larva of coelenterates in general form but, although they have a solid two-layered construction, the inner cells are all reproductive cells, with a single type of ciliated cell forming an outer layer around them. Their life-cycle is always complicated, and usually seems to involve an alternation of asexual and sexual generations. There are three types of mesozoans, differing mainly in the details of their multiplicative history; these are the dicyemids, the heterocyemids, and the orthonectids. The least rare forms are dicyemids found in the excretory organs of cephalopod molluscs. Adult dicyemids show cell constancy: one species always having 30 outer cells around a single long axoblast. Asexually multiplying vermiform larvae (each 33 cells) are produced until crowding in the infected cuttlefish kidney brings about a shift to production of infusiform larvae (each 28 cells, including four germinal cells) which are passed with the urine and which can infect another cuttlefish. In the stage most usually found, the core reproductive cells are termed axoblasts, and functionally correspond to the agametes which are formed when protozoa reproduce by schizogony. Such agametes are found in no other many-celled animals. The orthonectids (e.g., *Rhopalura*) are also rare parasites of various invertebrates and are most remarkable at one stage of their life-cycle. They form an asexual, multinucleate, amoeboid plasmodium which spreads through the tissues of the host, and may even carry out parasitic castration (in some clams and brittle-stars). A "female" of the genus *Rhopalura* is illustrated in Figure 5-4.

Some distinguished workers have suggested that the similarities to protistan structure and function in Mesozoa are real, and that it is improbable that they could arise by degeneration of the flatworm body form. If this could be demonstrated to be so, then the question would arise whether the Mesozoa are (like the Parazoa) independently evolved from protistans, or whether they are in

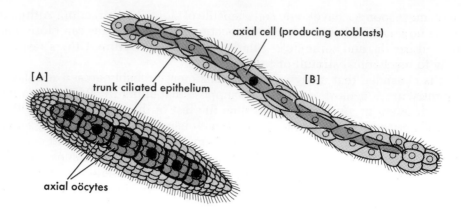

FIGURE 5-4. Representatives of the minor and aberrant phylum Mesozoa. **A:** An adult female of the orthonectid genus *Rhopalura,* parasite of brittle-stars. **B:** The vermiform stage of a dicyemid mesozoan, parasitic in the kidneys of cephalopod molluscs.

some way related to intermediate forms between Protista and true Metazoa. If this last is true, then the group is of great phylogenetic significance. However, recent biochemical work on the nucleic acids of dicyemid mesozoans suggests a closer relationship to *ciliate* protozoans and thus an origin totally distinct from both Parazoa and true Metazoa. In any case, development and life-cycle in the group merit more modern investigation.

6 THE EXTANT PROTOZOANS: SURVEY OF SUCCESS WITHIN A SINGLE CELL MEMBRANE

WE are concerned in this chapter with the animal protistans, or the phylum Protozoa, animal organisms in which all living functions are carried out within a single cell membrane. As we noted in Chapter 1, there are both semantic and fundamental biological difficulties involved in attempting diagnostic descriptions of this group. Difficulties concerned with the interrelationships among the primary divisions into which we place living organisms are best discussed at the end of this chapter after we have gained some substantive knowledge of the diversity of protozoans. For the present, it suffices to note that distinctions among the Protozoa, the Protophyta, and the lower many-celled plants are very difficult to make; that the distinction between the Protozoa and the Metazoa, or many-celled animals, is possible by a moderately complicated definition; but that the distinction between all of these groups and such organisms as bacteria is both semantically easier and biologically much more significant. Bacteria and certain other simply organized organisms, such as blue-green algae, are described as procaryotic; protozoans, along with the cells of higher plants and animals, are eucaryotic. Structurally, a distinguishing feature is the presence of a membrane, closely similar to the bounding cell membrane, surrounding the nuclear material in eucaryotic cells. This diagnostic feature is paralleled by other features of functional morphology; for example, eucaryotic cells can be considerably more compartmentalized with membrane-bounded structures—organelles—responsible for particular cell functions. There are also fundamental differences between procaryotic and eucaryotic organisms in the mechanics of the processes of cell multiplication, particularly as concerns division of nuclear material, which in eucaryotic cells involves the DNA of the chromosomes and the RNA of the nucleolus. Of course, all of the "invertebrates" mentioned in this book, both protozoan and metazoan, are eucaryotic. The definition we use here to separate protozoans from many-celled animals is that the protozoan body never has any specialized parts of the cytoplasm under the sole control of a nucleus. In some protozoans, there can be two, very few, or even many nuclei, rather than one, but no single nucleus ever has separate control over any part of the protozoan cytoplasm which is specialized for a particular function. In contrast, in metazoans there are always many cases of nuclei, each in control of cells specialized for particular purposes: for contraction, for secretion, or for conduction—that is, as muscles, as gland-cells, or as nerves—even if occasional metazoan tissues are syncytial (multinucleate without cell boundaries).

Another semantic difficulty, inflated by a few textbook writers into the status of a controversy, concerns the description of the Protozoa as acellular (not divided into cells) or as unicellular (single-celled). A brief review of some history of these descriptive terms may be useful. The early British microscopist Robert Hooke first used the word *cell* to define the tiny units he had observed. Many kinds of observations were made over the next century and a half, and in 1839 two German scholars, the zoologist Theodor Schwann and the botanist M. J. Schleiden, proposed the concept that the cell is the basic unit of life. A second important basic concept, that cells come only from pre-existing cells, was proposed in 1859 by another great German scientist, Rudolph Virchow, almost at the same time that Charles Darwin outlined his theory of natural selection. Subsequently, and well into the twentieth century, the majority of biologists would define the protozoans as *unicellular* animals. By the 1920's, however, many authors, including the great British protozoologist Clifford Dobell, had pressed for the usage *acellular* on the basis that the body of a protozoan could be conceived functionally to be the equivalent of the entire metazoan organism rather than of any one of its component cells. There are still a number of texts in use which argue this point very strongly. However, over the last three decades, evidence from studies by electron microscopy has clearly shown that, in the majority of their ultrastructures, the individual protozoan cell appears to be strictly comparable to an individual generalized cell of a metazoan. It seems best to state that individual protozoans are both whole organisms *and* individual cells. Accordingly, in this book, uninhibited use will be made of the terms unicellular and single-celled in describing conditions in the protozoans and in making comparisons with many-celled animals.

In some ways, a far more significant feature of the description of protozoans is that they are almost all minute, ranging downward from about 1 millimeter in length to about one-hundredth of that. Many, if not most, of the features of functional morphology that mark off the protozoans from the many-celled invertebrates can be attributed to their small size alone, and to the relatively high surface:mass ratios which result. It was D'Arcy W. Thompson, that twentieth-century polymath and seemingly perpetual professor of zoology in the University of St. Andrews, Scotland, who first pointed out that the majority of shapes among protistan organisms could be explained as a result of the forces of surface tension between fluids. In fact, he claimed that the form of all very small organisms is independent of gravity and mainly produced by surface-tension interactions. It is clear that the high surface:mass ratios in protistans make some kinds of specialized surfaces, for example, those for respiration, unnecessary. To anticipate, it will be clear in many cases discussed later in this book that the evolution of higher metazoans has in many cases been circumscribed, or even channeled, by the need for larger organisms to have acquired in evolution a differential increase of certain surfaces of physiological importance. For the protozoans, it is important to emphasize again that all the more complex functions of locomotion, of food capture, and of mating, which in the many-celled invertebrates involve special tissues or organs, are here combined with the normal intracellular functions, involving the endoplasmic reticulum, the mitochondria, and one or more nuclei. In protozoans, those cytoplasmic parts which are structurally differentiated for particular functions are termed *organelles*. The evolutionary implication of the name Protozoa, or "first animals," may still remain true, but recent work on the ultrastructure of and physiology in protozoans brings increasing evidence of considerable levels of adaptive complexity. Thus the early hope that study of the apparent simplicity of protozoans would more readily provide understanding of certain life processes than similar studies in higher animals, has proven largely illusionary.

In Chapter 1, we briefly noted four main body forms (Figure 1-10) and lifestyles in the phylum Protozoa which might have been encountered by the beginning biologist. It is now necessary to replace that overview of four "kinds" by a somewhat more sophisticated attempt at a classification of the between 25,000 and 60,000 species of extant protozoans.

Basic Architecture

Four major assemblages of protozoans are represented by our four main "kinds": flagellates, *Amoeba*-like forms or sarcodines, sporozoans, and ciliates. The flagellates, which include the most primitive of the protozoans, are closely allied in structure and in life-cycle pattern to the sarcodines. This is recognized in the major classification of this enormous phylum by their combination in the first of four subphyla, the Sarcomastigophora. Two of the other subphyla *do* correspond to the remaining two of our major kinds of body forms and lifestyles. The subphylum Sporozoa comprises the parasitic organisms whose adult stages usually lack locomotory organelles and have complex nuclear processes associated with complicated life-cycles and reproduction. The subphylum Ciliophora (which constitutes a single class, Ciliata) encompasses the most complex and most highly organized protozoans, the ciliates. The remaining and much smaller subphylum Cnidospora comprises parasitic forms which form spores with polar capsules which are multinucleate and bear some ultrastructural resemblance to nematocysts.

These four major subdivisions along with their more important superclasses and classes (in the case of the ciliates, the subclasses) are set out in Table 6-1.

Rather more than with other phyla of animals, classification at the higher levels of the Protozoa has not been stabilized and remains somewhat controversial. As my senior colleague Reginald D. Manwell has pointed out, comparative morphology, which has been so important in elucidating probable evolutionary relationships from homologies of structure in higher Metazoa, can provide much less information about the interrelationships of Protozoa. Table 6-1 is basically a simplified version of a scheme of classification drawn up by a distinguished committee of protozoologists in 1964. Like the King James Bible, it is one of the few documented sets of compromises drawn up by an academic committee which is relatively defensible. The notes below the table outline a few of the areas of disagreement and of honest biological doubt.

Before we pass to a survey of the architectural differences between these various protozoans, it is worth outlining the components of a generalized protistan cell. A much stylized representation appears in Figure 6-1 wherein we have the intracellular structures common to all cells combined with the organelles differentiated for particular whole-animal functions in the protozoans. Among the former are the cell membrane, the endoplasmic reticulum, the ribosomes, the Golgi apparatus, the mitochondria, and the nucleus with its nuclear membrane and nucleolus. The universality of most of these intracellular structures has

TABLE 6-1
Outline Classification of the Protozoa

Subphylum I   SARCOMASTIGOPHORA
  Superclass A   MASTIGOPHORA
    Class 1   PHYTOMASTIGOPHOREA
    Class 2   ZOÖMASTIGOPHOREA
  Superclass B   OPALINATA
  Superclass C   SARCODINA
    Class 1   RHIZOPODEA
    Class 2   ACTINOPODEA
    (one additional class, PIROPLASMEA, may belong here)
Subphylum II   SPOROZOA
  (only major class:) TELOSPOREA
  (two additional classes, TOXOPLASMEA and HAPLOSPOREA, may belong here)
Subphylum III   CNIDOSPORA
  Class 1   MYXOSPORIDEA
  Class 2   MICROSPORIDEA
Subphylum IV   CILIOPHORA
  (only class:) CILIATEA
    Subclass a   HOLOTRICHIA
    Subclass b   PERITRICHIA
    Subclass c   SUCTORIA (or ACINETARIA)
    Subclass d   SPIROTRICHIA

Notes: Three relatively minor groups of parasitic protistans, possibly rankable as independent classes, remain of doubtful systematic position: PIROPLASMEA, TOXOPLASMEA and HAPLO-SPOREA. The majority of systematic accounts earlier than 1963 make two separate subphyla of the MASTIGOPHORA (flagellates) and the SARCODINA (*Amoeba*-like forms). A few authorities would link these two groups with the SPOROZOA (that is, unite subphyla I and II above) in a single subphylum PLASMODROMA, with the subphylum CILIOPHORA as the only other major division of the phylum PROTOZOA. Other earlier accounts regard the OPALINATA as primitive members of the CILIOPHORA, and do not separate the CNIDOSPORA from the SPOROZOA.

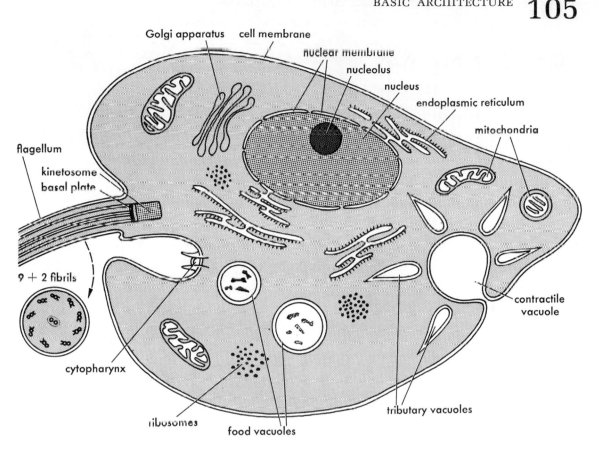

Golgi apparatus   cell membrane

nuclear membrane

nucleolus

nucleus

endoplasmic reticulum

mitochondria

flagellum

kinetosome

basal plate

9 + 2 fibrils

contractile
vacuole

cytopharynx

tributary vacuoles

ribosomes

food vacuoles

FIGURE 6-1. A generalized diagram of a mastigophoran (flagellum-bearing) protozoan. It shows not only the features of all eucaryotic cells (including the cell membrane, the nucleus with its nuclear membrane and nucleolus, the endoplasmic reticulum, the ribosomes, the Golgi apparatus, and the mitochondria) but also typical protistan organelles (including the cytopharynx, a flagellum with its kinetosome, a contractile vacuole with tributary vacuoles, and food vacuoles containing ingested material).

been established by electron microscopy, but perhaps the most fundamental and most significant element as a universal feature of all cells is the cell membrane which bounds the cell surface. At suitably high magnifications, it can be visualized as a two- or three-ply structure, and most authorities assume that irregular protein molecules penetrating from the outer and inner faces are associated with a regularly ranked bilayer of phospholipid (fat-like) molecules. A stylized representation of this bilayer construction is shown in Figure 6-2, but it should be noted that it is probably erroneous to conceive of this as a *static* array of molecular building blocks. Modern experimental studies of the functioning of these cell membranes are modifying the stylized visualizations derived from electron microscopy, and it seems clear that in the living functioning cell, the cell membrane is a dynamic system within which some molecular rearrangements are occurring continuously. There are still controversial aspects of the bioenergetics and the physical biochemistry of the functioning of cell membranes, and it would be inappropriate to try to summarize these here. However,

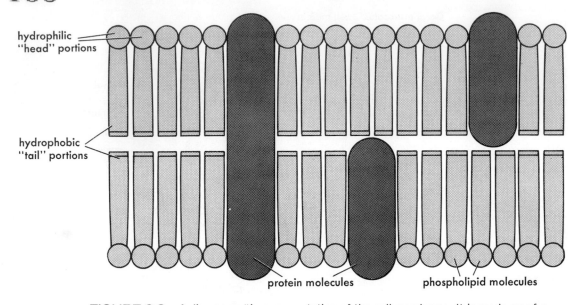

hydrophilic "head" portions

hydrophobic "tail" portions

protein molecules

phospholipid molecules

FIGURE 6-2. A diagrammatic representation of the cell membrane. It is made up of a regularly ranked bilayer of phospholipid molecules, with their hydrophilic "heads" forming the outer and the inner surfaces of the cell membrane, and less regular protein molecules penetrating or partially penetrating the membrane. The cell membrane is about 50 Ångstroms thick. For discussion, see text.

physiological functions can involve not only differential molecular permeabilities and energy-consuming *active transport* (against concentration gradients), but also free (passive) diffusion of certain ions and molecules, and facilitated diffusion of others. Temporarily or locally, membrane properties can be altered, apparently as a result of changes of metabolic state within the cell. Thus the cell membrane should be regarded not merely as a containing film, coating and bounding the cell, but as an essential part of many of the cell's functional mechanisms.

As in all eucaryotic cells, the nucleus in protozoans is enclosed in a nuclear envelope which is porous but involves two concentric membranes which are probably continuous by way of the endoplasmic reticulum with the cell membrane on the outside. Besides the DNA and proteins of the nucleoplasm which condense as the chromosome material at cell division, the nucleus usually contains a mass of RNA called the nucleolus.

In protozoan cells, we also have organelles. Figure 6-1, which is stylized on a mastigophoran basis, shows the root of a flagellum with its kinetosome and the basal organization of the fibrils. Locomotor organs—flagella, cilia, and pseudopodia—will be considered in more detail in the next section. Other organelles are concerned with protozoan nutrition and excretion. A set of four rather old-fashioned terms are useful in classifying the nutritional methods of protozoans. The most typically "animal" protozoans are termed holozoic, meaning that they ingest pieces of material from plant or other animal forms or else hunt and capture bacteria or other small protistans which are ingested whole. Plant-like protozoans capable of photosynthesis are termed autotrophic. Those free-living protozoans which absorb organic materials in solution and do not ingest solid food are termed saprozoic. Parasitic forms are normally holozoic or saprozoic, or may employ a combination of these methods. Both free-living and parasitic protozoans with any combination of the three principal nutritional methods are often termed heterotrophic. The cell body of holozoic protozoans always contains a series of food vacuoles which are membrane-bounded vesicles formed at the cell surface to envelope the food. As they migrate through the cytoplasm they undergo definite changes in pH corresponding to a sequential process of intracellular digestion, involving first enzymatic breakdown in an acid phase followed by an alkaline phase combining further enzymatic action and absorption of material from the vacuole into the cytoplasm. Undigested remnants are egested to the exterior of the cell in a thoroughly animal fashion. In some holozoic protozoans a cytostome (or cell-mouth where the food vacuole is formed) is detectable only during ingestion, while in others, including all the more complex ciliates, there is a cytopharynx with associated permanent structures including contractile and stiffening elements. In contrast, in the autotrophic protozoans, there are chromatophores of chlorophyll in the cytoplasm and these are the sites of energy-absorption and of photosynthesis. Many autotrophic protozoans also have light-sensitive structures termed eyespots or stigmata. Another membrane-bounded vesicle is associated with excretion and water control in protozoans and is termed the contractile vacuole. It is usually fixed at a definite site within the cytoplasm and may have contributory canals or other vesicles leading into it. It shows a rhythmic process of pumping, with a slow expansion by accumulation of fluid from the adjacent cytoplasm followed by a relatively fast contraction which expels the contents to the outside of the cell. Functionally, contractile vacuoles can be regarded as maintenance pumps which remove excess water from the cytoplasm. Their occurrence in protozoans has an obvious ecological significance. All freshwater protozoans have functioning systems of contractile vacuoles, while these are only sporadically distributed in marine and parasitic species. Obviously, regular pumping out of excess water for cell maintenance is universally necessary in freshwater circumstances where the cytoplasm will be hypertonic to the surrounding medium of the environment. Certain other features of the functional morphology of protozoan cells must be noted. Within the bilayer cell membrane there is an outer layer of the cytoplasm, usually in a gel state and termed ectoplasm, and this contrasts with the greater bulk of internal cytoplasm which is more fluid and termed endoplasm. As we shall see when we discuss sarcodine locomotion and the formation of pseudopodia, ectoplasm and endoplasm represent reversible colloidal states of the cytoplasmic materials. A variety of other intracellular structures in protozoans, particularly in the cortical or outer parts of the cell, are made up of compound bundles of fibrils. These are known as myonemes, rods, microtubules, axostyles, and axial rods. The myonemes and certain others are contractile, while some of the rods are clearly skeletal, or supporting, in function. A large number of protozoans also have external support, since they secrete nonliving tests or shells outside the cell membrane. These may be formed of calcium carbonate or of silica, or they may be of secreted and hardened (or polymerized) organic materials such as cellulose and chitin. A few protozoans also collect materials from the environment around them, such as sand grains, which they cement together to build shells in much the same fashion as do the larval stages of caddis-flies. Such protozoans are termed arenaceous and are found among both the sarcodines and the ciliates (see Figure 6-15).

We can now return to Table 6-1 and to the intermediate levels of classifica-

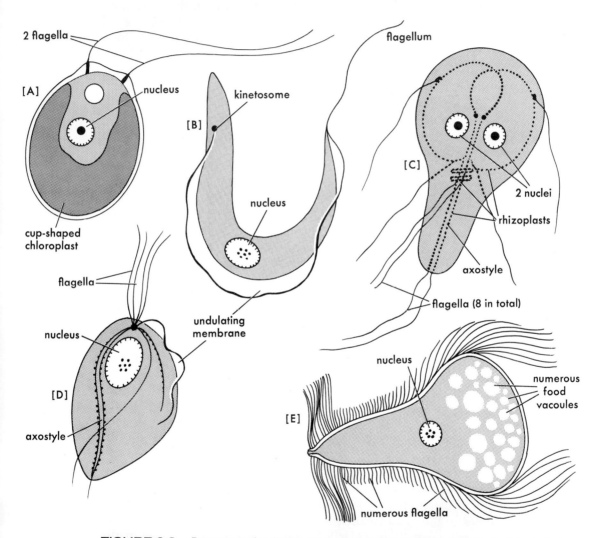

FIGURE 6-3.   Representative flagellates (see also Figures 6-1 and 33-2). **A:** A typical phytomastigine of the genus *Chlorella,* which is capable of photosynthesis and often is regarded as a representative of the "single-celled algae." **B:** The causative organisms of sleeping sickness in man and cattle are various species of this genus, *Trypanosoma.* They are almost all parasites of the bloodstream. The adult form has an undulating membrane associated with the long flagellum, as shown. **C:** *Giardia lamblia,* a common parasitic flagellate from the duodenal region of the human intestine. **D:** *Trichomonas vaginalis,* a moderately common parasite of human reproductive ducts. **E:** *Trichonympha,* a hypermastigine flagellate from the gut of termites, where it symbiotically assists in the digestion of wood.

tion in the phylum Protozoa. The diagnostic description of the subphylum Sarcomastigophora includes no spore formation, a single type of nucleus, and locomotion by flagella, by pseudopodia, or by both. The first superclass is the
Mastigophora (or flagellates) with one, few, or many flagella and asexual reproduction by a symmetric binary fission. The first class, the Phytomastigophorea,
have chromatophores and are normally autotrophic. Most of them are free-living, and they commonly have only one or two flagella (Figure 6-3). In addition
to forms like *Euglena*, this group includes many of the most important microscopic green plants of the world's oceans, including dinoflagellates and coccolithophores, the latter having a body covered with calcareous platelets. A few
forms like *Volvox* form spherical colonies. In textbooks of botany, many phytomastigophorans are classified and described as single-celled algae. The second
class of the flagellates is the Zoömastigophorea, which includes many free-living holozoic forms which are otherwise essentially similar to the chromato-
phore-bearing forms of the other class. There are also some parasitic forms including the trypanosomes, the causative organisms of human sleeping sickness
and of Nagana in cattle in Africa. As parasites of certain insect groups, there are
zoöflagellates with numerous flagella and multiple basal bodies to which these
are connected (Figure 6-3), and such flagellates represent a level of complexity
of cellular organization rivaling anything found among the ciliates. The second
superclass, the Opalinata, consists of forms like the genus *Opalina* (see Figure
6-7), which have many nuclei all alike and are covered with cilia-like organelles (considered to be flagella) in oblique rows like spiral rifling all over the
body surface. They are all parasitic in frogs and toads, and they have clearly
flagellate affinities. Although many earlier authorities regarded them as relatively primitive members of the Ciliophora, they lack the dimorphic nuclei and
the often elaborate cytostomes and associated structures of the true ciliates. The
third superclass is the Sarcodina, *Amoeba*-like organisms which move typically by pseudopodia and, since they lack a pellicle, usually are without fixed
shape. The class Rhizopodea includes forms like *Amoeba* itself, both free-living and parasitic; and there are also a variety of testate sarcodines which differ
in their external tests or shells but are in their intracellular organization closely
similar to *Amoeba* (Figure 6-4). Forms like *Arcella* have lobose pseudopodia
and a relatively simple case, while the abundant and important foraminiferans
have a many-chambered calcareous test with rather finer reticulate pseudopodia extending out through a series of pores (Figure 6-4D). Forams are abundant
in the oceanic plankton. Many are herbivores feeding on diatoms and green
flagellates, others are carnivores (Figure 6-4F), and still others are hosts to symbiotic zooxanthellae (Figure 6-4E) in a similar fashion to the soft-corals discussed in Chapter 4. In forams like *Globigerinoides* (Figure 6-4E), the green
symbionts move (or are moved) in a circadian rhythm so that they are contained within the test during the night and exposed in the outer parts of the
reticulate pseudopodia by day. The shells of dead forams form extensive areas
of marine sediments in today's oceans and have formed many fossil deposits in
the past. These fossil protozoans can be of immense economic importance,
since they are used by oil-geologists for the identification of particular strata
from drilling cores. Another subdivision of the Rhizopodea includes some of
the "slime molds" or Mycetozoa where amoeboid forms can develop into a

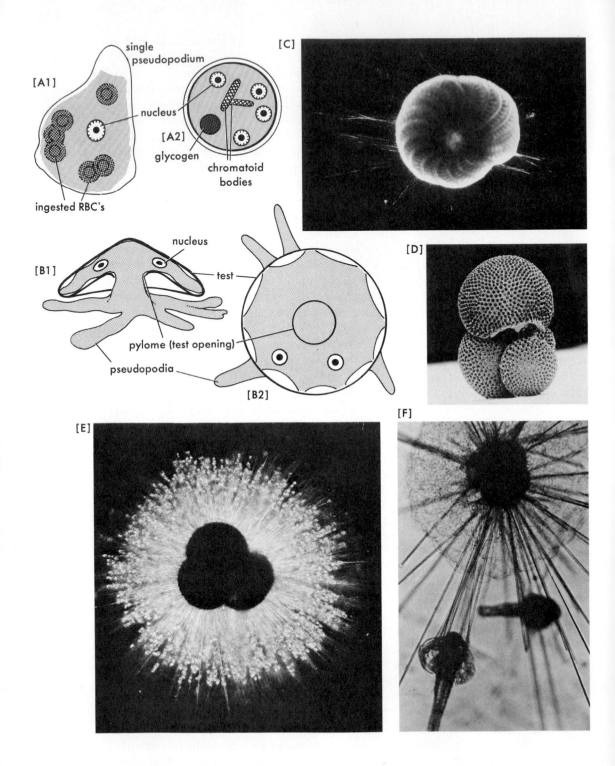

[A1]
single pseudopodium

[A2]

nucleus

glycogen

chromatoid bodies

ingested RBC's

[C]

[B1]
nucleus

test

[B2]

pylome (test opening)

pseudopodia

[D]

[E]

[F]

110

FIGURE 6-4. (opposite) Representative sarcodines (see also Figure 6-8). **A:** The parasitic amoeba, *Entamoeba hystolytica,* is the causative organism of amoebic dysentery in man. **A1:** Motile amoebic stage with the characteristic single pseudopodium and ingested red blood corpuscles; **A2:** a cyst from human faeces. **B:** A representative shelled or testate amoeba from moorland waters, *Arcella.* The "flying saucer" shell is largely proteinaceous, but also contains silica. **C:** A relatively large living foraminiferan, *Elphidium crispum,* with many characteristic fine and reticulate pseudopodia. **D:** Scanning electron micrograph of the calcareous test of a typical oceanic foram, *Globigerinoides sacculifer,* showing the numerous (and regularly arranged) tiny apertures through which the fine pseudopodia are extended. **E:** Light micrograph of a living planktonic foram, *Globigerinoides ruber,* with hundreds of symbiotic zooxanthellae in the fine reticulate pseudopodia. **F:** Another living planktonic foram, *Hastigerina pelagica,* engaged in feeding as a carnivore. Two *Artemia* larvae have been trapped by the radiating pseudopodia and are being drawn into the central "bubble capsule" for digestion. [**A** and **B** from the author's student notebooks prepared under the direction of Margaret W. Jepps in 1944 at the University of Glasgow; photo **C** © Douglas P. Wilson; photos **D–F** courtesy Dr. Allan W. H. Bé; **D** from A. W. H. Bé in A. T. S. Ramsay (ed.), *Oceanic Micropaleontology,* Academic Press, 1977; **E** and **F** from Bé, Hemleben, Anderson, Spindler, Hacunda, and Tuntivate-Choy, *Micropaleontology,* **23:**155–179, 1977.]

multinucleate plasmodium or even into multicellular aggregations in the course of moderately complex life-cycles. Many of them are saprozoic; they involve young stages with flagella and they include some important parasites of crop plants, including the causative organism of clubroot disease in cabbages. In a separate class, the Actinopodea, are placed certain sarcodines with slender radiating pseudopodia and a delicate and often very beautiful radial skeleton made of silica. These radiolarians are also important in oceanic bottom deposits and their fossils form a few rocks including the cherts, but these are never so thick or extensive as are the limestones formed from forams. One additional group of parasitic forms which may belong in the sarcodine superclass is the Piroplasmea or babesias, the causative organisms of Texas cattle fever (which is transmitted by ticks).

The second subphylum, the Sporozoa, and its only major class, the Telosporea, are parasitic protozoans that normally lack all locomotory organs and occur as internal parasites of a great variety of animals, often with an intracellular stage in their complex life-cycles. The "spores" of their name are somewhat different from the resistant forms of bacteria or the fruiting stages of fungi which are also called by the same name. In the Sporozoa, the term refers to daughter cells produced by a process of multiple fission or sporogony, which typically occurs after a process of sexual fusion. A few groups of sporozoans including the gregarines (Figure 6-5) are moderately large protists, with little in the way of internal organelles, which live as internal parasites of a variety of invertebrates. Much more important to man are the coccidian sporozoans which are mostly intracellular parasites of vertebrates. They include the malaria parasites which were, until relatively recently, the largest single cause of human mortality in the world's population each year. Their complex life-cycles, involving intermediate hosts (or vectors) such as mosquitoes, will be discussed later. Two additional classes may belong in this superphylum, the Toxoplasmea, intracellular parasites of a variety of mammals and birds, and the Haplosporea, parasites of invertebrates and fish with transmission by relatively simple spores. Some of the latter are hyperparasites living in parasitic flat-

112

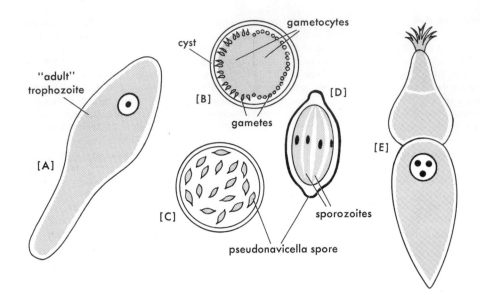

FIGURE 6-5. Representative sporozoans (see also Figure 6-12). Gregarine sporozoans are common parasites of invertebrates. **A–D:** Stages in the life-cycle of *Monocystis,* a parasite living in the male genital system of earthworms. **A** is the "adult" trophozoite; **B** and **C** two of the stages of sexual reproduction within the "association" cysts. In **B** the two gametocytes have already undergone multiple divisions to form two kinds of gametes, while in **C** these have associated in syngamy and formed many pseudonavicellae or spores. In every pseudonavicella (**D**), eight sporozoites are formed, each of which is capable of infecting a new worm and growing up to become a trophozoite. At **E** is shown the more complex trophozoite stage of another gregarine, *Corycella,* a parasite of water-beetles. [**E** adapted from Leger and Grell, **A–D** from the author's student notebooks, University of Glasgow, 1944.]

worms. A separate subphylum, the Cnidospora, has now been erected for certain related forms, formerly classified as sporozoans. They all have complex spores with complex polar capsules and filaments which resemble the nematocysts of the cnidarians (see Figure 6-7), which apparently assist in dissemination from host to host. None is known to parasitize man but a number of them cause diseases in species of economic interest including silkworms, honey

FIGURE 6-6. (opposite) Representative ciliates (see also Figures 6-14 and 6-15). **A:** The "classic" ciliate of elementary biology, *Paramecium* is structurally rather complex, with an elaborate ciliated groove (or vestibule) leading to a large buccal cavity anterior to the cytostome and cytopharynx, where food vacuoles are formed. **B:** Another common freshwater ciliate, *Spirostomum,* is one of the few protozoans visible to the unaided human eye since some reach lengths of about 1.5 millimeters. **C:** An attached ciliate of the genus *Stentor;* again some species are visible with the naked eye. **D:** A more complex attached ciliate, *Vorticella,* has a contractile stalk and elaborate peristomal and buccal apparatus. **E:** *Euplotes* is an example of the hypotrichid ciliates, where compound cilia form cirri, which are used for "walking," and the general body surface lacks cilia. **F:** A symbiotic ciliate from the gut of cattle, *Eudiplodinium,* with cirri and, alongside the cytopharynx, a complex skeletal plate of polygonal prisms, to which contractile fibers attach and in the cavities of which glycogen is stored. Such ciliates, along with the gut bacteria of cattle, play an essential part in the supply of beef and milk to man. "All flesh is grass," but grass requires fermentive digestion in the rumen by microorganisms.

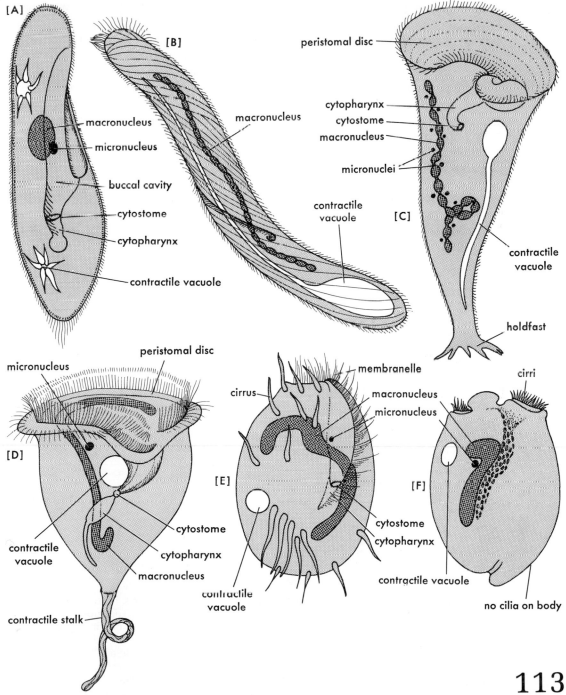

[A]

macronucleus

micronucleus

buccal cavity

cytostome

cytopharynx

contractile vacuole

[B]

macronucleus

macronucleus

contractile vacuole

peristomal disc

cytopharynx

cytostome

macronucleus

micronuclei

[C]

contractile vacuole

holdfast

peristomal disc

micronucleus

[D]

contractile vacuole

contractile stalk

cytostome

cytopharynx

macronucleus

membranelle

cirrus

macronucleus

micronucleus

[E]

cytostome

cytopharynx

contractile vacuole

contractile vacuole

cirri

[F]

contractile vacuole

no cilia on body

113

bees, and several fishes. Once again, a few species are hyperparasites occurring in parasitic flukes and even in other cnidospores.

The fourth and last subphylum is the Ciliophora (see Table 6-1), which comprises the most highly organized protozoans, all with a complex pellicle and subpellicular arrangement of fibrils, to which are attached the many cilia which give the group its name (Figure 6-6). They typically possess two types of nuclei and they are mostly free-living with holozoic or heterotrophic nutrition. Their sexual reproduction involves conjugation and exchange of nuclear material from the micronuclei. The macronucleus appears to be responsible for the control of general cellular functions and has been referred to as the vegetative control center, while the micronuclei seem to be essential for sexual processes of reproduction. The majority have compound ciliary organelles associated with the cytostome and cytopharynx and variations in this are important in the classification of ciliates. Some of the varieties of ciliate body form are shown in Figure 6-6. Of the four subclasses, the Holotrichia have relatively uniform ciliation on the body and include forms like the well-known *Paramecium* and one of the most studied of all protozoans, *Tetrahymena*, whose maintenance in continuous culture has allowed fundamental studies of basic nutritional requirements and provided an assay organism for several vitamins also required by man. The subclass Peritrichia are mostly sessile with stalks, or sedentary with attachment discs. They all have an anticlockwise spiral or ring of cilia,

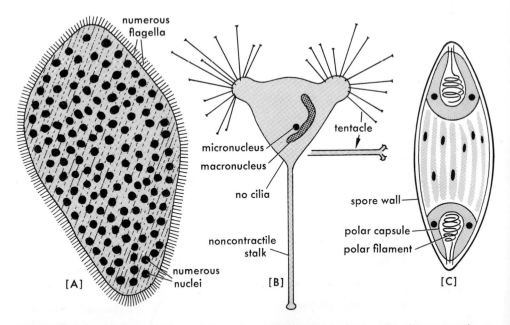

FIGURE 6-7. Representatives of three aberrant protozoan groups; note the comments on the systematic status of these forms in Table 6-1 and in the text. **A:** The superclass Opalinata is represented by *Opalina,* a common parasite of frogs and toads, multinucleate and with numerous rows of cilia-like flagella. **B:** The subclass Suctoria is represented by *Acineta* with two groups of sucking tentacles and no true cilia in the stalked sessile adult. **C:** The subphylum Cnidospora is represented by a myxosporidean spore of a fish parasite. Compare with the gregarine spore in Figure 6-5D, and with nematocyst structure in Figure 3-1.

differentiated around the mouth, to serve as a food-collecting organelle. *Vorti-cella* (Figure 6-6) is a typical representative of this group. The Suctoria are again sessile but their adult stages lack cilia and ingest their prey through suck-ing tentacles. Finally, the subclass Spirotrichia, including forms like *Stentor*, are a somewhat mixed group with complex organelles formed out of compound cilia. The buccal ciliature is conspicuous and consists of many membranelles winding clockwise down to the cytostome. A few other examples of the com-plexities of ciliates will be described later. Certain aberrant protozoan groups are illustrated together in Figure 6-7.

Basic Locomotion

Investigators of locomotion in protozoans have attempted to elucidate the pro-cesses at two distinct functional levels, and with somewhat different levels of success. First, there are the mechanics involved, including the descriptive ki-netics of the movements of the locomotory organelles, and what can be de-duced regarding the forces which they exert upon the environment. This me-chanical level of explanation—how protozoans as efficient machines move through their environment—has been the subject of extensive and moderately successful investigations and these will be briefly summarized here. Secondly, there is the biochemical and biophysical level concerned with the translation or "release" of the energy of molecular bonding as the kinetic energy of con-traction and of movement. There are still considerable uncertainties regarding these processes at the molecular level, and they will not be explored here. How-ever, the supply of energy to contractile structures in protozoans is believed to be based on the albeit universal process of energy supply involving the break-down of ATP (adenosine triphosphate) to ADP (diphosphate), but the subse-quent processes, whereby that available energy is involved in the contraction or sliding or bending of filamentous protein molecules in those fibrils revealed by electron microscopy, remain uncertain.

The principal locomotor organelles of protozoans are pseudopodia, flagella, and cilia. The naked sarcodines like *Amoeba* move by lobose pseudopodia, and it is relatively easy for a student to observe this with a low-power microscope. Unlike the movements produced by flagella and cilia which normally occur *through* a fluid medium, pseudopodial movement of this kind is essentially lo-comotion over a surface. Even with ordinary microscopic illumination, one can see that an *Amoeba* flows into one or a few lobes, apparently pulling along a temporary "rear end" or uroid. A little closer observation will show that the central, more fluid protoplasm, the endoplasm, is flowing toward the extending pseudopodium. In many cases there will appear to be a thicker, somewhat smoother layer of gel ectoplasm at the advancing tip of each pseudopodium and in contrast the uroid is of irregular shape (not smoothly lobed) and ap-parently sticky with debris clinging to it. In Figure 6-8 are shown some of the movements which can be observed within the cytoplasm during the simplest locomotion of an *Amoeba* with a single pseudopodium. The trailing uroid and the leading hyaline gel cap can be seen, along with the flow of the central endo-plasm in the direction of movement. As demonstrated many years ago by S. O.

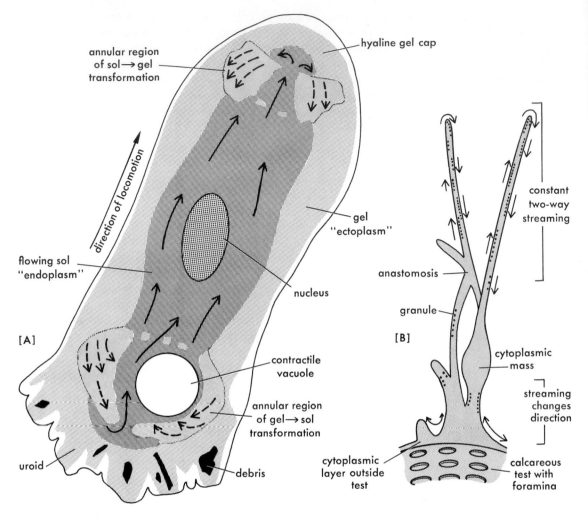

FIGURE 6-8.   Locomotion by pseudopodia. **A:** Locomotion in a stylized specimen of *Amoeba* with a single large pseudopodium. The cortical gel "ectoplasm" contains a streaming mass of sol "endoplasm" moving in the direction of locomotion. At the front an annular zone of sol → gel transformation lies behind a smooth convex hyaline gel cap while, posteriorly, a zone of gel → sol transformation is trailed by an irregular, flattened, sticky uroid. The central position of the nucleus in a wider region of sol conditions and the posterior position of the contractile vacuole nearly surrounded by cortical gel cytoplasm at the zone of gel → sol transformation are both characteristic and related to the hydromechanics of locomotion. **B:** Part of the reticulate pseudopodia of a foram (compare with Figure 6-4C, D). The delicate radiating pseudopodia are sometimes called filopodia, and each shows a continuous two-way streaming of the cytoplasm with granules being carried along out and back, while the cytoplasm masses nearer the foraminiferan test show streaming which is less consistent in direction.

116

Mast of Johns Hopkins and Woods Hole, around the hyaline cap is an annular region of sol to gel transformation, while just anterior to the uroid is a similar ring-zone of gel to sol transformation. These observations of pseudopodial mechanics are accepted by all, but there are somewhat controversial theories concerning their causal mechanisms. One theory involves a "fountain" streaming and continually adding to the anterior end of the gel tube of the pseudopodium which pulls the central endoplasm in its sol state forward. An alternate theory involves a posterior contraction pushing the central sol core forward by hydraulic means.

Even greater uncertainty concerns the mechanics of the more delicate pseudopodia of other sarcodines. Figure 6-8 also diagrams a branching pseudopodium in a foram, where there is a constant movement of cytoplasm in opposite directions on the two sides of the pseudopodial filaments as they lie extended out across a surface. The importance of this adaptation to the feeding of forams is obvious, since it provides a conveyor belt system for particulate food, but it is somewhat difficult to explain in terms of folding changes in molecules. In the even finer radial axopodia of the Radiolaria, which have a central axial fila-

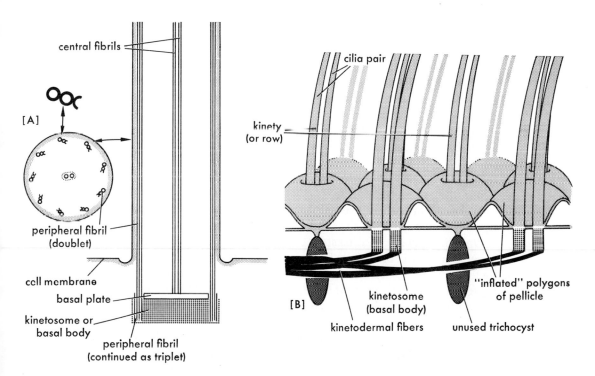

FIGURE 6-9. A: Basic structure of a flagellum. An extension of the cell membrane encloses the characteristic **9 + 2** organization of fibrils, and the nine peripheral doublets are continued as triplets in the basal body or kinetosome. Cilia, flagella, and sperm tails from the majority of animal phyla all are essentially similar in this **9 + 2** arrangement of fibrils. [Adapted in part from Sleigh and from Satir.] **B:** Stylized reconstruction from electron microscopy of the infrastructure of a small number of cilia in a protozoan like *Paramecium*. Paired cilia are arranged by row (or kinety), with alternate rows of trichocysts. The polygon "cushions" of pellicle (forming basal cups for the cilia) are also in parallel arrays. Kinetosomes of the cilia appear to be linked by kinetodesmal fibers, again by row.

ment, we are even further from a complete elucidation of the mechanics at the molecular level.

The other organelles of locomotion in protozoans, flagella and cilia, are of basically similar ultrastructure, and the nine-plus-two unit structure (shown in cross section) is also found in the sperm tails of Metazoa, and in certain sensory structures, as well as in their flagella and cilia. Each consists of an outer membrane continuous with the bilayer membrane of the cell surface enclosing a matrix in which there are eleven sets of fibers, two single central ones with nine double fibers forming in sections a circle around the central pair. In protozoans, each flagellum arises from a basal body or kinetosome, while cilia have their basal bodies linked by complex filaments (or kinetodesmata) in a complex infraciliature (Figure 6-9). Apart from this potentiality for the coordination of large numbers of cilia, and from the fact that each individual flagellum is usually proportionately longer, the main differences between the two kinds of locomotory organelles concern the mechanics of their action upon the fluid environment. In brief, a flagellum normally moves the fluid medium at right angles to the surface of its attachment. Further, the flagellum may have several waves of bending passing along it at any instant, and thus the movement of fluid can be relatively continuous. In contrast, cilia move fluid in a flow parallel to the surface of their attachment and each cilium has a clear active stroke followed by a recovery phase. Of course, smooth movement by cilia is achieved by many of them being coordinated in a pattern of metachronal rhythm. In this, each row of cilia are slightly out of phase with those of the next row and a wave appears to pass over the surface in the opposite direction to the effective stroke of the cilia (often described as resembling the movement of wind over a field of wheat). Examples of both flagellar and ciliary movement are shown in Figure 6-10. In their simpler forms, both kinds of organelles can only act on fluids. Obviously, when they are attached to the surfaces of free-living protozoans, they will cause the protozoan body to move in relation to the fluid medium. When they are extended from a sessile, attached protozoan (such as *Vorticella* or *Stentor*), or from a fixed surface in a many-celled animal, then they will set up a movement of the fluid medium relative to their bases.

One final class of ultrastructures should be mentioned in relation to the movements of protozoans. Particularly in the more complex ciliates, there can be fibrils which have been called endoplasmic myonemes. In forms like *Stentor* there is good evidence that these are contractile organelles concerned in retraction and in other changes of shape. Phase microscopy on living animals shows that these myonemes become markedly thickened when that part of the ciliate contracts. There are similar myonemes in the stalks of vorticellids and also, perhaps surprisingly, in the large sporozoans called gregarines.

Some Life-Cycles

In all protozoans, some form of asexual reproduction occurs. In some species of flagellates and of sarcodines no other mode of reproduction has ever been observed.

The simplest kind of asexual reproduction involves binary fission, and this is

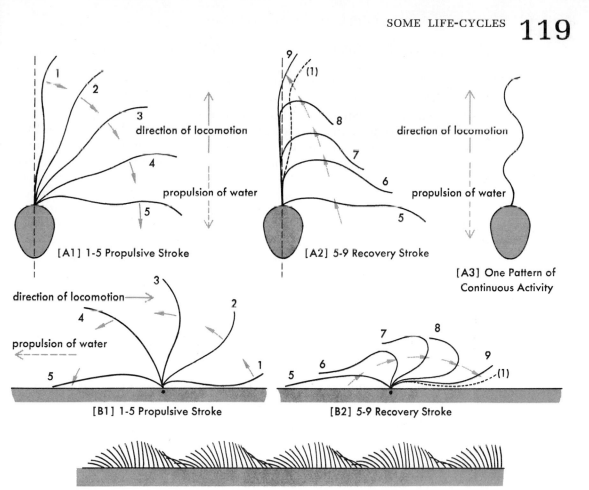

FIGURE 6-10. Activities of flagella and cilia. Propulsive stroke (**A1**) and recovery stroke (**A2**) of a simple flagellum. Note that water is moved at right angles to the surface of attachment of the flagellum. Some flagella (see **A3**) may have several waves of bending passing along at any instant, and thus create relatively continuous movement. Propulsive stroke (**B1**) and recovery stroke (**B2**) of a single cilium. Note that water is moved in a flow parallel to the surface of attachment of the cilium. Cilia are usually coordinated in a pattern of metachronal rhythm (**C**). Each cilium is beating slightly later than the one in front of it, but in phase with its neighbors in adjacent rows: thus waves appear to pass over the ciliated surface *in the opposite direction* to the effective stroke of the cilia. [Partly adapted from Krijgsman and from Gray.]

always preceded by appropriate nuclear division. At least in the functional result, this nuclear division is similar to mitotic cell division in many-celled plants and animals. In other words, the daughter cells that result from successive processes of binary fission have identical nuclei, and thus form clones possessing identical genetic material. Binary fission in mastigophorans is usually longitudinal and symmetric, involving equivalence of the kinetosomes and so on as well as of the nuclear material in the daughter cells. Binary fission in ciliates is usually transverse and is preceded by divisions, not only of macronuclei and micronuclei but also of the cytopharyngeal structures. In a few sessile ciliates and certain mastigophorans, processes of asexual budding produce daughter cells much smaller than the parents. Here again, the preceding nuclear division is of a sort that ensures that the daughter cells are genetically identical with the parent, despite the differences in cytoplasmic mass.

In a number of protozoans there are also processes of multiple fission or schizogony where a number of daughter cells are produced simultaneously. A simplified version of this occurs in some parasitic flagellates; for example, in trypanosomes there is essentially a repeated budding within a parental membrane. The more typical process of schizogony is found among the Sporozoa. A stage termed a mature schizont is formed by successive binary fissions of the nucleus occuring without any division of the cytoplasm, until the characteristic number of nuclei (eight or sixteen, or as few as four, or as many as some hundreds) for the schizont in that particular species has been reached. Then, after a varying period of time, multiple fission of the schizont will give rise simultaneously to the appropriate number of daughter cells. In sporozoans, the term sporogony is applied to the multiple fission of a zygote cell resulting from sexual reproduction.

If we exclude for the moment the relatively complex processes of conjugation in the ciliates, sexual phases of a relatively simple type have been described for the life-cycles of all sporozoans, most phytomastigophorans, several kinds of zoömastigophorans, and many sarcodines. Gametes are produced by appropriate nuclear divisions (involving reduction of nuclear material) preceding the division of single cells, or preceding special processes of unequal cell division (like those of asexual budding). The gametes may be identical in appearance (isogametes) or differ in size and motility (anisogametes), like the eggs and sperms of many-celled animals. The gametes fuse in the process called syngamy and the resultant nuclear fusion restores the appropriate level of genetic material, though now of mixed origin from two "parents." In a few forms of phytomastigophorans and sporozoans, there is evidence that the meiosis or reduction division of the nucleus does not occur in the process of gamete formation, but is carried out after the formation of the zygote and of the zygote nucleus.

In the subphylum Ciliophora, we find conjugation as a method of exchanging genetic material that is peculiar to the group. In this process of temporary union, individuals of complementary mating types adhere to each other and their micronuclei undergo a constant number of successive nuclear divisions (usually three), which involve at least one meiotic stage. The macronucleus in each individual then degenerates as do all but two of the daughter micronuclei in each sexual partner (Figure 6-11). These surviving nuclei are termed pronuclei, one being migratory and the other stationary. Each migratory micronucleus then moves across the zone of adhesion into the other partner and fuses with the stationary nucleus. The product of this nuclear union is called the synkaryon, and it undergoes a series of mitotic divisions as the partners separate. These divisions restore the correct micronuclear number and one micronucleus in each cell becomes enlarged to form the new macronucleus. This enlargement involves a multiplication of the genetic material involved (if the micronucleus is regarded as containing one gene set, then the macronucleus will con-

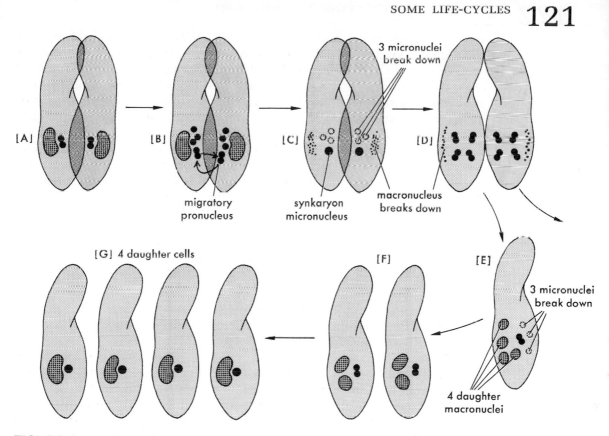

FIGURE 6-11. Sexual reproduction in ciliate protozoans. In this stylized sequence of events in a form like *Paramecium caudatum,* complementary mating types adhere in a temporary union, their micronuclei subdivide (**A**), three micronuclei and the macronucleus in each break down and are resorbed, a further division of the remaining micronuclei yields stationary and migratory micronuclei, and the latter are exchanged and fuse to form the synkaryon (**B** and **C**). Each synkaryon nucleus then undergoes three mitotic divisions (**D**), and of the eight daughter nuclei, four become daughter macronuclei, three break down and are resorbed (**E**), and the remaining micronucleus then subdivides (**F** and **G**) to yield the four micronuclei of the daughter cells.

tain from twenty to forty sets or "genomes," depending on the species of ciliate). As first discovered by H. S. Jennings, and more recently elaborated by T. M. Sonneborn, species of *Paramecium* are divided into an elaborate series of mating types and groups, some of which can involve lethal interactions. The adaptive significance of these series of antigenic types, which involve both cytoplasmic and nuclear controls, remains somewhat obscure.

In terms of the life-cycles of protozoans, it is the processes of asexual reproduction that can result in rapid population increases. Reginald D. Manwell points out that, in small ciliates like *Colpoda,* binary fission every three or four hours can result in 64 individuals from a single isolated ciliate after one day, and 2^{42} would represent the potential progeny after one week. The life-cycles of some amoebae would seem to consist of nothing but an endless series of binary fissions, without any sexual reproduction intervening. However, in most groups, a period of asexual multiplication is followed by a sexual phase. In some shelled sarcodines like forams, schizogony within the fully grown foram shell results in a large number of daughter units which emerge naked and then begin their own shells. Forams and many flagellates have sexual phases intervening after a restricted number of asexual multiplications. This seems also to be true in certain stocks of ciliates, although it is complicated by the elaboration of mating types and races (including stocks which are preferentially in-breeders and others which are normally out-breeders). As with many-celled animals, recombination of genetic material can only occur with sexual reproduction. As will be discussed below, the life-cycles of certain parasites, including those which cause malaria, involve sustained asexual reproduction in one host and sexual reproduction in the other.

Efficient Parasites

As we will discuss in certain metazoan cases of parasitism, the most efficient protozoan parasites are those which are never lethal to their hosts, and thus never lethal to the habitat housing that population of parasites. Obviously, the most extensive biological investigations have concerned those protozoan parasites of man which produce marked pathological symptoms or death. The much larger numbers of relatively "benign" parasites remain comparatively undetected in individual humans and underinvestigated by human science.

In the alimentary canals of metazoan animals of all phyla reside many relatively harmless protozoan parasites: amoebae, flagellates, and gregarine sporozoans. Typically, their life-cycles involve some resistant stage, possibly a passive cyst which, if passed out of the gut of the host with faeces, can then lie passively in the soil or on vegetation until it is accidentally taken into the body of another host. Parasites of the circulatory system and of the blood, and of other internal organs and tissues, require more elaborate means of transmission from host to host, and associated with this requirement are successive multiplicative stages which produce enormous numbers of potential propagule cells. In itself, this need for rapid multiplicative phases makes such parasites potentially more pathogenic.

In man there are many relatively harmless protozoan parasites. These include flagellates, such as *Giardia lamblia* living in the duodenum, *Trichomonas buccalis* living in the mouth, and *T. hominis* living in the colon. Among the sarcodines are two species of the genus *Entamoeba: E. gingivalis* living in the mouth between the gums and teeth and *E. coli* living in the colon. Surveys of the incidence of these parasites yield figures ranging from 2% of populations infected with *Trichomonas hominis* to 15% with *Giardia,* and to possibly 30% with *Entamoeba coli,* with even higher percentages of children and older persons supporting *E. gingivalis.* It is particularly important that both *Trichomonas buccalis* and *Entamoeba gingivalis* are not pathogenic, since both are effectively and regularly transmitted from one human to another by kissing.

It is of some evolutionary interest that closely related forms both of human amoebae and of human trichomonads have become harmful. *Entamoeba histolytica* is the cause of amoebic dysentery and can occur all over the world. This species burrows in the wall of the colon and destroys cells and causes ulceration. In severe infections, it becomes difficult to separate the effects of the amoebae from those of the bacteria which accompany them into the tissues as secondary invaders. Infections readily spread in areas of poor sanitary standards, since the cysts which pass out in the faeces can survive for long periods of time and pass via flies to human food, or more directly via contaminated water supplies. Even in Western Europe and the United States, local outbreaks of amoebic dysentery can occur. Somewhat similarly, a flagellate species closely related to the harmless ones, *Trichomonas vaginalis,* can cause a persistent, and irritating rather than harmful, vaginitis in women, although infected males seldom show any symptoms.

Much more harmful among flagellate parasites are the trypanosomes, two species of which cause human sleeping sickness, while other species are responsible for a variety of diseases in domestic animals. *Trypanosoma gambiense* is the cause of the chronic and regularly fatal form of African sleeping sickness widespread in West Africa. A more virulent species, *T. rhodesiense,* occurs in a limited area of southeastern Africa. Both forms of this disease are transmitted by blood-sucking tsetse-flies of the genus *Glossina,* and there is a multiplicative stage in this intermediate host. Appropriate stages for reinfection make their appearance in the salivary glands of the flies sometime after an infected meal of blood containing trypanosomes. Transmitted to a new human victim, the trypanosomes live and multiply in the bloodstream and, in later phases of the disease, invade fluids around the central nervous system including the cerebrospinal fluid. The sleep-like stupor and disturbances of postural motor control, which characterize the later phases of the disease, are produced in this way. Combined use of insecticides to control tsetse-flies and of certain trypanosome-killing drugs to treat the disease are currently reducing its incidence in Africa.

Malaria remains the most important disease caused by protozoans. Once worldwide in warm temperate and tropical countries, it has virtually disappeared from the southern United States and from such European countries as Italy, in areas of which, even in the early twentieth century, it killed many thousands of humans annually.

The causative organism is a sporozoan of the genus *Plasmodium.* Many species of this genus and certain allied genera occur extensively as intracellular parasites of the blood cells of birds, while a few occur in mammals, especially rodents and primates. Four species of *Plasmodium* occur in man, of which *Plasmodium falciparum* is the most extensively pathogenic. The red blood corpuscles of a malarial patient are the sites of a rhythmic process of asexual reproduction by multiple fission for the parasite cells. Each cycle begins with an invasion of a fresh group of host red blood corpuscles by tiny *Amoeba*-like parasite cells which first grow into trophozoites of characteristic ring-form within the corpuscle. When the trophozoite fills the blood cell, it commences nuclear division to become a schizont, the stage which will reproduce by schizogony. When the daughter cells, or merozoites, are fully formed they break out of the remains of the corpuscle and each then goes on to infect a new blood cell. In *Plasmodium falciparum* there are usually about twelve merozoites so formed. It is the release of the toxic metabolic products of the parasites' growth and schizogony, released at the same time as each new crop of merozoites, that causes the recurrent fever so characteristic of the disease (Figure 6-12). In *P. falciparum* and in two other species of human malaria (tertian fevers), each synchronized cycle of asexual reproduction by schizogony in the red cells takes 48 hours, while in the fourth species, *P. malariae* (quartan fever), it requires 72 hours. The malarial patient's fever thus occurs at the same time of day, at two-day or three-day intervals.

Periodically some of the parasites which invade the blood cells grow into specialized gametocytes, and begin the sexual or sporogony cycle which in-

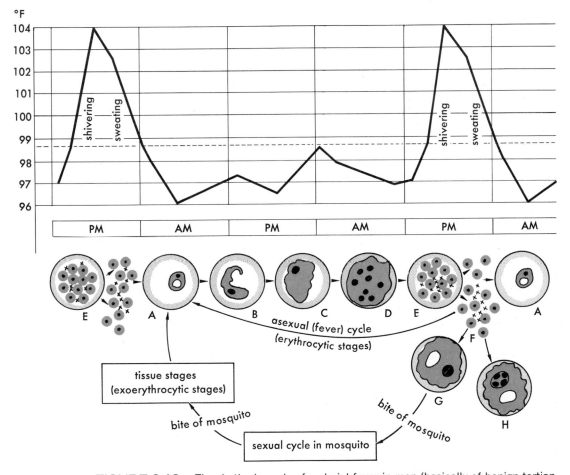

FIGURE 6-12. The rhythmic cycle of malarial fever in man (basically of benign tertian malaria, as caused by *Plasmodium vivax*). Each cycle begins with a new invasion of host red blood corpuscles, each infection forming the trophozoite stage (**A**), which grows to fill the blood cell (**B, C**). By nuclear division, the trophozoite becomes a schizont (**D**); by asexual schizogony the schizont gives rise to daughter cells, or merozoites (**E**), which break out (**F**), releasing toxic metabolic products, and go on to invade new corpuscles. For the parasite, this cycle continues as an asexual process of rhythmic proliferation in more and more blood corpuscles; for the patient, as a recurrent fever at the same time of day, every second day. Some merozoites give rise to gametocytes; if ingested by the appropriate mosquito, gametocytes begin the sexual cycle, which ends with many sporozoites (ready for injection into a new human host). Introduced sporozoites do *not* directly invade red blood corpuscles (see text).

volves stages in the intermediate host or vector. In the case of all the human malarias, this is an anopheline mosquito which ingests some gametocytes along with its blood meal. In the gut cavity of the mosquito, male and female gametocytes give rise to male and female gametes which fuse to form a zygote that invades the mosquito's gut wall to form an oöcyst. This in turn undergoes a multiple fission to form many sporozoites which migrate to the salivary gland of the mosquito, where they form the infective stages ready for injection into a new human host. The first few cycles of asexual reproduction in the new host are exoerythrocytic, occurring not within the red blood corpuscles but in cells of the lymphoid (endothelial) system. These exoerythrocytic stages are sometimes known as cryptozoites (the "hidden" stages) but from them some of the daughter cells move over into the bloodstream and start the classical rhythmic proliferation in red cells which is malarial fever.

As noted earlier, antimalarial drugs and mosquito control, including the extensive use of insecticides like DDT, have markedly reduced the incidence of malaria throughout the world in the last quarter-century. However, as the World Health Organization has recently pointed out, processes of economic development and social progress in tropical countries may sometimes bring about a local and tragic recurrence of malaria as an endemoepidemic disease. One such area is the Gezira near the confluence of the White Nile and the Blue Nile in the Sudan. Completion of dams and irrigation schemes in the 1930's and 1940's turned what had been semiarid savannah into a prosperous and densely populated agricultural region. For a time, when cotton was the almost exclusive crop, mosquito control was easy since all irrigation ditches were regularly dried out. However, as a more varied agriculture, including millet, peanuts, and even rice, became developed for the increasing population, the periodical drying out of irrigation channels was abandoned and anopheline mosquitoes became abundant. By the middle 1950's, malaria affected more than half of the population of this prosperous and economically well-developed area. By 1975, malaria was still endemic in much of the region, and the long, costly fight against it has been beset by certain difficulties, including local stocks of anopheline mosquitoes which have developed resistance to DDT. According to a 1975 communication from the World Health Organization, tropical diseases still stand as major obstacles to social and economic improvements in the underdeveloped countries of the tropics.

Although it is difficult to get biologists to agree about any evolutionary matter, perhaps the majority would concur with the view that such exceptionally pathogenic parasites are the less successful ones. The most successful protozoan parasites of man may be the undetected stocks of forms like *Entamoeba gingivalis* and the flagellate *Trichomonas buccalis* which live undetected in so many mouths.

Limits of Complexity

Members of the subphylum Ciliophora show not only the greatest complexities of external shape but the most complicated arrays of internal ultrastructures that are found in *any* living cells. They represent the greatest complexity of organization possible within a single cell membrane.

Most of the more complex ciliates are relatively large for protozoans. A few have lengths of 2 or 3 millimeters, and, as beginning biologists studying *Paramecium* know, many are visible with the naked eye. Some of their superficial complexity comes from the formation of organelles by the fusion of groups of cilia. These may form structures like flat paintbrushes, or membranelles (Figure 6-13), or be fused into the pointed, relatively large, tentacle-like appendages called cirri. In some genera of surface-crawling ciliates, these cirri function as limbs for walking and even for jumping. Their constituent cilia no longer move in fractionally out-of-phase metachronal rhythm but bend *together* in a powerful levering action. Some of the complexities of the cytostome and cytopharynx (shown in Figure 6-14) are the basis for some ciliate classification. Their complexity is founded on the elaboration of organelles formed from fused cilia and from the modified fibrillar elements of their infraciliature. Especially in peritrichs and spirotrichs, many membranelles of compound cilia provide the collecting mechanisms that pass food particles and prey organisms of suitable size into the cytostome. Recent studies of ultrastructure in ciliates, particularly those carried out by scanning electron microscopy (Figures 6-13 and 6-14) have revealed more of the fantastic complexity underlying the organization of the cirri as "limbs" in a walking form like *Euplotes*, or in the food-collecting membranelles of a sessile form like *Stentor*.

A few heterotrichs and the whole group of tintinnids among the subclass

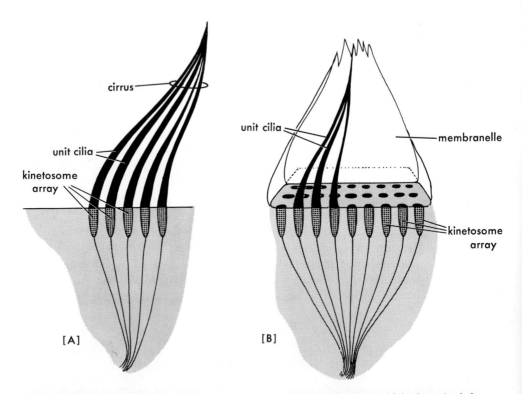

FIGURE 6-13. Compound ciliary structures, in stylized diagrams which show the infrastructure. **A:** A "walking" cirrus from a ciliate like *Euplotes*. **B:** A membranelle such as is borne on the peristomial disc of *Stentor*, and on the buccal and vestibular regions of other ciliates.

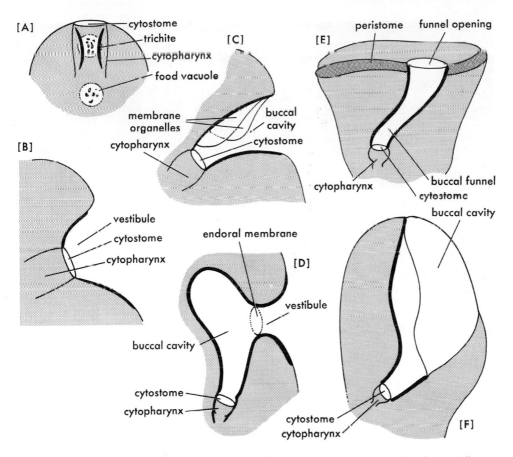

FIGURE 6-14. Levels of oral complexity in ciliates. Increasing complexity of organelles (mostly formed of compound cilia) as found in: (**A**) gymnostomes, with trichites and food vacuoles being formed; (**B**) trichostomes; (**C**) hymenostomes (including *Tetrahymena*); (**D**) peniculine hymenostomes (including *Paramecium*); (**E**) peritrichs (including *Vorticella*); and (**F**) hypotrichs. [Adapted from Corliss, Manwell, and others.]

Spirotrichia, and a few forms in the subclass Peritrichia, construct tests or "houses" of mineral particles cemented together. In the case of the tintinnids, their tests (or loricas) are somewhat better known than the contained organisms and when suitably magnified may have an aesthetic appeal to the protozöologist (Figure 6-15).

These more complex ciliates show the greatest elaboration of functional morphology that is possible wihin a single cell membrane. However, as was noted at the beginning of this chapter, single-celled organisms are necessarily restricted in size, and no protozoan—even the most elaborate ciliate—can have a mass of more than a few milligrams.

Protista and Metazoa

Most biologists accept that the Metazoa or many-celled animals were evolved from protozoans. The majority view is that they were derived from stocks of flag-

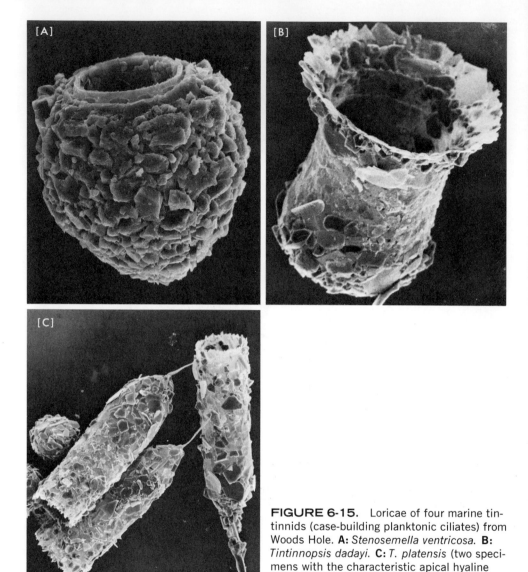

FIGURE 6-15. Loricae of four marine tintinnids (case-building planktonic ciliates) from Woods Hole. **A:** *Stenosemella ventricosa.* **B:** *Tintinnopsis dadayi.* **C:** *T. platensis* (two specimens with the characteristic apical hyaline horn) and *T. radix* (on the right). [Photos courtesy of Dr. Kenneth Gold.]

ellates like the present-day Mastigophora, and that this evolutionary process may have involved (as an intermediate stage) a colonial association of individual flagellate cells such as occurs today in *Volvox*. Thus both many-celled animals and the Metaphyta (or many-celled green plants) may derive from a common protistan stock. An alternative minority opinion is that metazoans were derived from some kind of multinucleate protozoan in which there were many nuclei in a continuous mass of protoplasm. In this case the intermediate stage involved the isolation and specialization of parts of that protoplasm by the interpolation of cell membranes. Whatever was the case, it is clear that only on a many-celled basis could animal organisms of a larger size and complexity evolve. Discussion of the interrelationships of metazoan phyla will be deferred to Chapter 35. In the meantime, two useful working hypotheses are: first, that coelenterate-like forms were the most primitive of many-celled animals and, secondly, that the next more complex forms to evolve from them resembled certain simpler flatworms alive today. The point as regards size, and the surface:mass ratio mentioned early in this chapter, cannot be overemphasized. Even the relatively simply organized many-celled animals already discussed in Chapters 3–5 can be many orders of magnitude more massive than even the largest protozoans.

Another point to be reiterated here is that the phylum Porifera (or subkingdom Parazoa, the sponges—see Chapter 5) were evolved independently from the rest of the Metazoa. It is most reasonable to suppose that either they evolved from a totally different group of flagellates, or else they diverged very early from the major metazoan stem. It is tempting to assume that the Porifera evolved from forms like those in the choanoflagellate order of the class Zoömastigophorea where the individual organisms have a single anterior flagellum surrounded by a delicate collar and closely resemble those important component cells of sponges, the choanocytes.

Returning to the question of surface:mass ratios, one could claim that the increasing complexity of functional morphology in the higher animal phyla (and even the complementary increase in independence of the environment or in levels of homeostasis) remains basically related to increasing size. To crystallize that argument one need only state that a sarcodine protozoan like *Amoeba* of larger size—say with a mass of a few grams—could not possibly exist in any environment on earth. It would not have a sufficient surface area for the uptake of oxygen, far less sufficient area of cell membrane to form sufficient food vacuoles or to carry out any of the other surface-based functions. Maintenance of a suitable ratio of surface:mass with increasing size is simply impossible in a single-celled organism. In the simpler metazoans we see some deceptively simple, but ultimately limited, solutions to the problem. In such cnidarians as corals and jellyfish a thin layer of cells is spread over a larger mass of nonmetabolic material: the calcareous skeleton of the coral, or the "nonliving" organic gel of the mesogloea in the jellyfish (see Chapter 4). This solution is paralleled in many higher plants where the proportion of inert, nonrespiring lignin and cellulose in the wood of trees, for example, allows an increase in total bulk to be accompanied by disproportionate increase in the living surfaces of the enveloping cells. Another limited solution is provided by the flatworms (Chapter 7) where the shape allows a condition where no living cell is more than a few microns from an exchange surface with the environment. Of course, many other kinds of metazoan animals do not have limitations of form imposed on corals, jellyfish, and flatworms by their means of maintaining appropriate levels of surface:mass ratios. The majority of these many-celled animals have achieved the necessary differential increase of surfaces by increasing the complexity of their multicellular tissues and organs, most obviously by complex patterns of folding or of branching. For example, it is a commonplace of vertebrate evolution that smaller, more primitive air-breathing forms, such as newts, have lung walls whose folds are simple (if, in fact, present), while the largest mammals and birds have complicated systems of folded folds or of branching tubes, respectively, to achieve appropriate areas of respiratory surfaces.

Along with the potentialities and liabilities of increased size in the Metazoa has emerged increasing functional specialization of the parts of the many-celled body. This specialization in turn involves an increased interdependence of parts. Much of the later chapters in this book is concerned with this interdependence, as a central feature of functional morphology in the higher metazoan phyla, but three obvious needs (only barely foreshadowed in the protozoans) are worth noting here. First, there is a need for skeletal structures both to support the mass of many-celled structures and to provide a basis for a more sophisticated use of contractile tissues. Secondly, the need arises for communication between different organs and tissues and coordination of their activities in an efficient machine. Coordination involves the development both of nervous or conducting tissues and of endocrine and neurosecretory cells which produce chemical messengers. Thirdly, there is the need for the development of an internal medium (to use the concept first promulgated by the great French physiologist Claude Bernard over a century ago), an internal pseudomarine environment, to supply basically aquatic cells. This involves the development of various body fluids and spaces, most importantly blood and circulatory systems. It can be said that this third need arose from the increasing isolation of some cells far within the bodies of metazoans, but its development (probably initially in organisms still living in the sea) proved an important preadaptation allowing stocks of larger many-celled animals to move out of the sea into the harsher environments provided by fresh waters and by the land surface. Once again, these possibilities were not open to single-celled animal organisms. No matter whether the colonial view or the syncytial view of the origin of metazoans from the protozoa is preferred, it is clear that the potentialities and penalties of increasing size, and particularly of the surface:mass ratio, have shaped the evolutionary limits and possibilities of metazoans.

Differing theories of metazoan origins are but one part of the background for competing schemes of classification above the level of phyla. Even on the basis of a few aspects of function in single-celled organisms, it is obviously difficult to adhere to the simple plant kingdom and animal kingdom of early nineteenth-century naturalist-biologists. Early in this century, Clifford Dobell and others designated a third kingdom, the Protista, in whch they included bacteria and algae as well as protozoans, and which, as we noted earlier, they regarded as acellular organisms. By the middle years of this century it was clear that the differences between bacteria and protozoans were greater than those between protozoans and many-celled animals, and a fourth kingdom was proposed, usually termed the Monera, to include the procaryotic cells, bacteria and blue-green algae. This four-kingdom system still involves considerable difficulties of definition, particularly as regards the higher fungi which are clearly a totally separate stock from the other Metaphyta or many-celled green plants, and as regards the truly acellular viruses. A few years ago, R. H. Whittaker proposed a broad classification in five kingdoms, which has found general acceptance. These are the Monera, Protista, Metaphyta (termed Plantae by Whittaker), Fungi, and Metazoa (called Animalia by Whittaker). To these five, some classifiers have added a sixth kingdom to encompass the viruses, but this is based upon a misconception of viral organization as being that of independent parasitic microorganisms. Discussion of the significance and value of schemes of classification at the level of kingdoms need not detain us here, although it should be noted that our treatment of the protozoans and the sponges (phylum Protozoa and phylum Porifera, respectively) as merely two phyla among the other thirty of invertebrate animals is very questionable. As noted elsewhere (see pages 17 and 92), a higher ranking as subkingdom Protozoa and subkingdom Parazoa, both of equivalent status to a kingdom or subkingdom Metazoa, would be more appropriate. Extended discussion of classification of living organisms at these higher levels would not accord with the purposes of this book. It is sufficient to state that, like the artificial term "invertebrates" itself, they have the same sort of practical value that major directional arrows can have at the entrances to groups of stacks and levels in a very large library. A systematic matter of considerably greater phyletic and chronological significance concerns the possible interrelationships between the various higher phyla of many-celled animals. Aspects of this topic will be taken up in Chapters 27 and 35.

7 PLATYHELMINTHES: THE FLATWORMS

No matter how one prefers to believe the many-celled animals have evolved, the flatworms must at least resemble one central basic stock. These dorsoventrally flattened worms include one group (class Turbellaria) of free-living forms, found in fresh waters, in the sea, and in damp soil, and two parasitic groups. These latter are the class Trematoda or flukes (mainly internal parasites of higher animals) and the class Cestoda, the tapeworms, which show replication of parts and are mainly parasites of the alimentary canal of higher animals. The flatworms are certainly the simplest—and probably the most primitive—animals that can be characterized as triploblastic and as having true bilateral symmetry. They are the only animals with these characters which lack a definitive anus and a body cavity or coelom. Their diagnosis then as "triploblastic acoelomate bilateria without a definitive anus" involves terms which require, if not definition, at least some further remarks on their biological significance. For reasons discussed earlier (see page 38), the term triploblastic is acceptable if it is taken to mean that in the Platyhelminthes, as indeed in all the other many-celled animals except coelenterates, there is a true mesoderm which arises from, or with, the endoderm and forms a third tissue layer lying between the outer epithelium and the endoderm lining the gut. The flatworms can also be said to have reached the organ system level of complexity: in other words, they show a higher degree of interdependence of parts than do the coelenterates. Not only are their tissues specialized for various functions, but two or more types of tissue cells may be combined to form an organ of specific function.

There is no extensive cavity within the body except the gut, which is in the form of an endoderm-lined blind sac. The simple opening to the exterior must serve both as mouth and as functional anus. In one group of small Turbellaria, which may or may not be primitive, there is not even a differentiated gut cavity, and in one highly modified group of flatworms, the Cestoda or tapeworms, there is no gut at all, nutritive uptake being through the general body wall from the host's gut cavity. In no flatworm is there any other internal space, and certainly nothing which corresponds to the body cavities of more complex animals, the coelom and the pseudocoel (see pages 157–158 and Figure 8-5). Flatworms are thus acoelomate. They are also the simplest animals (if one excludes the planula larvae of some coelenterates) to have a definite front end and rear end, upper surface and under surface, and therefore left and right sides, and true bilateral symmetry. This symmetry they share with all the other more complex Metazoa excluding only adult echinoderms. It should be noted that the presence of a definite front end in flatworms does not mean that there is a head, although there are some beginnings of cephalization in Turbellaria with anterior concentrations of nerve and sense cells.

TABLE 7-1
Outline Classification of the
Acoelomate Flatworms

Phylum MESOZOA
Phylum PLATYHELMINTHES
  Class TURBELLARIA
      Order Acoela
      Order Rhabdocoela
      Order Alloeocoela
      Order Tricladida
      Order Polycladida
      Order Temnocephalida
  Class TREMATODA
    Subclass MONOGENEA
    Subclass ASPIDOBOTHRIDEA
    Subclass DIGENEA
  Class CESTODA
    Subclass CESTODARIA
    Subclass EUCESTODA
Phylum RHYNCHOCOELA (or NEMERTEA)
    Subclass ANOPLA
      Order Paleonemertini
      Order Heteronemertini
    Subclass ENOPLA
      Order Hoplonemertini
      Order Bdellonemertini
Phylum GNATHOSTOMULIDA

Free-Living Flatworms

Many of the characteristic features of form and function in the trematodes and cestodes are clearly adaptively related to their parasitic habit. Accordingly, it is among the Turbellaria that we have to look for the characteristics of the archetypic flatworm. Unfortunately, there is considerable diversity of form among the presently living turbellarians and, perhaps because it is difficult to distinguish primitive simplicity from regressive simplicity, the classification of the group is subject to controversy and is unsettled. Accordingly, it is best for us to survey the main subdivisions, and then use as our archetypic flatworm an admittedly evolved pattern which has shown considerable ecological success. The subclassification of the Turbellaria adopted here is a compromise one and involves five main groups, based on the organization of their gut, and one minor group of ectocommensal Turbellaria (see Table 7-1).

The order Acoela consists of small exclusively marine flatworms with a mouth but with no gut cavity. The Rhabdocoela are small freshwater or marine turbellarians with a sac-like intestine without branches. This order is probably unnatural and the group polyphyletic. The third order is the Alloeocoela with a similar simple gut but usually with a more complex pharynx leading into it.

The fourth order is the Tricladida, which includes larger and more successful Turbellaria, marine, freshwater, or terrestrial in humid habitats, with a complex eversible pharynx leading into a characteristic three-branched intestine (Figures 7-1 and 7-2). The fifth order is the Polycladida, again a group which includes relatively large and successful Turbellaria, with a complex pharynx

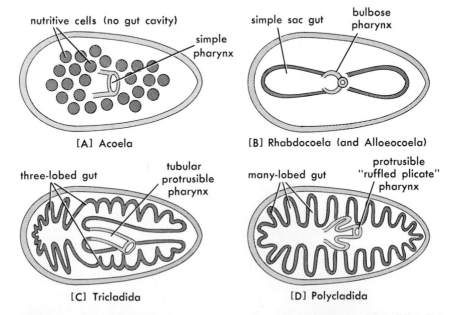

FIGURE 7-1. Gut and mouth structures in the five principal groups of free-living flatworms. Compare with Figure 7-2.

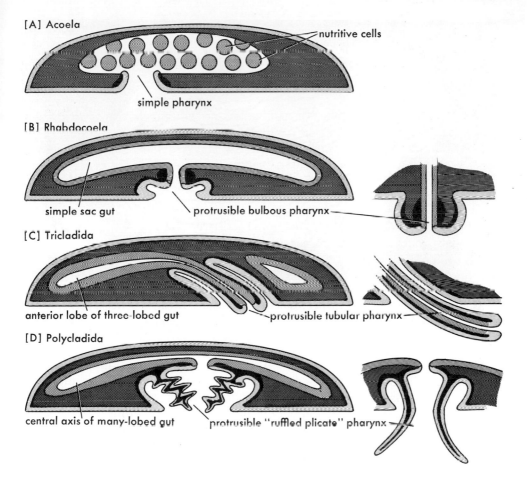

[A] Acoela

nutritive cells

simple pharynx

[B] Rhabdocoela

simple sac gut

protrusible bulbous pharynx

[C] Tricladida

anterior lobe of three-lobed gut

protrusible tubular pharynx

[D] Polycladida

central axis of many-lobed gut

protrusible "ruffled plicate" pharynx

FIGURE 7-2. The oral organization of free-living flatworms. Pharyngeal apparatus of increasing complexity from the simple pharynx (**A**), with only a single ring of muscle fibers, to bulbous (**B**), tubular (**C**), and plicate (**D**) forms all capable of protrusion and all with specialized masses of muscles. (Although part of the contiguous mesoderm, the pharyngeal muscles are shown in solid black in these diagrams for clarity.) Anterior ends of flatworms to the left in all cases. Note that, unlike the gut-lobe arrangements of Figure 7-1, the levels of oral organization are not diagnostic for the four groups in which they occur. For example, many genera of the Rhabdocoela have a simple pharynx, different genera of Alloeocoela have different pharynxes encompassing all types, and several genera of Polycladida have a tubular pharynx (like that typical of Tricladida but arranged for protrusion anteriorly). [Adapted in part from Hyman, and from Jennings.]

leading into a main intestine from which numerous branches radiate, Polyclads (Figure 7-3) are all marine. A sixth order is the Temnocephalida, a group of ectocommensal freshwater Turbellaria, obviously related to the Rhabdocoela but with adhesive discs at the posterior end and "cephalic" tentacles. They have some features which suggest an evolutionary history paralleling that of primitive trematodes. Both microanatomy and physiology are best known in the triclad Turbellaria. Typical genera such as *Planaria*, *Dugesia*, and *Dendrocoelum* live in fresh water and are all about 1–2 centimeters in length.

FIGURE 7-3. Living specimen of *Prostheceraeus vittatus,* a marine polyclad flatworm. [Photo © Douglas P. Wilson.]

The outside is generally covered with a ciliated epidermis (Figure 7-4). The cilia are usually better developed on the underside of the flatworm, and there are also numerous mucous glands discharging onto the sole. One of the types of locomotion in Turbellaria is dependent on these glands laying down a slime trail along which the worm crawls, or rather swims, by its cilia. Also embedded in the epidermis are unusual hyaline bodies called rhabdites, whose function is still imperfectly understood. Some workers have claimed them to be defensive structures whose discharge results in a protective sticky coat around the flat-

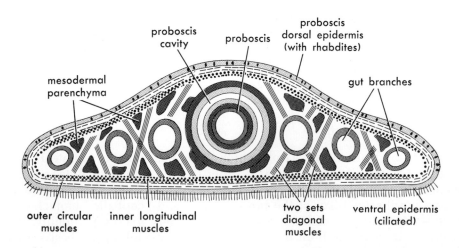

FIGURE 7-4. Cross section through an archetypic triclad flatworm. Compare with Figures 8-1 and 8-4.

worm, while others have considered them excretory in function. They may well be both. Circumstantial evidence for the defensive properties of rhabdites includes the fact that in certain polyclads, where defensive epidermal acid secretion (pH 1.0 achieved with sulfate anions) is highly developed, rhabdites are relatively uncommon. Immediately below this epidermis lie the muscular layers of the flatworm.

A layer of circular muscles is underlain by longitudinal muscles (Figure 7-4). There are also two series of diagonal fibers which run at right angles to each other. In triclads the longitudinal muscles are usually better developed on the ventral side of the animal. Watching a triclad moving, one can readily see that a major part of the locomotion is brought about by peristaltic waves of contraction, which correspond to alternate contractions of circular and longitudinal muscles in sections of the body. When, in one region of the triclad, the circular muscles contract, that part becomes thinned and elongated, with the longitudinal muscles, which are relaxed, being correspondingly stretched (Figure 7-5).

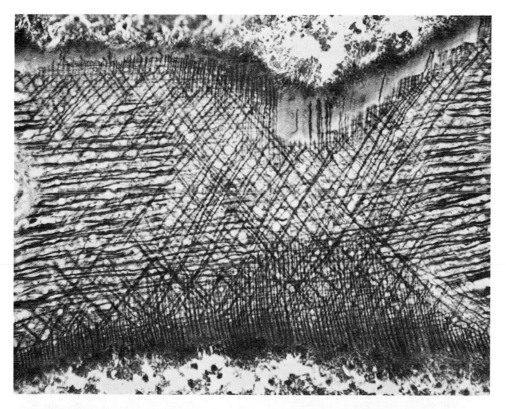

FIGURE 7-5. A horizontal section through one locomotory wave in *Dugesia tigrina,* stained to show the muscle strands. This section was prepared by a freeze-substitution technique of fixation which allowed the locomotory wave of muscle contraction to be "caught." There is epidermis at the top and bottom of the photo, and the muscle strands between are arranged in four sets: two oblique, one longitudinal, and one transverse (or circular). For further discussion, see text. [Photomicrograph, section, and original preparations by Drs. Ned Feder and Richard L. Sidman of Harvard University.]

In contrast, when the longitudinal muscles of a section of the body contract, that part of the flatworm becomes thick and shortened and the circular muscles are correspondingly stretched. Essentially, the regions where the longitudinal muscles are more contracted (that is, where the worm is fattest) form the temporary points of attachment to the substratum. Contraction of the diagonal muscles seems to stiffen the worm (increasing its internal turgor), and, when a turbellarian lifts up its anterior end from the dish and looks around, this is probably achieved by alternate contractions of the longitudinal muscles on the right and the left side of the "head," carried out while the whole body is stiffened by tonic contraction of the diagonal muscles.

Style of locomotion in flatworms is largely determined by size. The smallest flatworms, mostly acoels and rhabdocoels, use ciliary action both in swimming and in movement over mucus-covered surfaces. Turbellarians of intermediate size both swim and crawl by the peristaltic waves of muscle contraction described above and illustrated in Figure 7-5. The few genera of terrestrial tri-

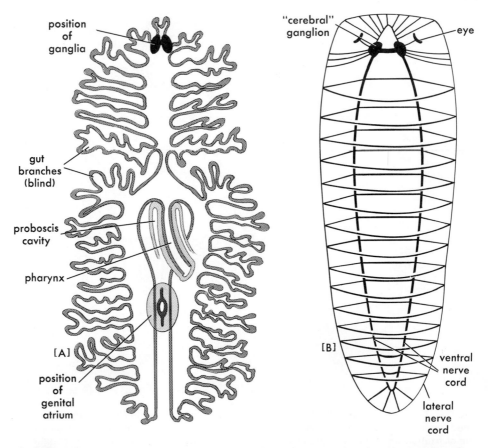

FIGURE 7-6 Organ systems in an archetypic triclad flatworm. **A:** Alimentary system. Protrusible pharynx leads to threefold system of blind branch sacs. **B:** Simplified diagram of major trunks in the nervous system; note that the fine network of the delicate subepidermal plexus and many cross connections of the submuscular plexus are omitted.

clads all move in this fashion along slime trails. Certain larger flatworms, notably freshwater triclads, can crawl rapidly over submerged surfaces by a
leech-like looping (see Chapter 11) which involves coordinated contractions of
circular and diagonal muscles throughout the body in antagonism to all the
longitudinal muscles, rather than sequential waves of muscle contraction. Special mucous glands at the ends of the body are used to provide fixed points for
this looping.

Below and between the muscle layers there is packing (or parenchymatous)
tissue in which both the genital structures and the alimentary canal wall are
embedded. It should be noted that muscles, parenchyma, all genital and excretory structures are mesodermal in origin. As described earlier, however elaborate in pattern, the alimentary canal is always a blind sac with no anus—the
branching merely serves to increase surface area, and there is no differentiation
into regions of particular function (Figure 7-6A). Water regulation and perhaps
some nitrogenous excretion are carried out by flame-cells lying at the inner ends
of an extensive branched duct system of convoluted protonephridial tubes discharging to the exterior through a series of nephridiopores (Figure 7-7). Each
flame-cell unit involves both a flagellar propulsive system (Figure 7-8) and,
as electron microscopy has clearly revealed, an ultrafiltering meshwork in the
wall. Turbellarian protonephridia and their flame-cells are clearly osmoregula-

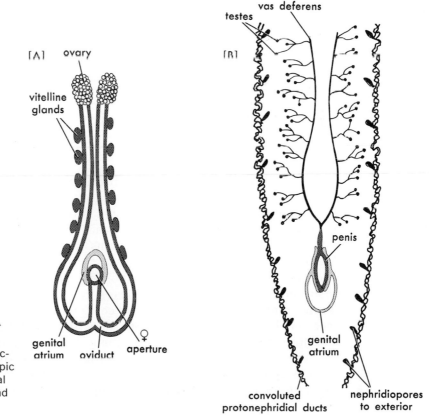

FIGURE 7-7. Reproduc-
tive systems in an archetypic
flatworm. **A:** Female genital
system. **B:** Male genital and
excretory systems.

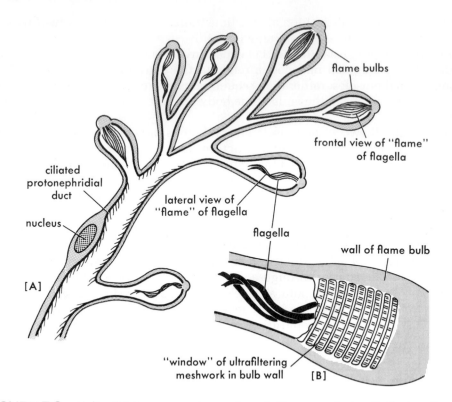

FIGURE 7-8. Units of flatworm excretory system. **A:** Diagrammatic longitudinal section through a "twig" of the protonephridial branches (main "trunks" in Figure 7-7), showing a group of flame bulbs each with a group of internal flagella. The "flicker" of these flagellar groups in living tissues has given rise to the name *flame* bulb. **B:** A reconstruction based on electron micrographs of the meshwork window in the lower wall of each flame bulb, which is thought to be involved in ultrafiltration of the hypotonic excretory fluids which are continuously driven down the protonephridial ducts (in freshwater turbellarians) for excretion to the exterior through the nephridiopores.

tory in function and are best developed in freshwater forms. They are absent or much reduced in marine and parasitic flatworms, where there is no need for continuous pumping out of water. Most nitrogenous excretion and all respiration (that is, gas exchange) are carried out over the entire body surface. In these physiological respects flatworms are *thin* organisms like coelenterates (Figure 7-3).

In addition to a nerve net like that of coelenterates, there are also well-developed central nervous elements. There is a knot of nerve cells termed the "brain" with two longitudinal nerve cords which are connected to the general nerve net by fine branches and between which are delicate transverse connections forming the ladder-like arrangement shown in Figure 7-6B. There are a variety of sensory cells spread over the surface of the body and connected to this nervous system. Many of these are chemoreceptors or mechanoreceptors and are not greatly differentiated from the epidermal cells around them. There are two groups of sensory cells known as auricular organs at the sides of the "head," and these are supposed to be specialized taste receptors. There are typically two eyes consisting of light-receptor cells lying within cups of pigment cells. Behavioral studies have proved that these eyes allow the animal to discriminate between light intensities and to determine the direction of illumination and orientate in relation to it. Much of this behavior is only possible because of the bilateral arrangement of the pair of pigment-cups.

Apart from the more elaborate pharyngeal structures of the "higher" Turbellaria, the most complicated organs in flatworms are reproductive structures. In a typical triclad, which like almost all Turbellaria is hermaphrodite, there is a common genital atrium in the posterior of the ventral surface of the flatworm. Into this opens a vaginal duct formed by the union of two oviducts leading from a pair of ovaries in the anterior part of the body past a series of vitelline glands which supply food to the eggs. There is also a series of scattered testes in the lateral margins of the body which drain through a series of fine ducts to a pair of *vasa deferentia* which unite to form a muscular penis before discharging in the genital atrium. It is probable that cross fertilization is usual in most flatworms. The eggs which are produced by triclads are laid in groups enclosed in a capsule which seems usually to be secreted by the atrial part of the system. Some freshwater flatworms apparently produce two types of eggs: summer eggs where the capsule is thin and development relatively rapid, and winter eggs. In the latter, the capsule is thicker and the egg, which is resistant to desiccation and lowered temperatures, is involved in a lengthy diapause before development. In this way the freshwater flatworms have physiological adaptations to the environment which parallel those we have already discussed in *Hydra* and in freshwater sponges. Although the eggs of these freshwater forms hatch as miniature adult flatworms, some marine polyclads produce a ciliated larva, known as Müller's larva, which is free-swimming as a temporary member of the marine plankton. It has eight or more ciliated lobes which may be involved in feeding as well as locomotion during larval life. After it settles, the lobes are resorbed and growth leads to the adult polyclad form. The ciliated larva is developed from the polyclad egg by a process of spiral cleavage. The significance of such embryonic and larval development will be discussed in Chapter 27.

A few Turbellaria reproduce asexually by fission and, particularly in triclad flatworms, great powers of regeneration are present. Usually, any piece about a tenth the size of an adult flatworm will regenerate into a complete worm, with the front end of the regenerate developing at the anterior or inner end of the fragment, and the tail at the posterior or outer end of the fragment. A great deal of work has been done on this apparent polarity of regeneration. By appropriate manipulation, two-headed or two-tailed "monsters" can be regenerated. In natural circumstances of winter starvation, triclads can "degrow;" adults becoming not merely smaller but functionally juveniles again, with undifferentiated rudiments of their once fully developed reproductive organs. Under certain experimental conditions of partial starvation, it appears that any processes of senescence in individual triclads can be prevented and "potentially immortal" juvenile flatworms can be produced.

Reproductive Patterns

It is worth noting that the increased complexity shown in the gut and pharyn-geal structures by the series of turbellarian groups from Acoela through Rhab-docoela and Alloeocoela to the most complex Tricladida and Polycladida is paralleled by increasing elaboration of the auxiliary genital structures. In the most primitive Acoela there are no female genital ducts, the eggs simply rup-turing to the exterior through the body wall. There are male ducts ending in a muscular portion termed a penis, but the claim that fertilization results from hypodermic impregnation has not been confirmed by the most recent investiga-tions. Some other acoels have a short blind female duct serving for the recep-tion of sperms, but none of the group have oviducts. Within the Rhabdocoela and Alloeocoela, we find increasing complexity of the female organs, involving the separation of distinct yolk glands from the ovaries in more advanced forms. In the "higher" turbellarians (Polycladida and Tricladida), special copulatory canals and bursae, and even uteri for the accumulation of ripe eggs, are devel-oped. The male organs show a similarly evolving complexity, with groups of glands and vesicles forming an elaborate prostatic apparatus. Notably in the Polycladida, there are additional male reproductive structures termed muscu-loglandular organs, a ring of which may serve as additional stimulatory struc-tures in copulation. With their relatively large and yolky eggs, internal fertil-ization is the rule among flatworms. As hermaphrodites, they might be termed "simultaneous" in that eggs and sperm are ripened at the same time in each individual. (In this they contrast with certain hermaphrodite molluscs which show "consecutive" sexuality—see Chapter 23.) Thus copulation is mutual and sperm are exchanged in cross-fertilization. At least in most polyclads and triclads self fertilization fails to occur even in isolated individuals. It has been postulated that some chemical process occurring during copulatory ejaculation is necessary to induce motility and fertility of the sperm masses.

If the Acoela represent the most primitive flatworms, then it is important to re-alize both their similarities to and differences from the planula larva of coelen-terates. The similarities include the ciliated epithelium covering the outside, the solid mass of nutritive cells in the inside, and the lack of any gut lumen. The differences include the possession by acoels of a mouth (though it does not open into a gut cavity but simply into spaces between these nutritive cells) and the occurrence of muscle cells and gonadal cells derived from mesoderm.

Parasitic Flukes

The class Trematoda consists of the parasitic flatworms called flukes, whose general body form is not unlike that of the free-living Turbellaria. The group divides clearly into three subclasses, two of which, the Monogenea and the Aspidobothridea, are relatively minor groups of parasites (Table 7-1). The Monogenea have a fairly simple life-history and are basically ectoparasites living on the skin and gills of fishes, although some have become "internal" parasites of the mouths and urinary bladders of amphibians and turtles. The Aspidobothridea, distinguished by a huge and elaborate adhesive sucker (often as massive as the rest of the body), are mainly endoparasites of molluscs living in the pericardial or renal cavities, although a few are also found in the guts of fishes and turtles. Since these aquatic vertebrates probably ingest infected mol-luscs, this may suggest the way in which two-host and more complicated para-sitic life-histories have been evolved. Neither of these minor groups causes extensive tissue damage to their hosts. In contrast, the larger and more important subclass Digenea are more pathogenic endoparasites with a much more com-plicated life-cycle, living in the gut, the bloodstream, or the actual tissues of their hosts. A typical life-cycle will involve stages in three kinds of animals as hosts: the sexual adult parasite occurring in the animal designated the "pri-mary" host, the other stages in "intermediate" hosts. In the adult stage, the body form of most flukes is essentially that of a turbellarian, though the mouth is anterior instead of midventral and there are two suckers for adhesion.

  Clonorchis sinensis is a typical digenetic fluke, found in China, Japan, and Korea, which lives as an adult in the ducts of mammalian liver. Man is the usual primary host to the Chinese liver-fluke, though adults are often found in dogs and cats. The adult fluke is a little over 1 centimeter long and is a her-maphroditic organism with both male and female gonads (Figure 7-9). Nor-mally two individuals cross-fertilize each other, and the ootype serves as an assembly point in the center of the complex system of genital and accessory

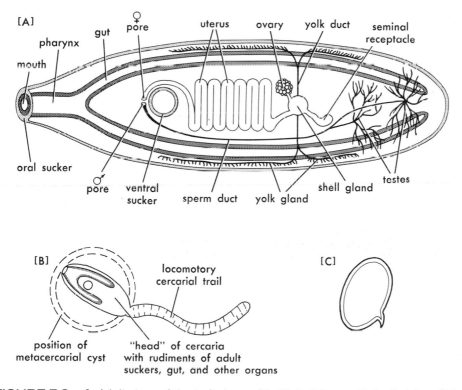

FIGURE 7-9. **A:** Adult stage of the typical parasitic "fluke" *Clonorchis* (= *Opisthorchis*). Note the relatively simple gut and the complex hermaphroditic genitalia. **B:** The cercaria larva of such a fluke (about 250 microns long). **C:** The egg of a blood-fluke, *Schistosoma mansoni*, viewed in optical section.

sexual organs. In that space, an egg cell from the ovary of the individual, a sperm from the seminal receptacle (into which it was passed from the other individual), yolk food material, and then secreted material for a shell come together. All this adds up to a fertilized egg with a shell, which is then stored in the uterus of the fluke for the initial stages of its development. In enormous numbers, these fertilized eggs are passed out continuously from the fluke into the liver ducts of the host, whence they pass via the bile-duct to the host's gut, and thence out the anus with faeces. In suitable conditions of temperature and water, the eggs hatch out as miracidia—the first of a series of larval stages. Each miracidium is a many-celled ciliated organism which swims to bore into the tissues of the freshwater snail *Bithynia*. In *Clonorchis* the eggs may occasionally be swallowed by a snail before hatching, and then miracidia bore through the snail's gut wall, but this is a relatively unusual process in trematodes.

Once within the tissues of the snail (which is thus the first intermediate host) the larva grows into a bag-like sporocyst which is nourished at the expense of the snail's tissues. The sporocyst grows rapidly, continually budding off asexually further larvae called rediae. They themselves reproduce asexually in many trematodes, and several generations can result in the host snail containing an enormous number of larvae. Finally, in the snail, the rediae give rise to the more complicated cercariae which break out to the exterior and swim through the water. These cercariae larvae (Figure 7-9) normally have a ciliated locomotory tail propelling a head portion in which rudiments of the adult structures such as sucker, gut, and reproductive organs can be distinguished. The cercariae usually swim through the water to the second intermediate host and, in *Clonorchis*, this is normally a fish such as a carp or trout. Upon contact with the second intermediate host, the cercaria bores under the scales of the fish and encysts as a metacercaria. To complete this life-cycle, the fish must be eaten by the primary host, and if this occurs (and the cyst has not been destroyed by cooking) the metacercarial cyst wall is dissolved away by the digestive enzymes of the host and a young fluke emerges into the gut cavity. The juvenile fluke then migrates from the gut to the liver by way of certain blood vessels; arrived there, it grows for three weeks and then is ready to reproduce in its turn.

A larger and somewhat similar liver-fluke, *Fasciola hepatica*, which occurs mainly in sheep, is often described in textbooks as a typical trematode. However, its size (up to 4 centimeters long) is atypical, and its life-cycle is unusual in omitting the second intermediate host. The first intermediate host is a snail commonly found in swamps and marshy places, *Lymnaea truncatula*, but the metacercariae are formed on grass blades whence they are eaten by the sheep (without the intervention of a second intermediate host).

Other typical flukes are intestinal or occur in the lungs or the livers of humans and various other mammals, but there is one genus of considerable medical importance, *Schistosoma*, which lives in human blood vessels. These blood-flukes are peculiar in having separate sexes as adults with considerable dimorphism and in having a shortened life-cycle (with usually Planorbid snails as intermediate hosts). The cercariae, which are forked-tailed, bore directly into the primary host's skin where it is exposed to infected water, and there is no metacercarial stage. Paired together in permanent copulation, the adults move in the larger vessels of the bloodstream and then migrate (depending on species) to veins around the rectum or to veins around the bladder where they cause local hemorrhage as they lay their spined eggs. Schistosomes are essentially tropical, two species occurring in Africa and one in Southeast Asia. As malaria has become controlled over the last few years, the diseases caused by blood-schistosomes—called schistosomiasis or billharziasis—have increased in importance as the tropical diseases taking the greatest toll of man. Reduction of these diseases from their present endemoepidemic importance will come only where a threefold medical and biological control operation is tried. This will involve improved sanitation, chemotherapy to control adult flukes, and control of snail populations, particularly in the water flowing over rice paddies and other working places. There are difficulties associated with this threefold approach to control. Some of the antischistosomal drugs, many of which contain antimony, can cause serious side-effects, especially in individuals already emaciated by chronic urinary or intestinal schistosomiasis. Other difficulties arise from the fact that snail populations are not so easily controlled as those of mosquitoes. It is not generally realized that schistosomiasis is a tropical disease often associated with economic development and with processes of agricultural improvement. Artificial lakes and irrigation canals can be important sources of infection. The building of the famous high dam at Aswan has increased the occurrence of schistosomiasis in upper Egypt and Sudan, and similar increases have occurred in northern Nigeria and in Ghana. Even in the desert margins of the Arabian peninsula, irrigation projects funded by oil revenues have resulted in the spread of fluke disease to regions where it did not previously exist.

It is worth noting that many populations of pulmonate snails in lakes of North America contain large number of rediae and cercariae larvae of blood-flukes, for which the primary hosts are water birds such as ducks and waders. Human skin exposed to such waters, at times when cercariae have broken out of the snails, will be subjected to attempts to penetrate by the larval swarms. Although there is no possibility of their getting into the bloodstream and breeding as adult blood-flukes in man, the "wrong" species, they do invade the epidermis and can produce the local tissue reactions or more extensive allergic responses known as "swimmer's itch." In some areas of the northern United States, management of lakeside swimming beaches has involved somewhat ineffectual attempts at local snail control.

Parasitic Tapeworms

The tapeworms, class Cestoda, are also endoparasites. The majority of tapeworms have a rather standardized body form and life-cycle. However, one small and obscure group, the subclass Cestodaria, do not form the characteristic strobilating tapes as adults, but instead superficially resemble adult trematodes. They are found as coelomic or intestinal parasites in certain primitive (ganoid) fishes or in aberrant shark-like fishes. Most cestodes are placed in the larger subclass Eucestoda, or true tapeworms, with the eponymous body form as adults and a two-host life-cycle. Usually, the adult tapeworm is found in the alimentary canal of vertebrates. It produces eggs which give rise to larvae which when fully developed form "bladder-worms" in the tissues of a vertebrate or invertebrate secondary host, and this secondary host has to be eaten by the primary host for the bladder-worm to give rise to the adult tapeworm again. Man is host for several adult tapeworms and for a few bladder-worms of other species. Our gut can house the adult stages of *Taenia solium*, the pork tapeworm, whose bladder-worm occurs in the flesh of pigs, and *T. saginata*, the beef tapeworm, with bladder-worm stages in cattle. Both species can grow to about 5 meters in length, and at least one *T. saginata* has been recorded at 21 meters. The adult form is characteristic (Figure 7-10) and shows a complete lack of any digestive organs. There is an attachment scolex, then a neck region of budding growth, then a strobila of many similar units termed proglottids. These worms could be regarded as showing a sort of segmentation, but it is more usual to term them polyzoic adults, since each proglottid has a complete set of male and female reproductive organs just like an individual trematode.

In a fully formed tape, behind the scolex and budding zone, there is a line of immature proglottids, followed by a series of fully mature proglottids, followed by a line of proglottids which are gravid (that is, full of fertilized eggs), followed by a series of proglottids which are dropping off as bags filled with already partially developed eggs (Figure 7-10). The developing eggs (either loose or enclosed within the remains of the proglottid) pass to the exterior in the faeces of the host. They must then be taken up by the appropriate intermediate host. If this occurs, then the hexacanth or oncosphere larva, after being freed

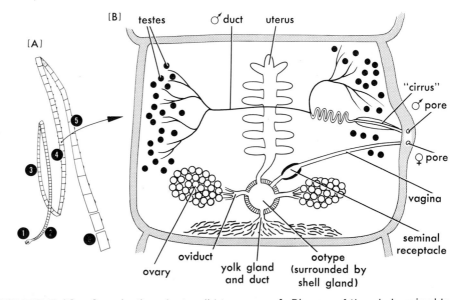

FIGURE 7-10. Organization of a taeniid tapeworm. **A:** Diagram of the whole animal to show scolex (1); the several series—budding (2), immature (3), mature (4), gravid (5)—of proglottids in the tape; and the detached proglottids (6) which are reduced to bags of partially developed eggs. **B:** Diagram of a mature proglottid to show the complex hermaphroditic genitalia. Compare with Figure 7-9A.

from its membranes, penetrates the intestinal wall and moves out into the muscle tissues of the pig or cow. It there forms another resting stage, termed the cysticercus or bladder-worm. Within the bladder-worm can usually be discerned an involuted scolex. If the cysticercus is ingested by the primary host, the scolex everts and begins to strobilize—growing up to the adult tapeworm in the gut of the primary host. *Taenia saginata* is probably the most common tapeworm of civilized countries, *T. solium* being rarer. There is another large tapeworm, *Diphyllobothrium latum,* which is relatively common; infection with it results from eating improperly cooked freshwater fish. This species has a more complicated life-cycle involving two intermediate hosts—first copepods and then freshwater fish. It was formerly common in human populations living around the Great Lakes in North America, and still persists here and in those Baltic states where lightly smoked fish is commonly eaten. Although public health inspection of meat has made the taeniid tapeworms relatively rare in urban communities, their incidence still reflects culinary habits. Human tapeworms do not occur where meats are *not* eaten rare.

Echinococcus granulosus lives as an adult as a relatively minute tapeworm in dogs. However, its bladder-worm stage can occur in cattle, in sheep, or in man. Its medical and veterinary importance comes from the fact that the bladder-worm is capable of considerable growth and asexual multiplication of its contained scolices. A single bladder-worm can grow into a hydatid cyst of more than 15 centimeters in diameter, which can be removed only by surgery. Human hydatid cysts occur mainly in sheep-raising countries, but the impressive pathology of this form should deter dog-lovers from any oral contact— whether direct or indirect—with their pets. It should be remembered that the minute adult is often present in dogs and completely escapes detection, since it causes no ill effects or symptoms.

Many kinds of cestodes have the polyzoic adult form and type of life-cycle described above. However, one aberrant group—the Cestodaria—and a few scattered forms within other groups of cestodes are monozoic, that is, do not proliferate a series of proglottids in the adult stage—rather, each has a single set of male and female organs. It is possible that one or other of the monozoic types of cestodes resembles an ancestor intermediate between modern tapeworms and a rhabdocoel type of turbellarian. All the available evidence tends to suggest that the cestodes and trematodes were independently evolved from free-living turbellarian ancestors, probably of the rhabdocoel type.

Some of the most interesting recent work on the physiology of parasites has been carried out on cestodes both *in vivo* and *in vitro,* and stems largely from the absence of all digestive structures in tapeworms. Every component needed for growth and metabolism, from water and oxygen through glucose to amino acids and fats, is taken into the growing tapeworm through its general body surface. The recent work has not only revealed some of the basic biological principles regarding the physicochemical nature of the sites of uptake of specific amino acids, but it has given zoologists new insight into the nature of the environment provided within the gut cavity of higher vertebrates. In brief, vertebrates do not merely absorb amino acids and suchlike from their guts, but maintain there particular ratios of concentrations of them, where necessary, by secretion. It is this maintained internal environment of the gut fluid which is parasitized by the tapeworm rather than the food as it is being broken down—and this results in the fact that as regards its amino acid needs, though not as regards its sugar needs, a tapeworm can successfully parasitize a starving vertebrate host. In general, cestodes can be regarded as extremely efficient parasites with (in normal circumstances) little or no pathological effects on their vertebrate hosts. No dog, cat, fish-eating bird, or human ever dies of infection with a species-appropriate tapeworm, though it may cease to appear "well-fed." Notoriously, this "efficiency" was exploited in the 1930's by unprincipled medical advisors of Hollywood actresses, who administered live cysticercal stages of appropriate *Taenia* spp. as a "slimming pill" of guaranteed effectiveness. Totally without guts—in fact lacking all specialized organs of digestion, respiration, and circulation—cestodes have all components essential to their growth and reproduction provided in abundance and in appropriate proportions by their primary hosts. As my senior Woods Hole colleague, Horace W. Stunkard, has pointed out, tapeworms impose "a graduated income tax" on the host vertebrate. Under conditions of partial starvation or stress from other disease in the host, the cestode cuts back on its own uptake and growth and remains inactive only until the host vertebrate is again well-fed and vigorous. As Stunkard puts it, with no need to expend energy for protection against environmental changes or for the acquisition of food, the tapeworm "can get down seriously to the business of reproduction, sexual and asexual."

Both trematodes and cestodes illustrate several general features of parasitic adaptation. In their endoparasitic stages they show reduction of sense-organs and indeed of organs of locomotion. To a large extent they are not homeostatic animals, relying on the host's powers of regulating the internal environment in which they live. The elaborate life-cycles, involving one or more intermediate hosts, reflect the difficulties of certain and widespread transmission from primary host to primary host. Finally, and perhaps most obviously, the fantastic reproductive capacities of flukes and tapeworms have obviously arisen as an adaptation to their parasitic habit. For example, a single beef tapeworm. *Taenia saginata,* living in a human, can shed up to ten proglottids daily, each proglottid being a bag containing 80,000 eggs. Thus per day, one infected human with a number of adult tapeworms can provide several million chances of infecting cattle. Less obvious, but equally important, are the increased chances of transmission resulting from the buildup of parasite tissues in the intermediate host in trematodes. One miracidium infecting a snail can, if successful, result in hundreds of thousands of cercariae emerging from that snail to infect the next stage of host. Recent work on trematode life-cycles has revealed that the sequence of larval stages is to some extent plastic, and that increased reproduction of sporocysts or increased numbers of redial generations can occur under appropriate circumstances.

Phylogeny of Flatworms

Returning to a general consideration of the flatworms as a whole, there seems little doubt that the parasitic groups were evolved from free-living forms like the Turbellaria. It is generally supposed that the first flatworms were derived from planuloid ancestors, not directly from any adult coelenterate as we know them, but from a motile gutless larval stage with bilateral symmetry. That is to say, the coelenterate or precoelenterate ancestor of the flatworms had a definite anterior and posterior end, dorsal and ventral surfaces, and no gut—like no adult coelenterate but like planulae larvae. It should be noted that, no matter whether one adheres to a colonial theory or a syncytial theory for the origins of many-celled animals from protozoans, some kind of turbellarian flatworm must at least resemble the stem-stock of bilateral triploblastic animals. Its planuloid (coelenterate) ancestor, having taken up a crawling life on the sea bottom, evolved differentiated dorsal and ventral surfaces and subsequently both a ventral mouth and the left and right symmetry of both sense-organs and contractile tissues necessary for orientation and directional locomotion. Some authorities regard the Acoela as the most primitive of living turbellarians and believe the first flatworms to have lacked gut-sacs and have had a relatively simple pharynx leading into a solid internal mass of nutritive cells (like the "core" cells of planulae larvae in representatives of all the major cnidarian groups). Another more widely supported theory would regard certain simply organized forms like *Microstomum* and *Macrostomum* (here classified as primitive rhabdocoels) as being closest to the stem-group of flatworms. In this case the simple sac-like intestine, lined with a ciliated epithelium and lacking any diverticula, would correspond to the blind gastrovascular cavity of any less complex hydrozoan coelenterate. Almost all zoologists would agree that the more complex turbellarians, such as triclads and polyclads, represent more advanced levels of organization derived from forms like primitive rhabdocoels, rather than the reverse. It is worth emphasizing that particular relation of specific turbellarian groups to particular coelenterate groups, for example, comparison of polyclads to ctenophores, is rather ridiculous, similarities between organ systems being superficial rather than resulting from common origins. In other words, attempts to derive all Turbellaria through polyclads from ctenoplanid ctenophores, or all coelenterates from flatworms in the opposite fashion, as J. Hadži has done, are futile.

On the other hand, there is little doubt that the Platyhelminthes form a single phylum (into which a few authorities would even incorporate the Mesozoa). Within the phylum, the parasitic trematodes and cestodes have clearly been derived from free-living progenitors like the Turbellaria. Even among the turbellarians, there are a few forms (anatomically unmodified from their free-living relatives) which have become ectoparasites living on molluscan, crustacean, and echinoderm hosts. Evolution of more specialized parasites such as trematodes and tapeworms has involved loss of sense-organs and replacement of locomotory structures by adhesive or attachment organs; progressive simplification of the digestive organs leading to total loss of the gut; and, to increase the chances of infection from host to host, both elaboration of life-history patterns with intermediate hosts and a fantastic increase in reproductive capacities. From an evolutionary viewpoint, it is probably significant not only that the simplest flatworm parasites are found in ancient lines of invertebrates, such as molluscs and echinoderms, but also that the complicated life-histories of the great majority of the more specialized endoparasites begin with the infection of an invertebrate "intermediate" host. Obviously, we can have no fossil evidence of the phylogeny of trematodes and tapeworms, but it is tempting to conclude from the circumstantial evidence of life-cycles that the various groups of parasitic flatworms first evolved in the Cambrian (600 million years ago—see Chapter 34) with primitive molluscs, arthropods, and echinoderms as hosts and acquired vertebrate final hosts (such as mammals and birds) and more complex life-histories at a much later date (perhaps only 60 million years ago). If this interpretation is correct, then two evolutionary advantages were conferred with the acquisition of a higher vertebrate as a habitat: first, wider dispersal and longer life for the flatworm parasite and, secondly, a more homeostatic environment for sustained parasitic reproduction.

8 RIBBON-WORMS (RHYNCHOCOELA) AND OTHER MINOR GROUPS

CLOSELY allied to the flatworms is the minor phylum Rhyncho-coela, sometimes called the ribbon-worms or Nemertea. Nemertines differ from flatworms in several ways, but resemble them in possessing a mesoderm without a general coelomic cavity and in the structure of their nervous and excretory systems. However, they have a tubular alimentary canal with both mouth and anus, a circulatory system, and an eversible proboscis which is enclosed in a special cavity (the rhynchocoel, lying just above the anterior part of the gut). There are about six hundred species of nemertines. The great bulk of these are marine scavengers living between tidemarks, under stones or seaweed, or in burrows in muddy sand. A few species are commensal, a few parasitic, and a few have managed to invade freshwater and tropical land habitats. They show bilateral symmetry, but are usually very elongate, occasionally *extremely* elongate. Cylindrical members of the group occurring on the seashore are called "bootlace-worms"; others, more flattened, resemble living ribbons. On our Atlantic coast the former kind is represented by species of the genus *Lineus* which, though only a few millimeters in thickness, may be over a meter in length, and the latter by *Cerebratulus* which can grow to a pinkish-white ribbon several meters long. In summary, nemertines (or nemerteans) can be defined as acoelomate worms which have achieved great length without segmentation.

151

Nemertine Functional Organization

The outside of the body is covered with a ciliated columnar epithelium. Smaller nemertines move by their cilia, just as do small flatworms, exuding mucus as they go. Although they are true acoelomates, like the flatworms—that is, have a solid construction with all the space below the epidermis filled with parenchyma and muscle cells—the body wall of nemertines is much more highly organized than that of flatworms. Structurally, there are four basic patterns of muscle layers (as illustrated in Figure 8-1), and these are used in the standard classification of the group. It should be noted, however, that there is great similarity of structure throughout the phylum, and it is generally regarded as consisting of a single class and these four divisions, based on the

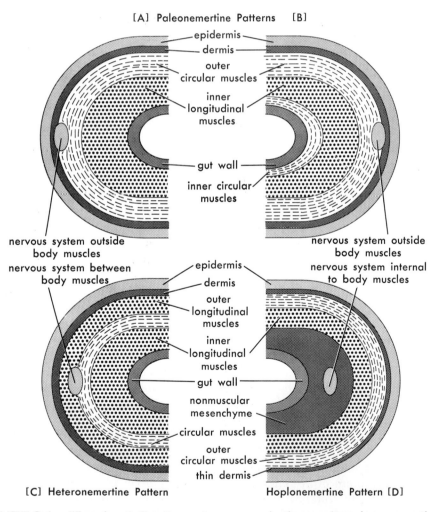

FIGURE 8-1. "Experimental" patterns of worm organization, as shown in cross sections of various nemertines (acoelomate Rhynchocoela). For further explanation, see text, and compare with Figure 11-1.

musculature of the body wall and other characters, are recognized as orders within two subclasses. In general, the nemertines can be thought of as a series of relatively unsuccessful experimental patterns of worm-like form. They have achieved length without segmentation or development of a body-cavity. The various patterns of musculature tried are all mechanically less successful than the more stereotyped plan which we will find in the true segmented worms— the Annelida—where metameric segmentation is complete, and where there is an extensive coelom which acts as the basis of a hydrostatic skeleton. In nemertines, the skeletal system through which the various muscles can act as antagonists to each other consists of a remarkably thick basement membrane lying beneath the epidermis and consisting of fibers which run in sinistral and dextral spirals around the body in alternate layers. These fibers are themselves inextensible and nonelastic and, for mechanical purposes, may be regarded as being attached to each other where they cross. This meshwork can be deformed in much the same way as a trellis fence with diamond-shaped openings. The mechanical properties of such a system were elucidated some years ago by J. B. Cowey and R. B. Clark, who found theoretically that the maximum volume contained within such a system would occur when the fibers were inclined to the worm's longitudinal axis at about 55 degrees. It proved possible to compare theoretical equilibrium positions with those assumed by the worms when anesthetized with all the body-wall muscles relaxed. In general there was good agreement between the actual performances of each different worm and predictions based on a mechanical analysis involving the ratio of its greatest to its

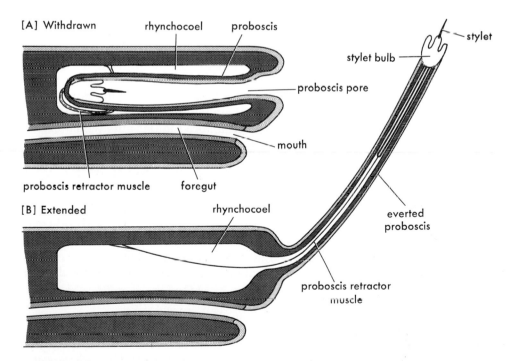

FIGURE 8-2. The rhynchocoel and its contained proboscis in a nemertine worm (actually in a hoplonemertine; see text).

smallest length. This ratio in turn depended on the difference between its actual volume and the maximum capacity of the fiber system, and thus in turn upon the degree of flattening of the cross section of the worm when in the relaxed equilibrium position. The details of this need not concern us here, but it is important to remember that, in general, as a nemertine elongates the angle between the fibers decreases, and the fibers become more nearly longitudinal. When it contracts or shortens, the fibers tend to become more circumferential and the angle between them increases. This is of some importance in relation to the mechanics of locomotion in more highly organized worms such as annelids.

The most distinctive diagnostic feature of all nemertines is the proboscis apparatus, which is used in the capture of small prey or for browsing on carrion. It is completely separate from the tubular gut, its opening (or proboscis pore) generally lying anterior and dorsal to the mouth opening. The proboscis itself lies in a proboscis cavity surrounded by the proboscis sheath. When the muscles of this sheath contract they exert pressure on the fluid in the cavity forcing out the proboscis, which everts as shown in Figure 8-2. It is pulled in by the proboscis retractor muscles in most forms. However, in a few common species such as Cerebratulus lacteus, the longitudinal retractor muscles are apparently wanting. Many of the larger nemertines (Hoplonemertini) are armed with one or more stylets on the proboscis, and it is thought that some also have toxic secretions to aid its use. Food material captured or collected by the proboscis must pass in through the mouth to the gut. In some ways, digestion is similar to that in the coelenterates, being begun extracellularly in the gut of the nemertine and then continued intracellularly in its endodermal phagocytic cells. Food material is moved through the gut by ciliary action since there are no separate muscles for the gut and, indeed, no true gut wall. The precise relationships of gut and proboscis (Figure 8-3) again vary in the different orders. Again we do not see the "ground plan" uniformity of structure and function so characteristic of successful phyla like Annelida and Arthropoda, but rather a series of relatively unsuccessful experiments in length without body cavities or segmentation. In functional terms, however, all are tubular guts which extend from an anterior subterminal mouth to a terminal anus and thus allow for a one-way traffic of food materials along a mechanical and chemical disassembly-line. This allows for a digestive efficiency much greater than in the blind gut-sacs of coelenterates and turbellarian flatworms.

The reproductive organs are usually simple masses of cells scattered along each side of the body, and in many cases merely rupture through the body wall to the exterior. Nemertines are dioecious, and fertilization is external. Spiral cleavage results in a peculiar larval stage called a pilidium, which is not comparable to the larva of any other invertebrate group, and which must undergo a

FIGURE 8-3.   (opposite) Proboscis and gut relationships in stylized longitudinal sections in four principal groups of nemertine worms (phylum Rhynchocoela): a paleonemertine (**A**), a heteronemertine (**B**), a hoplonemertine (**C**), and an ectocommensal bdellonemertine (**D**). Note that the cell layers of the proboscis (solid black) are not shown in detail. The proboscis pattern of epithelial and muscle layers corresponds exactly with those layers in the body wall (see Figure 8-1) for each major group, being present in *reverse* order in the wall of the retracted proboscis.

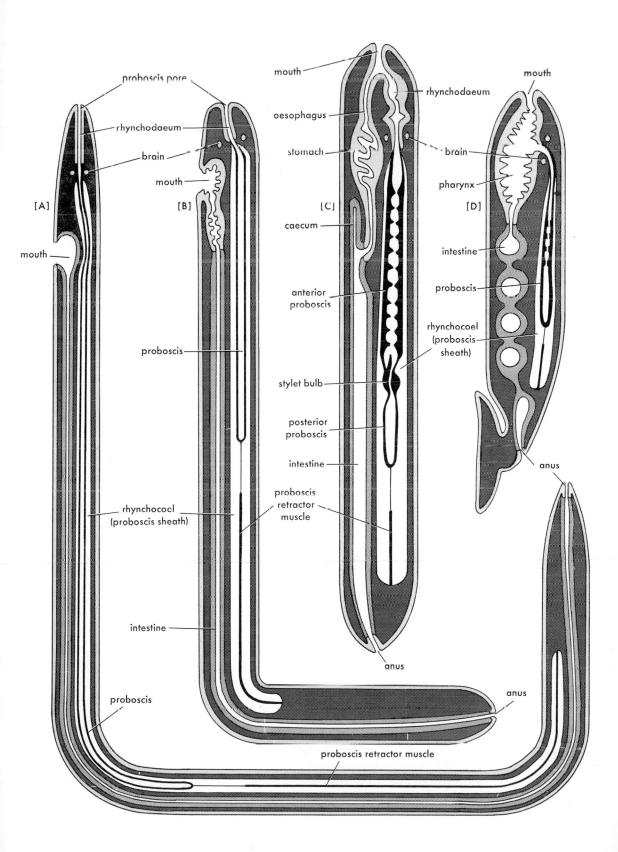

[A]

proboscis pore

rhynchodaeum

brain

mouth

mouth

[B]

proboscis pore

rhynchodaeum

brain

mouth

proboscis

rhynchocoel
(proboscis sheath)

intestine

proboscis

[C]

mouth

rhynchodaeum

oesophagus

stomach

caecum

anterior
proboscis

stylet bulb

posterior
proboscis

intestine

proboscis
retractor
muscle

anus

[D]

mouth

rhynchodaeum

brain

pharynx

intestine

proboscis

rhynchocoel
(proboscis
sheath)

anus

anus

proboscis retractor muscle

considerable metamorphosis before settling down as a miniature adult nemertine. In a few genera of nemertines there are viviparous species where the eggs are retained within the female worms until they develop.

The nervous system is essentially similar to that of the more complex Turbellaria. The excretory system, although based on the same sorts of tubules and flame-cells as are found in flatworms, is considerably more complex in the nemertines and may involve large numbers of ducts and nephridiopores to the exterior. In larger nemertines it is associated with the simple circulatory system. Nemertines are the only acoelomate animals to have true blood contained within a closed circulatory system. In its simplest form the system consists of only two vessels, one on either side of the gut-tube, connected by a series of spaces in the head and posteriorly around the anus. In some of the larger worms, this pattern has been elaborated by the addition of a middorsal vessel and numerous circular connections either around the gut or around the rhynchocoel. Although the large lateral vessels are contractile, there is no definite circulation, blood being merely stirred to and fro within the system by local contractions, with reversals of direction occurring at intervals.

Several of the larger nemertines are known to reproduce asexually by spontaneous fragmentation followed by the regeneration of the dozens of pieces each into a whole worm. Less experimental work on regeneration has been done on nemertines than on Turbellaria. A few common intertidal nemertines apparently reproduce throughout spring and summer by fragmentation, and turn to sexual reproduction only when water temperatures become lower in the fall.

To summarize the systematics of the phylum Rhynchocoela, the four orders are placed in two subclasses. Order Paleonemertini, with either two or three muscle layers (Figure 8-1) and a superficial pair of lateral nerve cords just below the epidermis, and order Heteronemertini, with three muscle layers (inner and outer longitudinal) and nerve cords just within the outer muscle layer, are placed together in the subclass Anopla (diagnosed as having a more superficial nervous system, a proboscis pore, a ventral mouth posterior to the brain, and a proboscis without stylets—Figures 8-1 and 8-3). The other subclass Enopla, with the nervous system internal to all the body musculature and a terminal anterior mouth through which the proboscis is everted (Figure 8-1D), comprises not only the order Hoplonemertini, wherein the proboscis is armed with stylets (Figure 8-2) and the gut considerably more complex than in the Anopla (Figure 8-3), but also a minor group of smaller ectocommensal rhynchocoels, the order Bdellonemertini, which have a posterior sucker, and proboscis and oesophagus opening together into a complex buccal chamber or pharynx. As with all nemertines and many flatworms, our knowledge of the functional morphology (particularly of the gut) in these ectocommensal bdellonemertines has been greatly enhanced by the detailed physiological and histological studies of J. B. Jennings and his associates. Some of the commonest intertidal nemertines of temperate seashores (Figure 8-4), including genera like Cerebratulus and Lineus, belong to the order Heteronemertini; while the freshwater and tropical land nemertines, as well as a variety of marine genera, belong to the Hoplonemertea.

Doubtfully allied to flatworms and nemertines is a minor phylum of minute triploblastic acoelomate worms; the Gnathostomulida (see Figure 1-12). With a

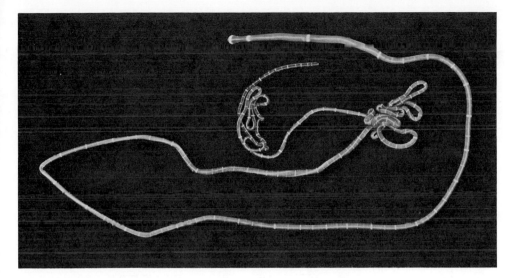

FIGURE 8-4. Live specimen of a rhynchocoel worm, *Tubulanus annulatus* (a paleonemertine, nearly 1 meter in length), from muddy sand in the marine intertidal zone. The geometrically arranged bands and stripes are superficial skin pigmentation and do *not* correspond to any internal structures. Length without segmentation is characteristic of all such nemertines, and muscle bands in the pattern of Figure 8-1B run the full length of the body in *Tubulanus* without any segmental breaks. [Photo © Douglas P. Wilson.]

blind gut, hermaphroditic sex organs, and a ciliated epithelium all resembling those of the smaller turbellarians, they differ in having the mouth armed with a tooth-bearing ventral plate and with a pair of lateral jaws moved by discrete strands of muscle. Mostly less than 1 millimeter in length, they live interstitially between sand grains in bottom deposits in the shallow sea. Since their first complete description in 1956, worldwide studies including those of P. Ax, W. Sterrer, and R. J. Riedl have enumerated at least ninety species.

Nature of Body Cavities

In spite of the complications of their muscle layers, their proboscis apparatus, and their circulatory structures, the nemertines can be regarded as derived from Platyhelminthes by the development of a tubular gut system. Much histology and physiology is essentially the same in the two phyla. Nemertines are the most complex acoelomate animals, and all the remaining phyla of animals have a body cavity of one form or another (Figure 8-5). The great majority of successful invertebrates are coelomate, that is, have a body cavity or coelom within the mesoderm. The internal organs are slung on mesodermal mesenteries and there is a mesodermal covering to the gut-tube endoderm as well as a mesoderm lining to the outer body wall (Figure 8-5C). The only other type of body cavity is that termed a pseudocoel which (as seen in Figure 8-5D) occupies a space between the mesoderm of the body wall and the endoderm of the gut. Thus, in this plan, there are no mesenteries suspending the internal organs and no mus-

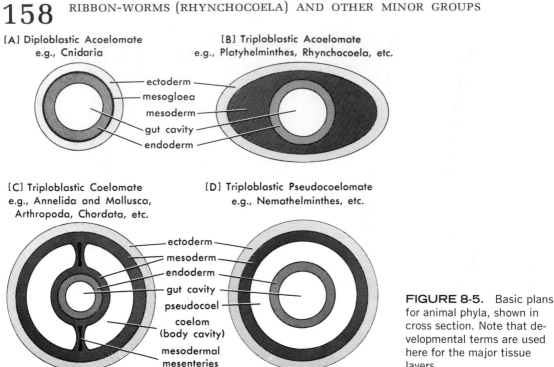

[A] Diploblastic Acoelomate
e.g., Cnidaria

[B] Triploblastic Acoelomate
e.g., Platyhelminthes, Rhynchocoela, etc.

ectoderm
mesogloea
mesoderm
gut cavity
endoderm

[C] Triploblastic Coelomate
e.g., Annelida and Mollusca,
Arthropoda, Chordata, etc.

[D] Triploblastic Pseudocoelomate
e.g., Nemathelminthes, etc.

ectoderm
mesoderm
endoderm
gut cavity
pseudocoel
coelom
(body cavity)
mesodermal
mesenteries

FIGURE 8-5. Basic plans for animal phyla, shown in cross section. Note that developmental terms are used here for the major tissue layers.

cle layers surrounding the gut. This condition has the functional implication that in no pseudocoelomate animal does muscular peristalsis move food through the alimentary canal.

Only one "major, successful" phylum of animals is built on the pseudocoelomate plan: the nematode worms. However, there are also six minor phyla whose body organization is built in this way, and these will be briefly considered now (see Figures 1-13 and 1-14).

Entoprocts and Rotifers

The first of these six minor phyla is the Entoprocta or Calyssozoa consisting of about sixty species of small sedentary animals, which look superficially like hydroid coelenterates that have developed some mesodermal structures. They are all marine except for one genus, and all are ciliary filter-feeders (Figure 8-6). At one time they were associated with the phylum Bryozoa-Ectoprocta in the "Bryozoa" or "Polyzoa," but these are now regarded as totally distinct sessile ciliated organisms, being coelomate and more closely related to brachiopods and other coelomate groups. The Entoprocta, on the other hand, are probably more closely allied to the Nematomorpha and some other pseudocoelomates. The cup-shaped body bears an encircling crown of ciliated tentacles and is carried on a stalk. The gut is U-shaped and ciliated with both the anus and the mouth within the cup. There is parenchymatous tissue between the ectoderm of the body wall and the alimentary canal, but embedded in it are two excretory

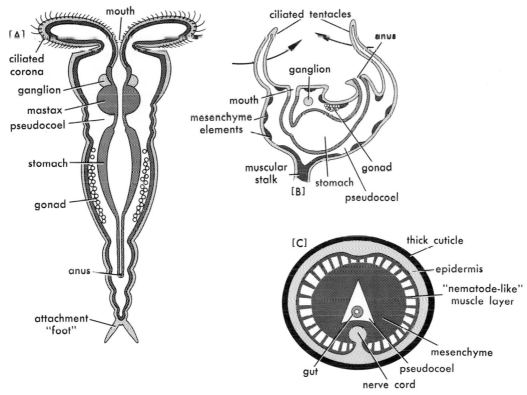

FIGURE 8-6. Some representatives of minor phyla. A: Horizontal section through an ar chetypic rotifer. B: Vertical section through an individual zooid belonging to the phylum Ento-procta. C: Cross section through the extremely elongate body of an adult horsehair-worm (phylum Nematomorpha). Note that the phyla Rotifera, Entoprocta, and Nematomorpha are all constructed on the "triploblastic pseudocoelomate" plan.

organs each consisting of a single flame-cell with a ciliated duct. There is a well-developed nerve ganglion and the gonads have specialized ducts leading to the exterior. Some species are dioecious, others are hermaphrodite, and it is thought that in most cases fertilization is internal. A ciliated, free-swimming larva hatches and is expelled, and is believed by some authorities to bear a re-semblance to the trochosphere larvae which are found in molluscs and in anne-lids. A complex metamorphosis is involved after settlement. Asexual reproduc-tion by budding is much more important and entoproct colonies (usually attached to rock surfaces or the shells of other animals) can grow rapidly or regenerate damaged groups of individual zooids. Budding, in some species, takes place on the calyx at the base of the feeding tentacles, and in others, from the attachment stalk as a stolon. They are all ciliary filter-feeders using the cilia of the tentacular crown both to move water and to filter bacteria, flagellates, and diatoms from it. Water is driven (Figure 8 6) centripetally from outside to in-side the crown, and thus the anus discharges into an exhalant water current. Filtered food particles pass down the ciliated inner faces of the tentacles and through the mouth into the U-shaped gut.

The phylum Rotifera is a somewhat more successful group numbering about 1500 species living mainly in fresh water, although there are a few marine species. They are mostly microscopic animals, having about the same dimensions as the larger ciliate protozoans, and only a few species reach lengths in excess of 1 millimeter. The head bears a ciliated crown which is used both in locomotion and in feeding and there is a spherical or cylindrical body ending in a bifurcated attachment structure called a foot. In most species, the ciliated corona or wheel-organ which gives the group its name drives fine, particulate food into the mouth where there is a pharynx armed with a mastax which functions as a chewing organ. A few have abandoned such suspension-feeding and become predaceous carnivores. There is a glandular stomach where the food is digested and a short intestine to the subterminal anus (Figure 8-6). The outside of a rotifer is covered with a distinct cuticle, which may be sculptured or ornamented and often looks as though it is segmented (although this is not true articulated jointing as occurs in the arthropods). There are a pair of nephridial tubules draining a series of flame-cells and leading to a bladder which discharges to the exterior. The system is undoubtedly osmoregulatory in such relatively minute freshwater organisms with a high surface:volume ratio, and the bladder usually pulses at intervals of a few seconds. However, little is known of the physiology of regulation in these animals. The nervous system includes a bilobed ganglion, dorsal to the pharynx, resembling that of flatworms, and there are complex sense-organs (tactile, visual, and chemoreceptory) arranged in pairs around the body. The pseudocoel lies within these organs of the body wall, and is often subdivided by the branching arms of amoeboid cells. The anatomy of rotifers shows an intriguing level of cell-number constancy with subadult growth involving no sequence of cell divisions. Thus, for example, in one freshwater species of rotifer, in newly hatched individuals, as in older adults, there are always 39 cells in the stomach, 22 cells in the circular muscles of the body wall, 183 cells in the bilobed central ganglion or "brain," and so on! Similar cell-number constancy occurs in the pseudocoelomate Nematomorpha, but also in the totally unrelated minor metameric phylum Tardigrada (see Chapter 18). Rotifers have separate sexes and, in many forms, the females are capable of parthenogenesis. Much reproduction in rotifer populations is by this method, with diploid eggs being produced which do not require fertilization. The egg-to-adult cycle in a typical parthenogenetic rotifer is about seven to ten days. In some species, apparently, when haploid eggs are produced, if these are not fertilized they hatch as males. The haploid males remain very small, mature rapidly, and can fertilize other haploid eggs to produce the so-called resting eggs, which usually overwinter. These resting eggs are capable of resisting desiccation and freezing; when they hatch in the spring, they produce only females which immediately give rise to more young by parthenogenesis. Once again, the success of a relatively minor group of minute animals in temperate fresh waters depends on a modification of the life-cycle to suit the peculiar conditions of that environment. Most of the freshwater rotifers live on submerged vegetation or as epizoites on other animals. A number of them are planktonic, swimming continually by means of the corona of cilia, and many of these undergo seasonal changes in body form (or cyclomorphosis), which changes seem to be related to changes in mode of reproduction. A few rotifers have become parasites and a number are termed terrestrial, though they really live as aquatic organisms in the water films on mosses and other marsh-dwelling plants. In some of these "land" forms, and also in more typical freshwater rotifers, there are species where the adults can go into diapause in cysts. They are then capable of resisting desiccation and freezing in a similar fashion to that of the resting eggs.

Four More Minor Phyla

The phylum Gastrotricha is another group of about 150 species of minute animals living in fresh waters and in the sea. They are usually less than 1 millimeter in length; they are ciliated and not unlike small rhabdocoels, but additionally bear bristles and sometimes scales and spines. They have a straight alimentary canal with a muscular pharynx at the anterior end and some have protonephridia. Their gut is characteristically like that of the nematodes, but their ciliation and excretory system link them to the rotifers. They have a ganglion and nerve cells in contact with the endoderm and no trace of metameric segmentation. Little is known of their physiology, but some freshwater gastrotrichs show parthenogenetic reproduction like rotifers.

A fourth minor pseudocoelomate group is the phylum Echinorhyncha (sometimes called the Echinodera or Kinorhyncha), comprised of minute marine animals without cilia and with a regularly segmented covering of cuticle. The body always consists of thirteen or fourteen "segments" which bear hollow, movable spines. Kinorhynchs have protonephridia and sac-like gonads with distinct genital ducts. The sexes are separate, but little is known about reproduction. Kinorhynchs are found in offshore silts and muds in all parts of the world, and apparently feed on detritus ingested from the mud through which they burrow. The anterior part of the gut is lined with cuticle and leads to a tubular intestine and a terminal anus. In some there are two pairs of digestive glands of unknown function. Little general physiology is known, but one striking feature is that subadult and adult kinorhynchs molt their cuticle at regular intervals like arthropods.

A fifth minor phylum, the Nematomorpha, is obviously related to the last groups, though somewhat distantly. The nematomorphs are very elongate, totally unsegmented worms, which are free-living as adults but parasitic in arthropods as larvae. The adults are the so-called hairworms, or "living horsehairs," which may be 1 meter long by about 1 millimeter in diameter, and usually black or dark brown in pigmentation. In rural districts, they regularly give rise to rumors of "living horsehairs" in the water supply. There are probably less than 250 species placed in genera such as *Gordius* or *Paragordius*, living in fresh waters, and in one marine planktonic genus *Nectonema*. In some typical freshwater forms, the larvae pass through two arthropod hosts in turn, but the adult stage is always free-living, with a degenerate gut and two long cylindrical gonads. The sexes are separate, and there seem to be sexual differences in behavior, the males being much more active. Eggs are laid through the cloaca, and the larvae which hatch out have boring organs which enable them to penetrate arthropod hosts such as grasshoppers, chironomid larvae, or beetles. The parasitic gordioid larva (which resembles an acanthocephalan or pria-pulid) grows through many molts inside the host, before emerging as a near-adult hairworm lacking only fully mature gonads.

The last (considered here, see Table 1-1) of the minor pseudocoelomate phyla is the Acanthocephala. There are about three hundred known species of these endoparasites, which bear a proboscis armed with hooks which gives them their name. They have no gut, and food is absorbed directly through the body wall from the host. They have protonephridia and a simple nervous system of longitudinal cords linked to an anterior ganglion. The life-cycle involves larval stages in the haemocoel of an invertebrate host, usually arthropodan, followed by the adult in the gut of a vertebrate, such as a fish, bird, or mammal. They occur throughout the world in terrestrial, freshwater, and marine hosts. As adults they are usually small, rarely more than 2 centimeters in length, but are often present in enormous numbers. A typical genus is *Echinorhynchus;* the larval stages live in the freshwater crustacean *Gammarus,* and the adults in the intestines of ducks, often many hundreds in a single duck. Other species occur in freshwater fishes, and some may have more complex life-cycles involving one invertebrate and two vertebrate hosts.

These minor phyla are often grouped together in various ways. Some modern textbooks use the superphylum Aschelminthes to include most of them (except the Acanthocephala) along with the nematodes. Older texts incorporated them in a superphylum Gephyrea, which also included some obviously coelomate phyla such as the Sipunculoidea. The case for these superphyla is not a good one. Such composite groups would be far looser in their relationships than the major phyla, and therefore it has seemed best to separate them and regard them all as separate phyla—the minor pseudocoelomate phyla.

9 SUCCESSFUL PSEUDOCOELOMATES: NEMATHELMINTHES

ONE phylum of relatively simply organized pseudocoelomate worms has been enormously successful. The phylum Nemathelminthes consists of worms built to a relatively uniform functional plan (so much so that they are regarded as constituting one class, the Nematoda), which stands apart without obvious relationship to any of the other phyla. These "roundworms" are triploblastic and completely unsegmented. They have a pseudocoel or perivisceral cavity, a tubular gut which runs through them from mouth to anus, and a thick, continuous cuticle. None of their cells in any organ or tissue is ever ciliated. They are found in all environments: in the sea, in fresh waters, on land, and as parasites of all kinds of animals and plants. It is difficult to say just how successful the nematodes are. Some authorities place the number of species between 10,000 and 20,000, while other authors have estimated between 400,000 and 500,000 species (less than one-twentieth of which have been described). In this book (see page 13 and Table 1-1) a compromise figure of 80,000 has been adopted. This is admittedly a subjective estimate. Although they range in size from microscopic forms to some parasites 2 meters long, they are all structurally alike. Their systematics are very unsettled and involve considerable inflation of categories. Free-living microscopic forms are often exceedingly abundant: it is said that one spadeful of good garden soil usually contains about a million nematodes. All

163

species of higher plants and animals have at least one species of nematode parasite, and most vertebrate animals are hosts to several kinds of roundworms. About fifty species of nematodes are known to occur in man, though only about a dozen of these are true parasites of pathogenic importance. Probably *less than* 2% of the readers of this book have *never* been hosts to a nematode infection; more than 70% of the readers are at this present moment hosts to adults, or eggs, or encysted stages of these worms. It is arguable that such parasites are the most successful of their kind: that is, in being nonlethal to their hosts, they can be claimed to show much greater efficiency as parasites than those which cause serious diseases. Such parasites are unobtrusive and often ignored.

Whatever estimated number is finally accepted for the species of nematodes presently living, the great majority of them will prove to be microscopic forms which are free-living: either in terrestrial soils or in marine bottom deposits. Such microscopic, free-living nematodes usually have relatively simple life-cycles. Thus the species of the genus *Ascaris,* are somewhat atypical, being parasites, about 30 centimeters long, of the intestine of mammals, with a rela-

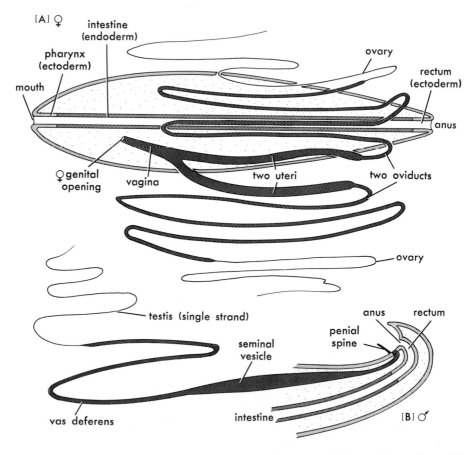

FIGURE 9-1. Anatomy of nematodes: diagrammatic representations of dissections of female **(A)** and male **(B)** ascaroid nematodes. Note the *paired* genital structures in the female and the *single* strand of gonad and ducts in the male.

tively complex life-cycle. Several lines of similar parasitic forms have evolved independently from the main groups of free-living nematodes. However, it is still justifiable to use features of *Ascaris* to build archetypic form since, as stressed earlier, in structure and physiology all nematodes are more or less similar. *Ascaris lumbricoides*, found in the intestine of pigs and occasionally in man, and *A. megalocephala*, in the intestine of horses, are the "typical" species. Their structure can be understood with reference to Figures 9-1 and 9-2. The thick cuticle is underlain by a syncytial ectoderm, and this in turn by the mesodermal cells which are the muscle cells. These are peculiar to nematodes and consist of a protoplasmic part containing the nucleus and a contractile part usually arranged as a sort of gutter along the inside of the body wall. In some ways these muscle cells do not show very much advance, in the efficiency of their contractility, beyond the myoepithelial cells of coelenterates. They are almost all arranged so that their contractions are longitudinal to the nematode, and they do not produce changes in shape but rather flexures of the whole body. This activity seems to be relatively uncoordinated, and in general nematodes have poor powers of locomotion. The muscles can do little more than flex the whole body rather violently from side to side. As a result nematodes are usually unable to swim where there is a lot of free water. They move much better in films of water over surfaces, and many plant nematodes can only invade plant tissues where a capillary layer of water of suitable dimensions lies over the surface to provide leverage for boring. Most nematodes can move more

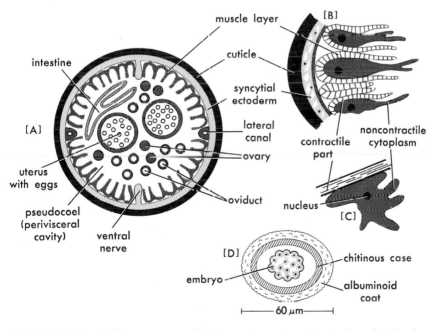

FIGURE 9-2. **A:** Transverse section through a mature female *Ascaris*, showing pseudocoelomate organization with many cross sections of reproductive structures. Compare with Figure 9-1A. **B** and **C:** Transverse and longitudinal sections through the peculiar muscle cells characteristic of nematodes. **D:** Embryonated egg of a typical ascaroid nematode, as extruded.

readily through the interstitial fluids of soils, or of tissue cells, than through the open sea.

It is worth considering the mechanics of the system in a little detail. In all other worm-shaped animals (except perhaps in the Nematomorpha) a local contraction of some longitudinal muscles will result in a thickening or dilation of that part of the body. This is not so in nematodes. The local contraction of longitudinal muscles involved in one flexure will result in the stretching of the longitudinal muscles in all other parts of the body. That is to say, the localized contraction is antagonized by the rest of the musculature. The forces involved in stretching the other musculature are transmitted in two ways. There is a high internal hydrostatic pressure. There is also, in the cuticle itself (that is, *outside* the ectoderm), a complex system of collagenous fibers forming a spiral basket-work around the body, rather like that already described in the basement membrane of the nemertines. These fibers are themselves nonelastic and inextensible, but their arrangement is again such as to allow certain kinds of limited deformation of the cuticle. Actually, in *Ascaris* the spiral fibers are found to be inclined at an angle of 75 degrees to the longitudinal axis of the worm, and it is this sort of arrangement, combined with the high internal hydrostatic pressure, which allows the nematode (possessing only longitudinal muscles) to undergo reversible changes of length of its parts and the typical flexures of its swimming.

It is not usually realized that growth in nematodes involves molting of this thick cuticle. Although the molts of early larval stages have been extensively studied in some parasitic forms, we still lack the sort of knowledge which has been accumulated on the arthropods about control of molting and the different physiological effects of the molt-cycle.

Within the body wall is the perivisceral cavity surrounding the gut-tube and the genital organs. This is, of course, not a true coelom since it is not lined by mesodermally derived cells. The gut is a simple tube of endoderm running through from end to end. There is an invagination of ectoderm to form a pharynx at the anterior end and another to form a rectum at the posterior end. These may be muscularized, but the main gut-tube is not. Many nematodes have a two-chambered pharynx in which alternate contractions serve as a pump to push food through the length of the gut-tube. The anus is normally sealed shut by the high hydrostatic pressure in the pseudocoel and must be temporarily opened by dilator muscles to permit a squirt of faecal material. It could almost be considered as a physiological axiom concomitant on the pseudocoelomate condition that the gut is unable to carry on peristalsis and food must be moved through the gut-tube by forces external to it. Actually the perivisceral cavity is traversed by strands of protoplasm from various cells of the muscular body wall and also by wandering amoebocytes which move around the gut and probably play some part in excretion. There are longitudinal, lateral canals in the body wall which may be excretory, but there are no flame-cells or cilia and no internal openings of any sort typical of excretory organs. There are usually no obvious sense-organs except for tactile papillae around the mouth, which may be arranged in groups of three or six in a radial fashion. In larger nematodes there is a nerve ring round the pharynx and six longitudinal nerves, of which the two dorsal and ventral nerves are usually larger. Not much neurophysiological work has been done on the system. In *Ascaris* spp., as in most nematodes, the sexes are separate. In the female, there are two thread-like ovaries which lead by two oviducts and two uteri to a single vagina and thence to the female genital opening in the anterior third of the body. In the male, a single strand-like testis leads by a vas deferens to a storage vesicle which discharges, near the anus, alongside a penial seta or setae used in copulation. Since there are no cilia in nematodes, even the sperm are amoeboid. Fertilization is usually internal and, in the case of *Ascaris*, a fertilized female will continue to extrude eggs, at a rate of about 15,000 a day for many months, which then pass out of the host with its faeces.

In most nematodes, an egg which passes out of the female has already begun development and contains an embryo. In the case of *Ascaris*, this has only begun to cleave, but in other nematodes the embryo may be a fully developed miniature worm before the egg leaves the mother. In *Ascaris* there is a double case around the egg; a chitinous capsule is surrounded by an albuminoid coat. In general, what hatches out of a nematode egg is a larva which is almost indistinguishable from the adult in its morphology, though it often exhibits a special larval behavior pattern. In the case of *Ascaris*, once out of the host, the embryonated egg develops to a small, coiled larva; then development stops, and the egg can survive in this state for over five years. Nothing then happens unless it is taken into the gut of the correct host with accidentally swallowed soil or in contaminated water. For example, if the eggs of *A. lumbricoides* are ingested by pigs or by humans, they hatch in the small intestine and the larvae (like miniature adults) bore into the blood vessels of the intestinal wall. They are then carried by the circulating blood to the liver and move through its tissues to other blood vessels, whence they are carried to the lungs, where they break out of the blood vessels into the cavity of the lungs. When they arrive there, the young *Ascaris* are about 3 millimeters long. They then have to be coughed up into the mouth of the host and swallowed; they then pass down the gut again to the intestine, where they grow to maturity. Infections with adult *Ascaris* may be of little importance pathologically, unless they are so numerous as to obstruct the intestine. On the other hand, if numerous eggs are ingested at one time, the migration of the young may damage the lung wall and be responsible for ascaroid pneumonia, which is, of course, much more common in pigs than in any other animals. In swine, this is termed "thumps." In humans, there may be an allergic reaction internally, or externally if *Ascaris* is handled. There seems to be considerable human variation in sensitivity to the secretions of *Ascaris*.

Other Human Parasites

As stressed earlier, all nematodes are similar in structure, no matter whether they are free-living or parasites of plants or of animals, and they never vary much in form, even in their larval stages. Most higher animals are hosts to a dozen or so nematode species living in different parts of the body, and man is no exception. Among the best-known human nematodes are the hookworms, *Necator* and *Ancylostoma*, of the New World and Old World, respectively.

These are unusually harmful nematodes. Their buccal cavities are armed with hooks by which they browse on the intestinal wall, feeding on the blood produced and causing an anemia (which may in some cases be fatal, but which always involves continual internal hemorrhage). The eggs pass to the exterior and must have a warm, damp place to hatch out. They hatch as boring larvae which penetrate exposed skin (usually of the foot or the hand) and thence migrate by the bloodstream to the gut. They are widespread throughout the tropics, wherever soil is warm enough for the development of the larvae, and where sanitation is poor.

Perhaps the commonest human nematodes of colder and temperate civilized countries are the pinworms of the genus *Enterobius* (sometimes called *Oxyuris*). The eggs pass to the exterior and infect a new host if swallowed. They develop directly, and there is no larval migration. The females, when they are laying their eggs, cause an irritation of the anal region, and an infected child's scratching of the region will contaminate his or her hands. The eggs may then be conveyed to the mouth and thus autoinfection, or self-reinforcement of the infection, is common in children. Pinworms have long been known to occur as an occupational hazard for preschool and kindergarten teachers. Given the human behavior patterns of the "permissive societies" of Europe and America, it is not surprising to find *Enterobius* also occurring in recent years in certain social and age groups as a widespread but relatively harmless venereal disease.

A more dangerous parasite is *Trichinella spiralis*, found in rats, in pigs, and in humans—a peculiar nematode in that both larvae and adults live in the same host. The adults live in the small intestine, are about 3 millimeters long, and do not have much effect, even if present in considerable numbers. However, after copulation in the gut, the fertilized female bores out into the bloodstream where she retains her eggs until they are fully developed. That is, she is viviparous, and she produces about a thousand young at a time. The young bore out of the bloodstream into muscles where they encyst inside the tissue, forming cyst walls which gradually calcify. There is no further development until these cysts are ingested by a new host. Only then do the cysts break down in the gut; the worms escape, grow, and copulate, and the females bore out again. Cysts are important pathologically if large numbers are ingested at one time, since females may cause hemorrhage upon leaving the gut and then rheumatic-like pains during migration. Larvae can even cause degeneration of muscle tissue if enough of them encyst, and smaller numbers in nerve tissues will impair neural function. A single ounce of measly pork (pork infected with *Trichinella*) can contain 80,000 cysts and, if it is improperly cooked and then ingested, a nearly simultaneous hatching of 80,000 worms will occur in the gut. Subsequently, the fertilized and fecund females can liberate in the bloodstream 40 million larvae, which then move through the muscles of the new host before encysting. Meat inspection at abattoirs and slaughterhouses is intended to control this danger, and, indeed, the most usual cause of rejection of pork is the presence of *Trichinella*. Nevertheless, a small quantity of measly pork does get marketed (for example, in some "country" pork sausages) and, as with cestode contamination, thorough cooking is necessary to avoid infection.

In Chapter 7, we noted how a minute and harmless tapeworm of dogs could, with suitable oral infection, give rise to a massive hydatid cyst on the side of a human lung—a "dead end" (uncompleted life-cycle) for the parasite but a pathological condition requiring surgery for the human. Within the last few years, somewhat similar cases concerning a relatively harmless canine nematode, *Toxocara canis*, have been reported in Western Europe. Once again, eggs from an undetected infection in a pet dog may be conveyed to the mouth, particularly by children. They hatch in the human intestine, and the larvae then move out through the tissues where they can cause serious disease by tissue degeneration of such organs as the liver or lungs. A more tragic result for children is that the larvae can move into the eyeball and cause such extensive areas of degeneration in the retina that permanent and irreversible blindness results. In the past, some such cases involving degenerated areas of the retina have been mistaken for malignant optic tumors, and the unnecessary operation of removal of the entire eyeball (optic enucleation) has been urgently performed. The enormous increase in urban and suburban dogs in the western world (about 28 million in the United States in 1976) has increased for children and adults alike the incidence, not only of hydatid cysts and toxocaran lesions, but also of a number of dog-transmitted diseases beyond the scope of this book. These range from the potentially lethal, such as rabies and tetanus, to the merely embarrassing, such as the fungus infection, incorrectly called ringworm, which causes hairless patches on the scalp. Transference of the dog's territory-marking-by-urination behavior to urban circumstances not only has resulted in erosion and collapse of lampposts in London (as reported in *The Times* in 1976) but also has created the risk of leptospiral hepatitis (appearing as jaundice) for any mechanic who removes large numbers of automobile tires from their rims.

Another dangerous nematode family is the Filariidae, species of which live in the tropics in the bloodstream of man. The larvae, which hatch out in the blood, are transmitted to new hosts by various means. In *Filaria* (*Wucheria*) *bancrofti*, the larvae move to the peripheral blood vessels at night and are transmitted by night-flying mosquitoes. They thus pass to new hosts and rapidly grow to mature adults, each about 20–30 centimeters long, which in turn produce larvae ready for transmission. Other genera and species belonging to the same family are transmitted by other biting flies. Heavy infections of certain filarian worms cause a disease called elephantiasis, in which some parts of the body (such as legs, scrotum, or breasts) become swollen to an enormous size. The swelling is largely due to the obstruction by the worms of circulation through blood vessels and lymphatics being followed by a reaction of the human tissues in hyperplastic growth. Some forms related to filarians are not transmitted by insects. These include *Dracunculus medinensis*, an elongate nematode (up to 1.3 meters long), which occurs as a subcutaneous and connective-tissue parasite of man in Africa and Asia. Larvae are discharged into water through a skin ulcer, formed by the gravid female dracunculid. For successful transmission, the microscopic larvae must then make their way into the haemocoel of the freshwater copepod *Cyclops*, usually after boring through the crustacean's gut wall. Development continues in *Cyclops*, which is thus a true vector. If the copepod has been infected for about five weeks and is then swallowed in drinking water, a new human infection can occur.

Filariasis in its various forms is one of the three most widespread tropical diseases, the others being malaria and schistosomiasis. In 1976, the World Health Organization estimated that some 250 million people have diseases caused by filarial parasites. One of the most dramatic forms is onchocerciasis (or "river blindness") caused by the nematode worm *Onchocerca volvulus* which occurs in some areas of tropical Central and West Africa. This nematode is transmitted by a black-fly of the genus *Simulium*. The larvae of such black-flies live in relatively eutrophic, fast-flowing streams, and anglers who seek such localities in many parts of North America and of Asia can suffer many hundreds of bites from the adult female black-flies. In tropical Africa, however, when a female black-fly feeds on human blood, it can also transmit the larvae of *Onchocerca volvulus*. Once in a new human host, the nematode larvae develop into long, thread-like worms which can live in their host for up to fifteen years and produce millions of new larvae. These larvae move through the dermal capillary blood vessels and, if ingested along with blood, can be transmitted by a black-fly to further human hosts. The adult worms have little pathological effect beyond causing characteristic nodules in the tissues under the skin; but the enormous numbers of larvae moving through the most superficial layers of the skin also invade the eyes and the victims gradually but inevitably become blind. Thus in some regions of tropical Africa, such as the upper basin of the Volta river, a common sight is a young child leading a string of adult men through a village to their work in the field. Most likely, the child is already infected with *Onchocerca* and will become blind in turn as an adolescent. River blindness affects the Ivory Coast, Ghana, Mali, Togo, and Upper Volta; the World Health Organization has estimated that there are over a million onchocerciasis sufferers in Upper Volta alone.

One additional aspect of the pathology of nematode infections occurs in all species where larval worms migrate through human tissues. Thus it does not concern the forms with simple life-cycles such as the common pinworm, *Enterobius*, but is involved in the tropical filarian worms and all such forms as *Trichinella* and *Toxocara*. In these cases the young nematodes may carry within *their* tissues viruses from the human gut to much more susceptible human tissues such as the nerve cells of the spinal cord and brain. It is possible for the virus of poliomyelitis to be transmitted in this way.

There remain a few general points to be made about the parasitic Nemathelminthes. As in the other parasitic worms, large numbers of eggs are produced to increase the chances of infection of new hosts. Just as in the other worms, there are elaborations of the life-cycle which ensure that transmission from primary host to primary host has a chance of success. It is important to note that the most efficient parasites are those, such as *Enterobius*, which are never lethal to the host and thus never lethal to the population of parasites inhabiting it. Another matter is worth raising. Some of the characteristics of the free-living nematodes have been important as preadaptations to the parasitic way of life. Besides the type of locomotion, there is an unusual capacity for sustained diapause in eggs and other stages. Eggs, encysted larvae, or encysted adults (both of free-living forms and of parasitic forms) can remain inactive for periods of years and take up active metabolism again only under suitable conditions. The physiology of these diapause processes is still imperfectly understood.

10 SEGMENTAL ORGANIZATION: ANNELIDA–ARTHROPODA

No two phyla of the more complex animals share more features of structural and developmental organization than the annelids and arthropods. The phylum Annelida encompasses the segmented worms of the sea, of fresh waters, and of terrestrial soils. The phylum Arthropoda, the enormous phylum of the most successful invertebrate stocks (see Tables 1-1 and 10-1), comprises the insects, crustaceans, spiders, and other groups, all with a jointed, chitinous procuticle which serves both as a protective armor and as an exoskeleton.

The structural homologies *common* to the two groups are mainly concerned with early development and with the nature of metameric segmentation. The patterns of functional interdependence, while somewhat stereotyped within each group, clearly separate the two phyla. Functional integration in the efficient machines which are annelid worms differs from that in arthropodan machines, and the differences are chiefly those associated with the nature of the integument: differences in the mechanics of growth, of respiration, of excretion, and of loco motion. If we consider, for example, only the last aspect of physiology in the two metamerically segmented groups, the distinction is clearcut. Annelid worms are soft-bodied animals, using longitudinal and circular muscles in the body wall for locomotion, the forces generated by contractions being transmitted by pressure changes in the worm's body fluids. Thus, the

TABLE 10-1
Outline Classification of the
Metameric Coelomates

Phylum ANNELIDA
    Class Archiannelida[a,e]
    Class Polychaeta
    Class Myzostomaria[a]
    Class Oligochaeta
    Class Hirudinea
Phylum ECHIUROIDEA[a]
Phylum ONYCHOPHORA[a]
Phylum ARTHROPODA
    Class Trilobita[b]
    Class Merostomata[a]
    Class **CRUSTACEA**[c]
    Class **ARACHNIDA**[c]
    Class Pycnogonida[a]
    Class **INSECTA**[c]
    Class Pauropoda[a,d]
    Class Chilopoda[d]
    Class Diplopoda[d]
    Class Symphyla[a,d]
Phylum TARDIGRADA[a]
Phylum LINGUATULIDA[a]

Notes:
[a] Minor groups with relatively few living species.
[b] No living species.
[c] The three major "successful" groups of arthropods.
[d] The four groups commonly termed "centipedes and millipedes" or "myriapods."
[e] This minor group should probably be suppressed (see pages 209 and 222).

efficient propulsion of an annelid through its environment involves antagonistic sets of muscles acting on each other by way of a hydrostatic skeleton and producing either undulatory movements or peristaltic waves of alternate elongations and bulgings which deform the entire "soft" body of the worm. In contrast, arthropods have a rigid exoskeleton and, eponymously, jointed limbs with internal muscles arranged antagonistically as flexors and extensors of each joint. The arthropod is propelled through the environmental medium, or over the substrate, by the movements of the interacting systems of levers which are its jointed appendages. The differences in the mechanics of locomotion in the two phyla obviously involve the structural arrangements and the functioning sequences of the muscles and of their motor innervation. Much more, however, is involved through the necessary functional integration within each type of animal machine. Not merely muscles and motor nerves, but the integrative parts of the nervous system, the sensory receptors (both external and internal), the circulatory system and heart, and even the functional organization of the alimentary canal must differ along with the pattern of locomotion. Similar interdependence would characterize a consideration of the different patterns of respiration in the two phyla.

Six terms of morphology can be applied to both Annelida and Arthropoda: triploblastic, coelomate, bilaterally symmetrical, with a tubular gut running from mouth to anus, with a ventral nerve cord, and with metameric segmentation. The implications of the first three characteristics are discussed elsewhere (see page 11). The functional significance of a tubular gut running from a subterminal mouth to a terminal anus through an elongate body is, of course, that it allows for one-way traffic of food material past a series of specialized tissues for trituration, secretion, digestion, and absorption. In other words, annelid worms and arthropods, although differently organized, have a functional alimentary layout similar to that in the gut of vertebrates, which layout is likewise capable of forming a disassembly-line for food materials. The last diagnostic character—that of metameric segmentation—implies that the body has its muscles, nerves, and internal organs subdivided in sets each of which makes up a metameric segment.

Metamerism

The essence of metamerism is the serial succession of segments, each containing unit subdivisions of the several organ systems. Not only the skin, the muscles, and nerves of the body wall are involved, but also, in more primitive segmented forms, such internal organs as those of circulation, excretion, and reproduction. Functionally, this type of organization is important in two ways: first, during development, in the nature of morphogenesis, and, secondly, with regard to the muscle systems and hydrostatic skeleton used in locomotion in soft-bodied forms. In the truly segmented animals such as annelids and arthropods, metamerism appears very early in embryonic development. Indeed, in many, as soon as mesoderm is formed, it is organized into mesoblastic somites. Although many of the obviously segmental features are ectodermal in origin, including the parapodia in worms, the limbs in arthropods, and the segmental ganglia of the nervous system in both, segmentation appears to be basically a phenomenon of mesodermal organization and proceeds from the inside of the embryo outward. It is important to note that, in the more highly evolved worms and arthropods, the exact numbers of somites or segmental structures in the adult can often be difficult to determine, but demarcation is always clearer in the embryo or in the larval stages. Later, fusion, loss, or modification of somites may obscure the regularity of repetition of the several organ systems or of the serial succession of segments themselves. It is in the least specialized annelid worms—presumably more primitive—that we find adults showing closest adherence to a serial pattern of identical units.

The second significant feature of metamerism in the physiology of soft-bodied segmented animals is its importance in locomotion. It is possible for acoelomate, nonsegmented animals to use body-wall musculature arranged in longitudinal and circular elements to carry out various forms of locomotion (see Chapter 8, on the locomotion of ribbon-worms). However, the evolution of a coelomic cavity has allowed a hydrostatic skeleton to be used in arranging the antagonistic action of these, so that faster and more powerful forces may be used to propel the coelomate animal through its environment. Further, the evolution of metameric segmentation, including the subdivision of the muscular units into a repeated series and the subdivision by septa of the coelomic cavity itself, has allowed not only even greater locomotory efficiency, but also considerably greater capacity for local changes of shape along the elongate body. The significance of this is immense, not only in burrowing, but also in all other types of locomotion. Compared with a ribbon-worm, an annelid worm (coelomate and segmented) is capable of faster and more efficient crawling and burrowing and of more powerful and more sophisticated responses to environmental dangers or to predators.

Archetypic Development: Annelids

In the most primitive marine annelid worms, the sperms or eggs produced by the ripe gonads either rupture through the body wall itself or pass out through the segmental excretory organs, the nephridia, or coelomoducts to meet in an external fertilization in the sea. The zygote thus formed undergoes a peculiar pattern of division known as spiral cleavage (see Chapter 27) to form a blastula and then, by invagination of a gut pouch, a ciliated gastrula. In a relatively primitive annelid, such as the genus *Polygordius*, this develops into the ciliated larva known as a trochosphere, which then drifts in the plankton. As shown in Figures 10-1A,B and 10-2A, the trochosphere has two ciliated bands, one above the mouth and one below the mouth, which are used both for locomotion and for feeding. There is also a characteristic apical sense-organ. Subsequent development of this larva takes place largely in the area around the anus. It is in this region that metameric segments are first divided off (Figure 10-1B). The first few segments may appear more or less simultaneously, or they may even appear to be budded off in sequence from the main body of the ciliated larva. However, the establishment of the main budding zone in the segment immediately in front of the anus (Figure 10-1C) is soon set up, and thereafter the segments form in an anterioposterior sequence so that the segments in front are older than the segments behind, the youngest segments always being the penultimate ones immediately in front of the anal somite, which may or may not be specially modified. In subsequent larval development the segmented portion of the body grows at the expense of the original trochosphere part and the little worm-like larva develops.

At about this stage, the young worm settles out of the plankton to the sea bottom and takes up a more or less adult way of life. At this stage (Figures 10-1C and 10-2B) the two head somites of the adult worm, the prostomium and the peristomium, are clearly seen to be derived from the structures of the original trochosphere: the apical sense-organ which gives rise to the prostomium with its contained supraoesophageal ganglia (or brain) and the rest of the original larva which gives rise to the peristomial segment around the mouth. The remainder of the adult worm consists of the metameric segments first divided off in the planktonic larva. If such developmental stages are examined more

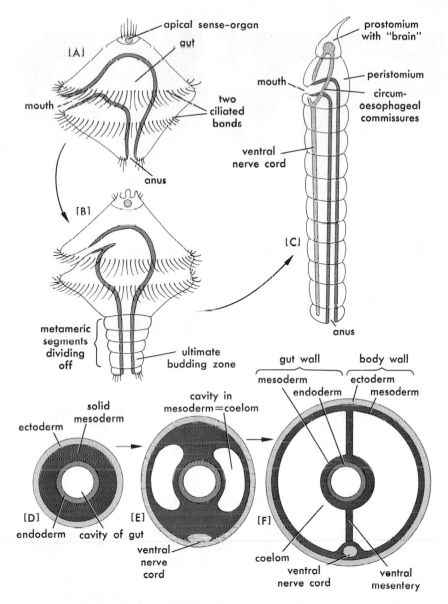

FIGURE 10-1. **A–C:** Development of metameric segmentation in a primitive annelid. **A:** Planktonic trochosphere larva of *Polygordius*. **B:** Later planktonic larva with metameric segments being budded off. **C:** Still later stage of juvenile segmented worm about to settle out from plankton. **D–F:** Development of mesodermal structures and of coelom, as shown in cross sections through successive ages of budded segments.

closely, it is seen that each segment, when it is originally budded off, consists initially of a solid walled tube of three layers (Figure 10-1D). Paired cavities appear as splits in the mesoderm in each segment, and these form the segmentally divided pairs of coelomic cavities. As shown in the figure, these cavities enlarge until, in each segment, they are separated only by the dorsal and

[A]   [B]

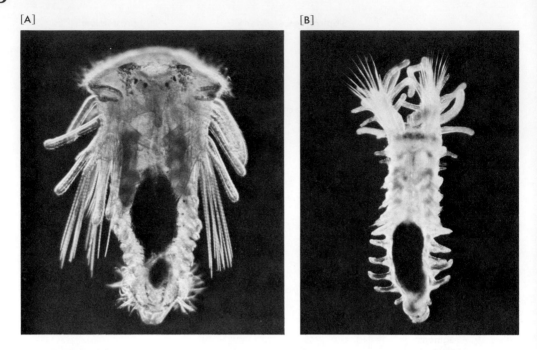

FIGURE 10-2.   Live planktonic and newly settled stages in polychaete development. **A:**
The late planktonic larval stage of *Lygdamis muratus,* a sabellarian worm which builds permanent tubes of sand grains and small pebbles in a secreted matrix. Like chaetopterids, these
related sabellarians are highly tagmatized worms. This stage corresponds to the diagram in
Figure 10-1 (B), where the trochosphere larva has begun to bud metameric segments posteriorly. **B:** A recently settled and metamorphosed juvenile worm of *Lygdamus muratus* removed from its initial tube. The adult elaboration of prostomium and peristomium characteristic of such worms is already being developed. [Photos © Douglas P. Wilson.]

ventral mesenteric slings of the gut and by similar thin, double septal walls
between adjacent segments. Thus, in a young developing worm of twenty segments, although about 75% of its volume is fluid-filled coelomic space, this
space is not continuous but is subdivided transversely between each pair of adjacent segments by a thin septum, and vertically within each segment by the
dorsal and ventral mesenteries, so that there is a series of forty fluid-filled coelomic cavities in a twenty-segment worm.

The budding of a series of sets of segmental organs from the budding zone
immediately in front of the anal segment in developing worms is paralleled by
the capacity of many worms to regenerate lost segments. This will be discussed
in Chapter 12.

Archetypic Development: Arthropods

Most arthropods have eggs containing massive amounts of yolk. This affects
early embryonic development, and cleavage is usually relatively superficial. A
form of spiral cleavage has been reported in a few arthropods, but, even where
it occurs, the pattern of mosaic development which follows differs significantly

from that of annelids (see Chapter 27). Primitive aspects of early larval development are bound to be similarly obscured in the terrestrial habitat of most insects and arachnids. Thus it is among the largely aquatic crustaceans that a larval pattern occurs which shows acceptably archetypic features.

In some representatives of each of the many diverse groups in the class Crustacea (see Table 14-1), the larva which first hatches from the egg is a nauplius.
This is the simplest crustacean larva, and the least differentiated of all arthropodan larvae. The nauplius has three pairs of jointed limbs which are usually
known by the names of the adult appendages into which they develop, though,
as shown in Figure 10-3 and 14-8, the naupliar limbs are not specialized as
head appendages but serve principally for locomotion and also for feeding.
There are usually several molt stages during which the young crustacean grows
but retains the nauplius form (see Chapter 13 for the physiology of the molting
process). In external appearance, the nauplius is a three-segmented larva with
three pairs of appendages, but internally there is, in some cases, evidence that
embryonically there are four segments involved. The first pair of appendages
(corresponding to the adult first antennae) are always uniramous, while the
other two pairs (corresponding to the adult second antennae and mandibles)
are biramous and have stout bristles in their basal parts forming gnathobases
(see page 232).

Then at one molt, a fourth pair of appendages appears posteriorly (corresponding to the first maxillae of the adult crustacean), and the larva becomes a
metanauplius. Although the nauplius shown in Figure 10-3A is that of a copepod, nauplii of many diverse groups of crustaceans are closely similar in form.
It is only after the metanauplius stage that the different group characteristics
begin to appear in the larval development (see also Figure 14-8). The later addition of more segments posteriorly is best seen in its simplest pattern in a relatively primitive form of branchiopod crustacean like *Artemia*. Newly added
segments may not develop limbs until a subsequent molt. A later metanauplius
(Figure 10-3B) and an even later twenty-segmented larva of a cephalocarid
(Figure 10-3C) serve to show the process of metameric addition. The process in
copepods is illustrated in Figure 14-8.

It should be noted that, once again, the segments form in an anterioposterior
sequence so that the segments in front are older than the segments behind.
Thus, in the developing arthropod, as well as in the developing annelid worm,
the youngest segments are always the penultimate ones, the main zone for the
development of new metameric segments being immediately in front of the
anus (Figure 10-3). In a number of crustaceans, the development of the paired
limbs lags behind that of the segments which bear them, so that during the
process of addition there may be, from anterior to posterior, segments with
fully formed paired appendages, then well-formed segments without appendages, then poorly differentiated segments, and then the anal parts. This can be
seen most clearly in forms like anostracans and cephalocarids; these relatively
undifferentiated and probably primitive forms show slow addition at each molt
stage (Figure 10-3). Even in some other, more highly specialized, crustaceans,
there is often a stage late in the larval development when there is a region in
front of the anus which is relatively undifferentiated and shows incomplete
segmentation. For example, this occurs in the preadult copepodite stages of copepods (see pages 256–257).

FIGURE 10-3. Archetypic development of metameric segmentation in crustaceans. **A:** Generalized nauplius larva (basically of a cyclopoid copepod) with three overt segments bearing three pairs of appendages. Although these will become specialized head appendages—antennae and mandibles in the adult—they are unspecialized and used both for propulsion and for food-gathering in the larva. **B:** A later metanauplius (actually of the anostracan *Artemia*) with twelve segments and seven pairs of appendages, including two pairs of trunk limbs. **C:** A still later metanauplius (actually of the cephalocarid *Hutchinsoniella*) with twenty segments and eight pairs of appendages, including three pairs of trunk limbs. (This larva corresponds to the stage 8 or stage 9 "nauplius" in the studies of cephalocarid development by Howard L. Sanders and Meredith L. Jones.) Note the mode of addition of segments and paired appendages and compare with the annelid pattern of metameric development illustrated in Figure 10-1. [**C** modified from Howard L. Sanders, *Memoirs Conn. Acad. Arts Sciences,* **15:**1–80, 1963, with most of the setation and slighter spines omitted for clarity.]

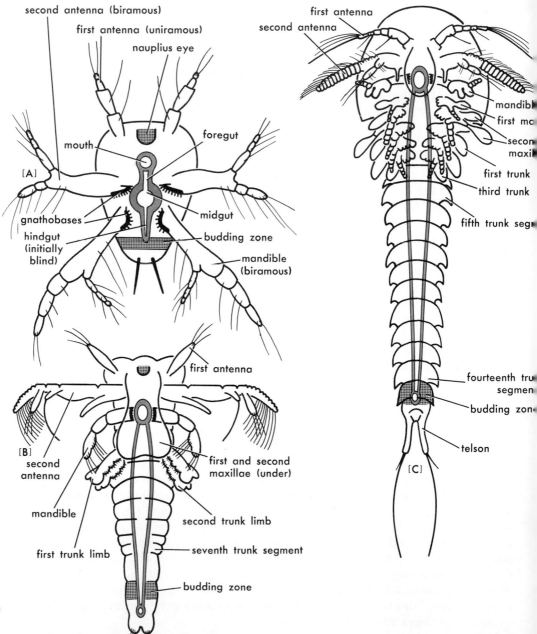

It is worth noting that it is extremely easy to rear the larval stages of bran-chiopod crustaceans like *Artemia* from their desiccation-resistant "eggs" in a beaker of salt water without elaborate apparatus. The processes of locomotion and of feeding in nauplius and metanauplius larvae are easily studied under low magnifications, and more detailed and sequential observations can yield a real understanding of many aspects of arthropodan physiology.

The most anterior of the head structures of the adult crustacean are clearly seen to be derived from the segments and appendages of the original nauplius. The adult mouth is somewhat more posterior, lying between the mandibles, and in all but the most primitive crustaceans the adult head is formed of the first five appendage-bearing segments of the larval series (that is, the head is equivalent to the larva at one appendage past the first metanauplius stage). The remainder of the body of the adult crustacean consists of the metameric seg-ments and their appendages which were first divided off during the larval de-velopment. Further features of crustacean larvae will be discussed for each major group later. It is of some significance that within the likely stem-group of the Arthropoda, the fossil class Trilobita, the sequence of larval stages is known in some detail and appears to have been closely similar to that just described for *Artemia*, although the first hatched larva, the protaspis, was of five seg-ments. Additional segments in trilobite development were added, once again, from a budding zone in front of the anus. To recapitulate, in spite of the major structural and functional differences between the annelids and the arthropods, the developmental processes by which segments are formed and added—the morphogenetic processes of metamerism—are closely similar in the two phyla.

Tagmatization

As mentioned earlier, in more highly evolved worms and arthropods, the adult pattern and numbers of somites or segmental structures can often be difficult to determine. Some part of this difficulty arises from the organization of tagmata (singular tagma). In relatively unspecialized annelid worms and in some elon-gate arthropods like centipedes, the pattern of serial succession of identical units is most clearly seen. In these forms the structures and functions of all seg-ments are almost uniform, all serial organs being equally developed in all seg-ments and many functions being carried out by every segment. In more highly organized forms we find groups of segments—the tagmata—structurally marked off from other groups and specialized to perform certain functions for the whole organism. The basis of the regional anatomical differentiation of a series of tagmata is the physiological division of labor among them. The results of tagmatization during development are most often discussed in relation to the arthropods, and it is not usually realized that the same process takes place in annelid worms to some extent. Indeed, the concept in its simplest form, of re-gional specialization and interdependence of regions or tagmata, is perhaps easiest to understand where it is least developed: in a series of marine bristle-worms (Annelida, class Polychaeta, see page 209).

A primitive polychaete like *Nereis* (Figure 10-4A; see also Figure 1-5A) can be considered as showing no tagmatization. Such a worm is a long series of like

[A]

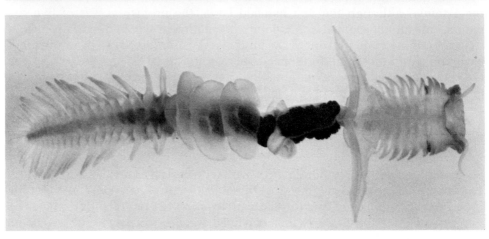

[B]

FIGURE 10-4. Contrasting segmental organization in two forms of polychaete worms. **A:** Anterior end of *Nereis diversicolor,* with a long series of uniform segments behind the head and no groups of segments (or tagmata) modified for special functions. **B:** A specimen of *Chaetopterus variopedatus* removed from its tube. There is a high degree of tagmatization involving structural and functional specialization of seven groups (tagmata) of segments. [Photos of living worms © Douglas P. Wilson.]

segments, uniform in structure and in their functional contribution to the whole *Nereis*-machine. A simple case where tagmata are marked off would be that of the burrowing lugworm, *Arenicola* (see page 202). In this, the prostomium, peristomium, and first body segment have a number of features in common and can be considered to form the first tagma of 3 segments. They are followed by the next 6 body segments which bear only setae externally, and which, internally have mixonephridia as excretory organs. The third tagma consists of 13 segments which bear externally both setae and gills and which internally lack mixonephridia. Then finally, there is a fourth tagma of 21 posterior segments which do not have setae or gills or mixonephridia. Thus, for *Arenicola,* we have a total of 43 segments arranged in four tagmata. At this simple level of specialization, the interdependence is obvious. The 13 segments of the third tagma carry out most of the respiratory exchange for the whole worm, while the 6 segments of the second tagma are the segments solely responsible for excretion and water control. More complex cases of tagmatization occur among polychaete worms. The tube-dwelling worm *Chaetopterus* (Figure 10-4B) lives in a secreted membranous tube and is a suspension-feeder utilizing mucous filtration. The anatomical and functional details need not concern us now but the 43 segments in *Chaetopterus* fall into seven clearly distinguishable tagmata.

In the great majority of arthropods the metameric segments are grouped into clearly defined tagmata. In fact, each major group of the arthropods has a nearly diagnostic pattern of tagmatization which is retained even in the most diversely adapted forms within each group. The insects show three tagmata—head, thorax, and abdomen—with a stereotyped pattern of segments involved. The insect head consists embryonically of six segments, the thorax consists of three segments which bear the only adult walking legs (hence the alternative name Hexapoda), and the abdomen has typically eleven segments without locomotory appendages. The crustaceans also show threefold tagmatization, though the segments involved do not correspond to those in the insects. The crustacean head has five obvious segments (but six embryonically) all of which bear appendages, including the two pairs of antennae anterior to the mandible and adult mouth. In the most typical crustacean pattern there follows an eight-segmented thorax, all segments bearing limbs, and a six-segmented abdomen, all segments of which can bear limbs. The class Arachnida (spiders, mites, and their allies) have the body divided into a prosoma and an abdomen or opistho soma. Fusion of segments has occurred in both tagmata. The prosoma seems solid and nonmetameric in most adult arachnids though it consists developmentally of three "head" segments, two bearing appendages, and the four segments which bear the eight legs. In most arachnids the opisthosoma is a single tagma of twelve completely fused segments, but in primitive arachnids, like scorpions, there are two tagmata: a preabdomen of seven segments and a postabdomen of five segments. The body of the fossil trilobites was divided into three tagmata: a cephalon, a thorax or trunk composed of a varying number of separately articulated segments, and a posterior pygidium. (The name, however, refers to a threefold lateral division by a pair of anterioposterior furrows which involves all segments.)

To sum up, only in unspecialized and presumably more primitive forms in both phyla, do we find adult animals with a strict metameric pattern of serial succession of identical units. In more highly evolved forms, fusion, loss, or modification of somites may obscure the regularity of pattern, as do specialized groupings of segments or tagmata. Tagmatization is in general more extensive in arthropods than in annelids. This is understandable because in the organization of soft-bodied forms like annelids, the adaptive value of metameric segmentation to locomotion is higher. With the acquisition of an arthropodan integument as an exoskeleton, segmented animals no longer required fluid-filled body cavities for the hydrostatic purpose of muscle antagonism. Thus in arthropods, tagmatization can be much more comprehensive, involving all organ systems and functions in the regional specialization of segments. However, it remains probable that the original evolution of metamerism, and indeed of the coelom or body-cavity, was governed by its value with regard to the muscle systems and hydrostatic skeleton used in locomotion. The only other major functional significance lies in the nature of segmental morphogenesis during development. Taking the hypothesis a little further, it might be concluded that metameric segmentation could only have arisen in a relatively elongate soft-bodied animal with a body-cavity.

11 ANNELID MECHANICS

THE segmented soft-bodied worms—of the sea, of fresh waters, and of terrestrial soils—make up the phylum Annelida. While not so vast in numbers as the allied phylum Arthropoda, annelid worms number nearly nine thousand known species, and the phylum is also a major one in terms of the ecological measure used in this book. The diagnostic features of the group have been set out earlier, and functionally the most significant is metamerism. This longitudinal division of the worm into a series of segments, each containing unit subdivisions of the several organ systems, is important both in development and in the mechanics of locomotion. The muscle systems, the nervous elements that control them, and the hydrostatic skeleton provided by the coelom through which they act, are all metamerically arranged.

As we have seen, it is possible for acoelomate, nonsegmented animals, such as flatworms and more notably rhynchocoels, to use body-wall musculature arranged in longitudinal and circular elements to carry out various forms of locomotion. However, the evolution of a coelomic cavity as a hydrostatic skeleton and the evolution of metameric segmentation, have allowed not only even greater locomotory efficiency, but also considerably greater capacity for local changes of shape along the elongate body. The significance of this is immense, not only in burrowing but also in all other types of locomotion. In other words,

183

although a ribbon-worm (acoelomate and nonsegmented) is capable of crawling, of burrowing, and of reacting to environmental dangers, a true annelid worm (coelomate and segmented) is capable of faster and more efficient crawling and burrowing and of more powerful and more sophisticated responses to environmental dangers or to predators. An analogy with man-made machines can be outlined. In both ribbon-worms such as *Lineus* (see Chapter 8) and in annelid worms such as the ragworm *Nereis* or the earthworms, the basic structures capable of contraction—the motors of the machine—are circular and longitudinal muscles. The more complex mechanical design of the annelid, however, allows a greater force to be exerted against the environment per unit volume of muscle than in the ribbon-worm—the more sophisticated design allows a markedly greater output from motors of the same total size.

Before we discuss the details of locomotion in annelid worms, it is best to relate the examples we will use to the major systematic subdivisions. As set out in Table 10-1, the phylum Annelida is usually divided into five classes, only three of which, Polychaeta, Oligochaeta, and Hirudinea, are of major importance (see Figure 1-5). The Polychaeta comprise the marine bristle-worms and use parapodia bearing numerous setae in a variety of locomotory activities: swimming, walking, and burrowing. Earthworms are included in the class Oligochaeta, characteristically without parapodial flaps and with fewer setae. The leeches form the class Hirudinea in which setae are entirely absent, and anterior and posterior attachment suckers are used as fixed points in locomotion. In order to begin with simpler muscle and skeletal mechanics, we will treat locomotion in these three classes in reverse order: we will consider first leeches, then earthworms, and lastly the more complex and eclectic movements of a variety of marine polychaetes.

Segmental Muscles and Nerves

In all three major classes of annelid worms, movements are based on an essentially similar layout of muscle layers and nerves around the coelomic spaces (Figure 11-1). The body wall always consists of a layer of circular muscles lying below the ectoderm and cuticle, and a layer of longitudinal muscles lying within them, as shown in the oversimplified cross section (Figure 11-1B, C). There are also muscle layers surrounding the endoderm of the gut, but these are not concerned in locomotion. In some polychaetes, there are oblique muscle strands in the position indicated by the dotted lines in the sketch, and in all leeches there are vertical columns of muscle—the dorsoventral muscles—whose contraction tends to flatten the leech's body. The body wall surrounds a true coelomic cavity (defined as lying between layers of mesoderm at all stages in its development—see Chapter 10) which in the adult worm is always fluid-filled. The polychaetes and oligochaetes have the coelom divided internally by septa. The Hirudinea have lost most of these divisions but have a meshwork of fluid-filled tissue occupying most of the coelom. However, the leech coelomic meshwork functions mechanically in the same way as an open, fluid-filled coelom. As noted earlier, leeches also differ in having two suckers for attachment to the substratum and in totally lacking setae. Both polychaetes and oligo-

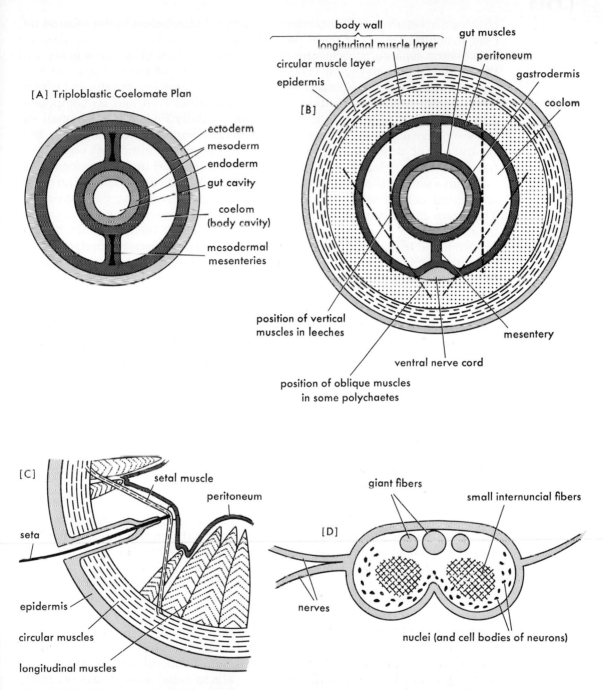

[A] Triploblastic Coelomate Plan

ectoderm
mesoderm
endoderm
gut cavity
coelom
(body cavity)
mesodermal
mesenteries

body wall
longitudinal muscle layer
circular muscle layer
epidermis
[B]
gut muscles
peritoneum
gastrodermis
coelom

position of vertical
muscles in leeches

ventral nerve cord

mesentery

position of oblique muscles
in some polychaetes

[C]
setal muscle
peritoneum

seta

epidermis
circular muscles

longitudinal muscles

giant fibers
small internuncial fibers
[D]

nerves

nuclei (and cell bodies of neurons)

FIGURE 11-1. The "efficient" pattern of worm organization. **A:** The basic triploblastic coelomate plan in cross section, as found in phyla Annelida, Arthropoda, Mollusca, and Chordata (see Figure 8-5). **B:** Cross section through an archetypic annelid, showing the layers of the body wall and the positions of additional musculature in certain polychaetes and in all leeches. **C:** Enlarged partial cross section through the body wall of an earthworm, showing one seta with its muscles. **D:** Cross section through the ventral nerve cord of an earthworm.

185

chaetes use setae to provide temporary points of attachment to the substratum. One structural feature of these setae should be noted here. It is that they are attached by separate muscle strands to the circular part of the muscle layer in the fashion shown in Figure 11-1. The gross anatomy of the nervous system is again similar in all three classes with segmentally arranged ganglia and a bilaterally symmetrical organization of the longitudinal cord and its component nerves. In the discussion of any worm, one usually finds a pair of small supraoesophageal ganglia linked by two circumoesophageal commissures to a ventral pair of ganglia which are part of a double cord running back on the ventral side of the coelomic cavity with a ganglion (or pair) in each segment. The transverse section shown in Figure 11-1 is through the ventral cord of an earthworm, but is reasonably representative of annelid central nervous structure. The nerve fibers making up the bulk of the cord usually have a maximum length of two or three ganglia, while the giant fibers, which are syncytial, can run from end to end of the worm's body. It can be demonstrated that the velocity with which a nerve impulse is passed is dependent on the diameter of the fiber, all other things being equal. In the earthworm, some of the ordinary fibers in the nerve cord with diameters of about 4 microns have impulse speeds of about 50 centimeters per second: the median giant fiber, about 50 microns across, conducts at 30 meters per second (that is, at about sixty times the velocity in the ordinary fibers).

The details of the smaller nerves in the annelid nerve cord are best known for forms like *Nereis*, from the detailed neuroanatomical studies of J. Eric Smith. As shown in Figure 11-2, the ganglia of the ventral cord each lie a little anterior to the segment which they control. Each gives rise to four pairs of nerves of which nerves II and IV are the largest. II is the main parapodial nerve, and IV is concerned with the muscles and receptors on the septum posterior to each segment. All four nerves are mixed, that is to say, they contain both motor fibers running out to activate muscles and sensory fibers transmitting inward to the central nerve cord from various kinds of receptors. Within the nerve cord, in addition to the giant fibers (providing the "through trunk" routes), there are six longitudinal tracts of fine fibers and it is from these that stimuli pass to the motor neurons in normal locomotion. Although it is much more difficult to map the fine fibers of the internuncial neurons from segment to segment, there is abundant functional evidence about the way in which they provide reflex arcs linking inputs in particular ganglia to the cell bodies of motor nerves in other ganglia. In the case of *Nereis*, it is clear that many of these fine fibers provide connections which span (in overlapping fashion) sets of six, eight, or twelve segments. Some of these reflex arcs are contralateral, that is they provide connection from proprioceptors or touch receptors on the left-hand side of one segment to the motor elements on the right-hand side of a segment eight units behind it. The total number of nerve cells in annelids is always relatively small, but they show rather complex interconnections repeated within each ganglion. Obviously, these can integrate a moderately complex sequence of movements, but it is important to realize that while these are centrally controlled, they can be functionally autonomous for any group of say twenty segments. It is not just that a 2-centimeter section of an earthworm or a marine ragworm accidentally chopped by a digging spade will still wriggle. It is that each

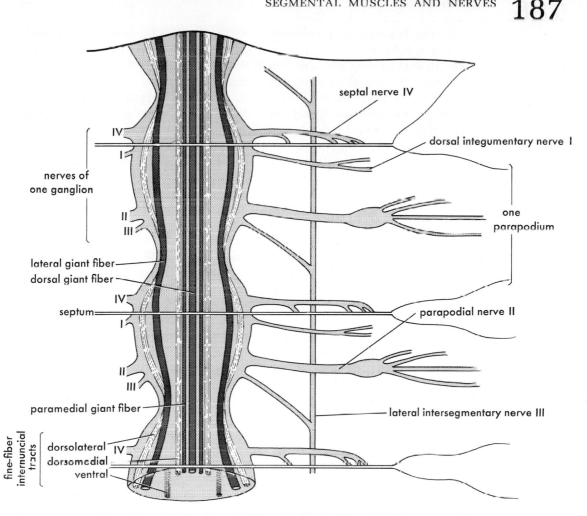

FIGURE 11-2. Neural organization in a primitive polychaete. Diagram of two segmental ganglia of the ventral nerve cord in an unspecialized polychaete like *Nereis*. The four pairs of segmental nerves are shown (II being the parapodial nerve, and IV the septal nerve). Within the ventral nerve cord run five giant fibers (one pair laterals, one pair paramedials, and a single dorsal), and six bundles of fine-fiber internuncial nerves. Two pairs of these internuncial tracts (dorsolateral and dorsomedial) are indicated by broken lines on the diagram. [Derived, in large part, from various publications by J. Eric Smith.]

of these sets of detached segments retains all the appropriate series of neural connections to allow it to carry out a coordinated series of muscle contractions passing along the small set of segments in the same fashion as the locomotory muscle contractions would have passed in the entire animal. Coordination in ordinary swimming, crawling, or burrowing locomotion in annelids is not dependent upon any "brain" in the worm, but rather upon the metamerically repeated segmental ganglia of the ventral nerve cord.

As with other aspects of development and of function in metamerically segmented animals, every segment is a unit comprising a complete set of muscles and nerves, but each unit is closely coordinated with those other segments adjacent to it, both anteriorly and posteriorly. In terms of nerves, the segments are linked in overlapping series of sets. It is the existence of these patterns of neural connections that allow rhythmic series of locomotory activities to be carried out in the worm and, as we have noted, to persist in isolated pieces of worm consisting of a dozen or so segments. Some early investigations of the neurophysiology of annelid worms were much concerned with the question as to whether the rhythmic patterns of contraction of particular muscles (for example, the contraction together of the left-hand longitudinal muscles in every tenth segment of a slowly wriggling *Nereis*) were generated *within* the nerve cord or resulted from a series of reflexes established from peripheral sensory stimulation. It must be emphasized that there are several kinds of sensory information continually being fed into the center in a moving annelid. First, there are sense-cells monitoring internal conditions in the worm, such as proprioceptors and stretch receptors, which detect the condition of the major muscle blocks and relay this information into the central nervous system. Secondly, there are arrays of tactile receptors concerned with external monitoring, in particular relaying information about those parts of the annelid which are in contact with surfaces or burrow walls. Perhaps the most important information in this latter, externally derived class, concerns the temporary points of attachment to the substratum through which the forces of locomotion are exerted. These are provided in polychaetes and oligochaetes by setae and in leeches by suckers. Although in these worms, we have not yet reached the sort of mechanical efficiency we will find in the arthropods (see Chapter 16, page 291), where the slender-jointed limb with its pointed tip allows the propulsive force to be applied through a single *point d'appui* on the substrate, the temporary points of attachment provided by setae or suckers represent a considerable advance in mechanical efficiency over the ciliary areas in surface contact for flatworms and the like. As we shall see, in a variety of forms of annelid locomotion, information coming in from the sensillae around the setae, or in the sucker-discs of leeches, is essential to coordinated locomotion. Thus the rhythms of locomotion do not originate *within* the nerve cord, although the internuncial nerve cells of the cord do serve to translate a stream of information coming in from the peripheral sense-organs (particularly the exteroreceptors associated with setae and suckers) into a rhythmic motor output which goes to the two basic sets of musculature working antagonistically to each other. As we shall see this alternate excitation by motor nerves will, in leeches and earthworms, be controlling the alternate contractions of longitudinal and of circular muscles within each metameric unit, while in polychaetes like *Nereis,* the alternate messages will be going to the right- and left-hand sides of the longitudinal muscle blocks.

Patterns of Antagonism

The mechanics of locomotion in annelids are most readily understood if two relatively simple facts are fully accepted. First, worm muscles—like the muscles of any other animals—can exert force only by contracting. (No mechanism which involves muscles in exerting force by "expanding" or "stretching" can possibly exist in animals.) A muscle is normally caused to contract by a stimulus from a nerve or nerves. As soon as the stimulus from the nervous system ceases, the muscle will become limp again, but the unstimulated muscle does not regain its original precontraction length until it is extended by some other force outside itself. In most animal mechanisms this force is provided by the contraction of another muscle or set of muscles, which are said to be antagonistic to the first. An antagonistic pair of muscles must of course act upon skeletal elements which transmit the contracting force of one to cause the stretching of the other (unstimulated) one to its precontraction length. The second simple point to remember is that in annelid worms there are no rigid skeletal elements, but instead a hydrostatic or hydraulic skeleton provided by the fluid in the coelomic space or spaces. The essential feature of the hydrostatic skeleton provided by the coelom is that it is of constant volume (watery body fluids can be regarded as totally incompressible by such forces as we are concerned with). In broad general terms, contraction of any muscle in the body wall of an annelid worm causes an increased pressure in the hydraulic skeleton which, in turn, causes the stretching of any muscles which are not receiving a nerve stimulus and are therefore flaccid. Actually, the coelomic fluid in any annelid is usually under a slight positive pressure: this amounts in a quiescent lugworm (*Arenicola*) to a positive pressure of 14 centimeters of water, in a totally anesthetized lugworm to about 3 centimeters of water, and in an actively burrowing lugworm to about 30 centimeters of water. Positive pressures of more than 90 centimeters of water can occur in some annelids and have been recorded in leeches which were standing rigidly upright on their posterior suckers. In this rather special case it is probable that most of the muscles of the body wall were simultaneously in a state of slight contraction. Normally, however, contraction of one set of muscles in an annelid means stretching of another.

The Leech as a Hydrostatic System

Although structurally more complex and undoubtedly more specialized than other annelid worms, leeches are mechanically simpler, and we shall consider locomotion in them first. Septa are missing and the coelom is functionally continuous: that is, mechanically, a leech is one fluid-filled bag. Thus, contraction of the circular muscles in a leech, which then elongates, causes a stretching of the longitudinal muscles. Contraction of the longitudinal muscles, which shorten and fatten the leech, cause stretching of the circular muscles. The leech can attach to the substratum by suckers at either end and the sequence of events which can be readily observed in a living leech is roughly summarized in Figure 11-3. When the hind sucker is attached to the ground a stimulus causes the circular muscles to contract; the volume of the leech remaining constant, the longitudinal muscles are thus stretched and the whole leech elongates. If the front sucker is then fixed to the ground (Figure 11-3B) the circular muscles are no longer stimulated. The posterior sucker is caused to detach and the longitudinal muscles stimulated to contract. This shortens the leech (Figure 11-3C) and thickens the body as the circular muscles are stretched. The hind sucker is

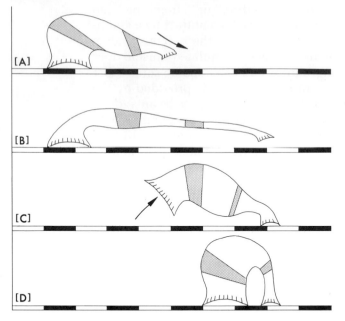

[A]

[B]

[C]

[D]

FIGURE 11-3. Locomotion in an artificially marked leech. When the leech is at its most elongate and thinnest, the circular muscles are fully contracted, and when the leech is shortest and thickest, the longitudinal muscles are fully contracted. The leech is a fluid-filled bag of unchanging volume, and the longitudinal and circular muscles act antagonistically around this hydrostatic skeleton. For further details, see text. [Adapted from J. Gray, H. W. Lissman, and R. J. Pumphrey, *J. Exp. Biol.,* **15:**408, 1938.]

drawn forward and, at the full contraction of the ventral longitudinal muscles, fixes down next to the front sucker. The front sucker is then freed and the circular muscles contract again, and so on, and so on. Apart from the obvious antagonistic arrangement of longitudinal and circular muscles, there are clearly also reflexes which involve tactile sensillae in the suckers sending signals to the central nervous system about their condition. For example, contraction of longitudinal muscles normally takes place only when the larger posterior sucker is detached from the substratum.

A few very simple experiments can be carried out by a student with a living leech. If a leech is suspended from a thread tied lightly round it, the posterior sucker will readily attach to any surface such as a glass coverslip presented to it. If then, another coverslip is brought into contact with the anterior sucker, it will attach, and the posterior sucker will then let go its coverslip. This replicates, of course, the series of steps which result from the alternate attachment and release of the suckers. But the sensory signals from the tactile receptors in the suckers do more than signal to each other; they originate reflex arcs which can either stimulate or inhibit contraction in the two main sets of muscles. When the posterior sucker of the slung leech is attached to a coverslip, contraction of the circular muscles, and thus elongation of the leech, will result. When the anterior sucker is presented with its coverslip, the longitudinal muscles will contract, but only after the posterior sucker has released its hold, and the slung leech will shorten and thicken. The reflexes involved in annelid locomotory controls and rhythms are rarely as simple as this, and most involve some integration of input from more than one peripheral sensory source. For example, we already noted that attachment of the larger posterior sucker in the leech inhibits contraction of the longitudinal muscles. In looping locomotion, the leech reaches out (Figure 11-3B) to its maximum length (circular muscles contracted) while the posterior sucker is attached. James Gray and his colleagues showed that a headless leech (lacking both anterior sucker and the most anterior ganglia of the nerve cord—the suboesophageal ganglia) could not show contraction of the circular muscles and thus could not elongate in this stage of the locomotory rhythm. However, the posterior sucker still can attach and, when it does so, the longitudinal muscles are relaxed (that is, are inhibited from contraction by a signal from the posterior sucker). If the posterior sucker lets go (or is detached) then the longitudinal muscles *can* contract. If a healthy, whole leech is moving from place to place, there is an obvious rhythm in the alternation of attachment of the two suckers and in the alternate contraction of the longitudinal and circular muscles. The frequency of the rhythm in natural conditions seems to depend upon environmental factors affecting the energy expended by the leech. The rhythm will be slower in a leech proceeding up a slope or through aquatic vegetation. Here again, the elegantly simple experiments of Gray and his colleagues first showed us that the attachment of small weights to the leech will always slow the rhythm by way of reflex signals from proprioceptors in the muscles themselves, which "inform" the central nervous system of the state of contraction. It is especially clear that the frequency of the rhythmic phases is inversely related to the time that it takes for complete isotonic contraction to be reached in the longitudinal muscles. The signals emitted by proprioceptors in these muscles are delayed by the additional weighting, and this imposes a pattern of slower "steps" on the entire locomotory sequence of the leech. Once again, it is worth emphasizing that the rhythmic motor signals from the central nervous cord do not originate there, but are merely transduced by the interconnections of the central neurons from the incoming information, derived from the tactile sensillae of the suckers and the proprioceptors in the muscle blocks.

Mechanics of Other Locomotion

In more typical polychaetes and oligochaetes, the essential difference in the mechanics of locomotion stems from the fact that the coelom is divided into segments by septa. Each segment forms essentially one hydrostatic skeletal unit within which the muscles can act antagonistically against each other. In the more elongate bodies of errant polychaetes and of earthworms, it is also possible for different parts of the body or different series of segments to be employed mechanically in different ways at the same time. Consider the case of an earthworm crawling on a surface: in it, the longitudinal muscles of each segment are antagonized by the circular muscles of that segment and a local grip on the substratum can be maintained by the action of the setae. When the longitudinal muscles of a segment are contracted, the setae of that segment are most completely protruded. When the circular muscles contract, the setae are withdrawn.

The sequence of events in locomotion is illustrated in Figure 11-4 in a diagrammatic earthworm consisting of only twenty segments. In those segments in which the longitudinal muscles are contracted the segment as a whole is at

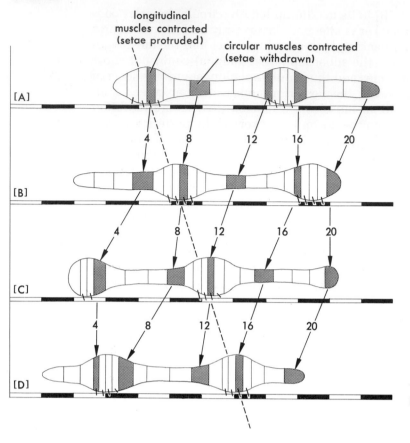

longitudinal
muscles contracted
(setae protruded)

circular muscles contracted
(setae withdrawn)

[A]

4   8   12   16   20

[B]

4   8   12   16   20

[C]

4   8   12   16   20

[D]

FIGURE 11-4. Locomotion in a twenty-segment earthworm. Every fourth segment is marked and linked by numbered lines; the broken line links a "region of contact." For further details, see text. [Adapted from J. Gray and H. W. Lissman, *J. Exp. Biol.*, **15**:506, 1938.]

its thickest and shortest with setae protruded and in contact with the ground. Since the volume of individual segments remains constant around each unit of coelomic-hydrostatic skeleton, those segments in which the circular muscles are contracted are at their most elongate and thinnest, have their setae withdrawn, and are not in contact with the ground. Neurally controlled waves of muscle contraction pass anteriorly along the worm. This means that segments are added to the posterior of each region of contact, and taken away anteriorly. As a result the region of contact (that is, the region wherein the longitudinal muscles are contracted and the setae protruded) moves back slowly in relation to the ground. However, the body of the earthworm moves forward more rapidly in relation to the region of contact so that there is a net forward movement. In spite of misleading statements in several textbooks, the bulges in an earthworm move *backward* during *forward* locomotion of the earthworm as a whole. In other words, an earthworm does *not* move like a caterpillar tractor. This can be readily verified by any student watching a live, healthy earthworm on which a few segments have been marked or stained to allow their identification as in Figure 11-4. The main coordination of this type of locomotion, which also serves for normal burrowing of earthworms through soft soils, is carried out by the relatively short fibers called internuncial neurons in the ventral nerve cord.

Among other elements involved, there are stretch receptors in the longitudinal muscles of each segment which are stimulated by the contraction of the circular muscles and these are connected through the short internuncial neurons to the motor neurons controlling the muscles of a more posterior segment.

The giant fibers in the earthworm nervous system are concerned with the control of the escape reactions which are used by the worm. Such reflex reactions are provoked by gross stimulation, either mechanically or by increased light intensity, at either the head or the tail end. There are two separate systems. Stimulation of the head causes an impulse in the median giant fiber which causes all the longitudinal muscles to contract but only the setae of the tail part to be protruded. This results in a fast withdrawal of the head of the worm. Similarly, "painful" stimulation of the tail causes an impulse in the two lateral giant fibers which again causes a contraction of all the longitudinal muscles together but, in this case, accompanied by a protrusion of the setae in the head region alone and, therefore, results in a withdrawal of the tail of the worm.

Similar giant nerve fibers are very important in the withdrawal reflexes of tube-dwelling polychaetes, many of which have exceedingly delicate structures used for feeding at the head end. On the basis of observations under unnatural laboratory conditions, some textbook authors have claimed that one form of locomotion in earthworms involves the attachment of the head using the mouth as a sucker followed by a contraction of longitudinal muscles pulling the body up to this point of attachment. This is not a normal method of locomotion in nature. The process is, in fact, part of the normal feeding behavior of earthworms whereby they pull dead leaves into their burrows and normally involves a simultaneous longitudinal contraction like that in the first escape reaction described above but occurring after the worm has taken hold of part of a leaf with its mouth. Incidentally, quiet observation of a moderately small songbird attempting to pull a large earthworm out of its burrow should convince any reader of the truth of the statements regarding the increased efficiency of muscles arranged in a segmental pattern around a hydrostatic skeleton.

At this point, it may be worth recapitulating some fundamental aspects of the use by most annelids of the coelomic fluids as a hydrostatic skeleton (and, in some forms, as a hydraulic one where fluid movements transmit the forces generated by the effector muscles to parts of the animal distant from their contraction). As already stressed, muscle fibers can only contract. It would be inappropriate to discuss here the recently elucidated biophysics of movement in myofibrils and the functional dependence upon a reversible "sliding" association of myosin and actin molecules, but the fact cannot be overemphasized that the *only* expression of kinetic energy which can occur results from *contraction* of the complex myofibrils. Thus the only forces which can be generated for use in locomotion are those resulting from contraction of certain elements or, to put it another way, tension-generating mechanisms. The "pulling forces" of tension-generating systems must usually, as Garth Chapman has pointed out, be converted to "pushing" ones if they are to be used in any simple locomotory mechanism. In the soft-bodied worms, as we have seen, this is done through the hydrostatic or hydraulic use of an incompressible fluid, where a tension generated in one set of muscles can be converted into a stretching of other tissues (including some other muscles) and also simultaneously exerts a thrust against the environment. Structurally, the earthworm was an example of an annelid worm with relatively complete septate divisions between a series of metameric coelomic cavities. In certain other worms, to be considered below, the septa between segments are structurally incomplete; these worms can be regarded, from a hydraulic standpoint, as nonseptate. A diagrammatic representation of some of the differences in the functioning of the fluid skeleton in nonseptate and in septate annelid worms is provided in Figure 11-5. In a nonseptate worm, the fluid of the hydraulic skeleton is free to move from one part of the body to another part. The forces exerted by contraction of circular muscles in one region and longitudinal muscles in another create temporary pressure changes which are dissipated and transmitted to other parts of the worm's body wall. In contrast, in a septate worm, the coelomic fluid is not free to move and thus the increase in fluid pressure caused for example by contraction of the circular muscles in one region creates a localized pressure change without dissipation. Similarly, a contraction of longitudinal muscles will fully stretch the local circular muscles to their precontraction length and create that region of thickened (and shortened) body segments where, as in the earthworm, the body wall can exert a force against the environment. As we have seen in the locomotion of earthworms, it is this segmental partitioning of the coelomic fluid into a series of hydrostatic skeletal units which allows different parts of the body or different series of segments to be employed mechanically as separate units with, at any specific point in time, different relationships to the substrate outside the worm. As important as the unit independence, which originates in this segmental separation of hydrostatic units, is the greater pressure which can be developed within them when there is no dissipation by fluid movement. As already noted, the internal hydrostatic pressures in active annelids like leeches and earthworms can amount to positive pressures of from 20 to 35 centimeters of water. Recently, M. K. Seymour has shown that the forces exerted radially by the body wall of a *Lumbricus* can amount to about two and a half times that which can be ascribed to the internal hydrostatic pressure. This implies that the rigidity developed in the body wall itself by contraction of the longitudinal muscles represents an important component to be added to the force measurable as pressure within that coelomic unit. Earlier estimates by Garth Chapman of the forces exerted against the environment by the divided mechanical systems of segmented worms can illustrate this point in a different way. Measurements of coelomic pressures show that, in the burrowing earthworm, a pressure of up to 8.5 grams per square centimeter can be applied to the posterior face of the first septum. This force is transmitted to the soil ahead of the earthworm by the prostomium which has the shape of a rounded and somewhat truncated cone (see Figure 1-5). The tip of this conical wedge will have a diameter of close to 1 millimeter and thus the pressure exerted through it can amount to more than 1 kilogram per square centimeter. In discussing the mechanical efficiency of the lever system of the limbs of arthropods (Chapter 16), we will use the term *point d'appui* for the single point on the substrate through which the propulsive force of each slender walking limb with its pointed tip is applied. The same term was used by writers of earlier centuries on military strategy to denote certain parts of fortification systems de-

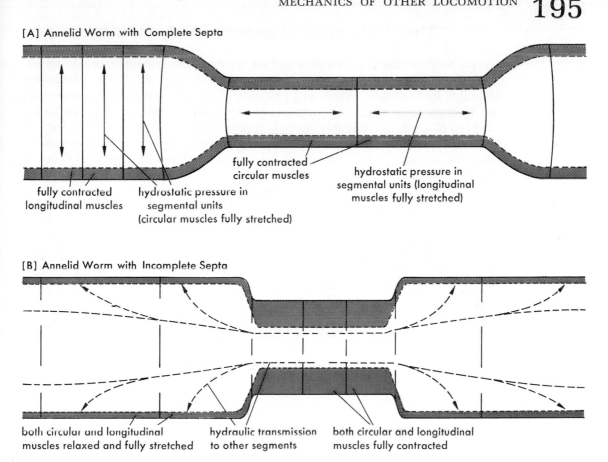

[A] Annelid Worm with Complete Septa

fully contracted
circular muscles

hydrostatic pressure in
segmental units (longitudinal
muscles fully stretched)

fully contracted
longitudinal muscles

hydrostatic pressure in
segmental units
(circular muscles fully stretched)

[B] Annelid Worm with Incomplete Septa

both circular and longitudinal
muscles relaxed and fully stretched

hydraulic transmission
to other segments

both circular and longitudinal
muscles fully contracted

FIGURE 11-5. Fluid skeleton in septate and nonseptate worms. **A:** Longitudinal diagram of the body wall of an annelid worm (like an earthworm; see Figure 11-4) which has complete septal partitions between segments. Each isolated coelomic space functions as a hydrostatic skeletal unit so that, when its longitudinal muscles are fully contracted, its circulars are fully stretched. Note that this compensation within each segment means that the body wall as a whole (combining circular and longitudinal elements) remains more or less of uniform thickness no matter whether the segment is at its thickest and shortest, or at its most elongate and thinnest. **B:** Longitudinal diagram of the body wall of an annelid worm (like the polychaete *Polyphysia;* see Figure 11-8) which does not have any complete structural septal divisions between segments. In such a nonseptate worm, the fluid of the hydraulic skeleton is free to move from one part of the body to another. The diagram shows one of the possible patterns of muscle mechanics which can result. Both circular and longitudinal elements can contract *together* in certain segments, and the forces generated by such joint contraction can be transmitted to many other segments to stretch both kinds of muscles (both circular and longitudinal in these other segments) *simultaneously.* Note that the body wall in the actively contracting segments can appear four times thicker than the body wall in the hydraulically stretched segments.

finable in terms of potential attack. The prostomial tip in the actively burrowing earthworm represents a true *point d'appui* in both senses, with that kilogram per square centimeter of pressure available to punch a way through the soil derived from the contraction of the segmentally arranged units of circular muscle behind it in the worm.

The Protean Mechanics of Polychaetes

Locomotory mechanics in leeches and earthworms has been discussed first because these animals are mechanically simpler. Although polychaete worms are undoubtedly less specialized and probably more primitive than earthworms and leeches, their modes of locomotion—even in untagmatized errant forms like *Nereis*—are considerably more complex. Indeed, the great versatility of movements in other more advanced or specialized polychaetes involves, in some cases, simplifications of errant mechanical action which parallel those of leech locomotion.

In general, the various patterns of locomotion shown by the more primitive polychaetes, such as *Nereis,* depend upon much more complex systems of reflex arcs that link the muscles of each segment. There are essentially three kinds of locomotion possible, but they are all basically dependent upon a different pattern of muscle antagonism from those we have just discussed. In this case, in any segment, the longitudinal muscles of one side can be the antagonists of the longitudinal muscles of the other. In other words, when the longitudinal muscles on one side of a segment are fully contracted, the longitudinal muscles of the opposite side are relaxed and stretched fully (Figure 11-6). In this way, the whole worm body can be thrown into a series of waves. The three modes of locomotion in *Nereis* are slow walking, rapid crawling, and swimming. The structural elements, in particular the effector muscles, on which nereid locomotion is based, are considerably more complex than those in leeches and earthworms. There we had essentially two concentric cylinders of circular muscles and of longitudinal muscles forming continuous layers along the length of the body, even although they were arranged in segmental units. In nereid worms we have additional smaller muscles *within* the parapodia, while the circular muscles only form complete bands on either side of the septa between the parapodia, and the longitudinal muscles are arranged, not in a continuous tubular layer, but as four blocks in each segment (see Figure 12-1B). The longitudinal muscle blocks are in two pairs arranged in bilateral symmetry, with the dorsal pair more massive and more powerful. The intrinsic parapodial muscles can be used along with the segmental oblique muscles inserting in the parapodium to protrude the setae and generally stiffen the parapodium for its effective stroke, either as a lever in walking locomotion or as a paddle in swimming. Despite this complexity, the main power for faster locomotion in all modes in nereid worms comes from alternate contractions of the right and left blocks of longitudinal muscle in each segment, particularly the dorsal blocks.

Slow walking involves a metachronal rhythm of action in the parapodia as a series of levers with the opposite sides of each segment exactly out of phase. Every fifth or sixth parapodium on one side of the worm is at exactly the same stage in the cycle of forward recovery stroke and effective backstroke, and controlling the rhythm there is certainly a relay system of reflex arcs connected in sixes running through the central nerve cord. During each effective backstroke the whole parapodium is protracted and turned down toward the ventral side. The setae and the two acicula (see Chapter 12; and Figures 11-1 and 12-1) are protruded at this time, thus creating a series of temporary *points d'appui*. At the same time, the parapodium on the other side of the segment is retracted,

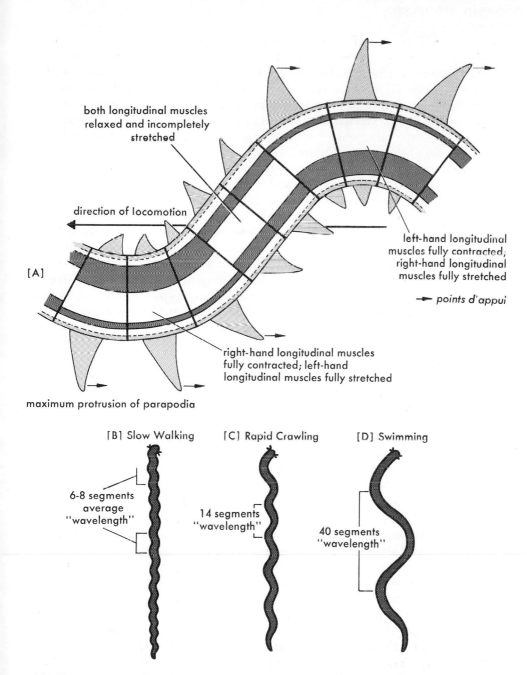

FIGURE 11-6. Locomotion in a primitive polychaete. **A:** Diagram of the dorsal aspect of the longitudinal muscles and parapodia in part of a rapidly crawling specimen of *Nereis*. Although segmental divisions are complete, here the left-hand longitudinal muscles of each segment are the antagonists of the right-hand longitudinal muscles. In segments where the longitudinal muscles of one side are fully contracted, the longitudinal muscles of the other side are completely stretched and the parapodium is at maximum protrusion (providing a temporary *point d'appui*). Waves of contraction pass anteriorly, in the *same* direction as the worm's locomotion. In rapid crawling (**A** and **C**), the interval between segments in exactly the same phase of contraction is usually from eleven to fifteen segments in length. In slow walking (**B**) the interval is between six and eight segments long, and in swimming (**D**) the "wavelength" increases to about forty segments.

197

raised toward the dorsal side, and then carried forward in a recovery stroke. The major longitudinal muscle blocks do not contribute much power in this kind of locomotion, although slight contraction of the blocks on the side of the retracted parapodium in its recovery stroke aids hydrostatically in the stiffening and extension of the opposite parapodium for its power stroke in levering the worm along. Metachronal waves of activity appear to pass forward and, with a phase interval of only five or six segments, appear numerous in relation to the length of the worm (Figure 11-6B).

Rapid crawling is much more dependent on contralateral waves of contraction of the longitudinal muscles, waves which again pass forward. As James Gray and his colleagues first showed, the two sides of any marked segment (Figure 11-6A) move forward alternately, with the longitudinal muscles most fully contracted on the side opposite the crest of each crawling wave. After an intervening stage, as it were "between waves," with left and right longitudinal muscle blocks of the marked segment both flaccid and incompletely stretched, the opposite side will contract, stretching the longitudinal muscles of the other side and causing the maximum protrusion of the parapodium on the crest of the wave. A series of *points d'appui* are established at the wave crests by the protrusion and downward pressure on those parapodia, and the power for locomotion is almost entirely derived from the contracting longitudinal muscles of the opposite sides pulling against these *points d'appui*. In rapid crawling, the interval between segments in exactly the same stage of contraction is usually from eleven to fifteen segments—phases move metachronally—and once again an appropriate relay system in the internuncial neurons of the central nervous system must be responsible for the moving series of reflex arcs involved.

Swimming is basically similar to rapid crawling, but in it the waves of muscular contraction are fewer and their amplitude and frequency greater (Figure 11-6D). The waves still pass forward, and swimming in *Nereis* and its allies is clearly mechanically inefficient since forward locomotion depends upon the power strokes of the parapodia. To some extent, however, the body movements create a flow of water in the opposite direction to the backward flow created by the paddling of the parapodia. Any student with access to a living specimen of *Nereis* can verify all these statements regarding the three modes of locomotion and in particular, can readily appreciate the swimming inefficiency. Observed against a measured background (for example, a piece of graph paper under a dish), the progress of the ragworm in a forward direction seems extraordinarily slow, compared with the frequency and speed with which the propulsive waves pass along the body. In a series of mixed metaphors, the apparently galloping level of activity "full of sound and fury" achieves only a slow forward gain as though being achieved against the retrograde motion of an invisible treadmill. Nereid worms can also move water in another fashion. This water-propulsion occurs when they are living in tubular burrows or placed in appropriate-sized tubes by investigators. It involves an antagonistic contraction of the dorsal blocks of longitudinal muscle causing stretching of the ventral blocks in each segment and vice versa. This results in dorsoventral undulations which propel water through the tube largely for respiratory, but also occasionally for feeding purposes. Certain other polychaetes are much better adapted for this propulsion of water past a sedentary worm in a tube.

Earlier, we noted the lack of tagmatization in forms like *Nereis*, where there is a long series of like segments, uniform in structure and in their functional contributions to the worm-machine. As we have just noted, the locomotory organization of *Nereis* is similarly unspecialized. Thus it is not surprising to find that some polychaete worms are markedly more efficient at slow walking, and others at rapid swimming, than is *Nereis*. Among the most efficient walking forms are the polynoids or scale-worms, which are not uncommon on hard substrata in the littoral (Figure 11-7). In polynoids, the longitudinal muscle blocks of the trunk are reduced and have no locomotory function; rather steps are taken by the stout, conical parapodia as a result of differential action of their intrinsic muscles and of the oblique muscles in the segments. Regular protrusion of the acicula is important in the power strokes. In some ways walking movements of scale-worms resemble those of millipedes, which we will describe in Chapter 16. It is important to note that such scale-worms and "sea-mice" cannot swim.

[A]   [B]

FIGURE 11-7. Live specimens of two more massive polychaete worms. **A:** *Aphrodite aculeata*, the so-called "sea-mouse." **B:** *Halosydna gelatinosa*, a typical "scale-worm." Both are intertidal, *Aphrodite* from sandy shores and *Halosydna* from near low water mark on boulders. Both are relatively massive for annelid worms and have bodies with few, stout segments; *Aphrodite* reaches nearly 10 centimeters in length, *Halosydna* about 5 centimeters. [Photos © Douglas P. Wilson.]

Truly efficient swimming locomotion is shown by forms like *Nephthys*, whose parapodia are modified as paddles (see Figure 12-1C) and whose body segments are relatively short, making the parapodia more closely set than in true nereids. The more powerful swimming of *Nephthys* results largely from the fact that the effective parapodial stroke (or paddling propulsion of water to the rear) takes place more rapidly, on the crest of each lateral undulatory wave, but occupies a much smaller proportion of the total locomotory cycle of each segment than in *Nereis*. A most significant observation, which can be made by any student, is that in *Nephthys*, as the speed of swimming is increased, the number of undulations into which the body is thrown is also increased so that a larger proportion of the parapodia are executing effective paddling strokes at any one time. Obviously, the shift in patterns of neural coupling with increased speed is precisely the opposite of that which occurs in *Nereis*. Once again we can note that *Nephthys* is not very good at slow walking over a surface; the flap-like parapodia do not provide levers which can be rigidified by parapodial muscles as in *Nereis*. When *Nephthys* burrows in sand, it first drives its relatively pointed head into the sand by swimming and then uses the active protrusion of the proboscis, acting against a set of *points d'appui* provided by the protruded setae of the anterior segments, to create a hole into which the worm moves by further "swimming" movements. We noted in discussing the locomotion of earthworms that the waves of contraction were metachronal or retrograde; that is, they moved anteriorly along the body in the direction opposite to that of locomotion. We also noted that because of the means of providing *points d'appui* by a body-thickening contraction of longitudinal muscles, this method of locomotion was suitable for movement through a burrow as well as over a surface. A contrasting mechanism for burrowing locomotion in soft marine muds by the polychaete *Polyphysia* has recently been elucidated by my former colleague Hugh Y. Elder. In its anatomy, *Polyphysia* is largely nonseptate; therefore, its coelomic fluids form a low-pressure hydraulic system rather than a relatively high-pressure hydrostatic one. Viewed externally, *Polyphysia* excavates its burrow and proceeds through sublittoral muds by a direct series of peristaltic constrictions rather than a retrograde series (Figure 11-8). In this case, the *points d'appui* are provided by longish sections (only one or two in the whole worm body) of segments which are dilated by a relative relaxation of both circular and longitudinal muscles. Along these regions, the setae are protruded as incomplete rings of backwardly directed fans. Against these anchoring rings, direct peristaltic constrictions (in which both longitudinal and circular muscles contract, thickening the body wall while narrowing the segment) create an advancing movement. As Elder has shown, the nonseptate condition (and the consequent low-pressure system) has here been turned to advantage since hydraulic deployment of the coelomic fluid converts the periodic progression of small groups of trunk segments into a continuous advance by the head through the soft mud. There are obviously considerable adaptive advantages for an animal living in soft substrata which result from having some three-quarters of the body surface in contact with the burrow wall at any one time. This allows the application of a relatively low pressure over a wide area (in this case more of a "zone of contact" than a series of *points d'appui*).

Earlier studies by such workers as G. P. Wells, Garth Chapman, and E. R.

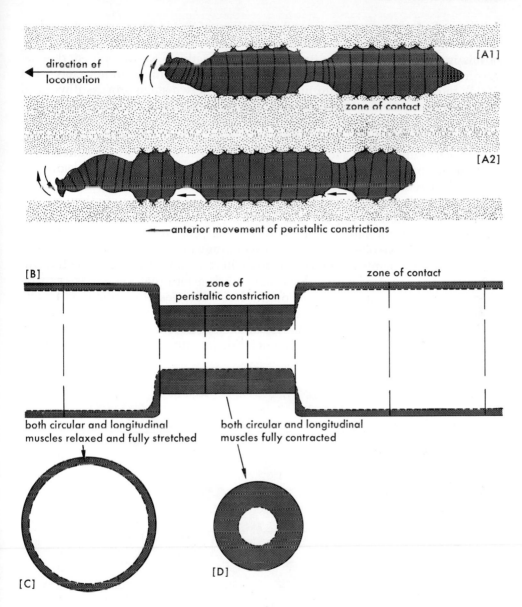

direction of
locomotion

[A1]

zone of contact

[A2]

anterior movement of peristaltic constrictions

[B]

zone of
peristaltic constriction

zone of contact

both circular and longitudinal
muscles relaxed and fully stretched

both circular and longitudinal
muscles fully contracted

[C]

[D]

FIGURE 11-8. Locomotion in a nonseptate worm. **A:** Two stages in forward locomotion in a nonseptate burrowing polychaete, *Polyphysia,* viewed dorsally. Apart from side-to-side movements of the horned prostomium, locomotion is principally carried out by the *forward* peristaltic movement of zones of contraction of *both* longitudinal and circular elements together. Much of the body surface is stretched and provides this low-pressure system with long zones of anchoring contact rather than successive *points d'appui.* **B–D:** Diagrammatic sections to show the changing dimensions of the body wall. Note that, despite the reduced overall diameter, the body wall in the contracted segments is four times thicker. Compare with Figures 11-4 and 11-5. [Adapted in part from papers by, and personal communications from, Dr. Hugh Y. Elder.]

Trueman had given us some understanding of locomotory and of feeding movements in such forms as the lugworm, *Arenicola,* where the eversible proboscis is important. As noted in Chapter 10, *Arenicola* shows a moderate degree of tagmatization, and it is essentially a worm which cannot crawl, but which burrows and moves through the sand by a two-stage process illustrated in Figure 11-9. After an initial entry for the anterior part of the body is achieved by repeated protrusions of the proboscis to loosen the sand, the anterior segments are distended by contraction of the longitudinal muscles in them and in nearby segments to form a terminal anchor. The rest of the worm is pulled up to this anchor by further longitudinal contractions in more posterior segments. Then, in a second phase of burrowing, the anterior segments are formed into projecting annular flanges, which form the so-called penetration anchor against which *points d'appui* the contraction of anterior circular muscles pushes the head forward and again everts the proboscis. During this second phase, the attentive observer of a burrowing lugworm will notice that some of the posterior segments are temporarily displaced backward out of the burrow. After the head and proboscis have advanced, a terminal anchor is then reformed, and contraction of the longitudinal muscles throughout the body shortens the worm and pulls some more segments into the burrow. Internally *Arenicola* is nonseptate in its anterior tagmata, but this open body cavity is a relatively high-pressure system, and coelomic pressures amounting to 90 centi-

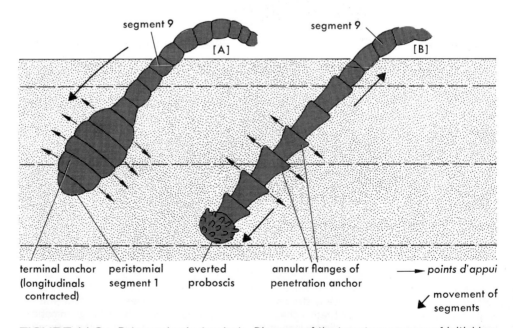

FIGURE 11-9.   Reburrowing in *Arenicola.* Diagrams of the two-stage process of initial burrowing in the lugworm *Arenicola.* In **A,** a terminal anchor is formed by contraction of all longitudinal elements and the rest of the worm is pulled down toward it. This phase alternates with a phase (**B**) in which projecting annular flanges are formed in the anterior segments and the head (peristomium) is pushed forward and the proboscis everted. Note that some segments (including the ninth in the diagrams) are temporarily displaced backward out of the burrow. [Modified from Trueman, 1966.]

meters of water can be generated (even though this is a hydraulic system with considerable movement of fluid into and out of the head region). It is clear that lugworms are adapted for burrowing and, with this organization of their muscles and coelom, can have no capacity for swimming and almost none for crawling over the surface of the substratum. (The earthworm is obviously *less* specialized in this respect since it has, because of the distribution of its setae and setal muscles, retained some capacity for local application of thrust to a surface.)

To obtain nutrition, lugworms are sand-swallowers. When *Arenicola* is living in its U-shaped burrow, it uses rhythmic movements of the proboscis, both to irrigate its burrow from the hind end forward and to swallow loosened sand and particulate organic matter sifting down from the surface of the sand, in a steady piston-like pumping (Figure 11-10). This is continued in long bursts of a regular rhythm alternating with pauses to move backward in the burrow to defecate, which are precisely timed. This rhythmic behavior has been studied in some detail by G. P. Wells, and there is good evidence that it is controlled by an internal pacemaker within the central nerve cord. Such control is extremely significant, since it is the only case in all the locomotory and other mechanical activities of annelids we have so far discussed which is not reflex behavior (no matter how complex) dependent upon a system of internuncial neurons within the central nervous system to link input signals from sense-organs on the outside or inside of the worm with the motor neurons (or efferent pathways) which stimulate the various muscular elements to contraction. We are commenting here merely on the fact that the rhythmic activity of burrow irrigation and phasic defecation in *Arenicola* is not a matter of reflexes but, at least in the short run, seems to be purely endogenous or evoked by a spontaneous innate rhythm *within* the cells of the central nerve cord. We need not be concerned here with the even broader question of whether this rhythm is endogenous in the longer time sense of development. The whole question of "animal clocks," as well as the relative importance of endogenous and exogenous (including subtle geophysical) factors in the control of such rhythmic activities in animals, remains a matter of some controversy. In the case of *Arenicola*, the pacemaker appears to be located in the ventral nerve cord and will continue to function even if the relatively small anterior "brain" or supraoesophageal ganglion complex has been removed experimentally. It is also important to note that the usual kinds of exogenous signals which control rates of water currents used by other worms for respiratory or nutritive purposes (signals such as low oxygen or lack of food) do not operate to alter the rhythms in the pumping of *Arenicola*.

As already noted, an even more highly tagmatized worm, *Chaetopterus*, lives in a somewhat more permanent U-shaped burrow and uses a temporary mucous bag in a specialized filter-feeding mechanism. Segments 14, 15, and 16 bear the notopodial "fans" which create a water current through the burrow and through the filter-bag. The mechanics of water pumping by *Chaetopterus* have recently been investigated by my former colleague Stephen C. Brown. As shown in Figure 11-11, the musculature consists of thin sheets of fibers running in radial, transverse, axial, and circular directions, almost all located in a relatively thin, membranous wall surrounding an open parapodial cavity which is the hydrostatic unit. Certain remotor bundles of muscle fibers origi-

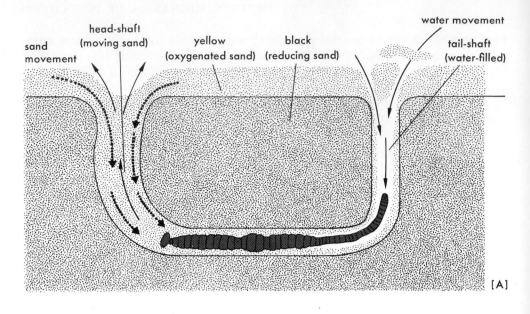

sand movement

head-shaft (moving sand)

yellow (oxygenated sand)

black (reducing sand)

water movement

tail-shaft (water-filled)

[A]

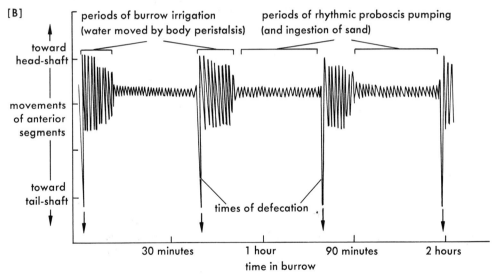

[B]

periods of burrow irrigation (water moved by body peristalsis)

periods of rhythmic proboscis pumping (and ingestion of sand)

toward head-shaft

movements of anterior segments

toward tail-shaft

times of defecation

30 minutes   1 hour   90 minutes   2 hours

time in burrow

FIGURE 11-10. Feeding and burrow irrigation in *Arenicola*. **A:** Diagram of *Arenicola* in its burrow. The sand in the head-shaft falls continually inward and downward, as a result of the pumping activities of the proboscis and the continuous ingestion of the fool-laden surface sand. Water flow through the U-tube is in the opposite direction: in by tail-shaft from above the faecal sand coils, from tail to head of the worm, and out by the head-shaft (against the slow flow of surface sand). **B:** Simplified sequence of the feeding and irrigation cycles in *Arenicola*. Each complete cycle shown takes about 42 minutes from defecation (when the worm is mostly in the tail-shaft) to next defecation. After each defecation there is a period of rapid peristaltic movements of the body irrigating the U-tube from tail-shaft to head-shaft, followed by a longer period of proboscis-pumping and feeding. With individual live lugworms, there may be considerable variation in the proportions of time allocated to feeding and to irrigation, and much more overlap between them than is shown in this simplified trace. However, each lugworm maintains the time interval of its complete cycle (for example, 37 minutes for specimen A, 45 minutes for B, and so on) rather precisely over periods of many hours. [Modified from G. P. Wells, and others.]

204

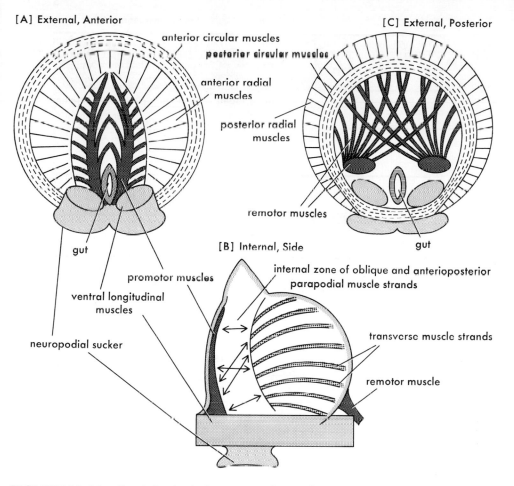

[A] External, Anterior

anterior circular muscles

posterior circular muscles

anterior radial
muscles

posterior radial
muscles

[C] External, Posterior

remotor muscles

gut

[B] Internal, Side

gut

promotor muscles

ventral longitudinal
muscles

neuropodial sucker

internal zone of oblique and anterioposterior
parapodial muscle strands

transverse muscle strands

remotor muscle

FIGURE 11-11. Specialized polychaete musculature. One of the three segments of *Chaetopterus* (14, 15, 16), which are modified for pumping water through the tube in diagrammatic views externally from the anterior **(A)** and posterior **(C)** faces and laterally **(B)** as a transparent object to show some of the internal muscle strands. The segment and its parapodia form a thin-walled balloon-like "piston," power and recovery strokes of which (see Figure 11-12) are accomplished by a sequential program of muscle contractions. The main muscles are superficial: anterior and posterior circular and radial elements, along with promotor and remotor "fans" of fibers (the latter originating in the neuropodial sucker of the next segment to the posterior). [Simplified from Stephen C. Brown, *Biol. Bull.*, **149**:136–150, 1975.]

nate in the ventral neuropodial sucker of the segment next to the posterior, and it is these neuropodial suckers which provide fixed points on the secreted tube lining. The pump segments form a series of three reciprocating pumps, at least one of which is always in its power phase (Figure 11-12). Such pumping is obviously much more efficient than tube-irrigation currents created by undulatory body movements or by peristaltic contractions of the body as in other tubicolous annelids. As Brown has pointed out, the pumping system in *Chaetopterus* is closely analogous to a well-designed reciprocating man-made pump in the compound single-acting category. The three single-acting units of

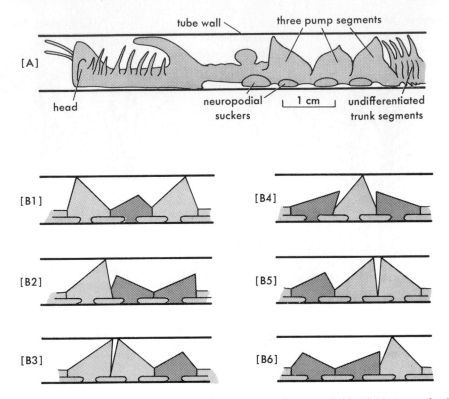

FIGURE 11-12.   Water pumping by *Chaetopterus*. **A:** Posture of this highly tagmatized worm in its tube during active pumping. Compare with Figure 10-4. **B1–B6:** Successive strokes of the three-unit pump. Light tone indicates "piston" segments in the power stroke (that is, driving water to the right), and darker tone indicates segments in the recovery phase. [Reproduced, with minor modifications, from Stephen C. Brown, *Biol. Bull.*, **153**:121–132, 1977.]

Chaetopterus arranged in series achieve a continuous flow with dampened pulsations, just as do commercial pump units arranged in compound with overlapping power strokes. It is of interest that in commercial pump practice, *three* piston and valve units are regarded as the minimum number to yield a continuous flow in a compound single-acting pump. As we will note in Chapter 12, the need for a continuous water flow in *Chaetopterus* may be related to the need to make optimal use of the fine-meshed, thimble-shaped bag-filter in feeding. Brown has computed that most of the "work" performed by the triple pump in *Chaetopterus* is directed toward accelerating the fluid through the tube, and relatively little is used to overcome "frictional" resistance to flow. There is, of course, no "lift" component in the work done since the inhalant and exhalant ends of the U-tube are at the same height and pressure. The system is clearly of high hydrodynamic efficiency.

   In the next chapter, we will consider other aspects of the diversity of form and function in polychaete worms. It is worth recapitulating the basic theme, which may have become somewhat obscured in this survey of the versatility of mechanical systems—hydrostatic and hydraulic—in polychaetes. The point is that the mechanical efficiency of the muscles in a worm-shaped body is undoubtedly increased by the organization of both the fluid skeleton and the muscles of the body wall around it into a series of metameric segments. In addition to the developmental significance of metamerism discussed in Chapter 10, the functional significance of segmentation is that it allows a relatively simply organized system of muscles to carry out complex and sophisticated movements. All things considered, this is a major reason why the Annelida form a numerous and successful phylum, while the Rhynchocoela remain a relatively unsuccessful one of minor ecological significance.

12 ANNELID DIVERSITY

THE phylum Annelida is subdivided into three major and two very minor classes. The most numerous and most diverse group encompasses the bristle-worms, the almost entirely marine class Polychaeta, comprising more than 5300 species. Polychaetes are mostly dioecious, and spawn their eggs and sperms into the sea so that fertilization is external and early development involves ciliated planktonic larvae, like those described for *Polygordius*. Among the bristle-worms, there are several unspecialized families which show little or no tagmatization, including some which are probably closest to the most primitive annelid pattern. These are families of actively crawling worms (such as the Phyllodocidae) with a long series of similar segments, each bearing a pair of parapodia with numerous setae which are used in swimming, walking, and burrowing. The next largest group is the class Oligochaeta, numbering more than three thousand species of mainly land and freshwater worms. Earthworms and freshwater oligochaetes have fewer and smaller setae on the segments, never have flap-like parapodia, and usually have head and sense-organs reduced if compared to the polychaete pattern. A third group is the leeches, class Hirudinea, with a more specialized and much more uniform pattern of body involving attachment suckers and a modified mouth and gut. There are about three hundred known species of leeches in the sea, in fresh waters, and on land—the last largely in the tropics. In leeches, setae are entirely absent: fixed points for locomotion are provided by the suckers. In both the Oligochaeta and Hirudinea, the majority of species are hermaphroditic and some form of copulation results in cross-fertilization, usually internally. As a result, the eggs produced in these two classes can be relatively large, and the developing young upon hatching are not usually ciliated larvae but rather miniature adults. There are two other minor groups of annelid worms, each numbering about fifty species of little ecological importance: the Archiannelida and the Myzostomaria. The former are a heterogeneous collection, the genera of which may be primitive survivals or may show secondary degeneracy. The Myzostomaria are relatively specialized parasites, almost without internal segmentation as adults, but undoubtedly derived from some polychaete stock.

The class Polychaeta, besides the least specialized forms, encompasses the greatest diversity of form and function among the annelids. By contrast, within the earthworms, or within the leeches, both anatomy and physiology are somewhat stereotyped.

Form and Feeding in Polychaetes

Phyletic discussion of functional morphology in polychaetes is not possible. The structural and functional adaptations associated with certain specific habits and habitats in polychaete worms are often known to be polyphyletic. Each ecological and physiological pattern appears to have been independently evolved in several stocks of bristle-worms. This is revealed by the state of systematics in the group. There are about fifty distinct families of polychaetes, accepted as such by most systematists, but there is no acceptable grouping of these into orders or agreement on any other hierarchy of interrelationships among them. Apart from differing degrees of tagmatization, much of the functional morphology concerns the sensory and feeding structures of the prostomium and peristomium and variation in the parts and proportions of the parapodia. Perhaps the adaptive value of these structural differences is most easily understood in relation to the ecology of the different worms. However, it should be remembered that ecological similarity need not mean close relationship. For example, microphagous polychaetes which construct fixed tubes have apparently been evolved independently in several stocks and thus reflect convergence rather than close relationship.

As already stressed, archetypic polychaetes can be considered as medium-sized worms with many undifferentiated segments, almost certainly detritus-feeding, and possibly living in muddy, offshore bottom deposits. Both *Polygordius* and the least differentiated phyllodocids, although the latter have a more muscular pharynx which can allow macrophagy, can be regarded as independent survivors of such stocks. (It should be noted that the remaining archiannelid genera, other than *Polygordius*, are certainly unrelated and obviously not primitive. A better understanding of polychaete phylogeny would probably merge the Archiannelida into a few families of the Polychaeta.)

Several groups of larger worms—carnivores and macrophagous scavengers—have evolved from these, the principal modification being the de-

velopment of eversible buccal and proboscis structures along with small cuticular teeth, or stout jaws, or both. Such free-living worms used to be designated
as the "Errantia" but involve many distinct and clearly unrelated families including the Syllidae, Eunicidae, Nephthyidae, and Nereidae. Ragworms or
clamworms like *Nereis* show the characteristic features and are readily available. There is a high degree of cephalization (see Figures 10-4A and 12-1A), the
prostomium bearing four eyes, two tentacles, and two palps, while the peristomium has four pairs of tentacles in addition to the oral apparatus. The pharynx,
shown everted in Figure 12-1A, bears stout jaws and many paragnaths or small
cuticular teeth. In *Nereis* and similar forms, there are about eighty segments
behind the head, all of which are uniform except for the slightly modified first
and last. The structure of a regular segment is shown in Figure 12-1B, which, in
addition to the internal arrangements of muscles, circular, longitudinal, and
oblique, shows the structure of a relatively undifferentiated pair of parapodia.
Among the many setae of each parapodium are the stouter acicula which provide a skeletal stiffening, and there are sensory outgrowths called cirri. Most
undifferentiated polychaete parapodia are biramous, consisting of a dorsal no-
topodium and ventral neuropodium (see Figure 12-1B–F), and variations in
their proportions allow these paired appendages to be used as swimming paddles, or for creating water currents in tubes, or for crawling or burrowing. In
many diverse groups of polychaetes there are also additional lobes, conspicuously blood-filled, which act as gills.

In contrast, the dorsal surface of the segments themselves forms the vascularized respiratory surface in scale-worms and their allies (see Figure 11-7B). Typical scale-worms, such as *Polynöe,* have an overlapping "armor" of dorsal
scale-plates or "elytra" covering the surface and, in the sea-mouse *Aphrodite*
(see Figure 11-7A), this is augmented by a felted mat of setal threads from the
notopodia. Respiratory currents pass *below* these dorsal protective layers. As
noted in Chapter 11, scale-worms and sea-mice are efficiently stepping walkers
and crawlers but nonswimmers.

The evolution of a different pattern of eversible proboscis characterizes such
forms as lugworms like *Arenicola,* where the "proboscis" so formed is involved both in digging through the mud and in ingesting suitable organically
rich deposits. When lugworms are living in their characterstic U-shaped tubes
and feeding normally, rhythmic movements of the everted proboscis are used,
not only to shovel into the mouth the richer surface sand which flows down the
funnel at the head of the burrow, but also to pump water with suspended material into the gut. The extruded proboscis is also used extensively in the construction of each new burrow, with the rhythmic cycle of extension, expansion,
and longitudinal contraction effectively pulling the worm through the substrate. *Ophelia* and its allies form another family of similar stout-bodied worms
—though some are spindle-shaped—which live in sand and feed by using a
similar eversible proboscis.

A vast number of different polychaete groups have developed ciliated tentacles on the head which are used in various methods of particle collection and
feeding. The Ampharetidae and Terebellidae have ciliated tentacles on the prostomium which are used to gather materials to build permanent tubes or burrows as well as to collect food. If a terebellid is carefully observed, three

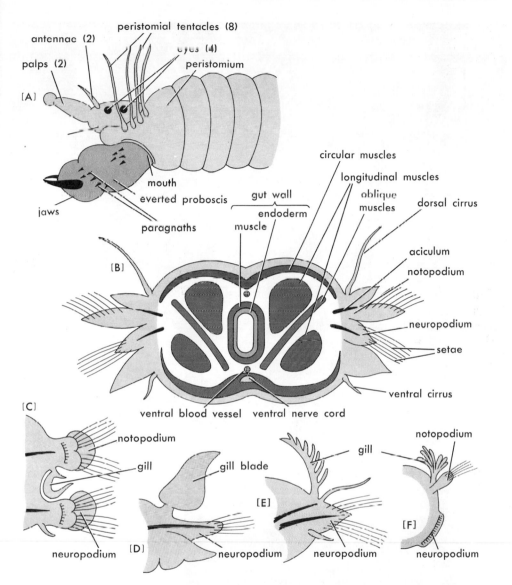

FIGURE 12-1. Organization of polychaete worms. **A:** Head structures—prostomium and peristomium—in a nereid polychaete, shown with the proboscis everted. **B:** Generalized nereid segment showing the characteristic arrangements of internal muscles and a pair of typical nereid parapodia. **C–F:** Stylized diagrams of other parapodia from a nephthyid (**C**), a phyllodocid (**D**), an eunicid (**E**), and an arenicolid (**F**).

methods of food transport are seen to occur. In a terebellid like *Amphitrite*, only one side of each tentacle is strongly ciliated, and this face is usually infolded to form a gutter-like groove in which the cilia beat toward the prostomium and mouth. Fine particles are transported by the cilia; medium-sized particles are maneuvered by the cilia along the groove but their passage back to the mouth is assisted by muscular peristalsis of the gutter (see Figure 12-3B).

[A]

[B]

FIGURE 12-2. Living fan-worms and tubeworms. **A:** Specimens of the fanworm *Sabella pavonia,* with the tentacles or "filaments" of the cephalic fan fully extended in active feeding. There is also a solitary sea-squirt, *Phallusia mamillata,* at lower right. **B:** Coiled calcareous tubes of *Spirorbis borealis* on the common brown seaweed, *Fucus serratus,* from a rocky seashore. Such tiny serpulid worms are extremely numerous in temperate and colder seas throughout the world, and, like sabellids, they feed using pinnate cephalic tentacles. [Photos © Douglas P. Wilson.]

The largest particles ingested are carried to the mouth by being enfolded in a single tentacle which is then almost entirely retracted.

The polychaetes have also developed true suspension-feeding forms, once again in several families and involving two entirely distinct types of mechanism. As already mentioned, Chaetopterus is a suspension-feeder using a mucous bag for filtration. The greatly differentiated body occupies a U-shaped burrow with a secreted lining (see Figure 10-5B). Segments 14, 15, and 16 bear the notopodial "fans" which create a water current through the tube as described in Chapter 11. Segment 12 manufactures the thimble-shaped bag in which particles are trapped, the bag being ingested every quarter-hour or so. This filtering mechanism can certainly collect organic particles as small as 1 micron in length, and some workers have claimed that protein molecules of about 40 Ångstroms can be retained.

A totally different type of filter-feeding mechanism is provided by the elaborate tentacles of the fanworms, the Sabellidae and Serpulidae (Figures 12-2 and 12-3A). In forms like Sabella, the crown of tentacles forms a wide funnel with the mouth at the bottom; cilia on the sides of the pinnules which line each tentacle cause a water current to pass centripetally into the funnel, food particles being trapped, not by a filtering on the outside of the funnel, but by eddies forming in front of, and between, the pinnules. Particles are passed by ciliary tracts toward the base of each tentacle where there are folds forming a sorting mechanism (Figure 12-3A2). In at least a few species of sabellids, this sorting is threefold: large particles which cannot enter the folds are carried to rejection tracts; medium particles pass along an intermediate groove and are transferred to a ventral sac to become the "graded" material used in tube-building; and lastly, the finest particles pass along the deepest grooves, the cilia of which are continuous with tracts leading directly into the mouth. All fanworms show some degree of tagmatization. This is perhaps most marked in the serpulids, which secrete permanent calcareous tubes (on rocks, dock pilings, ships' bottoms, and other hard substrata) and have part of the tentacular crown modified as an operculum which can close off the door of the tube when the serpulid is fully retracted.

Since direct uptake of dissolved organic materials from seawater has been somewhat more convincingly demonstrated in certain polychaetes than in any other marine invertebrates, it is appropriate to end this section on feeding in polychaetes with some account of the present status of this controversial field. Early in this century a marine biologist, A. Pütter, after calculating energy budgets for aquatic animals, concluded that dissolved organic matter is the main source of food energy for the majority of animals living in the sea and in fresh waters. Pütter's theory was generally refuted as regards uptake by the majority of physiological studies carried out on marine and freshwater invertebrates in the years between 1910 and 1940; and particularly important studies by August Krogh showed that the hypothesis by Pütter that dissolved organic materials were the main source of nutrition of freshwater animals was definitely erroneous. However, in the period 1957–1966, Grover C. Stephens and his associates clearly demonstrated by the use of radioactive labelling that both glucose and a variety of amino acids could be taken up from solution by a large number of marine species belonging to twelve invertebrate phyla. Unexpectedly they dis-

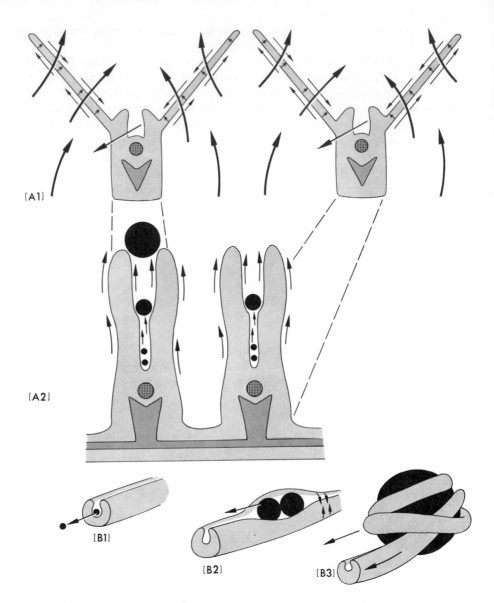

FIGURE 12-3. Feeding apparatus of sabellid and terebellid worms. **A1** and **A2:** Horizontal sections through two adjacent filaments of the feeding fan in a sabellid, distally (**A1**) showing the ciliary tracts on the pinnules of the filaments and more proximally (**A2**) showing the "size-sorting" apparatus and fused skeletal elements. For further explanation, see text. **B1–B3:** Methods of food collection exhibited by the tentacles in a terebellid like *Amphitrite*, including the ciliary action on small particles in the temporary "gutter" (**B1**), peristaltic action of the "gutter" on medium particles involving sequential contractions in transverse muscles within the tentacle (**B2**), and the wrapping action around large particles which is followed by contraction of longitudinal muscles in the tentacle (**B3**). [Adapted in part from E. A. T. Nicol, in *Trans. Roy. Soc. Edinburgh,* **56**:537–598, 1930; R. P. Dales, in *J. Mar. Biol. Ass. U.K.,* **34**:55–79, 1955; and unpublished work of Dr. Meredith L. Jones of the Smithsonian Institution.]

214

covered that no corresponding freshwater species could take up any significant amounts of any organic materials. Subsequently, some of this work was criticized on the grounds that many aquatic animals leach or release organic molecules *into* the medium, and that measurements of unidirectional influxes of various labeled compounds could not provide knowledge of any *net influxes* or "gains" to the animal. A further difficulty lay in the relatively low concentrations of the smaller organic molecules such as glucose and amino acids in natural seawaters, when compared to the concentrations which had to be used in the earlier experiments. Studies of net exchanges have now been carried out in a few marine invertebrates, and some of the transport systems have been described in terms of appropriate kinetics. It now seems that the net uptakes possible from unpolluted seawaters by most benthic and pelagic organisms (including the bivalve *Mytilus* and planktonic echinoderm larvae, both of which have appropriately high surface: mass ratios for active epithelia), could provide only small to insignificant fractions (mostly less than 1%) of the total energy requirements of these forms. However, the case of certain polychaetes, inhabiting sediments rich in organic material which is steadily being degraded by microorganisms, may prove to be different. Species of both bamboo-worms, *Clymenella,* and the thread-like marine bloodworms, *Capitella,* are able to cover a major part (perhaps 20–40%) of their maintenance energy requirements by uptake of dissolved amino acids. Appropriate concentrations of primary amines *are* present in the interstitial water of the sediments in which they live, as demonstrated recently by Stephens. Thus we can summarize the present status of Pütter's theory. Although uptake of dissolved organic materials from seawater *is* possible in a variety of marine animals, in the majority of cases, including such forms as bivalve lamellibranchs, the calorific gains are tiny in relation to needs. In a few infaunal polychaetes, a significant fraction of energy requirements can be gained from such uptake when the animals are living in suitably rich substrates. As will be noted in Chapter 31, the peculiar minor phylum Pogonophora probably resembles these polychaetes in this nutritional capacity.

Excretory and Circulatory Organs

Compared with other worm-like animals—mostly acoelomate or pseudocoelomate in organization—the body of annelids can be both more massive and more actively employed. Thus the organs of respiration, excretion, and circulation have occasion to be more highly organized than in the other "worms."

The archetypic circulatory arrangement can be best understood with reference to Figure 12-4A. The contractile elements consist of two longitudinal vessels: a dorsal aorta in which the blood is propelled anteriorly and a ventral aorta in which the blood is propelled away from the head. The circulation is a closed one, these two longitudinal vessels being linked by metamerically arranged sets of vessels which join them through capillary systems. Functionally, there are three capillary networks: one on the gut, where nutrients pass into the bloodstream; one in the parapodia where oxygen is taken up; and the third in the muscles where all the usual metabolic exchanges can take place. The afferent blood vessels to the capillaries of the parapodia and muscles leave the ven-

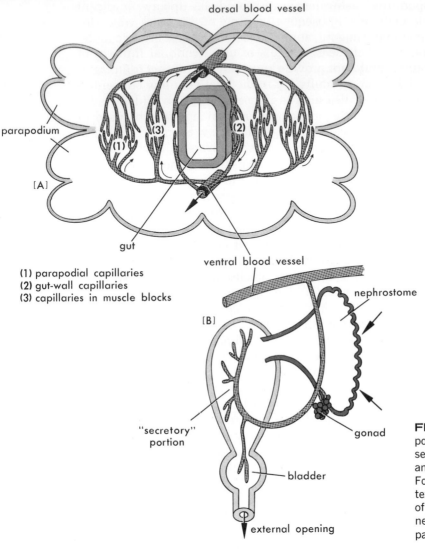

parapodium

[A]

gut

(1) parapodial capillaries
(2) gut-wall capillaries
(3) capillaries in muscle blocks

dorsal blood vessel

ventral blood vessel

[B]

nephrostome

"secretory" portion

gonad

bladder

external opening

FIGURE 12-4. A: One possible arrangement of segmental blood vessels in an archetypic polychaete. For further explanation, see text. **B:** One of the six pairs of excretory organs (mixonephridia) in *Arenicola*. Compare with Figure 12-5.

tral vessel in each segment and the flow is toward the dorsal side. In contrast, in each anterior segment the afferent supply to the gut-wall capillaries is from the dorsal vessel, the collecting vessels running to the ventral. However, to complicate archetypic concepts, directions of flow in some "errant" forms have been reported to be exactly contrary to the above. In forms showing a higher degree of tagmatization (see pages 180 and 203), specific groups of segmental circulatory arcs are greatly modified. In some of the more specialized annelids, there is a tendency for the segmentally arranged lateral vessels to become contractile "hearts." At the anterior end of earthworms like *Lumbricus*, there are five pairs of these which pump blood collected from the dorsal aorta through channels with valves to the ventral blood vessel which is, as in all annelids, the principal distributing vessel of the body.

Our present understanding of comparative anatomy, development, and evolution of excretory systems in many-celled animals is largely based on the detailed studies of E. S. Goodrich, which extended over half a century and were in large part based on the diversity of conditions found in various annelids. As Goodrich first elucidated, primitive coelomate animals, including some surviving annelid types, have two distinct sets of tubular organs connecting the coelom to the exterior. In metameric animals, these are arranged segmentally, archetypically a pair of each in every segment. First, the coelomoducts are mesodermal and grow outward through the body wall from the coelomic cavity. Their archetypic function seems to have been the transport of gametes from the gonads to the exterior. Secondly, the nephridia are developed centripetally from the ectoderm and are primitively blind at their inner or coelomic ends. Each nephridium was archetypically concerned in excretion: as is usual in animals, both in the removal of nitrogenous waste and in osmotic and ionic regulation. The structural distinction will be best understood with reference to Figure 12-5. There are two principal forms of nephridia: protonephridia, where the canals end blindly in either flame-cells or solenocytes, and metanephridia, where the duct system opens to the coelom. Protonephridia are assumed to be the more primitive, and besides being found in such polychaete families as the Phyllodocidae and Nephthyidae, are also the main excretory organs of flatworms and nemertines and are found in *Amphioxus*, though not in any other primitive chordates (see page 543). In one annelid family, the Capitellidae, the primitive metameric distribution of the organs is found, each segment having a separate pair of coelomoducts and metanephridia. In all other annelids and in many other phyla, the two sets of structures are combined in various ways, and, once again, the details of the different types of fusion which can occur were first deciphered by Goodrich. The more important combinations are illustrated in Figure 12-5C, including the protonephromixium as in Phyllodocid worms which involves combination of a protonephridium with the coelomoduct, and two different forms of combination of metanephridia with functional genital ducts—metanephromixia and mixonephridia. Mixonephridia, the last form of combined segmental organs, are found in such worms as *Arenicola*, where they are restricted to the six segments of the second tagma of the divided body. Functionally, they are concerned both with the passage of gametes to the exterior and with nitrogenous excretion and ionic and osmotic regulation (Figure 12-4B). They are obvious organs in a freshly dissected lugworm, each with a frilly funnel as an internal opening, a rich supply of blood vessels, and the gonad tissues in close proximity. All oligochaetes and leeches have metanephridia. In these groups, the coelomoducts with the gonads are restricted to a limited number of segments and are not usually connected with the nephridia. Obviously the functional importance of nephridia is greater in nonmarine annelids. In polychaete groups, where there are estuarine and freshwater forms, these have nephridia that are not only larger and more complex than the corresponding marine species but also have a richer vascularization. Physiological investigation has been most complete in earthworms. In most of them, the fluid discharged by the nephridia is hypotonic to the coelomic fluid (i.e., contains less dissolved salts), and has a higher proportion of such nitrogenous substances as ammonia and urea. At present, there is some controversy among physiologists about the functional processes in annelid nephridia.

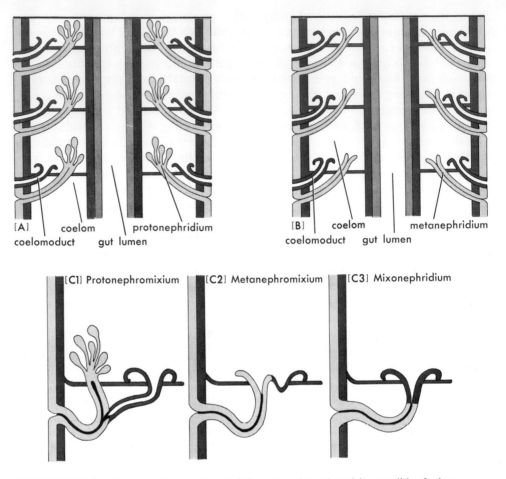

FIGURE 12-5. Segmental organs (nephridia and coelomoducts) in annelids. **A:** A presumed archetypic condition with separate pairs of protonephridia and coelomoducts in each segment. **B:** Another archetypic condition, actually found in some capitellid polychaetes, with separate pairs of metanephridia and coelomoducts in each segment. Note that, in both cases, the nephridia are ectodermal and develop centripetally while the coelomoducts are mesodermal and develop outward from the coelomic cavity. **C1–C3:** Three of the more important of the combinations of nephridia with coelomoducts which are found in various polychaetes: a protonephromixium, a metanephromixium, and a mixonephridium. For further explanation, see text.

Regeneration and Reproduction

The pattern of morphogenesis of segments in annelids, described in Chapter 10, is associated with a fantastic capacity for repair after injury by regeneration of lost segmental units. In some polychaetes and aquatic oligochaetes, fragmentation is combined with regeneration as a method of asexual reproduction. The development of specialized sexual individuals, each produced asexually, seems also to have arisen from the regenerative capacity.

Restorative (or reparative) regeneration can occur to some extent in all annelid worms and usually returns the injured worm to the characteristic number of segments of its adult form. The restorative process is easiest to understand in worms where all the segments are relatively uniform. It should be remembered that the actively growing zone in young worms is usually just in front of the most posterior segment. In many forms regeneration can take place headward or tailward, as appropriate; the wound surface first seals to form a blastema, which then proliferates an appropriate number of segments to restore the adult pattern. It is significant that in restorative regeneration of the anterior region, it is the head which is differentiated first, and in many elongate worms this leads to some imperfections in the restorative process. In such forms, if the head plus 3–5 segments is cut off, then the head plus 3 segments only regenerates; if the head plus 6–10 segments is cut off, then the head plus 4 segments regenerates; and if the head plus 20–30 segments is cut off, then only a head plus 10 segments is restored. In the elongate syllid *Autolytus*, a cut made between segments 5 and 13 will cause restoration of a head plus 5 segments; amputation between there and segment 42 results in restoration of a head only; and with a more posterior cut, no head can be regenerated. In several worms, there is a specific regeneration no matter how many segments are removed. For example, in *Syllis spongicola* the restorative regeneration is always of the head plus 2 segments no matter where the amputation. Perhaps surprisingly, even worms with a high degree of segmental specialization (tagmatization) are capable of restorative regeneration. Any number of segments up to 14 may be cut off the anterior end of the highly specialized *Chaetopterus* and complete restoration can occur. If 15 anterior segments are cut off, there is no regeneration. Fantastically, any two segments from the anterior tagmata will regenerate both anteriorly and posteriorly to form an exact copy of the original highly specialized worm.

Tube-dwelling fanworms like *Sabella* must naturally be rather subject to loss of the tentacular crown and anterior segments which are exposed from the tube during feeding. Restorative regeneration in *Sabella*, after the head structures and any number of segments are removed, always results in the head (prostomium and peristomium) plus one body segment. Subsequently, the existing segments are modified by metamorphosis and new ones added posteriorly to make up the number. There is considerable evidence that the restoration of the correct segment order in *Sabella* and similar forms involves secretions passing from the regenerated head structures posteriorly.

Competence in restorative regeneration, particularly the capacity to regenerate a lost head, is linked in many aquatic annelids with the development of asexual fragmentation. In the genus *Ctenodrilus*, worms simply periodically fall apart, and each group of one, two, or three segments then regenerates into a complete worm. In *Autolytus* and several other genera, a similar asexual process is more closely controlled, with epidermal ingrowth forming a macroseptum at the position where the break will occur and a head forming behind each macroseptum before fission takes place. A chain of stolons may thus be formed (Figure 12-6A) and can result in the budding off of nearly complete and fully grown "adults."

A special form of this budding of individuals is called epitoky. This involves the development by asexual means of a reproductive individual, called an epi-

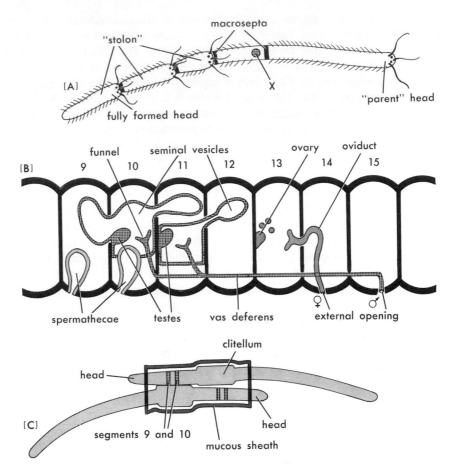

FIGURE 12-6. Reproduction in annelid worms. **A:** Stolonic budding in a syllid polychaete, *Autolytus*. After each macroseptum is formed by epidermal ingrowth, a new head is differentiated behind it before the asexually budded individual breaks off. **X** marks a region where implantation of cephalic neurosecretory tissue will prevent head differentiation. **B:** Schematic lateral view of the hermaphroditic sex organs in a typical earthworm. **C:** Characteristic arrangement of two earthworms in copulation.

toke, which differs from the nonsexual atoke usually, not only in the possession of mature reproductive organs internally, but also in the possession of modified parapodia to allow more efficient swimming. The ecological importance of epitoky is twofold. It can allow synchrony of spawning, and by providing a motile distribution stage, permit the scattered burrowing worms of the sea bottom to come together in swarms for reproduction.

Both in cases where epitokes are budded off, and in cases where the whole adult worms metamorphose to a sexual swarming form, both the structure and behavior of the sexual forms is markedly different from their normal, bottom-dwelling progenitors or predecessors. Structurally, the sexual individuals have reduced guts, enlarged eyes, and parapodia elongated and elaborated as effective swimming paddles. Behaviorally, sexual individuals are usually markedly photopositive being attracted to artificial lights in appropriate seasons. Scientifically, there are many palolo-worms, several genera and species occurring in different parts of the world. The commonest Pacific palolo belongs to the genus *Eunice,* and its spectacular swarms can be predicted with some accuracy since they show response to the lunar cycle. As a result, these or similar worms are netted as a source of human food in Fiji, Samoa, and certain parts of Japan. More minor, but still impressive, swarming of sexual individuals can be observed in many parts of the world. Two examples are well-known to students who have visited Woods Hole on Cape Cod. Both regularly appear in July two nights after the occurrence of full moon. In the case of sexual individuals of *Nereis succinea,* the males appear first and swim in rather loose circles. When females appear amongst them, the males form swirling vortices around each female and only then emit sperms. The sperms apparently elicit egg-shedding by the females, fertilization occurs, and both male and female worms survive the sexual aggregation. In *Platynereis megalops,* sexual swarming involves a unique copulatory process, regularly observed by investigators and students at Woods Hole, but so peculiar that a number of writers on the subject have doubted its occurrence. The eggs in this species are unfertilizable after 40 seconds of contact with seawater. As with the previous species, both male and female sexual individuals swim toward night lights held over the surface waters. On contacting a female, the male wraps his body around hers, inserts his anus into her mouth, and goes into a general contraction of the body muscles. Since in both sexes the gut wall is already degenerate, sperms pass from the male's anus directly into the coelom of the female where the eggs are fertilized and spawned via the female anus almost immediately. In this case the sexual individuals do not survive the process.

Apart from the complete metamorphosis of a bottom-dwelling polychaete into its sexual swimming form, there is great variation in the type of asexual division leading to epitoke formation. Thus some epitokes are long, of many segments, others short; some have heads and obvious sense-organs, others have none. There is considerable evidence, however, that some of the processes of sexual metamorphosis after epitoky can only take place after separation from the original parent head. It seems almost certain that neurosecretory cells in the head of the "normal" worm secrete a hormone which has an inhibitory effect on the development of the secondary sexual characters (including the enlarged parapodia and eyes) of the epitoke and may even control the onset of gonad maturation. It seems likely that a similar endocrine control could be involved in some of the processes of restorative regeneration discussed earlier. Neurosecretory cells and their products will be more fully discussed in the Arthropoda (Chapters 13 and 17), where their physiology has been more thoroughly investigated.

As already noted, the terrestrial annelids—earthworms and leeches—are hermaphroditic with internal fertilization and relatively large eggs. The functional organization is often complicated, but is essentially concerned with the prevention of self fertilization while accepting sperms from other individuals to fertilize the eggs. In an earthworm like *Lumbricus* (Figure 12-6B, C), there are essentially three sets of structures: first, two pairs of spermathecae which store the "foreign" sperm received from the partner in copulation; secondly, two pairs of testes, with their storage seminal vesicles and their modified coelo-moduct system to the exterior; and thirdly, a pair of ovaries and their modified coelomoducts (Figure 12-6B). In copulation, two such worms have their ventral surfaces in contact and secrete an enclosing sheet of mucus. Actual contact is between segments 9 and 10, the spermathecal segments, of one worm and the clitellum of the other. Sperm stored in each seminal vesicle pass down the vas deferens, out of the male opening in segment 15, and along the outside of the body to the clitellum where they enter the spermathecae of the other worm. The worms then separate, and the clitellum secretes a cocoon of hardened mucus which slides forward with the wriggling of the worm. As it passes the female opening on segment 14, eggs are extruded into it, and a little later it receives sperm from the spermathecae. Finally it is passed over the head and the ends are closed off; the zygotes develop inside the egg-cocoon to hatch out as minia-ture adults. The details vary in different terrestrial annelids, but the functional implications are always the same. They involve copulation and cross fertiliza-tion followed by direct development (without larval stages) from relatively large eggs.

Minor Annelids

The minor class Myzostomaria is a highly specialized group of parasites, al-most certainly derived from polychaetes. Myzostomarians appear as adults like tiny flattened discs with five pairs of reduced parapodia. They are all parasitic on echinoderms, the majority on crinoids. Although they are hermaphroditic, the development of their fertilized eggs leads through a typical free-swimming polychaete larva.

The other minor class in the phylum Annelida is the Archiannelida. As pres-ently constituted, it is certainly polyphyletic. Apart from *Polygordius,* the other archiannelid genera all have muscular buccal bulbs and may severally repre-sent stocks showing simplification (including absence of setae) from several polychaete families. Such genera as *Dinophilus* and *Nerilla* could possibly rep-resent small-scale and degenerate versions of ancient stocks of eunicids and syllids, respectively. The common interstitial habitat could well be significant. Even if some of the other genera, including *Polygordius,* genuinely represent surviving primitive stocks, there seems to be a good case to suppress the artifi-cial group Archiannelida and incorporate its parts into various subdivisions of the class Polychaeta.

Nonmarine Annelids

All available evidence from comparative anatomy and physiology suggests that the first oligochaetes were evolved from representatives of one of the less-dif-ferentiated polychaete stocks which acquired a capacity for osmoregulation and invaded fresh waters. The primitive oligochaetes of fresh waters gave rise to more completely adapted forms with internal fertilization, large eggs, and resistant resting stages as well as osmoregulatory capacity. The successful freshwater oligochaetes of the present day, including such ubiquitous families as the Naididae and Tubificidae which are ecologically important in lake bottoms, belong to this group. Both the land oligochaetes, or true earthworms, and the freshwater leeches must have been evolved from similar freshwater oligo chaete stocks which already possessed the full concert of structures and functions for nonmarine life. For example, evolution of internal fertilization and the production of large eggs preceded the development of earthworms or leeches.

It is worth noting that the evolution of earthworms must have taken place during and after the Cretaceous, in close correspondence (as with the insects) with the evolution of a land flora of flowering plants. Terrestrial soils, as we know them, evolved along with the seasonal and deciduous vegetation of the newly evolved angiosperms, and with the newly evolved land oligochaetes (see Chapter 34). Angiosperms, earthworms, and humus all have exactly the same geological antiquity.

Locomotion in earthworms involves neurally controlled waves of muscle contraction which pass anteriorly along the worm, adding posteriorly to each bulge where the longitudinal muscles are contracted and the setae protruded. Thus the bulges move slowly back, and the body of the earthworm moves somewhat more rapidly forward. (The process is fully described in Chapter 11.) This method of locomotion serves the earthworm both in crawling over the surface (usually at night) and in burrowing through soft soils. Along with the soil itself, earthworms ingest all kinds of organic material but especially leaves and other dead plant tissues. As was extensively investigated by Charles Darwin nearly a century ago, earthworms modify the soil structure by passing soil through their bodies and casting it toward the surface. Essentially, besides improving aeration and drainage of the soil, earthworms move organic material downward deeper into the soil and bring inorganic particles from the subsoil upward toward the surface. Earthworms occur in all parts of the world, under all types of vegetation, and in all climates except the most arid. They form an important part of the terrestrial biomass, apart from their ecological importance in restructuring soil. Under one square meter of pasture grassland, eight hundred larger earthworms, and ten times that number of smaller oligochaetes, can occur. There is probably a greater abundance in the soils of deciduous woodlands.

The most highly specialized annelids are the leeches (class Hirudinea). Although they are found in the sea as well as on land and in fresh water, it is probable that the marine forms, like the terrestrial ones, are evolved from freshwater leech stocks, which in turn were evolved from freshwater oligochaete stocks. They form a very stereotyped group in which the two suckers have replaced setae as the fixed points in locomotion and where the coelomic-cavity has been largely obliterated by the growth of fluid-filled cells. The mouth is in the center of the anterior sucker and may be armed with three piercing jaws. It leads into a muscular pumping pharynx and thence to a large storage crop which is functionally related to the irregular meals. Leeches, like earthworms, are hermaphrodites with the organs arranged for cross-fertilization. Most leeches are not true parasites since they remain attached to their hosts only for the short period of feeding. Many leeches are predaceous carnivores rather than blood-suckers and eat such invertebrates as earthworms, slugs, and insect larvae. As is well known, in the blood-sucking forms, the salivary glands secrete an anticoagulant which is injected to prevent clotting of the host's blood during feeding, and is responsible for the continuation of bleeding after the swollen leech has withdrawn.

A few leeches have a capacity for restorative regeneration of the head but there is no asexual reproduction in the group. This is probably connected with the highly stereotyped body plan; except for one genus, they have a body invariably composed of 33 segments, the last 6 of which form the large posterior sucker. The Hirudinea undoubtedly comprise the most specialized, and least diversified, group of annelids.

The possible relationships between other phyla of many-celled animals and the Annelida are discussed in Chapters 16, 17, and 35.

13 FUNCTIONAL ORGANIZATION OF ARTHROPODA

ON the basis of mechanical functioning and of inferred phylogeny, the characteristic features of arthropods can be divided into two distinct lists. First are those paralleled in the phylum Annelida, which mainly involve aspects of development and metamerism, and secondly, those more associated with the possession of a chitinous exoskeleton. The features on the first list would include their being triploblastic and coelomate and having perfect bilateral symmetry, with a tubular gut running from the mouth to the terminal anus, with metameric segmentation as we have already defined it, and with the central nervous system in the protostomous form (see Chapters 10 and 27), that is, as a chain of segmental ganglia on the ventral side with only one pair of ganglia—the supraoesophageal ganglia—lying anterior and dorsal to the gut. The second list would include the many features associated with the exoskeleton of chitin and epicuticle; such as the need for periodic molting or ecdysis during development and growth; the jointed limbs with their internal arrangement of antagonistic muscles as flexors and extensors; the restriction, compared to annelids, of suitable permeable surfaces for respiration, excretion, and so on; the contrasting capacity for resistance to water losses, which has made them the most successful group of terrestrial animals; tagmatization (as already defined) occurring to a much greater extent than in the annelids; an apparent lack of ciliated epithelia even in the linings of the internal duct systems; and the body cavity as a haemocoel involved in the circulation of the blood, rather than as the annelid coelom, which is totally separate from circulation and greatly involved in the mechanics of muscle antagonism.

Thus all the features truly diagnostic of the arthropods are connected with the exoskeleton—even the most unlikely diagnostic characteristic, "hearts with ostia," is related to the passage of blood from the haemocoel into the heart and the lack of a "venous" system, and thus to the decline in the importance of the hydrostatic function of the body-cavity fluids.

By any measure, the phylum is enormously successful, clearly the most successful invertebrate phylum, whose species numbers (see Table 1-1) outnumber those of all the other phyla added together. The arthropod fraction of the total animal biomass is likewise disproportionate: consider, for example, the composition of the faunas of terrestrial soils or of the marine plankton. Physiological aspects of this success are, once again, almost all associated with the possession of an exoskeleton of chitin and epicuticle. As noted in Chapter 2, the account of arthropod functional organization below and the chapters which follow are principally concerned with the biology of the arthropod class Crustacea. Since the other nine classes will be neglected until Chapter 16, and then dealt with somewhat synoptically, it is worth reviewing their relative importance now.

As outlined in Table 10-1, the forms included in the phylum Arthropoda in the strict sense fall into ten groups, here termed classes, though considered as subphyla by some authors. Of these, four are tiny groups of minor ecological importance: Merostomata, Pauropoda, Symphyla, and Pycnogonida. One relatively large group, the class Trilobita, is represented only by fossils, although it has considerable significance as a likely stem-group for many, if not all, the other arthropod classes. The classes Chilopoda and Diplopoda form, along with the Pauropoda and Symphyla, a moderately large-sized collection of forms (about 9500 species) usually termed centipedes and millipedes, but of relatively minor ecological importance except in some rather specialized terrestrial environments. This leaves us with three enormous classes: Crustacea (more than 28,000 species), Arachnida (more than 47,000 species), and Insecta (more than 700,000 species). The crustaceans are a largely marine group, with considerable representation in fresh waters and only a few genera on land. On the other hand, the arachnids and insects are primarily land animals with only a few forms secondarily returning to aquatic life, usually in fresh waters. It is this ecological distribution of the three large classes, and the occurrence of relatively primitive marine forms in several of the subclasses of the class Crustacea that makes a study of the more primitive crustaceans relatively more rewarding in terms of an understanding of the possible functional homologies and the course of physiological evolution in the arthropods as a whole. Their greater physiological diversity contrasts with rather stereotyped patterns of anatomy and physiology in the arachnids and insects. Since we cannot investigate aspects of physiology in the class Trilobita, the closest approximation to archetypic conditions of structure and function is likely to be found in the more primitive marine Crustacea.

Physiology of the Exoskeleton

Structurally, the layers of the arthropod exoskeleton are divided into two zones, and the functional significance is markedly different for each. The thin, outer nonchitinous epicuticle is responsible for the "chemical" properties of the integument. The much more massive chitinous procuticle provides the mechanical properties of this combined armor and skeleton. The procuticle consists of many fused laminae which can be grouped together on the basis of differences in composition or properties. Unfortunately, the nomenclature adopted by workers on insects does not correspond to that used by most workers on the higher Crustacea: the former divide the chitinous layers into an exocuticle on the outside next the epicuticle and an endocuticle below this, while the latter describe the whole procuticle as the endocuticle and divide it from outside inward into a pigmented layer, a calcified layer, and an uncalcified layer (Figure 13-1A,B). Both the epicuticle and the chitinous procuticle consist, of course, of nonliving layers secreted by the cells of the ectodermal epidermis which lie within and in close contact with them. This epidermis has typically a stout basement membrane to which muscles can be attached and is itself strengthened by intracellular fibrils which run from the basement membrane into the cuticle.

The epicuticle is responsible for the "chemical independence" of the environment which the exoskeleton confers on arthropods, for it is markedly hydrofuge (or nonwettable), and almost impermeable in a physicochemical sense, and provides protection against microorganisms such as bacteria. The details of the ultrastructure responsible for these properties, and the sequence of secretion of units, have been most thoroughly studied in insects, but it is obvious that similar organization of the epicuticle occurs in arachnids and in the higher Crustacea. In insects, the outermost layer of the epicuticle consists of a very thin protective cement layer of lipoprotein and is secreted last (actually after molting), the secretion being through pore canals from the epidermal cells which pass through the layers of the procuticle and the deeper layers of the

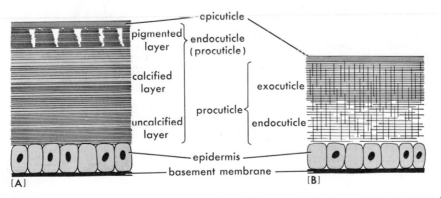

FIGURE 13-1. Structural patterns of arthropod integuments in decapod crustaceans (**A**) and in insects (**B**). The functional significance of each of the different layers is discussed in the text.

epicuticle. Just below this, lipids are laid down as a waterproof wax layer secreted immediately before molting. Below this again, and next to the outer face of the chitinous procuticle, is a layer of proteins—the cuticulin layer associated with polyphenols.

Below these epicuticular layers lies the great bulk of the exoskeleton, the procuticle, consisting of complex laminae of chitin and proteins. Chemically, chitin is a polysaccharide consisting of acetyl-glucosamine residues which are polymerized into long, unbranched molecules. Enmeshed with the chitin in the laminae of the procuticle is a protein component which has been termed arthropodin. Layers of protein and chitin which are not further hardened remain flexible but nonelastic, and such procuticle is found in all the joints of the arthropod limbs and body and, to a varying extent (Figure 13-2), in the innermost layers of the exoskeleton of higher arthropods. Other layers are hardened into a more massive armor or more rigid skeleton. In the crustaceans, this is accomplished by the deposition of calcium carbonate in the middle layers of the endocuticle. Another method of hardening—used to some extent by all arthropods—has been termed sclerotization and involves the so-called tanning of the protein component in the procuticular layers. This is accomplished by the cross-linking of the protein chains by orthoquinones, and this process

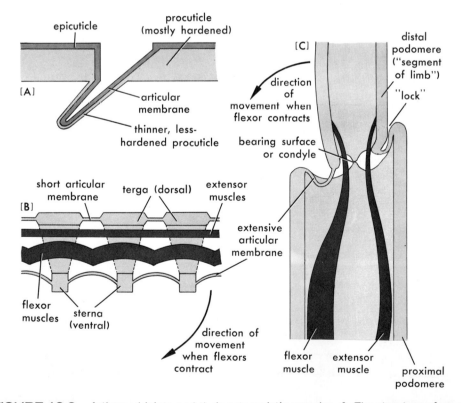

FIGURE 13-2. Arthropod joints and their antagonistic muscles. **A:** The structure of an articulation. **B:** Three segments of a schematic half abdomen from a decapod crustacean such as a lobster or crayfish, viewed laterally. **C:** A limb joint in a higher crustacean, split open at right angles to the axis of articulation. For further explanation, see text.

always involves, in addition to the proteins, polyphenols and polyphenol-oxidases. In insects, it is the exocuticular layers and in crustaceans, the pigmented layers of the endocuticle, which are most highly sclerotized.

Of course, the mechanical strength conferred by these hardening processes is not important merely in providing a protective armor. The development of an integument of pieces of hardened procuticle which can resist deformation linked by unsclerotized joints is the basis of the superbly efficient arthropod locomotory machines, all of which are constructed as interacting systems of levers. Just as metameric subdivision of muscles and hydrostatic skeleton can increase the mechanical efficiency of soft-bodied worms over similar forms with unsegmented muscle arrangements (see pages 174 and 207), so the development of arthropodan levers makes locomotion even more efficient in terms of energy expended. The nice application of force to the substrate or environmental medium is accomplished by the jointed appendages of arthropods as lever-systems, with less waste of energy and with considerably greater precision than the forces resulting from the contractions of the muscular body wall of soft-bodied animals. Once again, an analogy with man-made machines can be conceived. As always, the basic structures capable of contraction—the motors of the machine—are muscles: circular and longitudinal in the worm-like animals and arranged as flexors and extensors in the arthropod. The more sophisticated mechanical design of arthropods, however, allows a greater force to be exerted more precisely against the environment, per unit volume of muscle, than in worms. As a result, arthropod locomotion can be faster, more precisely directed, and much more efficient in terms of energy expenditure.

The properties of the exoskeleton as a whole are responsible for the enormous success of the arthropod body plan. To recapitulate, the epicuticle is responsible for the physiological separation of the arthropod's tissues from its external environment. The mechanical properties of the procuticle are basic to the precise and efficient movements which characterize the various forms of locomotion in arthropods. Both the physicochemical and the mechanical properties have obviously contributed to the success of arthropods as land animals. It is worth noting that no other invertebrate group is so successful on land, and further, that the greatest proportion both of terrestrial species and of terrestrial animal biomass is arthropodan.

Joints, Muscles, and Limbs

The characteristic joints of arthropod bodies and limbs depend on their being regions of the procuticle which are thinner, less hardened, and thus lacking the mechanical rigidity of the other parts of the integument. It should be noted that these articular membranes, as they are called, are never intrinsically elastic, although this is suggested in many textbook accounts. In general, the articular membranes are more extensive on one side of the joint than the other (Figure 13-2). Thus the abdominal segments of the crustacean in Figure 13-2B will be capable of greatest bending ventrally, while those of the joints in Figure 13-2C will bend to the left. The actual "bearing surfaces" are termed the condyles and the invaginated exoskeleton provides points—apodemes—for the attachment of the muscles. Mechanically, each podomere (segment) of a stenopodous limb (such as the walking legs of insects, spiders, or Crustacea—see page 291) is in effect a rigid tube linked by the articular membranes to the next rigid tube. A line through the condyles, where each tube actually touches the next, forms the axis of articulation of the joint, and normally there are muscles arranged across the joint on either side of this axis in antagonism to each other.

The functioning of the complex internal musculature of arthropods and the mechanics of arthropodan movements are most readily understood if two simple facts of muscle physiology are fully accepted. First, the muscles in any arthropod exert forces roughly proportional to their size as measured by the cross-sectional area of the contractile fibers. (The work that the muscles can perform is—even more roughly—proportional to the volume of the contractile fibers.) Secondly, arthropodan muscles—like the muscles of any other animals —can exert force only by contracting. (No mechanism which involves muscles in exerting force by "expanding" or "stretching" can possibly exist in animals.) A muscle is normally caused to contract by a stimulus from a nerve or nerves. As soon as the stimulus from the nervous system ceases, the muscle will become limp again, but the unstimulated muscle does not regain its original, pre-contractional length until it is extended by some other force outside itself. In arthropods, as in most other animal mechanisms, this force is provided by the contraction of another muscle or set of muscles which are said to be antagonistic to the first. The action of antagonistic muscles around the joints of the *internal* skeleton in vertebrate limbs is familiar, and the antagonism of circular and longitudinal muscles around the fluid-filled bag of a hydraulic skeleton in worm-like animals has been discussed elsewhere (see Chapter 11). In arthropods, the antagonists span each joint on opposite sides of the axis of articulation. In general in the great bulk of arthropod systems, the muscles—the flexors—lying on the side with the greatest extent of articular membrane are the more massive. The muscles crossing the other side of the joint—the extensors—are usually slighter. Again in a generalization to which there are some exceptions, it is the flexors in arthropod systems which do most work against the environment and the extensors are the "recovery" muscles. (Many of the exceptional cases, where extensors are propulsive, occur in land arthropods; see Chapter 16.) In very general terms, therefore, flexors tend to be distributed on the ventral, or posterior, or medial sides of the joints of individual limbs or the body and are normally the more bulky muscles. (In the analogous mechanical system of the muscles working round the elbow joint of the human arm, the biceps brachii—with which more work is done—is much more massive than the corresponding antagonist which straightens the arm, the triceps brachii.) In many higher crustaceans, like the crayfish, the abdominal segments are arranged as in Figure 13-2B. These crustaceans often show an escape reaction involving flexure of the "tail," which drives them backward away from any danger. This flexure is carried out by the more ventral set of longitudinal muscles in the abdomen (see Figure 13-2B) which are more massive, while straightening of the abdomen is accomplished by the slighter dorsal longitudinal musculature. Another characteristic feature of the functioning of the arthropod muscle systems is illustrated in Figure 13-2C, where on contraction of the extensor muscles, the rigid elements on that side of the skeleton, being separated

by much less extensive articular membranes, are so arranged that they lock together rigidly on contraction of their muscle. In a minority of arthropodan joint
systems, there are no extensor muscles in the strict sense, and the flexors are
stretched by other means. Elastic structures counter them in a few cases, while
in others (including the mouth parts of butterflies, the cirri in barnacles, and
the limbs of certain peculiar arachnids), extension is accomplished by hydraulic means. In these peculiar cases, bending of the structures is accomplished by
the flexor muscles but these are stretched again and the structures are extended
by blood being pumped into them. A few zoological textbooks claim this extension mechanism to be universally important in the mechanics of arthropod
movements, but this is not so. The usual antagonists of flexor muscles are extensor muscles. In the claws or chelae of crabs and lobsters, the closing muscles
are flexors and the opening ones extensors. A simple mechanical experiment
with a healthy crab or medium-sized lobster will verify the truth of the statement above regarding the greater force exerted by the flexors. It is relatively
easy for human fingers to hold such a claw closed, that is, to oppose the force
exerted by the extensor muscles, but much more difficult to hold the same claw
open, that is, to oppose the flexors.

Another aspect of the physiology of arthropod movement seems peculiar to
us as vertebrates. The nerve-muscle system seen in crustacean limb muscles
functions in an entirely different fashion from vertebrate somatic muscles. The
extent and nature of contractions in crustacean limb muscles is variable, but
the variation results from multiple innervation. Double, triple, and quadruple
motor innervation has been demonstrated in different crustacean muscles, and
van Harreveld and Wiersma demonstrated some years ago that a quintuple innervation of all fibers occurred in certain muscles in higher Crustacea. Experimentally, four of the five nerve fibers were found to be motor axons, each of
which elicits a contraction with different characteristics. The fifth fiber, when
stimulated simultaneously with any of the other four motor fibers, causes inhibition of the contraction. The muscles in which this quintuple innervation
was demonstrated occur in decapods and are the flexor muscles of the carpopodite; that is, they are more distal muscles in the walking legs which are important for the posture of the animal and the support of its body weight.

The limbs of primitive Crustacea—and possibly of all primitive arthropods—
are constructed on a biramous pattern. There is a basal protopodite bearing the
two rami, the exopodite and endopodite. Primitively, such limbs are generalized in function; they are involved in locomotion by swimming or walking, in
feeding both in the creation of a water current and in sieving, in respiration
(particularly if they involve flat laminae), and as sensory organs bearing receptors. In more advanced arthropods, the limbs are more specialized for one or
two of the four basic functions. In crustaceans there are two main lines of modification of the limbs: the phyllopodium and the stenopodium (see Figure
14-1B,C). The names of the individual podomeres, or joints of the endopodite
in the stenopodous walking limb, need only be remembered by arthropod systematists, but there are a few terms used in limb anatomy which are concise
and valuable in any discussion of functioning systems in Crustacea. Besides the
main terms protopodite, endopodite, and expopodite, it will be advantageous
to remember that additional lobes developed on the inside of the limb are en-

dites, and those on the outside exites. Endites on the protopodite form the gnathobase, usually used in food collection or trituration, while exites on the protopodite (proximal to the exopodite) are epipodites, which, in many Crustacea, are flattened lobes used as gills.

Two points of mechanical functioning in these two basic limb types are worth noting here. In the typical stenopodium, the axes of articulation vary in direction in successive joints. Thus, although each pair of podomeres can be moved in relation to each other only in a plane at right angles to the axis of articulation, the limb as a whole is capable of nearly universal movements. In some other systems of joints, for example in the sensory antennae of higher Crustacea, the condyles are not present and a specific axis of articulation is not defined. In these, movement is possible in any plane (in theory at least). In phyllopodous limbs, the principal musculature works around the joints of the protopodite and, to a much less extent, around more distal joints in the limb. In many primitive Crustacea, phyllopodous limbs are used both for swimming and for creating a feeding current. In these limbs, the joints between the more distal lobes—such as endites and the expodite—and the central parts of the limb are so arranged that they flex toward the animal until they lie at right angles to the plane of the central parts of the limb, but are arranged with extensor locks (as discussed above) which prevent them from being flexed anteriorly at all (see Figure 14-2B). Thus, purely passive movements involving these joints allow the limb, moved by the contractions of muscles across the joints of its protopodite, to execute an effective backward stroke and a recovery stroke and achieve effective forward locomotion. On the recovery stroke, the limb is moving anteriorly but all its peripheral flaps are flexed back in a manner analogous to the "feathered" blade of an oar in rowing. While this arrangement of jointing is essential for the locomotion of the animals concerned, it is also of fundamental significance in their feeding mechanisms. This will be discussed later (see pages 245–247).

Molt-cycles

Growth, in animals other than arthropods, can involve a gradual increase in size, accompanied if necessary by gradual changes in shape, until the size and form of the fully grown adult is reached. Possession of an exoskeleton prevents this in arthropods. Once an arthropod exoskeleton has been hardened, by sclerotization and calcification, no change in its external linear dimensions can occur. Growth in arthropods must thus proceed through a series of molts. Each must involve the secretion of a new cuticle and an ecdysis, or shedding, of the old exoskeleton. The actual molting, or ecdysis, takes place *after* secretion but *before* hardening of the new, and larger, cuticle. During the short period of hardening the new exoskeleton, the arthropod shows a rapid increase in bulk, which usually involves uptake of water or air into internal spaces. Then, subsequently, the arthropod grows new tissues to fill the new armor. Thus arthropod growth appears to take place in spurts (Figure 13-3). It is worth emphasizing the paradox involved in this. It is that while the apparent increase in size in arthropods occurs at the molts (the near-vertical curves of Figure 13-3), almost

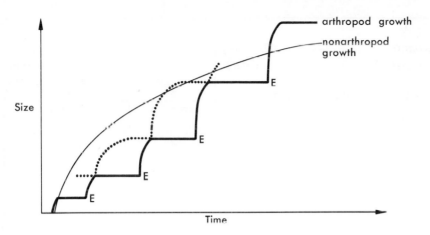

FIGURE 13-3. The characteristic pattern of arthropod growth (size being measured by any linear dimension on the exoskeleton). Rapid size increase follows each ecdysis or molt (E), and results in the growth steps which contrast with the smooth curve for growth in other animals. Paradoxically, almost all real tissue growth or increase in individual biomass (dotted line) takes place during the intermolt period when no change of size can be detected externally.

all of actual tissue growth, or increase in individual biomass of the arthropod, takes place during the intermolt periods when no change in size can be detected externally (the horizontal "steps" of the figure).

Many arthropods have changes of form at several ecdyses, which are termed metamorphoses. In others, including the crayfish, growth from juvenile to adult is accompanied by gradual shifts of proportions and there are no drastic changes at any one molt.

It should be noted that all surfaces covered with cuticle are renewed at each molt. There are many invaginations of the ectodermal epithelium, and with each is associated a cuticular lining, structurally and physiologically homologous with the exoskeleton. Such invaginations include the fore- and hindgut, the various apodemes for muscle attachments, and the tracheal tubes of air-breathing arthropods. The old cast cuticle of an arthropod, with all these attached internal processes, is convincing circumstantial evidence that the new cuticle was soft and pliable during the ecdysis.

The physiology of the molt-cycle has received most extensive and detailed study in certain insects and in higher Crustacea like crabs. For the latter, an accepted modern classification of the stages was developed as a result of the detailed studies of P. Drach (Figure 13-4). In summary, there are four functionally different stages covered in the detailed cycle from C4 to E to A1 and back to C4. In proecdysis, or premolt, calcium is removed from the exoskeleton and the calcium content of the blood rises as the new soft cuticle is laid down below the old one. After ecdysis, or the actual shedding of the old cuticle in the molt (Figure 13-5), the crab swells by uptake of water. Then, during metecdysis, the postmolt period, the exoskeleton is hardened and calcified. This is followed by the intermolt period when the animal is in normal condition with a hard exoskeleton. The intermolt stage (C4) may be long or short, depending in part on

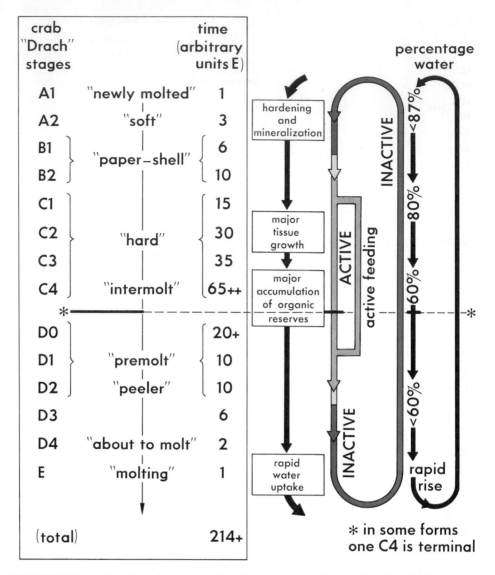

FIGURE 13-4. Functional aspects of the Drach stages of the molt-cycle in higher crustaceans like crabs. For further explanation, see text.

the rate of tissue growth and accumulation of organic reserves, this rate in turn depending on the rate of nutrition. In some higher Crustacea there is one intermolt (C4) which is terminal, and the adult animal in these cases can no longer grow. The Drach stages provide a practical classification with considerable detail, into which individual crabs can be fitted on morphological characters and condition of the exoskeleton without previous knowledge of each individual's molt history. Perhaps of greater significance is that Drach's detailed studies have emphasized the biological point: that molting is not a restricted act interrupting the "normal" life of the crab, but that all aspects of the normal physiology of the crab are continually changing along with the stages of the molt-cycle. For example, as shown in Figure 13-4, the internal water content is varied through the cycle by active processes of uptake and excretion. Similarly, the concentrations of organic food reserves and of mineral salts in the hepato-pancreas, and elsewhere in the animal, vary in a pattern corresponding to the Drach stages. Since Drach's work, it has been found that his stages are accurate in detail for all the Brachyura, and also for most other higher crustaceans with thick exoskeletons. Detailed application to smaller Crustacea is difficult but, in more primitive forms, it seems as though the Drach D stages are more extended in time and the C stages shortened.

In all arthropods, there are behavioral changes associated with the molt-cycle. Most go into hiding during the period when they are unprotected by the usual exoskeleton (for example, in higher Crustacea, during Drach E, A, and B stages). In many arthropods, this hiding behavior involves reversal of normal reflex patterns (for example, in light gradients). The molt-cycle even affects population dynamics in arthropods. In some forms, a survivorship curve for a population might resemble a mirror image of the arthropod growth curve (Figure 13-3), with a series of steps corresponding to the increased chance of death at each molt, in contrast to the more usual smooth concave curve of survivorship in most nonarthropod invertebrates.

It has long been known that the molt-cycle of arthropods is under hormonal control. Once again, the details are best worked out for insects and the higher crustaceans. It is convenient to postpone any discussion of insect hormonal and neuroendocrine mechanisms to Chapter 17, so that the following section will deal only with molt controls in the Crustacea.

FIGURE 13-5. The process of ecdysis or molting of the old exoskeleton in a swimming crab, *Portunus depurator*. The old integument is above and the newly molted "softshell" crab is emerging below. Size increase and hardening will follow. Photo © Douglas P. Wilson.]

Neurohormones and Other Crustacean Controls

Our present detailed knowledge of neurosecretory and other endocrine controls of molting and reproductive cycles in arthropods stems from an original discovery made early in this century. It was that in some crustaceans, extirpation of both eye stalks led to precocious growth and, in many cases, to more rapid and repeated molting. Detailed experimental work, particularly during the last 25 years, has revealed that three groups of structures are principally involved in higher Crustacea. A hormonal secretion from the Y-glands is responsible for the initiation of proecdysis and thence of the molt. The Y-glands lie either in the antennary or the second maxillary segments of the head. In those Crustacea with a terminal C4 stage in the molt-cycle, the Y-glands degenerate when the mature size is reached. Within each eye stalk is the X-organ which is neurosecretory, is molt-inhibiting, and also is inhibitive of the development of the ovaries to functional egg production. The third structure, the sinus gland, also lies in the eye stalk, and is misnamed since it is itself nonsecretory. It involves the endings of axons from nerve cells in the X-organ and elsewhere, receives a secretion from them, and releases this into the bloodstream where it inhibits the activity of the Y-gland. The sinus gland is thus best regarded as an endocrine reservoir.

An outline of these interactions, as they occur in higher crustaceans, is given in Figure 13-6. At least two secretions are involved. One is the molting hormone produced by the Y-gland, and the other is the neurosecretory material stored in the sinus gland but produced in the cells of the X-organ, which is a molt-inhibiting hormone. There are other organs and secretions concerned in the control of metabolism, of pigment distribution, and of the reproductive cycle in crustaceans. These include the pericardial organ (another neurohaemal organ), an androgenic gland formed from cells in the distal wall of the male

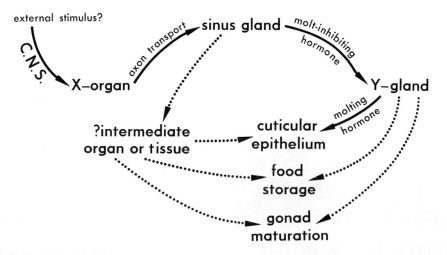

FIGURE 13-6.   An outline of the interactions of the neuroendocrine system controlling the molt-cycle in higher crustaceans. For further explanation, see text.

duct which regulates both spermatogenesis and the onset of male secondary sexual characters in higher Crustacea, and the ovary which secretes hormones directly regulating female secondary sexual characters.

In view of the importance of the neurosecretory material produced by the X-organs, it is worth outlining some aspects of the definition and demonstration of neurosecretory activity. Although neurosecretory cells form part of the central nervous system and are similar in structure to motor neurons, they do not have axons which innervate muscles or any other effector organs in the usual fashion. Their axons mostly terminate in close proximity to a blood vessel or sinus. The cell body often, and the axon at times, contain characteristic stainable droplets or granules which are not present in quantity in other nerve cells. Extraction and injection experiments demonstrate that they contain high concentrations of substances which if circulating in the blood act as hormones, that is, substances of high biological activity on certain other tissues. Thus there are three characteristics involved in their detection: their anatomical position and connections, their histochemical properties, and their biological activity if extracted.

It must be remembered that all nerve cells in metazoan animals are secretory —at their synapses. Therefore, it is often suggested that neurosecretion could have been evolved by a modification of certain motor neurons. A few investigators—notably R. B. Clark—have pointed out that the evolution of control mechanisms in animals could have taken the opposite course, that is, secretory control in animals could have preceded nervous control. According to this alternative hypothesis, neurohormones have not been evolved from the neurotransmitter substances characteristic of all synaptic integration and transmission, but the functioning of nervous systems in metazoan animals represents an evolutionary specialization of a more widespread (and more ancient) use of chemical messengers by living organisms. Irrespective of these theories, one temporal aspect of control mechanisms is highly significant. Neurohormones have a considerably *longer* "half-life" before being degraded by enzymatic action than do neurotransmitter substances, but a *shorter* period of effective action on appropriate target tissues and organs than do regular hormones.

Within the last decade, it has become clear that the control of molting in all arthropods involves chemically similar molting hormones. Since the most sophisticated experimental studies of hormonal coordination in invertebrates have been carried out in insects, a full discussion of this will be deferred to the section on insect hormones in Chapter 17. Meanwhile, it is sufficient to note that the molting hormone produced by the Y-gland in crustaceans is ecdysterone, or β-ecdysone, and this differs only by one additional hydroxyl group from the α-ecdysone which is the molting hormone of insects. Both have been isolated and chemically purified, and more recently both have been synthesized. It is now clear that either will induce molting in physiologically suitable arthropods of all groups. Another important control substance in insects is juvenile hormone, secreted from an endocrine gland termed the corpus allatum, which has no exact equivalent in crustaceans. In addition, the secondary sexual characters of insects seem to be directly determined by the sex chromosome mechanism and are not, as we noted in crustaceans, affected by hormones derived from other tissues.

Other Functional Systems

Before we move to a survey of functional morphology in the varied extant groups of the class Crustacea, it is worth outlining certain general features in the structure and function of the internal organs in the group. This is possible because there is considerable constancy in internal organization in crustaceans, and also in insects and arachnids. Unlike such animals as molluscs and echinoderms, where adaptive specialization of external structures usually involves considerable modification of internal organ systems as well, crustaceans have a fantastic variety of body forms and limb patterns but a rather stereotyped organization of the internal organ systems. The ecology and mode of feeding of any group of crustacean species are reflected in their limb patterns and body tagmatization, but not to any extent in their circulatory and excretory systems, as *is* the case in the Mollusca.

The central nervous system does show some variation within crustaceans. Its most primitive organization is as a ladder-like cord of ganglia, metamerically arranged. There is a greater concentration of ganglia in the more advanced crustaceans. For example, in primitive forms the supraoesophageal complex is made up of the antennular ganglia plus the optic lobes, while in the Malacostraca this is joined by the antennary ganglia. From this complex, the circumoesophageal commissures run to the suboesophageal ganglia, made up of the antennary units alone in most primitive Crustacea, but involving several segments in most of the more advanced groups. In crabs there is a posterior fusion of a large number of the thoracic ganglia. A variety of sense-organs occur in Crustacea, and some will be described in more detail later. They include compound eyes, simple ocelli, chemoreceptors (some on the distal parts of walking limbs), tactile and vibration-detecting mechanoreceptors, statocysts for gravity detection and response to turning movements, proprioceptors, and stretch receptors relaying both static information about limb posture and registering limb movements, and, in terrestrial forms, sensory receptors capable of detecting humidity changes.

In many crustaceans, the alimentary canal is a simple tube with a limited extent of endoderm restricted to the midgut and an ectodermal, and thus cuticularized, stomodaeum and proctodaeum. In the higher Crustacea we find the shortest midgut and the stomodaeum with various chitinous modifications including filters and triturating mills. The midgut may have diverticula, in some cases forming a hepatopancreas as the main site both of enzyme secretion and of absorption. The digestion is entirely extracellular, the foregut being concerned with mechanical processing, the midgut with chemical processing, and the hindgut with the formation of faecal pellets, sometimes formed sheathed in cuticular material.

The circulatory system varies, though functionally it always involves return of blood to the heart through haemocoelic spaces. The heart varies in shape but always has ostia and receives blood through them. It may, or may not, pump the blood out through a system of arteries, but there are never return vessels (veins) but only a series of sinuses. The rate of heartbeat in crustacean hearts varies greatly with environmental factors. In many forms, the rate of beat is doubled, and in some, tripled, by a rise in temperature of 10°C, and some crustacean hearts, in translucent forms, respond to changes in light intensity. In most, the heart is neurogenic; that is, the beat originates in nerve cells on the wall of the heart. In marked contrast to conditions in vertebrates, where they are antagonistic secretions, both acetylcholine and adrenalin, on perfusion through crustacean heart preparations, cause increased rates of beating.

There are usually only one pair of excretory organs, although there is considerable evidence that these are derived from a series of segmental coelomoducts (without nephridia). Adult Crustacea usually have a pair of maxillary or antennal excretory organs, but a few larvae have both, and some others change from one to the other in the course of the life-cycle. Since the excretory organs lie in the haemocoel, their duct system is always blind at the coelomic end. In most nonmarine crustaceans, the segmental excretory organs are larger and more elaborate, and this usually involves the addition of a long, intermediate convoluted tubule between the blind end-sac (with its attached labyrinth) and the storage bladder (with its short duct to the exterior). Such an addition probably corresponds to the evolution of the kidney tubules in nonmarine chordates. These crustacean segmental organs are clearly involved in osmotic and ionic regulation, and functional aspects have been best studied in freshwater crayfish, where urine hypotonic to the blood is produced. Samples taken from different parts of the antennal gland in *Astacus* have clearly demonstrated that the fluids in the end-sac and labyrinth are iso-osmotic with the blood and become hypotonic during their passage through the intermediate tubule but before reaching the bladder. As is the case with annelid segmental nephridia, the precise mechanisms involved in this process are still a matter of controversy among physiologists. It is possible that the functioning is closely similar to that in the vertebrate kidney tubule, with filtration in the crustacean end-sac and labyrinth being followed by reabsorption of ions in the intermediate tubule. However, the observed changes in ionic concentrations along the length of the crustacean excretory organ could result from secretion of water into the intermediate tubule, although this is less likely.

14 CRUSTACEAN DIVERSITY I

THE diagnosis of the class Crustacea need only read "arthropods with two pairs of antennae." Crustaceans have three somites in front of the mouth, the second and third of which each bear antennae, and this is true of no other arthropod group. The seeming triviality of this diagnostic feature again emphasizes the stereotyped structural patterns which characterize the major arthropod classes, in spite of their enormous species diversity. The probable relationships between the arthropod groups will be discussed later (see Chapter 16). Meanwhile, although some workers suggest that the crustaceans are polyphyletic, recent discoveries make this extremely unlikely. The available evidence suggests that all crustacean groups derive from a common ancestral type, and that that type bore certain close structural resemblances to the extinct arthropod stock, the trilobites.

As already noted, there is a tendency for increased tagmatization in the more advanced forms of Crustacea. As well as this specialization into series of different types with modification of limbs, there is also a tendency for a reduction in the total number of segments. There are up to 40 segments in some lower Crustacea, while the higher Crustacea have a consistent pattern of 19 fully developed segments plus one embryonic one, making a total of 20 segments. The segments of the primary tagmata in a primitive crustacean such as *Triops* run: head, 1 + 5; thorax, 11; abdomen, 22. For a higher crustacean, a typical decapod (such as a lobster or a crayfish), they run: head, 1 + 8; thorax, 5; abdomen, 6. This indicates both the reduction of total numbers and the process of increased cephalization.

The appendages borne on these series of segments consist primitively of one biramous pair in each segment, based on the protopodite plus exopodite plus endopodite pattern (Figure 14-1A). However, there are also three single appendages which can be developed: the labrum or upper lip, the labium or lower

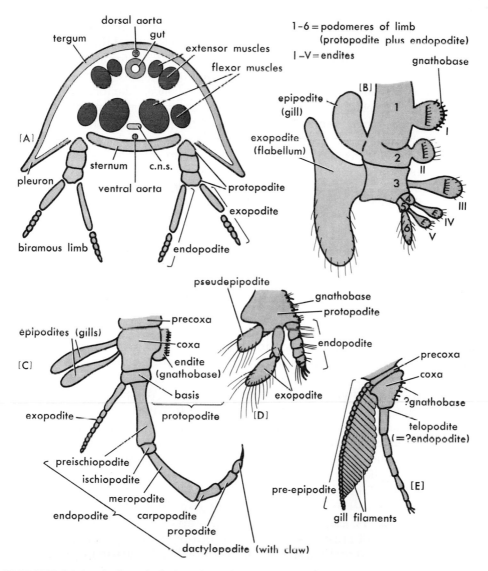

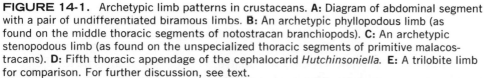

FIGURE 14-1. Archetypic limb patterns in crustaceans. **A:** Diagram of abdominal segment with a pair of undifferentiated biramous limbs. **B:** An archetypic phyllopodous limb (as found on the middle thoracic segments of notostracan branchiopods). **C:** An archetypic stenopodous limb (as found on the unspecialized thoracic segments of primitive malacostracans). **D:** Fifth thoracic appendage of the cephalocarid *Hutchinsoniella.* **E:** A trilobite limb for comparison. For further discussion, see text.

lip, and the telson or tail plate, which may be forked or entire and may itself bear claws or jointed rami. At the anterior end, the typical biramous appendages include the antennules or first antennae, which are clearly preoral and usually bear sense-organs, but may be used also in locomotion or in copulation. The antennae or second antennae are in front of the mouth but have nerve connections postorally; they are usually more clearly biramous than the antennules, but are also primarily sensory in function. The mandibles lack exopodites and each has the precoxa (part of the protopodite) developed as the chewing tooth, while the endopodite forms a tiny mandibular palp. These first three pairs of biramous appendages form the functional locomotory limbs of the nauplius larvae (see Chapter 10, and Figure 10-3). Behind the mouth are the first and second maxillae, both usually phyllopodous in form. The maxillary segments correspond to the end of the true head, though three further segments and pairs of appendages are added in decapod Crustacea like lobsters and crayfish. The trunk appendages on the thorax are generally called pereiopods, and those on the abdomen pleopods. The first to third pereiopods in the decapods and the first pereiopods in several other groups of crustaceans have become accessory mouth parts as maxillipeds. In *Triops*, there are 11 pairs of pereiopods and 22 pairs of pleopods. In the higher forms, there are five pairs of undifferentiated pereiopods and six pairs of pleopods. Particularly in these latter types, the trunk limbs can show great variation in form, being specialized either for feeding, for respiration, for walking or swimming, or for genital and other purposes.

An outline classification of the class Crustacea is presented in Table 14-1. Most textbooks divide the class Crustacea into six subclasses. Recent discovery of living representatives of two primitive genera has required the erection of two more groups: subclass Mystacocarida to include *Derocheilocaris* and subclass Cephalocarida for *Hutchinsoniella*. Some authorities have placed the latter as the most primitive order within the subclass Branchiopoda. A few authors consider that the Mystacocarida and Copepoda should be combined in one subclass. An old collective term for all the subclass except the Malacostraca (which constitute the higher Crustacea) was the Entomostraca. This collective term has no phyletic justification whatsoever. Within the Malacostraca, reinvestigation of the primitive forms dwelling in caves and hot springs has added a fifth major "division": the Thermosbaenida, and an additional order within the division Peracarida: the Spelaeogriphacea. In this book, reference to the "lower Crustacea" usually involves the Branchiopoda and the two "new subclasses," Cephalocarida and Mystacocarida. Similarly, reference to the "higher Crustacea" involves the Malacostraca only. Within this last enormous subclass, one order—the Decapoda—encompasses most of the familiar crustaceans: crabs, lobsters, crayfish, shrimps, and prawns. Of the more than 28,000 described species of crustaceans, over two-thirds fall in the Eumalacostraca, and all but about 200 of these species in the divisions Peracarida and Eucarida.

Living Archetypes

In some ways, it is relatively easy to construct a working archetype for the class Crustacea—one with an integrated concert of organs and functions. Apart from

TABLE 14-1
Outline Classification of the Crustacea

Subclass 1 CEPHALOCARIDA
Subclass 2 BRANCHIOPODA
        Order LIPOSTRACA (fossil only)
        Order ANOSTRACA
        Order NOTOSTRACA
        Order CONCHOSTRACA
        Order CLADOCERA
Subclass 3 OSTRACODA
Subclass 4 MYSTACOCARIDA
Subclass 5 COPEPODA
        Order CALANOIDA
        Order HARPACTICOIDA
        Order CYCLOPOIDA
        (and at least four orders of parasites)
Subclass 6 BRANCHIURA
Subclass 7 CIRRIPEDIA
        Order THORACICA
        Order ACROTHORACICA
        Order ASCOTHORACICA
        Order APODA
        Order RHIZOCEPHALA
Subclass 8 MALACOSTRACA
  Series I LEPTOSTRACA
        Order NEBALIACEA
  Series II EUMALACOSTRACA
    Division 1 SYNCARIDA
    Division 2 THERMOSBAENIDA
    Division 3 PERACARIDA
      Order MYSIDACEA
      Order SPELAEOGRIPHACEA
      Order CUMACEA
      Order TANAIDACEA
      Order ISOPODA
      Order AMPHIPODA
    Division 4 EUCARIDA
      Order EUPHAUSIACEA
      Order DECAPODA
        Suborder MACRURA–NATANTIA
        Suborder MACRURA–REPTANTIA
        Suborder ANOMURA
        Suborder BRACHYURA
    Division 5 HOPLOCARIDA (or STOMATOPODA)

involving the features archetypic to all arthropods, mainly those associated with the exoskeleton and its functioning, it would show, as already suggested, relatively little tagmatization or specialization of appendages. It would have a long series of similar segments, each bearing a pair of biramous appendages, probably phyllopodous, with all appendages being involved in many functions. It would necessarily be aquatic, with the limbs serving for respiration and the filtering of food particles as well as creating a water current for feeding and locomotion. On the other hand, the construction of hypothetical types is complicated by the actual survival, in certain living crustaceans, of structures and functions as primitive and unspecialized as any in our hypothetical arche-type. Such primitive structural features are displayed by the recently discovered cephalocarids. Further, the functioning of a series of unspecialized, multipurpose limbs can readily be investigated in detail in several anostracan genera, which include readily available animals of moderate size, more suitable for experimental work than the minute cephalocarids.

Probably the most primitive living crustaceans are the three species included in the recently erected subclass Cephalocarida. The first of these was discovered and described by H. L. Sanders in 1955 as *Hutchinsoniella macracantha*, the earliest specimens being obtained from an offshore mud-sand bottom in Long Island Sound. The results of a detailed investigation of microanatomy and life-cycle were published in 1957, but the significance of the many unspecialized features was never in doubt. These features include the possession of eight pairs of segmental organs in the thoracic segments (unlike the limitation to one or two pairs in all other Crustacea), the series of nearly uniform paired phyllopodous appendages which include the second maxillae as well as the eight pairs of thoracic limbs (see Figure 10-3C), the occurrence of gnathobases on all nine pairs, and the essentially triramous form of all these appendages (Figure 14-1D). Many authorities agree on the general similarity of these limbs to those of trilobites (Figure 14-1E). Among the more specialized features of *Hutchinsoniella* are the hermaphroditic condition (with separate gonads), and the organization of the ten abdominal segments which are somewhat cylindrical with pleural spines but no appendages. These features are connected with the nature of the habitat and the small size (about 3 millimeters long).

Obviously, the cephalocarids are closely related to the much more extensive subclass Branchiopoda, wherein are classified some of the other more primitive crustacean forms as well as some highly successful specializations built on the same ground plan. The evolution of this group has been based on the development of the phyllopodous multipurpose limbs, and the four living orders represent radiating lines of increasing specialization of the feeding limbs and reduction of their locomotory and respiratory functions. Throughout the four groups there is a tendency for increasing extension of a carapace enclosing the anterior segments. The first order of this subclass shown in Table 14-1 is the Lipostraca, represented only by fossils. The genus *Lepidocaris* includes forms, remarkably well-preserved from the Rhynie chert in Scotland, of Devonian age (see Chapter 34), which are similar in size and general organization to *Hutchinsoniella*. However, the thoracic limbs are in two sets, presumably having had differentiation of functions, and there is a pair of very large second antennae which were probably used for swimming as in the modern Cladocera. There is no carapace, and while in some other ways more primitive than *Hutchinsoniella*, *Lepidocaris* also shows more specializations.

The order Anostraca encompasses the least specialized living branchiopods, without any carapace and with the trunk limbs all similar and all used in both feeding and swimming. Genera include: *Artemia*, living in concentrated brine; *Branchipus*, living in brackish water; and *Chirocephalus* and other genera living in fresh waters and known as "fairy shrimps." Most of the species reach lengths of 2 centimeters, while some freshwater forms reach lengths of 10 centimeters as adults. There may be both ecological and evolutionary significance in the peculiar and highly restricted habitats frequented by these anostracan species, which involve them in characteristically ephemeral life-cycles. Natural populations of *Artemia* are found only in the world's limited range of salt lakes and brine pools. The freshwater genera and species are confined to temporary ponds, such as vernal pools formed by spring thaws or rains, where they can appear for a few years in some ponds but not others, and then somewhat unpredictably shift to a different pattern of local distribution. Ecologically, the fact that they are never found in larger, longer-established bodies of fresh water is probably related to the presence of predators such as fish which cannot live in temporary ponds. In a broader evolutionary sense, these are in some respects "relict" forms—forms more widely distributed early in the geological time-scale (Chapter 35), which now can thrive only in restricted habitats where the chance of competition from more advanced types (such as Cladocera and higher crustaceans) is greatly reduced. The continued existence of otherwise "primitive" (or, better, conservative) branchiopods like the Anostraca as "living archetypes" has depended upon their specializations of life-cycle, particularly of reproduction and dispersal, which allow them to colonize vernal freshwater ponds and brine pools.

Archetypic Feeding

Any student who hopes to understand archetypic organization, and particularly feeding processes, in crustaceans should spend some time making detailed observations of the limb movements, and the consequences, in living shrimps of a form like *Artemia* or *Chirocephalus*. Figure 14-2A shows the general body form and the crude movements of water, and of collected food particles, produced by the appendages. The paired limbs are similar (Figure 14-2B), each with seven endites bearing the filtering setae and with both exites and endites arranged to hinge backward as valve flaps. The shrimp swims on its back, and the same limb movements are used both to propel it forward and to collect particulate food. Anostracans are true filter-feeders, collecting and ingesting all suitably sized organisms (bacteria, diatoms, flagellates, etc.) suspended in the water and also fine organic detritus. No specialized movements of individual limbs are involved—the backstroke of the limbs is the effective propulsive stroke, while the sequence of events during the recovery, or forward, stroke allows for the filtration of the suspended food particles from the water. This will be best understood by reference to Figure 14-2C1, C2, C3, and D. The dual functioning depends on the metachronal rhythm of beating which passes along the series of segments. In general, water is drawn in toward the midline and expelled posteriorly and laterally. Meanwhile a food bolus is formed in the midline and pushed forward. The metachronal beat alternately enlarges and reduces the boxes formed between adjacent limbs. These boxes are created during the recovery stroke because of the manner in which the various flaps (exites, exopodite, and endites) are hinged to the main part of each limb. Therefore, as is shown in Figure 14-2D, when each limb moves forward toward the head, the exites are pressed back on the limb behind by the flow of water as the animal swims along, and this prevents water from entering from the outside

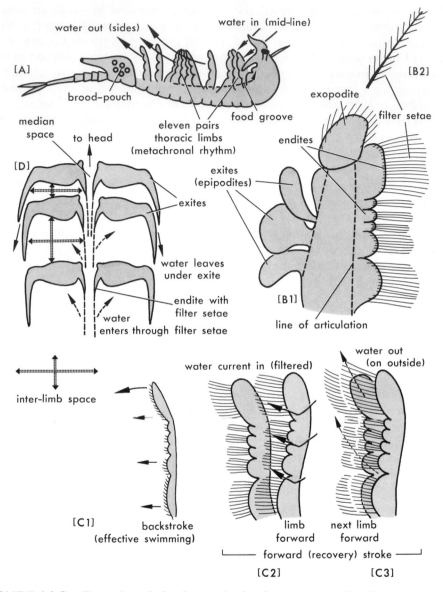

FIGURE 14-2.   The archetypic feeding mechanism in anostracans like *Chirocephalus* and *Artemia*. **A:** General body form, swimming posture, and water currents. **B:** One of the uniform series of thoracic limbs, with endites and exites arranged to hinge backward as valve flaps. **C1–C3:** Propulsive and recovery strokes of thoracic limbs as viewed from the midline. **D:** The same recovery strokes viewed in horizontal section, showing the sequence of enlargement of the interlimb spaces (water inflow from midline through endite setae) followed by compression (expelling water posteriorly under the exite flaps), which sequence results from the metachronal rhythm. For further discussion, see text.

edge. Thus water is drawn into the temporarily expanding interlimb space by way of the midline, passing from the median space through the filtering setae on the endites into the interlimb space. However, when the next limb in series presses forward in turn, water *can* be pushed *out* under the exites and add slightly to propulsion. It should be noted that the backstroke is, of course, the effective swimming stroke with all the "flaps" nearly in the same plane as the rest of the appendage, and that this stroke does not contribute to the collection and filtration of food.

The food particles are filtered on the endite setae, and thus come to lie in the median space (never in the interlimb spaces, as is suggested in some textbooks). The labrum secretes mucus which helps consolidate the food-string from the median space into a bolus between the maxillae, whence it is manipulated by chewing movements of the mandible into the mouth. It is worth stressing that in this feeding mechanism, as found in forms like *Artemia* and *Chirocephalus*, there are no specialized limbs concerned only in water propulsion or as filters. It is merely the dual action of the flaps as valves when the limb is being pushed forward in its recovery stroke that allows a series of exactly similar limbs to form a series of "temporary filtering boxes." Due to the metachronal rhythm, each interlimb space is enlarged in turn, water is drawn in from the midline to a region of reduced pressure temporarily produced, and that water is filtered on the setae of the endites. While this proceeds, blood is oxygenated within these limbs and the animal is propelled forward. This mode of action in a series of like phyllopodous limbs, allowing them to subserve the functions of feeding, respiration, and locomotion, probably represents the archetypic mode of action of crustacean limbs. It is tempting to adopt the hypothesis that the limbs of trilobites were moved in a similar fashion, that is, that the rhythm of beat of their appendages was metachronal.

Other Branchiopods

In some of their other characters the anostracans are somewhat more specialized. Many of them produce eggs which show a remarkable capacity to resist desiccation, and some, like those of *Artemia*, can be stored in diapause in a dry bottle on a laboratory shelf until needed. The sexes are separate and may show considerable dimorphism. The females have a brood-pouch carried on the last thoracic segment, and there are no abdominal appendages. The head of a male anostracan, shown in Figure 14-3A, demonstrates some of the more peculiar features of anostracan organization, including the stalked eyes characteristic of higher Crustacea—the naupliar eye retained in the adult suggesting primitive organization—and the peculiar structures of the male second antennae which are used to attach it to the female as it swims in copulation.

There are only two essentially similar genera—*Triops* (known in some older books as *Apus*) and *Lepidurus*—in the order Notostraca, and they are somewhat rare, occurring irregularly in temporary ponds all over the world. The carapace forms a dorsal shield which covers about one-half of the body and results in their common name of tadpole shrimps. There are numerous similar phyllopodous appendages used in swimming, feeding, and respiration. *Triops* and its allies, however, move along the bottoms of ponds, stir up mud, and are apparently capable of feeding on larger particles including detritus. All the thoracic limbs have gnathobases capable of chewing, and some of this is done between the posterior limbs before the food passes forward. There is some variability in numbers of segments and limbs, which is presumably a primitive fea-

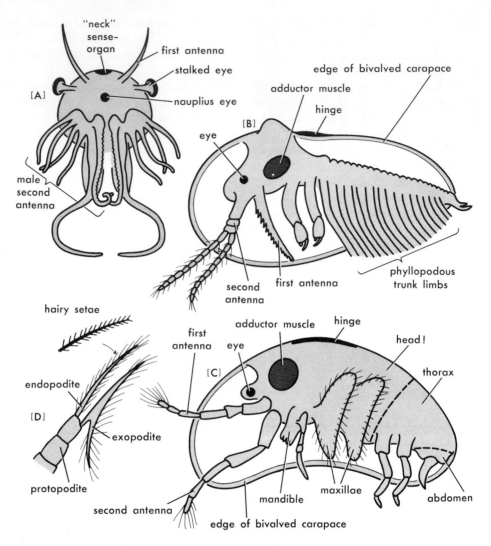

FIGURE 14-3. **A:** Front view of head structures in a male anostracan. The second antennae are modified for copulation. **B:** Diagram of a typical conchostracan with one valve of the carapace removed. **C:** Diagram of a typical ostracod with one valve of the carapace removed. **D:** Thoracic limb from *Argulus,* a branchiuran ectoparasite.

ture, but there are usually from two to five pairs of similar phyllopodous limbs on the abdominal segments and then five limbless abdominal segments to the telson. As already noted, this gives various species of *Triops* about 45 segments in all. The undifferentiated limbs are closely similar in form to those of anostracans. However, the first and second trunk appendages have modified endites of sensory importance and act as additional antennae, while the eleventh in females bears an egg-cup in which the developing eggs are retained. Most populations which have been discovered consist of parthenogenetic females, males only rarely being found. In summary, notostracans can be characterized as branchiopods somewhat specialized, in limb patterns and in reproduction, for life on the bottoms of temporary ponds.

The order Conchostraca is another smallish group of rare species, with a characteristic large carapace, hinged dorsally, which can enclose the whole animal. An adductor muscle (Figure 14-3B) is developed anteriorly and can close the two valves of the carapace. There are two series of similar phyllopodous limbs, differing in that the hind limbs do not have filtering setae but have strong spines which can be used like gnathobases to break up food particles. All species use the hind limbs of the series to chew up larger food particles, while using the front limbs for filter-feeding as in the Anostraca. An interesting reflex involves the clawed telson which, if a large particle gets jammed in the medial space, is bent forward and then rapidly straightened ventrally to kick the offending material out. In some species, the trunk limbs are no longer concerned with locomotion and the second antennae are used to produce jerky movements rather like some Cladocera. In some of their features, the conchostracans can be regarded as intermediate between the primitive branchiopod types (Anostraca and Notostraca) and the most specialized group—the highly successful order Cladocera.

More than half of the eight hundred living species of the Branchiopoda are cladocerans—the so-called water-fleas. The majority are filter-feeding forms like *Daphnia,* but there are a few predaceous carnivores which catch and feed upon other small crustaceans. The group is mainly freshwater with a few marine species. A carapace covers the entire trunk, but not the head, and the second antennae are the main means of locomotion, producing the series of jumps which gives the group its common name. There are usually only five pairs of trunk limbs (though some forms have four pairs or six pairs), and these form one functional filter pump. The abdomen is usually reduced and bears a claw-like telson. The great majority of cladocerans have direct development from big eggs and larval stages are omitted. The feeding mechanism, in typical forms like *Daphnia,* is totally unlike that in anostracans and can be best understood with reference to Figure 14-4. The single nearly watertight box of variable volume forms a very fast-acting pump filter. The box is closed on the dorsal side by the ventral groove of the trunk, laterally by the folds of the carapace, has an anterior wall formed of the protopodites of legs 1 and 2, a ventral wall formed by the exopodites and bristles of legs 3 and 4, and a posterior wall formed by the fifth pair of appendages. The third, fourth, and fifth pairs of appendages are those which move, alternately enlarging and reducing the volume of the box. Their forward stroke enlarges the volume, and food-bearing water is sucked in as in phase I (Figure 14-4B1) between legs 3 and 4. As the back stroke begins (phase II in Figure 14-4B2), these close together and the water is sieved through the interlaced hairs of their bristles. The volume is further reduced and the debris squeezed on the bristles toward the food groove in phase III, and finally the fifth legs turn from the transverse to the sagittal position and water is expelled posteriorly through the median cleft. This is phase IV of the feeding process. Meanwhile, the filtered detritus pressed up into the food groove is carried forward partly by the water current and partly by mechanical sweeping from the bristles. It is kneaded to form a bolus between the mandibles and then ingested. The rapid pumping of this filtering box forms a

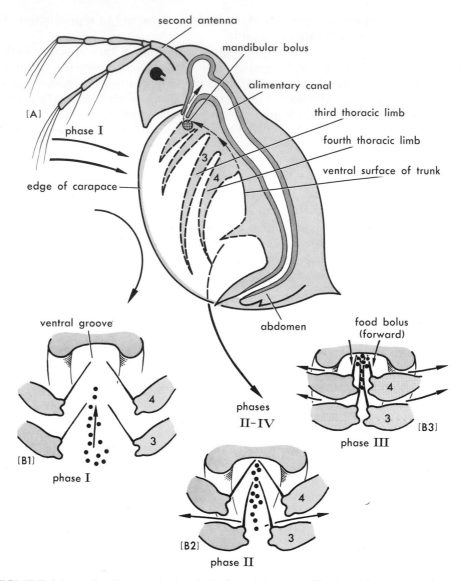

FIGURE 14-4. Feeding mechanism in typical cladocerans like *Daphnia,* where a single "box" of variable volume forms a fast-acting pump filter. **A:** General body form, swimming posture, and water currents. **B1:** Box enlarging as third and fourth thoracic limbs move ventrally (away from trunk) drawing in water and food particles in phase 1. **B2** and **B3:** Box being compressed in phases II and III by third and fourth thoracic limbs in active "propulsive" stroke toward trunk, with filtered food particles being pressed into median groove. For further explanation, see text.

very efficient food-gathering mechanism, so that *Daphnia* can be cultured with only bacteria as food. Although the individual limbs in cladocerans are not much modified from the branchiopod pattern, the functional pattern of their movements has been considerably modified to produce the sequence outlined above.

Some other aspects of cladoceran functional organization should be noted. The sense-organs are relatively well-developed, all forms having chemorecep-tors, mechanoreceptors, and eyes. The nauplius median eye is retained in the adults of some species, but there is also a large internal eye formed of a fused pair of compound eyes. This is said to be sessile but is moved by internal mus-culature. It is much larger and more complex in the predaceous forms of clado-cerans. For example, in *Podon*, a marine predaceous form, one-third of the bulk of the body is made up by this compound eye. The excretory organs are maxil-lary glands. The heart is modified from the primitive tube of the other branchio-pods, being shortened to a barrel-shape with two ostia. Since there are no blood vessels, the beating of the heart simply stirs the blood around the haemocoel. A number of freshwater cladocerans have haemoglobin as a blood pigment (or the potentiality of developing it under certain circumstances).

Sexual reproduction in cladocerans exhibits certain bizarre features. In gen-eral in freshwater populations, relatively few males are found, and partheno-genesis is common. In many cases the annual reproductive cycle appears to in-volve an alternation of parthenogenesis with ordinary sexual reproduction. The ecological aspect of this is that parthenogenesis can fill the pond with indi-viduals under favorable conditions, while fertilized eggs are desiccation-resistant and can persist through winter or through drought conditions. Such resistant eggs are also of enormous importance as a factor in the passive distri-bution of these forms. In a few species, all fertilized eggs seem to hatch as fe-males, and it can be shown that selective abortion during sperm maturation is responsible. The chromosome number remains constant, of course, throughout parthenogenesis, but when the unfavorable season approaches, among the parthenogenetically produced young are a few males. How this is achieved is still really unknown, but its adaptive significance is obvious. Simultaneously, eggs which need to be fertilized are produced by a reduction division. It has been suggested that substances are ingested from algal cells (which are already slowing their reproduction rate) and can induce the production of males and of meiotic eggs. As in several other groups of freshwater invertebrates, including rotifers and ectoprocts, a modification of reproductive life-cycle allows the cladocerans to exploit the peculiar conditions of the environment provided by temperate fresh waters.

Although the largest number of cladoceran genera are relatively cosmopoli-tan forms, similar to and feeding like *Daphnia*, there are also predaceous gen-era. *Polyphemus* and *Leptodora* in fresh waters, and *Podon* (Figure 14-5) and *Evadne* in the sea, are specialized carnivores with huge eyes, a reduced cara-pace, and reduced thoracic appendages. Of course, such carnivorous forms are never present in fresh waters in such enormous numbers as the species on which they feed. Such typical microherbivore genera as *Daphnia* and *Bosmina* can be the principal constituents of the animal plankton of fresh waters.

A typical nauplius larva occurs in the development of some members of all groups of the branchiopods, indeed, in some species in almost all groups of crustaceans. As already noted, the nauplius hatches from the egg and then there are molt stages with the nauplius form and three pairs of appendages, and then a molt to the early metanauplius with four pairs of appendages, and only after this do subsequent molts take the development into different patterns in

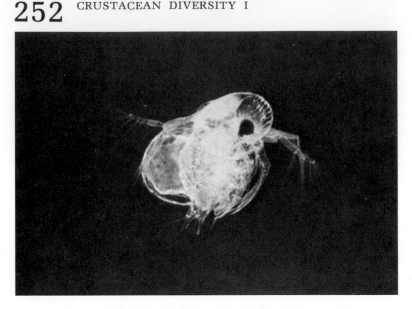

FIGURE 14-5. Living specimen of the tiny predaceous cladoceran *Podon intermedius* from the marine plankton. The single large compound eye is very characteristic both of *Podon* and of the closely related genus *Polyphemus,* which is a carnivore in many freshwater plankton communities. [Photo © Douglas P. Wilson.]

different groups. Later development, however, always involves the addition of other segments posteriorly (see Chapter 10). In several groups of crustaceans, a nauplius occurs in development but does not live a free-living planktonic existence, but is retained in a brood-pouch or even inside an egg membrane.

Of the six remaining crustacean subclasses, three groups have been enormously successful: the copepods, the barnacles, and the malacostracans, or "higher" Crustacea. The other three minor groups are treated together here merely for convenience, not to reflect any presumed relationships.

Three Minor Groups

The ostracods form a relatively stereotyped subclass. Conservative systematics would include only one order, the Ostracoda, with a small number of genera to encompass the two hundred or so cosmopolitan species, which are not uncommon in all kinds of fresh and marine waters. The development of the carapace into a bivalve shell with an adductor muscle parallels the case of the conchostracans, but in the ostracods only the rather stenopodous head appendages are important (Figure 14-3C). Some forms retain two pairs of trunk limbs, but others have none, and locomotion and feeding are principally carried out by the head appendages. Some aspects of their physiology, particularly of circulation and respiration, and even of their comparative internal anatomy, remain obscure. Many have two pairs of excretory organs as adults, antennary and maxillary, while a few have a third set associated with the first maxillipedal segment. They occur in a wide variety of marine and freshwater habitats: some in the plankton, some burrowing in mud with spade-like antennae, some climbing weeds with prehensile antennae, some ectoparasitic on higher crustaceans with the antennae modified as suckers, and even some species limited apparently to the pools of water in certain epiphytes or to other specific plants in the tropical rainforest. Their ecology has been little studied. A number of species are luminescent, mostly belonging to three fairly common marine genera. A grotesque claim to distinction is the possession in some forms of the largest sperms in the animal kingdom. In one species of *Pontocypris*, adult males 0.3 millimeter long have been reported to have sperm 6 millimeters in length.

The minor subclass Branchiura again consists of a single order and only a few genera, of which *Argulus* is typical. The 75 species of so-called carp-lice are ectoparasites of both marine and freshwater fish. They exhibit a number of features clearly associated with parasitism, including: suckers on the maxillae, a piercing spine in front of a suctorial proboscis, and marked dorsoventral flattening. However, they attach temporarily only, actively swimming from host to host; they have separate and equal-sized sexes; and they lay relatively large eggs singly on stones. Their relationships with other groups of crustaceans remain obscure, but most evidence suggests that they evolved (entirely independently of other crustacean parasites) from a fairly highly evolved free-living stock. Their pair of compound eyes and maxillipeds resemble those of the higher crustaceans, while the remainder of their thoracic limbs are biramous and stenopodous with hairy spines (Figure 14-3D), somewhat intermediate in character between those of the Copepoda and the barnacles. Of course, in the ectoparasitic branchiurans these limbs are used only for locomotion and have no feeding significance.

The third minor subclass is the Mystacocarida, which was set up recently for the unique genus *Derocheilocaris*, the first species of which was discovered and described in 1943 by R. W. Pennak and D. J. Zinn—the first specimens being found in intertidal sand at Nobska Beach on Cape Cod. Mystacocarids are minute crustaceans (about 0.4 millimeter long as adults) which live in the interstitial water between sand grains on certain littorals, and they have now been found in many parts of the world. Some anatomical features can be termed primitive, including the unspecialized biramous mandibles in the adult, the retention of a nauplius eye, and the retention of both antennal and maxillary glands as excretory organs. Other features—such as the long body of similar segments not grouped into tagmata—could involve the retention of ancient characteristics or could represent specializations, along with the minute size, for life in the interstitial habitat. Most authorities would agree that the mystacocarids are in several ways intermediate between the branchiopods and the big specialized group—the class Copepoda.

Copepod Success

The subclass Copepoda is of major importance, involving more than five thousand species in the sea and in fresh waters: many of them as members of the permanent zoöplankton, a few specialized for other habitats, and a large number modified as parasitic species. The free-living species—particularly those of the plankton—are relatively uniform in their functional anatomy. Ecologically, they are of supreme importance in the economics of the oceans and of larger bodies of fresh water.

Species of marine copepods in the genera *Calanus* and *Temora* are probably the most abundant animals in the world. They are almost certainly more numerous than the most numerous species of insects on land (see Chapter 17), and they may even outnumber the much smaller free-living soil nematodes (see Chapter 9) and soil protozoans. Of course, the whole populations of marine copepods are continually feeding, growing, and reproducing while, at any one time, a large proportion of the soil populations of both nematodes and protistans are in inactive diapause with "resting" metabolism. These copepods are microherbivores (ecologically of "the second trophic level") which feed directly upon the world's largest crop of primary producers—the flagellates, diatoms, and other single-celled green plants of the marine phytoplankton—a crop at least five times larger annually than that of all land vegetation, including human crops, added together (Figures 14-6B, 14-7A, and 14-11; see also Chapter 33). The single most abundant species—in terms both of numbers of individuals and of fraction of total animal biomass—is almost certainly *Cal-*

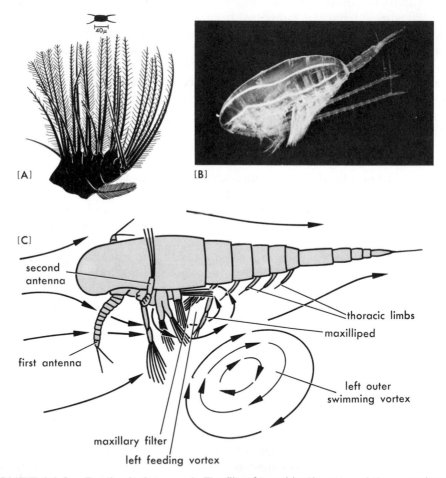

FIGURE 14-6. Feeding in *Calanus*. **A:** The filter formed by the setae of the second maxilla of *Calanus* with a food diatom, *Chaetoceros,* drawn to the same scale. **B:** Lateral view of a living *Calanus*. **C:** Diagram of a lateral view of *Calanus* while feeding. It shows the vortices created in swimming which bring food particles into the midline and through the maxillary filter. [**A** adapted from S. M. Marshall and A. P. Orr, in *J. Mar. Biol. Assoc. U.K.,* **35:**587–603, 1956; photo **B** © Douglas P. Wilson; **C** adapted from H. Graham Cannon, in *Brit. J. Exp. Biol.,* **6:**131–144, 1928.]

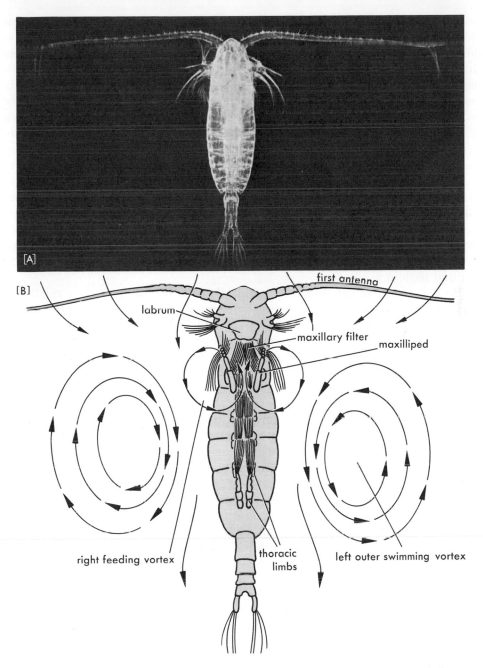

labrum

first antenna

maxillary filter

maxilliped

right feeding vortex

thoracic limbs

left outer swimming vortex

FIGURE 14-7. Feeding in *Calanus*. **A:** Dorsal view of a living Calanus. **B:** Ventral diagram of the vortices created in swimming which bring food particles into the midline and through *the maxillary filter*. [Photo **A** © Douglas P. Wilson; **B** adapted from H. Graham Cannon, in *Brit. J. Exp. Biol.*, **6**:131–144, 1928.]

anus finmarchicus or a closely similar form. Ecologically, such copepods provide the major part of the direct food supply of the world's largest animals—whalebone whales like the blue whale; some of the most abundant fishes—herring, menhaden, sardines, and mackerel; and the largest fishes such as the basking sharks (*Cetorhinus*) and whale sharks. From a human standpoint, copepods may hold one key to our future survival. As a species, our numbers are rapidly increasing, but about 45% of individual humans alive at present (1978) have insufficient diets, involving shortage of proteins. Marine copepods like *Calanus* constitute the world's largest stock of living animal proteins. Only a tiny fraction of this is now involved in human nutrition, and that mostly by indirect and energetically inefficient means. (For example, calanoid copepods of the Eastern seaboard of the United States are consumed by menhaden, among other fish species. A multimillion dollar fishing and processing enterprise catches menhaden and turns them into fish meal. This, in turn, supplements the diet of swine and broiler-fowl by being incorporated in the feeding-stuffs of "intensive" agriculture. Humans eventually ingest less than 1/10,000th part of the original protein.)

The body form in the planktonic and other free-living species is relatively uniform: the elongate body without a carapace is never more than a few millimeters in length. A nauplius eye is retained and slightly elaborated as the single central eye of the adult; the mouth parts and first maxilliped are used in feeding; the other five pairs of thoracic limbs are used for swimming. The stages in the life-cycle are always similar, even in parasitic forms (Figure 14-8D, E, F). There is a typical nauplius followed by a metanauplius with four pairs of appendages, followed by a series of copepodite stages resembling the adult but with only three pairs of biramous thoracic limbs, and finally molting to the adult copepod. The fact that the early copepodite stages have the posterior appendages and even the posterior segmentation underdeveloped is of course highly significant as evidence for the pattern of metameric morphogenesis earlier claimed as archetypic (see Chapter 10). The molt to the adult usually involves the addition of two more pairs of trunk limbs, reduction of the second antennae and great development of the antennules, and the development of long jointed rami on the furca of the telson. The genital openings are developed on the seventh trunk segment behind the last pair of limbs. The exact pattern of the adult varies in the three main orders (Figure 14-8A, B, C), the distinctions depending on where the flexible joint of the trunk occurs, and the biramous or uniramous nature of certain appendages. Separation of orders on such trivial anatomical features once again illustrates the stereotyped morphology (among a large number of species) which occurs when a successful pattern of animal machine is successfully exploited.

Locomotory and feeding functions are carried out in a similar fashion in the free-living forms of all three orders. They can swim smoothly by rhythmic beating of the trunk limbs, or, particularly in calanoids, by jerks of their large antennules. Most of them also use a sudden flexing of the body joint as an avoiding reaction. The thoracic appendages also serve as respiratory surfaces and create their own respiratory current. In several genera like *Cyclops*, the thoracic appendages are united transversely so that the four rami of each pair beat as one. Copepods can feed by seizing individual large particles or organisms in their

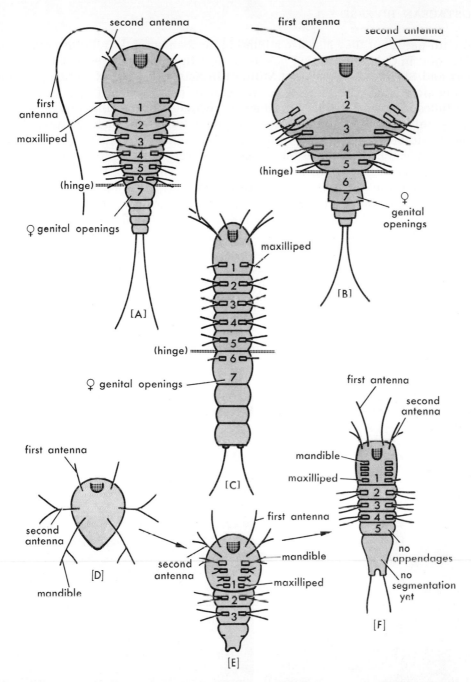

FIGURE 14-8. Copepod organization. **A:** Stylized calanoid copepod. **B:** Stylized cyclopoid copepod. **C:** Stylized harpacticoid copepod. Note the position of the flexible trunk joint in, and the minor limb differences among the three groups. (Mouth-part appendages are omitted.) **D–F:** Stages of larval development in copepods. Compare with Figure 10-4. **D:** Stylized nauplius larva (basically cyclopoid) with three pairs of limbs. **E:** Stylized late metanauplius with eight pairs of appendages including two pairs of thoracic limbs (basically calanoid; workers on zoöplankton would designate this stage "nauplius VI"). **F:** Stylized copepodite stage with three pairs of thoracic appendages and the fifth thoracic segment already differentiated.

257

mouth-parts, but they also use a filtering mechanism. Like all other aspects of function in *Calanus*, feeding has been most thoroughly investigated by A. P. Orr and Sheina M. Marshall at Millport in Scotland. When *Calanus* is swimming slowly, a series of vortices are created in the water. These bring small particles in toward the filter mechanism, which is of the single box type (but quite unlike that in *Daphnia*). Here the box is formed of the head appendages: the mandibular palps, maxillules, maxillae, and maxillipeds all being involved. Water is drawn in as the pressure is lowered by an outward swing of the maxillipeds with their fringe of long setae. The volume of the box is then decreased and the water expelled by the maxillules being pushed forward, the expelled water passing through the sieve of the maxillary setae (Figure 14-6A).

The particles thus filtered are removed from the "inside" of the sieve by being brushed forward by specialized long setae of the maxillipeds or by the endites of the maxillules. This brushing directs them between the mandibles and thence into the mouth. In most forms, including *Calanus*, the basket-like sieve of setae and the maxillae themselves are not moved during the filter-feeding process. However, in some species of *Acartia*, the setae of the maxillae can be spread apart and then drawn back together in a "clasping" movement.

As developed in calanoid copepods, the filter-feeding mechanism is superbly efficient for the retention of food organisms in the size range of from 10 to 30

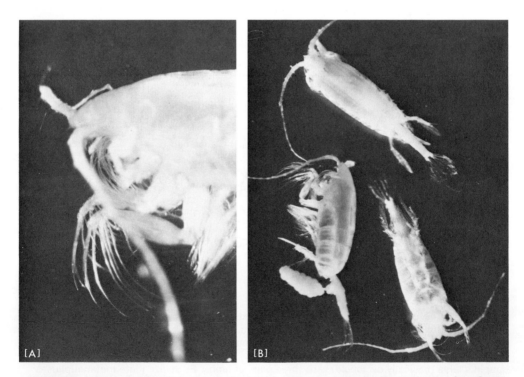

FIGURE 14-9.   A "giant" copepod, *Euchaeta,* collected from deeper cold water in a Scottish fjord, Loch Fyne, and preserved in formalin. **A:** Head with mouth-parts spread for food capture. × 14. Compare the maxilla with that of *Calanus* in Figure 14-6. **B:** Three adult specimens, including one female with eggs. × 4. [Photos by the author.]

microns. These same copepods also individually catch and ingest diatoms of considerably greater size. Almost all the dominant forms of the phytoplankton can be utilized as food by copepods like *Calanus*. Most investigators agree that the suspension-feeding is probably a secondary development within the group, copepods in general being primarily raptorial. The giant copepod *Euchaeta*, a form more frequently raptorial than *Calanus*, is shown in Figure 14-9A.

The seasonal life-cycle of *Calanus* in a Scottish sea-loch is illustrated in Figure 14-10 (see also Chapter 4 of my earlier book, *Aquatic Productivity*). There are three successive broods in such Scottish waters and off the New Jersey coast, contrasting with a single annual generation in the high Arctic. With the three-generation pattern of life-cycle in temperate seas, the biomass of *Calanus* in summer is of the order of a hundred times that in the same locus in the winter months. Such different life-cycle patterns in different latitudes allow populations of *Calanus* to exploit the changing seasons of plant productivity (Chapter 33).

As already emphasized, the calanoid copepods are dominant members of the marine zoöplankton (Figure 14-11), and *Diaptomus* species often occupy a similar position in the economy of large freshwater lakes. Cyclopoid copepods are common in bodies of fresh water of all sizes and include both benthic and planktonic forms. Harpacticoid copepods are mostly benthonic, living in and over bottom-deposits in both the sea and fresh waters. A few relatively unspe-

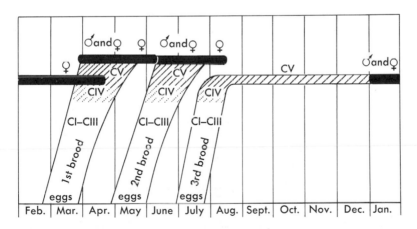

FIGURE 14-10. Life-cycle of *Calanus*. The diagram shows the annual succession of generations of *Calanus* in Loch Striven. The sizes of the stages are indicated by their positions in the vertical scale, with the eggs, copepodite stages, and adults indicated for each of the three broods. Note that the overwintering generation has a life-cycle lasting through 8–9 months and that the overwintering stage for that brood is the Vth copepodite, or preadult stage. In *Calanus,* the number of generations per year is related to the climatic and trophic conditions of the environment: a single annual generation in the high Arctic (for example, in the waters of the east Greenland fjords), two generations per year in the coastal waters of Norway and in the Gulf of Maine, three generations in Scottish waters and off New Jersey, and four generations in warmer and richer conditions such as parts of the English Channel. [From W. D. Russell Hunter, *Brit. Assoc. Handb.* (*Glasgow, 1958*), **5**:97–118, 1958, after Marshall and Orr, 1953.]

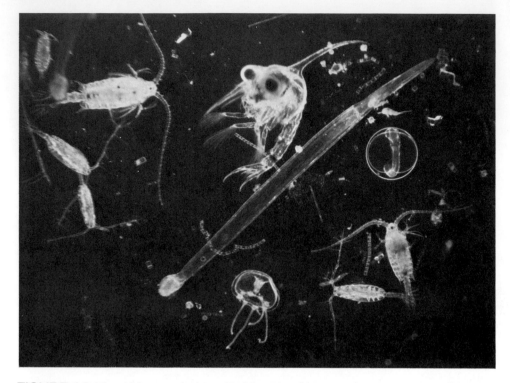

FIGURE 14-11. Living zoöplankton. Various calanoid copepods are seen along with an arrow-worm, *Sagitta,* diagonally across; a zoea larva of a crab above the arrow-worm, near center; a developing fish egg, below and to the right; a medusa; and, near the bottom, the fine chains of diatoms which serve as the basic food supply for all these forms of the animal plankton. [Photo © Douglas P. Wilson.]

cialized ectoparasitic forms occur in the cyclopoid and harpacticoid copepods, and it may be from forms like these that the more specialized parasitic orders have evolved.

Copepod Parasites

There are numerous parasitic genera in the subclass Copepoda, and they can be ranked as a series showing progressively greater degrees of modification for the parasitic habit. They range from commensal scavengers which can be occasional parasites through ecto- and endoparasites to extremely modified tissue parasites.

The simplest anatomical modifications may be those of the mouth: the labrum and labium form a sucking tube around the mandibles which become elongate piercing stylets. In more specialized forms there is a reduction of appendages and attachments to one host is permanent. In some, all parts of the attached adult become reduced except the gonads, and in many of these the males are shrunken, tiny forms permanently attached to the females. In the most modified forms, only the fertilized female survives as a bag of tissue—which is partly host tissue produced by reaction—enclosing the mature female gonads which enter a continuous and long-continued egg production. It is noteworthy that almost all species, even the most modified, show the entire series of larval stages: from nauplius to metanauplius to copepodite to adult copepod (Figure 14-8D, E, F).

The least modified copepod parasites fall into two groups. First, there are those in which only the mouth is slightly modified and the free-living organization of body and appendages is retained: such forms live in the lumen of the intestine of fishes, marine mammals, and large decapod crustaceans. Most of the species in such genera as *Botachus* and *Enterocola* are host-specific, although they usually feed only on faeces and rarely graze on the intestinal wall. Secondly, there is a group of ectoparasites of fishes, like *Ergasilus* (Figure 14-12A) and allied genera, which feed mainly on mucous secretions and have the antennae modified as clutching hooks. Related forms live on the gills and in the mouth cavity of marine and freshwater fishes and suck blood from gill lamellae and other exposed tissues.

Caligus (Figure 14-12B) and its allies show the further modification of the maxillae developed as suckers, grasping mandibles, and a generally clumsy build with an enlarged genital segment. These forms can still swim from host to host and are specific ectoparasites on the gills of various fishes and of *Nautilus*. The adult females of *Chondrocanthus* (Figure 14-12C) cannot swim and are permanently attached to the host after settling there. The appendages are reduced to blunt lobes, and the minute male is permanently attached to the genital segment of the female. The egg-sacs are in the form of a pair of long continuously produced filaments. Species of *Chondrocanthus* are permanent parasites on the gills of various fishes.

Both life-cycle and anatomical specializations are a little more complex in species of the genus *Lernaea*. In the copepodite stage, they are temporary parasites on fish gills, leaving the first host, which is usually a flatfish, after molting to adult copepods. They then copulate and males die, but the females attach to gadid fish, usually in the gill tissue. The body becomes vermiform with an enormously enlarged genital segment, and the anterior part of the head becomes modified into a series of branching roots which grow deep into the musculature of the fish (Figure 14-12D). Once again, there is continuous egg production for a long period of time.

Xenocoeloma is a genus of parasites of polychaete worms; the adults form mere sacs and feed by roots completely within the host. There is an extraordinary degree of host-collaboration, the sac wall being largely of host tissue, with even an invagination of the host coelom to provide a sort of false alimentary canal for the mass of parasite gonads. A normal nauplius is known, but some details of the life-cycle are controversial. Apart from the two long external egg-sacs, the adult form of *Xenocoeloma* is completely unlike a copepod.

In one group—*Monstrilla* and related forms—the late larvae are parasitic and the adults are free-living. Nauplius and metanauplius are free-living, but after this the host worm or mollusc is entered. The copepodite stages are passed as a long wormy parasite which builds up food stores living in the blood vessels of the host. The final copepodite then metamorphoses and breaks out as a rela-

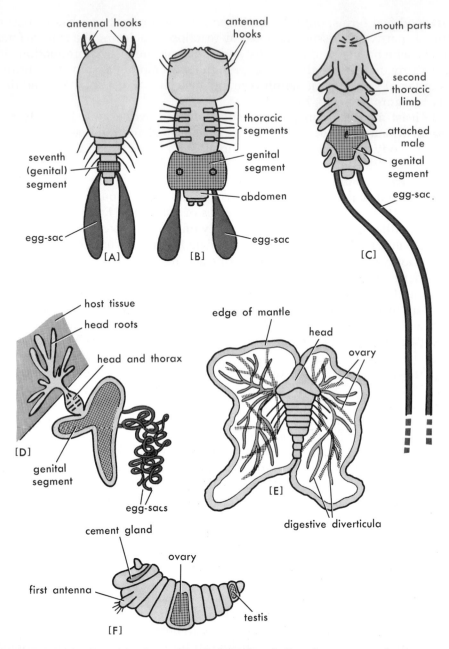

FIGURE 14-12. Parasitic copepods and cirripedes. **A:** *Ergasilus,* a copepod ectoparasitic on fishes. **B:** *Caligus,* another fish ectoparasite, with maxillary suckers and an enlarged genital segment, but retaining a basically copepod body form. **C:** *Chondrocanthus,* a permanently attached copepod parasite of fishes, with a minute male and continuous egg production. **D:** *Lernaea,* adult female copepod parasite, with branching roots embedded in the host fish's musculature. **E:** *Laura,* an adult parasitic cirripede whose host is an antipatharian coelenterate. **F:** *Proteolepas,* an unusual maggot-like cirripede, discovered by Charles Darwin as a parasite of stalked barnacles.

tively unmodified adult copepod lacking, however, any functional gut. The adult lives on its stored food and produces eggs, and thus acts as the dispersal stage in these parasites.

Barnacles

The major subclass Cirripedia is a widespread and successful marine group of nearly one thousand species. The majority are typical barnacles, specialized as adults for an entirely sessile mode of life, but there are also a series of parasitic forms which show progressively more extensive modifications. All have similar life histories. In characteristic barnacles (Figure 14-13 and 14-14)—all placed in the order Thoracica—the head becomes fixed down to the substratum by the antennules at metamorphosis, and the adult lives permanently attached using six pairs of biramous, thoracic limbs as a food-catching mechanism, which kick food into the mouth. The abdomen is reduced to a vestige, and most forms have become hermaphrodites. Adult acorn barnacles like *Balanus* (Figure 14-13A), with their calcareous shell cemented down to a rock surface, have no obvious resemblance to other crustaceans, and even Cuvier classified them with molluscs. (Early medieval myths had stalked barnacles like *Lepas* as development stages of "barnacle-geese.") Their true nature was elucidated (in

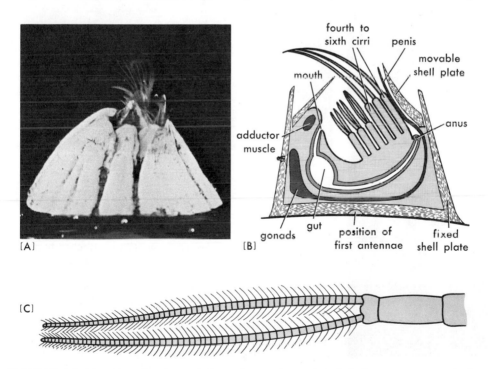

FIGURE 14-13. Functional morphology of acorn barnacles. **A:** *Balanus crenatus,* a typical acorn barnacle photographed alive while "fishing." **B:** The general anatomy of an adult acorn barnacle, showing the fourth, fifth, and sixth thoracic limbs as the elongate cirri used in food collection. **C:** A single cirrus, biramous, many-segmented, and fringed with stiff setae. For discussion of feeding methods, see text [Photo © Douglas P. Wilson.]

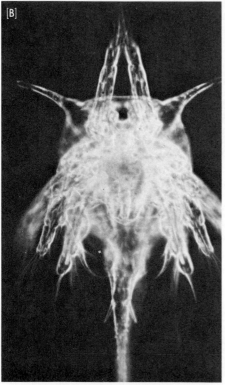

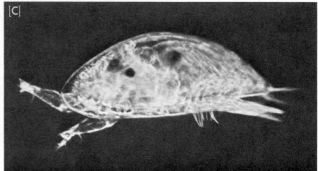

FIGURE 14-14. More living barnacles. **A:** A bunch of stalked or ship's barnacles, *Lepas anatifera,* attached to a floating bottle. Several individuals have the feeding cirri partially extended. Compare with Figure 14-13. **B:** The characteristic nauplius larva of *Balanus balanoides* from inshore marine plankton. **C:** The later presettlement "cypris" stage of *Balanus,* with the prominent blunt antennules which will provide the first attachment to a rock or ship's bottom before metamorphosis to the adult acorn barnacle (see Figure 14-13). [Photos © Douglas P. Wilson.]

1833) by J. Vaughan Thompson, a British army surgeon stationed in Cork, whose plankton studies first identified the free-living barnacle nauplius larva which, by way of the cypris-stage, settles and metamorphoses to the adult barnacle.

Cirripede nauplii have characteristic frontal horns and a dorsal spine (Figure 14-14B), which are retained through a series of molts to the last metanauplius. Then the mouth parts are reduced and the only head appendages are the antennules with their terminal suckers. A carapace forms a bivalve shell around this cypris-stage larva, which has six pairs of similar trunk limbs and a reduced abdomen. The cypris settles and then metamorphoses into the adult barnacle. The first attachment of settlement is by the antennules and then cement glands secrete while metamorphosis takes place by a twisting of the trunk (Figure 14-13B), the mouth with eye and adductor muscle moving away from the antennule. A fleshy mantle then replaces the cyprid carapace and begins to secrete calcareous plates. There are two major patterns of further development in the Thoracica: either toward a stalked barnacle like *Lepas* where the fleshy stalk supports a head with calcified plates, or toward an acorn barnacle like *Balanus*, where the conical limpet-like body is totally enclosed in calcareous plates secreted by the mantle. After the shell plates of the adult have been formed, they grow more or less continuously and totally independently of the molt cycle. In barnacles, molts really affect only the exoskeleton of the cirri, as the modified trunk limbs are called, and the surfaces associated with them. There is a very constant pattern of shell plates throughout the Thoracica, and we can trace evolution involving reduction of the number of plates in several lines.

Barnacles are a very successful group of sessile animals, dominating particular zones of the rocky littoral in most parts of the world and incorporating vast numbers of individuals in some species. They have the distinction that Charles Darwin worked for a number of years on their systematics and distribution. Over the last 25 years, there has been a great revival of interest in barnacles—particularly in aspects of their physiology and population ecology. This recent interest is partly spin-off from intensive applied research carried out during the Second World War. Research teams in both the United States and Britain were involved in preventive work on ship fouling. Barnacles are among the most important of marine fouling organisms, and, if not prevented, can add more than 15% to the fuel consumption of an average-sized merchant vessel or decrease by several knots the maximum speed of a warship.

Most species of both stalked and acorn barnacles are hermaphrodites. In relation to the sessile habit, many have long penes, and cross fertilization is achieved by their extension through the colony. A few forms liberate sperm which are taken in by the feeding mechanism to cross-fertilize. Some species of the stalked barnacle *Scalpellum* have dwarf, semiparasitic males attached to the mantle of the large females.

All barnacles catch their food particles with their cirri, three or six pairs being involved, but there are a number of different patterns of feeding activity. In all forms, each cirrus has two long, many-segmented rami bearing stiff bristle-like setae. The only other appendages in the adult are the relatively tiny mouth-parts. The methods of feeding which have been described can be classified into five main patterns, and a few species of barnacles can employ three of them. The first method utilizes the cirri as a passive net, deliberately turned to be held across the direction of any water current. The more posterior and longer cirri (three pairs in *Balanus* and its allies) are those used to form the net, and they may remain extended for periods of minutes before being withdrawn. Extension of the cirri is always relatively slow since it is due to a hydraulic

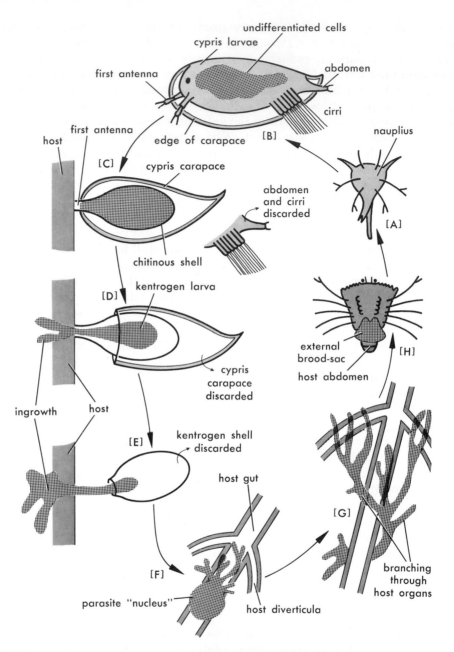

FIGURE 14-15. The life-cycle of the parasitic barnacle *Sacculina*. **A:** A typical barnacle nauplius is released from the brood-sac. **B:** A typical barnacle cypris larva, though lacking a functional gut, is the stage which follows the metanauplius. **C:** The cypris attaches to a host crab by the first antennae, and discards the cyprid appendages. **D:** A further molt of the cypris carapace gives rise to the kentrogen larva, as the undifferentiated parasite tissues begin to invade the host. **E:** The chitinous shell of the kentrogen larva is discarded in turn, as the remainder of the parasite moves internally. **F** and **G:** The ingrowing parasite tissue forms a "nucleus" on the host crab's gut and sends branches through all the visceral organs of the host. **H:** A brood-sac, containing developing eggs of the parasite, is formed externally under the host's reflexed abdomen. From this are released the nauplii of **A**.

mechanism, blood being pumped from the rest of the body into the lumina of the cirri by muscular action. Cirral withdrawal is rapid since it is performed by the flexor muscles which both roll up the cirri in an oral direction and pull them into the "mantle-cavity." The second pattern of feeding activity uses rhythmic movements of the cirri as a sweep net. The long cirri are extended, swept forward and downward in a scooping action, and withdrawn, each cycle taking about half a second. A third type of feeding behavior involves faster repetitions of the first part of the cycle, with the cirri being used again and again in sweeping movements and only being fully withdrawn at long intervals for the ingestion of the trapped food particles. A fourth feeding mechanism involves the short cirri, which are not normally protruded from the shell. Water currents pass through these each time the barnacle opens, and filtration of fine particles can occur even in the absence of activity of the long cirri. These four filtering mechanisms can apparently collect suitable food organisms ranging in size from 2 to over 800 microns in length. A fifth feeding mechanism—exhibited principally by *Lepas* and its allies—is not really filter-feeding. In these forms an individual cirrus may catch a small animal—for example, a copepod; that cirrus then rolls down, independently of the other cirri, and passes the captured organism to the mouth-parts. Local chemoreception is involved in such capture.

The functional systems of barnacles are relatively stereotyped. There is a U-shaped gut (Figure 14-13B) with an expansion which forms a simple stomach. There are no blood vessels or hearts; the body fluids are stirred around by the movements of the cirri and mantle, which organs also provide the respiratory surfaces. The nervous system shows some concentration in the trunk region but is relatively simple in organization. The adult excretory organs are maxillary glands.

Cirripede Parasites

Like copepod parasites, parasitic barnacles show a series of degrees of specialization. In this subclass, the parasitic habit probably first arose from the fact that many regular barnacles in the order Thoracica attach themselves as fouling organisms to other animals—for example, to big molluscs and decapod crustaceans, even to whales and turtles—and are thus transported about. Supplementing their nutrition by sending some roots into the host tissue was probably the first stage of ectoparasitism in such barnacles. Almost all parasitic cirripedes retain a typical barnacle life-history from horned nauplius through cypris stages, and it is usually the cypris larva which first attaches to the host organism.

In the order Acrothoracica, the least modified species live as adults as ectoparasites on whales, turtles, and sharks and retain a rough acorn-barnacle shape, but derive food through a root system in the host tissue. Others live as ectoparasites in the gill cavities of lobsters and crabs, while others bore into the shells of molluscs. A few allied species bore, by an unknown method, into coral or limestone rock and continue to use their cirri in feeding. The order Ascothoracica consists of species parasitic on coelenterates and echinoderms, which, though mostly embedded in the host tissues, often retain fairly obvious cirripede organization. Some show the head structures, six pairs of thoracic appendages, and a segmented abdomen, but in all the mantle is enormously developed as an absorptive sac containing diverticula of the parasite's gut and branches of ovarian tissue. The males are again minute. The genera *Synagoga* and *Laura* (Figure 14-12E) are parasites on antipatharian coelenterates. The order Apoda consists of a single genus, the maggot-like *Proteolepas* (Figure 14-12F), parasitizing stalked barnacles, which was discovered and described by Charles Darwin.

Finally, the order Rhizocephala, the species of which are almost all parasitic on crabs and lobsters, have claims to being the most highly modified parasites in the animal kingdom. The adult is merely a branching structure—rather like a fungus—growing through the host tissues, and at no stage in the life-history is there a functional alimentary canal. Externally, the nauplii and cyprid larvae are like those of typical cirripedes, but internally they have undifferentiated cells in place of a gut.

In *Sacculina*, the typical cirripede nauplius is retained in a brood-sac and liberated as a cyprid without a gut. The cyprid attaches to a bristle on the limb of the crab by its antennules, then moves to the thinner cuticle at an articular membrane. Its tissues bore in, and there is a loss of the cyprid exoskeleton (Figure 14-15), then the loss of a further larval skin. The motile tissue mass is carried in the bloodstream to the midgut of the crab, where it forms a nucleus from which parasitic tissues branch out through the animal in all directions. At a subsequent molt of the host, a brood-sac is formed externally in the reflexed angle of the crab's abdomen. It has long been known that all crabs parasitized by *Sacculina* look like immature females, and this is often attributed to the destruction of the gonads by the parasite. Thus it is quoted in many textbooks as an example of parasitic castration. If a male crab is infected with *Sacculina*, then, with successive molts, it develops a female appearance: the copulatory organs and other dimorphic limbs change, as does the shape of the abdomen. An infected female crab begins to look progressively less mature. The older texts suggest that these secondary sexual characters are altered when the animals are castrated by the parasites, but there has always been the difficulty that, in normal conditions, some of these secondary characters appear long before the gonads. Further, in parasitized specimens, the sex character change appears before the gonads are completely destroyed. It is now known that the androgenic gland (see page 236) controls the development of such characteristics in male crabs, and it appears that that endocrine organ is affected early in parasitization.

Species of *Peltogaster* are similar tumor-like parasites which live in hermit crabs. The genus *Thompsonia* on similar host species forms fungus-like ramifications capable of asexual reproduction. They also form numerous egg-sac bags at any available joints in the host exoskeleton, which bags appear regularly at every molt. The tissue of the egg-sac wall is apparently that of the host.

15 CRUSTACEAN DIVERSITY II

The numerous forms making up the remaining—and by far the largest—subclass of the Crustacea, the Malacostraca, are usually called the higher Crustacea. Once again, however, the most primitive members of the group are filter-feeders with relatively undifferentiated, relatively phyllopodous, limbs which are arranged in a uniform series with little secondary tagmatization.

The accepted phylogeny of the group is based on unusually extensive and detailed studies of comparative morphology, on a real understanding of archetypic function in a few critical types, and on some remarkably pertinent fossil forms. The scheme of interrelationships which has been deduced is therefore an atypically sound piece of exegesis but, unfortunately, a very complex one. Malacostracan evolution has *not* involved a single line of increasing specialization. There are at least four minor lines of "experimental" patterns, and two enormously successful lines of evolutionary specialization, parallel in some features but totally independent, both originating in rather undifferentiated filter-feeding types. Systematically, these are outlined in Table 14-1, in the two series, six divisions, and many orders and suborders of the subclass Malacostraca. As all "natural" classifications are intended to do, this attempts to portray the hierarchy of interrelationships which have been hypothesized. It is significant that, in this one, between class and order, three intermediate levels have to be invoked: subclass, series,

269

and division (or superorder). Further, the accepted classification of certain genera of crabs involves seven more intermediate levels between order and genus: suborder, section, subsection, superfamily, family, subfamily, and tribe. The student should realize that this elaborated hierarchy does not represent mere gamesmanship of museum zoologists, but an honest attempt to portray a complex series of interrelationships involving a highly successful group (of over 18,000 species) which has utilized a remarkably stereotyped basic pattern of anatomy.

The segments and tagmata have a remarkably constant arrangement throughout the Malacostraca. The head has five overt segments (almost certainly six embryonically), the thorax has eight (up to three of which may become associated with the head and carry maxillipeds), and the abdomen has six segments. The only exceptions are the species of *Nebalia* and two allied genera where there are seven abdominal segments, and these few species form the minor series Leptostraca. All others show the constant pattern: 5 + 8 + 6 (Figures 15-1 and 15-2B); throughout the series Eumalacostraca there is no variation in segment numbers. Within the Eumalacostraca, the divisions Syncarida (with 32 species), Thermosbaenida (with 4 species), and Hoplocarida (with 180 species—Figure 15-3) are small specialized groups off the major evolutionary lines. The two main stocks are the Peracarida and Eucarida, each of which numbers about 9000 species. Within each of these divisions is a series from filter-feeding forms—the mysids in the Peracarida and the euphausiids in the Eucarida—to highly modified carnivores and scavengers. These range from forms with a uniform series of phyllopodous limbs to those like crayfish, lobsters, and crabs, with chelae and other specialized limbs. A moderately good fossil record and the existence of undifferentiated living forms like euphausiids, in which limb function can be studied, makes the construction of a con-

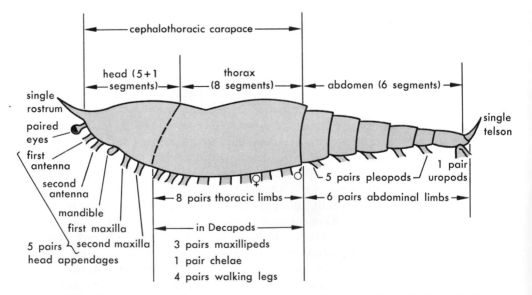

FIGURE 15-1. The archetypic pattern of the Malacostraca, as shown in the *caridoid facies* and in the successful decapods.

FIGURE 15-2. Three types of malacostracan crustaceans. **A:** *Mesodopsis slabberi*, a mysid. Compare with Figure 15-4D. **B:** *Palaemon serratus*, the "common prawn" of European waters. Compare with Figure 15-1. **C:** *Squilla desmaresti*, a hoplocarid or "mantis-shrimp." [Photos of living animals © Douglas P. Wilson.]

[A] [B]

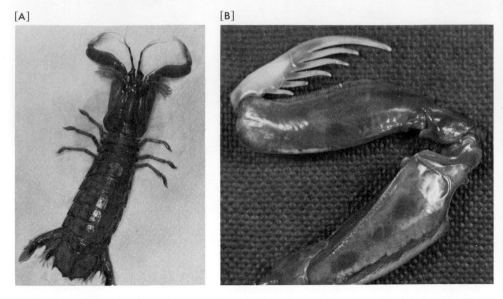

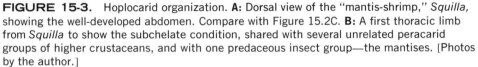

FIGURE 15-3. Hoplocarid organization. **A:** Dorsal view of the "mantis-shrimp," *Squilla,* showing the well-developed abdomen. Compare with Figure 15.2C. **B:** A first thoracic limb from *Squilla* to show the subchelate condition, shared with several unrelated peracarid groups of higher crustaceans, and with one predaceous insect group—the mantises. [Photos by the author.]

vincing archetype somewhat easier than is usual. In fact, the archetypic plan of organization in the Malacostraca has long been defined, and is termed the *caridoid facies*. In this pattern, the diagnosis of the three tagmata runs: head, with five pairs of appendages (two pairs antennae, mandibles, and two pairs maxillae); thorax of eight segments with uniform limbs, each with a walking endopodite, with an inwardly directed fringe of setae for filter-feeding, a swimming paddle exopodite, and a respiratory epipodite; and an abdomen of six segments, five with swimming pleopods and the sixth with uropods which, with the telson, form the tail fan.

Independent evolution of more specialized forms in the two groups—Peracarida and Eucarida—has resulted in parallel loss of the undifferentiated characters of the *caridoid facies* or, crudely, the characteristic general form of all shrimp-like relatives). The diagnostic features of the peracarid line include direct development, the thoracic segments not fused together with at least four being free from the carapace if present, and possession of a brood-pouch formed of the oöstegites on the thoracic limbs. The features of the Eucarida include a complicated larval development in most forms, a cephalothoracic carapace fused with the dorsal parts of all the thoracic segments, and no brood-pouch formed of oöstegites. Before we consider these two major stocks, the primitive relicts and other aberrant forms that make up the four minor groups of Malacostraca will be briefly discussed.

Experimental Patterns

In the series Leptostraca, there is a single order Nebaliacea consisting of three living genera. *Nebalia* and its allies have strong claims to being archetypic for the Malacostracan lines. Closely similar fossil genera are found more or less continuously back to the Cambrian (see Chapter 34). The eight thoracic segments bear uniform phyllopodous appendages, closely similar to those of *Hutchinsoniella* and to the limbs of trilobites. Other features are peculiar to the group: a two-valved carapace with an adductor muscle, a jointed rostrum, stalked eyes, a peculiar first antenna which may be primitive, and seven abdominal segments ending in a telson with rami. The last features are totally unlike the other malacostracans, as is the reported occurrence in some species of a series of eight pairs of segmental organs. The young show direct development.

Leptostracans are all detritus-feeders, using the thoracic appendages in a mechanism with resemblances to—but probably not directly homologous with —that of the more primitive branchiopods. As *Nebalia* moves through the fluid mud in which it lives, a food stream is drawn in anteriorly and the water passes out at the posterior of the carapace. The setae on the podomeres of the endopodites of each limb form the filtering mechanism (Figure 15-4A). The functional morphology in these forms is largely known as a result of the studies of H. Graham Cannon. Typical species of the group such as *Nebalia bipes* live characteristically in soft mud under stones in the littoral in many parts of the world. The genus *Nebaliopsis* lives in the oceanic plankton. It is almost certain that the typical mud-dwelling forms were evolved with some specialization of carapace and feeding mechanism from the free-swimming animals of the *caridoid facies*, and that *Nebaliopsis* represents a subsequent return to a planktonic habit from such detritus-feeding forms. In contrast to the 18,000 known species of the series Eumalacostraca, only 7 living species of leptostracans have been described.

Undeniably, the most primitive forms included *within* the higher crustaceans or Eumalacostraca are the 32 species of the division Syncarida. They all live in fresh waters in rather relict habitats and show most of the diagnostic features of the *caridoid facies*, although they do not have a carapace. There is little difference between the thoracic and abdominal segments; in this, they are more primitive than all the other malacostracans, and they have no specialized grasping limbs, either chelate or subchelate. Numerous fossil forms appear to be closely related to the extant syncarids, particularly from the Carboniferous period, and the presently living species are limited to geographically and ecologically peculiar habitats. This has probably an evolutionary significance similar to that of the distribution of present-day lungfishes. *Anaspides* is a medium-sized (reaching 5 centimeters long) shrimp-like form found in mountain springs around 4,000 feet in Tasmania; *Paranaspides* and *Koonunga* are modified forms living in larger lakes in Tasmania. They have direct development from large, single eggs. *Bathynella* is minute (about 2 millimeters long) and up to the late 1950's was known only from three cave localities: one in Wales, one near Prague in Czechoslovakia, and one near Kuala Lumpur in Malaya. It has recently been rediscovered as a member of the interstitial fauna of

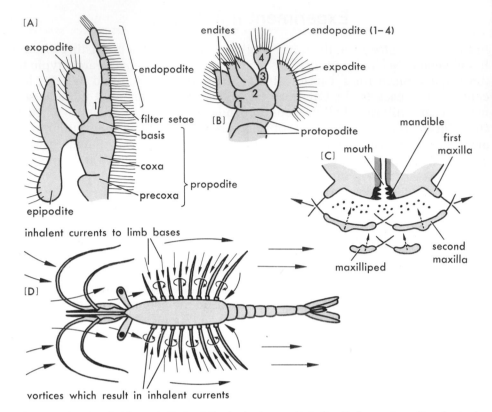

FIGURE 15-4. **A:** Thoracic limb of the leptostracan *Nebalia*. **B:** Second maxilla of a typical mysid. **C:** The maxillary filter press used in the filter-feeding of mysids in vertical cross section. **D:** The feeding and locomotory currents of a mysid, viewed from the dorsal side. Compare with Figure 15-2A. For further explanation, see text.

river sands in twelve more localities all over the world. The genus may be much more widely distributed than has been suspected, but it still shows a relict distribution: restricted to underground waters and to the peculiar habitat provided by the interstitial water in certain types of sand deposits in the beds of rivers having certain physical characteristics.

The second division of the series, the Thermosbaenidae, was recently set up for two genera, *Monodella,* including two Italian cavernicolous species, and *Thermosbaena,* from hot springs in Tunisia. *Thermosbaena mirabilis* cannot survive below 35°C and tolerates temperatures of 47°C (that is, close to the absolute upper limit for metabolism in metazoan tissues). They are clearly eumalacostracan animals, but cannot be fitted into any of the other groups.

The fifth and last division in Table 14-1, the Hoplocarida, consists of about 180 species placed in *Squilla* and related genera. These are not primitive forms, but have several features which separate them from both major stocks of malacostracans (Figure 15-2C). They are all marine, living in burrows, with a short carapace on three thoracic segments. The abdominal segments are unusually large and well-developed, with appendages for swimming and respiration. The head shows an anomalous division into segments, and the first five thoracic appendages are subchelate (catching prey like a closing jackknife—Figure 15-3). There is a planktonic larva of a peculiar type, the erichthus larva. In summary, the hoplocarids or Stomatopoda represent a specialized rather than a primitive group, but one which is well off the main lines of evolution in the higher Crustacea.

Peracarid Stocks

The division Peracarida encompasses a series of six orders, ranging from the mysids which retain many features of the *caridoid facies* through some intermediate groups to the Isopoda and Amphipoda. These last two orders are highly modified, bottom-dwelling scavengers and carnivorous macrophages. The more important characteristics of the division have been noted earlier, including the direct development and the possession of an oöstegite brood-pouch, but a few others are worth detailing. Only in Mysidacaea is there a carapace attached to the first four thoracic segments. Three segments are involved in the Cumacea, and two in the Tanaidacea, while the carapace is absent in the highly successful isopods and amphipods. Throughout the division, the first thoracic segment—and it only—bears maxillipeds, while the mandibles are often asymmetric and all have an articulated movable tooth called the *lacinia mobilis*. In the limbs, if there is any tendency for reduction of the number of podomere units to occur, it is the carpopodite and protopodite which are involved. (In the eucarid line, the ischiopodite and preischiopodite fuse if any reduction occurs.) The fact that these relatively trivial anatomical characters are entirely diagnostic of the more than nine thousand species in the division is highly significant. Once again, evolutionary success involves a stereotyped pattern.

The order Mysidacea consists of a very closely knit group of about 450 species, retaining many features of the *caridoid facies*. They occur in the marine plankton and over littoral sandflats, and there are a few forms in fresh waters which physiological investigations have shown to be relatively recent marine relict species. Some of the species in such genera as *Mysis* and *Praunus* are often very abundant as individuals, forming extensive and dense flocks in shallow inshore waters over sand and forming a major part of the food supply for such fish as flounders (Figure 15-2A).

The order shows the group characters already mentioned: the single maxilliped being followed by seven pairs of undifferentiated stenopodous, but bristly, biramous appendages. The pleopods are small, except for the last pair, which form the uropods of the tail fan (with the telson). Statocysts in these uropods betray their use as both rudders and elevators in steering the swiftly darting mysids. Mysids are excellent swimmers with two entirely different feeding mechanisms. First, largish particles of food can be seized by the endopodites of the thoracic limbs and placed between the mandibles for ingestion. Secondly, the second maxillae (Figure 15-4B) form a flat floor below the mouth and, along with the maxillule (or first maxilla) and maxilliped, act as a filter press—mechanically similar to that of *Daphnia*, though involving different appendages. In *Hemimysis* and related forms, each thoracic limb rotates, causing a vortex in the water which draws in particles to the midline (Figure 15-4D), whence they are transferred to the maxillary filter press (Figure 15-4C). In other forms the filter apparently receives only particles stirred up from the bottom by movements of the animal. From an evolutionary viewpoint, the first feeding method is that of the higher malacostracans, while the filter-feeding resembles that of some branchiopods.

Much of the biology of mysids requires further investigation. They have excellent sense-organs like predators and are good swimmers, and a number of planktonic forms show a diurnal, vertical migration, living on the bottom by day and swimming in the surface waters by night. In spite of the many primitive features in their functional morphology, it seems possible that many mysids could live as predaceous carnivores at night.

A similar body form, but modified for burrowing in sand, is shown by the more than five hundred species of the order Cumacea. The carapace is expanded laterally as a gill-chamber, the eyes are sessile, and the uropods filamentous. Such forms as *Diastylis* are still filter-feeders, using the setae of the maxilla. The pattern of the water current for feeding and respiration is somewhat peculiar: both inhalant and exhalant openings to the carapace-chamber being at the anterior end. Other forms including *Iphinoë* and *Cumopsis* have an unusual feeding mechanism. Essentially it involves scrubbing the organic matter off individual sand grains and using the first and second pairs of maxillipeds to tumble the grain and scrape the food material from its surface. Cumaceans show a cosmopolitan distribution, are all burrowing animals, and are found most typically in the lower littoral and sublittoral zones.

The order Tanaidacea is another smallish group (about 250 species) of burrow-dwelling or tube-building forms, which are in some ways intermediate between cumaceans and isopods in the degree of loss of the *caridoid facies*. The carapace is very small, the eyes usually sessile, and the uropods filamentous—all these features connected with burrowing. The thoracic exopodites are absent or vestigial, but some species have chelae on the anterior thoracic limbs. Both these last intermediate orders of the peracarid stock have shown increasing development of the endopodite and reduction of the exopodite in their thoracic appendages. Thus, they are better food catchers and graspers and increasingly poorer swimmers.

The recently described order Spelaeogriphacea, involving a single genus discovered in 1957 in caves in South Africa, may be related to the last two orders, but is more likely to prove an ancient (but physiologically specialized) offshoot of a mysid-like stock.

Peracarid Success

The two remaining advanced orders of peracarids—Isopoda and Amphipoda—are highly successful and numerous (4000 and 3600 species, respectively). The animals in both groups are poor swimmers, and highly modified as scavenging macrophages and bottom crawlers. They have in common such features as the lack of a carapace, sessile eyes, the first pair of thoracic limbs as maxillipeds, the remaining thoracic limbs stenopodous in form and without exopodites, the pleopods respiratory in function, and uropods not forming a tail fan. Diagnostic differences between the groups are concise and consistent. The isopods are dorsoventrally flattened, with all thoracic limbs similar and all abdominal limbs similar. The amphipods are laterally compressed, with thoracic and abdominal appendages, each arranged in at least two groups, differing in form and function, so that there are from four to six distinct series of cormopodites (or trunk limbs).

Perhaps the largest number of species in the order Isopoda fall into genera like Ligia, which live on the seashore as intertidal scavengers (Figures 15-5 and 15-6). These all show the group characteristic of having "all limbs similar": thoracic limbs with a single ramus of the endopodite for walking and foliaceous abdominal limbs for respiration. There are only one pair of maxillipeds at one end and one pair of uropods at the other, which are different from the two patterns. They have oöstegites on the bases of the thoracic limbs forming a brood-pouch, and they are usually dorsoventrally flattened (Figure 15-5A–C). The numerous species of Ligia, and allied genera such as Idotea, Oniscus, Porcellio, Gaera, and Sphaeroma, vary a little in shape and a great deal in size (adults range from microscopic species to forms about 22 centimeters in length), but are almost uniform in their structural organization. The only anatomical modification is a tendency for fusion of the abdominal segments.

There is a predisposition toward the terrestrial habit in several littoral forms. In many parts of the world the vertical zones of the seashore have specific isopods identified with them, the higher ones being progressively less aquatic in all aspects of their physiology, including respiration and excretion. Several genera, but notably Porcellio and Armadillidium, are completely terrestrial and are known as wood-lice or sow-bugs or slaters. The first pleopod is enlarged and extends posteriorly, covering the others, which are thus kept damp as respiratory surfaces in the air. In forms like Armadillidium, the cuticle on these limbs is invaginated into a ramifying system of tracheae. These obviously increase the efficiency of aerial respiration, even though the tracheae do not penetrate the body of the animal.

The development of tracheae in isopods totally unrelated to the other terrestrial arthropods is a classical example of independent evolution. Behavior in terrestrial isopods has been much studied, and shows all the taxes and kineses in response to such variables as light and temperature, as are appropriate to keep these imperfectly terrestrial animals from the dangers of water loss. Crudely, their behavior is such as to make them cryptic or nocturnal or both. A few genera of aquatic isopods have become euryhaline and invaded estuaries, and there are some completely freshwater genera including Asellus.

There is also a series of parasitic isopod species. The fish-louse, Aega, is a normal isopod in body form but with piercing mouth-parts and hooks on the thoracic limbs. Adult specimens of Bopyrus are permanently fixed parasites living attached to the gills under the carapace of decapod crustaceans. The sex of the adult parasite is apparently not determined until after its arrival in the gill chamber of the host. The first specimen of Bopyrus to invade a decapod inevitably becomes a female, and any later arrivals become males. If a young female is removed from a newly infected host and put into the gill chamber where there is a fully adult female, then the young one's sexual development is

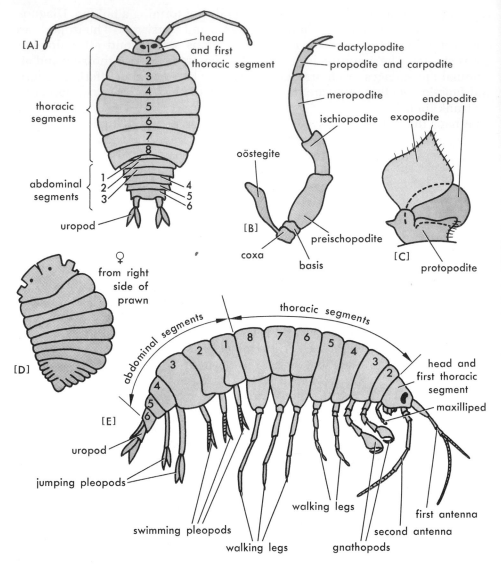

FIGURE 15-5. Peracarid organization. **A:** Body pattern in the dorsoventrally flattened *Ligia,* a typical isopod. **B:** Thoracic limb from *Ligia,* typical of isopods. **C:** Abdominal limb of same. **D:** A parasitic isopod, *Bopyrus,* from the gill chamber of a prawn. For further discussion, see text. **E:** Lateral view of an archetypic amphipod (actually basically gammarid), with characteristic lateral flattening of the body which bears seven types of body limbs in series.

retarded and reversed, and it will eventually become a male. Similarly, young males will develop into females if removed from the presence of females and put into the gill chamber of uninfected shrimps or prawns. The fully developed males are smaller and rather less modified than the females. A further peculiar feature is that the mature females show an asymmetry corresponding to whichever side of their host they live on (Figure 15-5E).

Other species of bopyrid genera are more modified tissue parasites, forming galls on the sides of various shrimps and crabs and often causing parasitic castration. An extreme case of specialization is shown by some parasitic isopods whose hosts are cirripedes of the order Rhizocephala. The isopod genus *Danalia* is hyperparasitic on *Sacculina*, while *Liriopsis* invades *Peltogaster* (parasite on hermit crabs), and these hyperparasites each castrate the parasites in turn. In the latter case, the hermit crab nourishes the fungus-like growth of the cirripede parasite, which in turn nourishes the reproductive cells of the isopod hyperparasite, and, of course, the eggs which are produced in the little external egg-sacs are those of the isopod.

The enormous and successful order Amphipoda is usually characterized by a lateral flattening of the body and almost always by division of functions among the varied limbs. There are about 3600 species, some of which are among the most abundant animals of the marine littoral. The gammarids—which include the abundant sand-hoppers or beach-fleas—have seven types of body limbs in series. The first of the thoracic appendages are maxillipeds, the second and third subchelate gnathopods, the fourth and fifth forwardly directed walking legs and the sixth, seventh, and eighth backwardly directed walking legs. Then there follow on the abdomen: three pairs of swimming pleopods, two for jumping, and the uropod pair. Thus in respect of limb differentiation, the amphipods represent the most extreme development of a peracarid stock from the *caridoid facies*. In the group as a whole, both chelae and subchelae are developed and used in some forms for climbing on seaweeds, burrowing in the sub-

[A] [B]

FIGURE 15-6. Representative peracarids. **A:** A living amphipod "skeleton-shrimp," *Caprella aequilibra,* climbing on the branches of a red seaweed, *Ceramium.* The upright stance is characteristic of caprellids. Some species show intraspecific aggression during which pairs of caprellids "arm-wrestle" and attempt to push each other out of their near-vertical posture. The "victorious" caprellid remains for a time immobile in its upright stance, while the "defeated" one crawls away. **B:** Dorsal and ventral views of a large intertidal isopod, *Ligia.* [Photo **A** © Douglas P. Wilson; photo **B** by the author.]

strate, or constructing dwelling tubes, as well as in handling food. All carry their eggs and developing young in a brood-pouch formed of oöstegites, and there is direct development, the young resembling miniature adults. The characteristic gammarids include species living in the marine littoral, in estuaries, and in fresh waters—there are even a few tropical land-dwelling forms. A few genera show some modification of the characteristic body shape. *Caprella*, the predaceous skeleton-shrimp living on littoral weeds and hydroids, has very elongate limbs and thorax, with a reduced abdomen (Figure 15-6). *Cyamus* is a characteristically flattened ectoparasite on whales. There are no greatly modified parasites in the order Amphipoda.

To recapitulate, the most highly successful forms in the division Peracarida are the isopods and amphipods, which are the two groups most completely—though independently—modified from the primitive malacostracan pattern.

Archetypic Eucarida

The major division Eucarida encompasses the most highly organized crustaceans and the largest living arthropods. They are highly successful animals with large numbers, both of species and of individuals. There are usually several distinct larval stages and metamorphoses in the life-history, and there is a cephalothoracic carapace fused to all thirteen of the segments of head and thorax. There is no brood-pouch, and no *lacinia mobilis* on the mandible. Once again, within the division we find a series of progressively more specialized forms ranging from close to the *caridoid facies* to the true crabs. Systematically, the group is divided into two orders (see Table 14-1), the archetypic euphausiids and the "advanced" decapod crustaceans.

There are only about one hundred species in the order Euphausiacea, but many of them are of great importance in the marine plankton. They are shrimp-like animals of moderate size (2–5 centimeters long), which, except for the eucarid carapace and the larval development, look very like mysids. All the thoracic limbs are similar and biramous with none greatly modified as maxillipeds. They have a single series of small gills at the bases of these legs, and this may involve a respiratory limit upon the size to which euphausiids can develop (see below). The young hatch as nauplii and then pass through zoea-like larvae and mysid-like forms. Euphausiids are certainly closer to the *caridoid facies* than are any decapod crustaceans. The feeding mechanism in typical genera such as *Meganyctiphanes* and *Euphausia* is principally by means of a conical basket which is formed by the inwardly developed setae on the endopodites of the first six thoracic limbs. These are held somewhat stiffly as the animals swim along; water is scooped into the conical basket and filtered through the setae. In many forms, including *Euphausia superba*, the bristles are arranged so as to form two or three distinct mesh sizes in different parts of the filtering walls of the basket. In *Meganyctiphanes*, there is a maxillary filter, like that of mysids, which is used in addition to the thoracic basket.

All euphausiids are marine and strictly animals of the deep-sea plankton. They can occur in relatively dense swarms many miles in horizontal extent, as does *Euphausia superba* in antarctic waters, and they have long been known to whalers as "krill," the principal food of whalebone whales in certain seas and at certain seasons. Over the last twenty years, a group of Russian marine biologists have investigated both biological and economic aspects of a direct fishery for euphausiids in the Antarctic, and Russian nutritionists have evaluated the protein-meal prepared from krill for supplementing the diets of both farm animals and of humans. On a size:energy basis, an important future contribution to global human diets is more likely to come from krill than from the vaster reserves of more usual plankton such as calanoid copepods.

Eucarid Success

The order Decapoda comprises more than 8500 species, of which 4500 are crabs (suborder Brachyura). It includes all the familiar forms: shrimps, prawns, crayfish, lobsters, and crabs, and all the biggest crustaceans. There are always three pairs of maxillipeds, and therefore five pairs of remaining thoracic legs (thus the group name)—usually as one pair of chelae and four pairs of walking legs (Figure 15-1). There can be up to three complicated series of gills—podo-

[A]

FIGURE 15-7. Peculiar crabs.
A: Lateral view of *Emerita talpoida*, an anomuran "mole-crab" from a surf-stirred sandy beach. B: Dorsal view of a young, but algae-encrusted, specimen of the brachyuran "spider-crab," *Libinia emarginata*, from among seaweed on a boulder beach. [Photos by the author.]

[B]

FIGURE 15-8. Male "fiddler-crab," *Uca pugnax*. The one hypertropied chela gives the common name; both chelae are small in the female. [Photo by the author.]

branchiae, arthrobranchiae, and pleurobranchiae—the details differing in various decapod groups. The exopodite of the second maxilla is modified as a bailing paddle whose continuous action pumps water through the gill series. They have the most highly organized central nervous system found in crustaceans and can show complex behavior patterns. Most have a series of larvae, though usually the nauplius stage is suppressed and the eggs hatch as zoea (see Figures 14-11 and 33-7). The chitinous lining of the stomodaeum is developed as a masticatory apparatus, and all the rest of the exoskeleton is increasingly impregnated with calcium carbonate. There is considerable controversy about the subclassification of this huge and successful group. The four suborders set out in Table 14-1, and used here, are a compromise arrangement and probably unsatisfactory as a complete reflection of phylogeny. In general, the macrurous groups are closer to the euphausiids and more like the *caridoid facies*.

The first suborder is the Macrura-Natantia, where the abdomen is relatively large and usually extended and there are well-developed pleopods used for swimming. The carapace (and whole body) are often compressed laterally, and the abdominal segments are approximately equal. This group includes all the familiar prawns and shrimps of such genera as *Palaemon, Penaeus, Crangon,* and *Palaemonetes* (Figure 15-2B).

The Macrura-Reptantia are the crayfish and lobsters, where the abdomen is large and usually extended but can be flexed under the cephalothorax, the walking limbs are well-developed and the pleopods are reduced and not suitable for swimming. The first abdominal segment is usually smaller than those posterior to it. Genera such as *Astacus* and *Cambarus* in fresh waters, and *Homarus*, *Palinurus*, and *Nephrops* in the sea are most extensively eaten by man. Lobsters, such as *Homarus americanus*, can reach weights in excess of 50 pounds, and are thus probably the most massive animals built on the arthropod ground plan. Obviously, such forms as *Homarus* do not have a C4 molt stage which is terminal.

In the third suborder, the Anomura, the abdomen is always small but somewhat variable in shape. The tail fan is usually less well-developed than in the macrurous groups and the abdomen is held flexed or is markedly asymmetrical. The group includes the hermit-crabs, such as *Eupagurus* and *Pagurus*; the squat-lobsters, such as *Galathea*; and the mole-crabs, such as *Emerita* (Figure 15-7). There are also a few forms clearly derived from hermit-crabs, but now free-living, which retain a markedly asymmetric abdomen. These include *Lithodes*, one of the stone-crabs (and the related king-crabs of commerce), and *Birgus*, the terrestrial coconut crab.

Finally, the Brachyura, or true crabs, have the abdomen reduced and carried permanently flexed below the thorax, the pleopods being greatly reduced. The carapace is massive and usually globose, or flat and laterally expanded (Figure 13-5). *Cancer*, *Carcinus*, and *Callinectes* are typical "crab-shaped" crabs; *Maia*, *Hyas*, *Libinia*, and *Macrocheira* are spider-crabs with pear-shaped bodies and long legs (Figure 15-7); and *Uca* and *Sesarma* are fiddler-crabs where one claw of the male is greatly enlarged (Figure 15-8). Specimens of the Japanese spider-crab, *Macrocheira*, are displayed in many museums: the fact that their spread legs can extend to 2.8 meters across makes them the world's largest living arthropods.

As already noted, there can be a full series of larval stages in the lower deca-

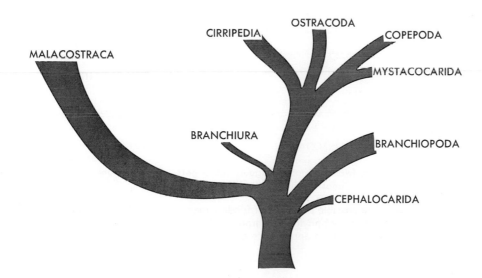

FIGURE 15-9. A compromise phylogeny of the Crustacea, partly modified from Howard L. Sanders. For discussion, see text.

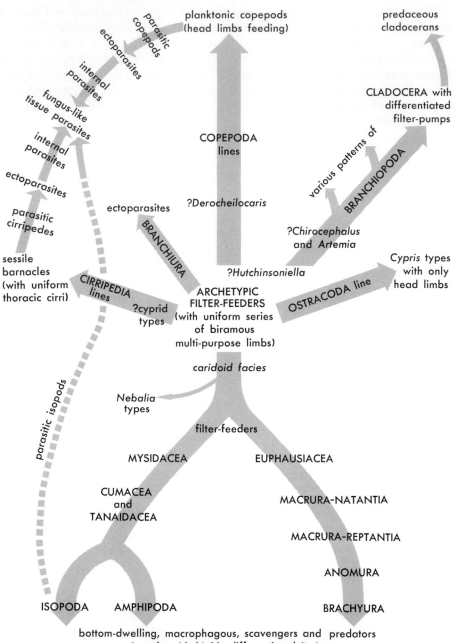

FIGURE 15-10. The possible evolution of feeding mechanisms within the Crustacea. This does not necessarily reflect phylogeny throughout. The feeding mechanisms characteristic of each "line" are discussed at appropriate places in Chapters 14 and 15.

pods, the ontogeny from nauplius via zoea (see Figures 14-11, 33-7) involving addition of segments posteriorly (see Chapters 10 and 33) and the replacement of foliaceous limbs by the stenopodous ones of the adult. In higher forms, the earlier larval stages may be suppressed. In cases where the larger eggs are carried by the females, there is of course no brood-pouch; the eggs are cemented to setae on the pleopods, giving the characteristic berried appearance.

Various phylogenies of the class Crustacea have been set out. The version presented in Figure 15-9 is a compromise one, but takes into account the evidence resulting from recent investigations on the cephalocarids and other primitive forms. In relation to one topic emphasized in the preceding chapters, we can attempt to set out the possible evolution of feeding mechanisms within the class, although this is not entirely phyletic. This is done in Figure 15-10. From primitive filter-feeders with uniform phyllopodous appendages can be derived the major successful groups of the "lower" Crustacea—the barnacles, copepods, and cladocerans—along with the parasites evolved from them and also (via the *caridoid facies*) the two major evolutionary lines of the malacostracan crustaceans. There is one additional aspect which is not brought out in this figure, and that is the markedly larger size of the successful decapod crustaceans (at the bottom right-hand part of Figure 15-10). The tendency of the eucarid line to move away from the *caridoid facies* by becoming macrophagous, bottom-crawling animals has—along with the development of more extensive gills and an efficient system for creating a respiratory current, and the greatly increased impregnation of the exoskeleton with calcium carbonate—allowed the group to evolve an ever-larger body size. Thus we have one of the most successful groups of predatory carnivores in the sea, which includes the largest living arthropods.

16 TERRESTRIAL ARTHROPODS AND OTHERS

THE other two major successful groups of arthropods are the classes Arachnida and Insecta (numbering more than 47,000 and more than 700,000 species, respectively). Unlike the crustaceans, arachnids and insects are primarily land animals, with their whole concert of structures and functions adapted for life in air. In typical winged insects, such physiological processes as aerial respiration, control of water losses, and methods of terrestrial locomotion show efficiencies (in energetic terms) unsurpassed by those of any other organisms. The only other animal groups which approach them in their degree of adaptation for life on land are the amniote vertebrates: reptiles, birds, and mammals. Although the most probable ancestors of presently living arachnids were aquatic arthropods (see page 303), among both arachnids and insects, the few nonterrestrial forms (almost all found in fresh waters) show *secondarily* acquired adaptations, both structural and physiological, for aquatic life.

Thus, the diversity which we have surveyed in the class Crustacea, both in terms of ecological distribution and of functional morphology, is not paralleled in either of the classes Arachnida or Insecta. Both groups are relatively stereotyped structurally and functionally, and both groups show the potentialities *and* limitations of highly specialized terrestrial animal machines. There are several good textbooks of entomology, largely concerned with describing systematically the enormous diversity of species which are built on the rather stereotyped insect plan. Perhaps as a consequence, many general accounts of invertebrate biology omit all mention of insects. A compromise treatment is adopted here, to fit with the book's purpose as a comparative survey of functional homologies throughout the invertebrates. In this chapter and the following one, insect and arachnid diversity are not extensively discussed. Instead, our compromise treatment is concentrated on those aspects of physiology which are peculiar to land arthropods, and the biology of insects is somewhat arbitrarily segregated in Chapter 17, all the other arthropods being dealt with in this chapter. Since insect physiology has, in general, received more detailed investigation in recent years, some topics such as hormonal controls must be deferred to the insect chapter. Somewhat less defensibly, locomotion on land, including that of insects, is treated in this chapter, while insect flight is discussed in the next.

Efficient Land Animals

Both the physicochemical and the mechanical properties of the arthropod integument contribute to the phylum's success in terrestrial environments (see Chapter 13). Both epicuticle and procuticle are separately indispensable. The epicuticle provides for the physiological separation of the tissues of a land arthropod from the aerial environment. An outermost layer of lipoprotein cement, middle waterproof wax layers of lipids, and an inner cuticulin layer of proteins together provide an impermeable, waterproof, and pathogen-proof coating over the outside of the jointed armor. The mechanical properties of the procuticle and of its articulations are basic to the precise and powerful levering actions which make locomotion on land—and flight—possible. The complex laminae of chitin and of differentially hardened proteins which make up the procuticle confer the essential mechanical properties on the arthropod exoskeleton as a whole. Functionally, this is not merely a jointed, protective armor. These limb-tubes of sclerotized procuticle which can resist deformation, linked by their flexible joints with unsclerotized proteins, provide the basis for the arthropodan lever systems, whose nice application of force to the substrate through single points provides the only energetically efficient method of locomotion on land (excluding that artifact of man, the wheel).

Apart from the fundamental contribution of the exoskeleton, there are some other physiological aspects important to arthropods living on land. The conservation of water is a determinant of patterns of respiration, excretion, and even of reproduction. A surprising number of insects lay their eggs so that the young larval or nymphal stages hatch and grow in fresh water, or in more humid circumstances (such as soils, or leaf litter, or *inside* living or dead plants) than in the truly aerial environments occupied by the adult forms. Other terrestrial arthropods, including some insects, have moderately large eggs which, since they are sealed within impermeable cases, are termed cleidoic (closed-box) in their development. All functioning systems of land arthropods are highly efficient in any bioenergetic assessment of their cost.

The success of the arthropods as land animals can be considered in terms of evolutionary preadaptation. In considering the functional morphology of the crustaceans, we noted that, although the more primitive forms had phyllopodous limbs and were filter-feeders, the more advanced forms were usually stenopodous and macrophagous. Although it is clear that the successful land arthropods, myriapods, arachnids, and insects were not derived from the higher crustaceans, they were undoubtedly derived from marine arthropods that already resembled higher crustaceans in being stenopodous and macrophagous. The importance of these two features becomes obvious if one considers certain contrasting aspects of marine and terrestrial environments. Except for some parts of the littoral zone, the marine environment is characterized by enormous areas of relatively uniform habitats and by having a major fraction of the world's primary plant production in the form of the microscopic single-celled organisms of the phytoplankton. In contrast, the terrestrial environment is much more patchy (with favorable microhabitats of limited extent separated by wide areas of unsuitable environmental conditions) and primary production consists largely in more massive, many-celled macrophytes. Obviously, the first land animals (probably in the Devonian period—see Chapter 34) had to be able to move rapidly from one suitable but limited habitat to another, and had to have the capacity to ingest meals by biting off relatively massive pieces of plant tissue from time to time rather than feeding continuously on finely divided food as they drifted or swam along. Thus the efficient levering action of the stenopodous limb, rather than the paddle-swimming of the flattened phyllopodous one, had to have been developed earlier in any marine ancestor before colonization of land. Similarly, the biting and chewing mouth-parts of macrophagy, rather than any head appendages fringed with bristles and used in filter-feeding were a required preadaptation in any marine arthropod stock poised to colonize land.

By all measures, insects—and as runners-up, arachnids—are the most successful land animals. Certain many-legged arthropods such as centipedes, millipedes, and some minor classes are not nearly so important ecologically, nor as diverse in species numbers, as are the two huge classes of terrestrial arthropods. These many-legged forms *do* also have stenopodous limbs and mostly have biting and chewing mouth-parts suitable for eating macrophytes. One aspect of the success of insects and arachnids in the land environment is undoubtedly their relatively small size. When animals first came on land and faced the "patchy" environment, small-sized animals, provided they could move relatively rapidly from microhabitat to microhabitat, would be at a distinct selective advantage. The terrestrial environment remains "patchy" today and small arthropods still have an advantage over most other forms of animal machine. The additional advantage conferred on insects by the evolution of flight will be discussed in Chapter 17.

Water Conservation

The problem of retention of water which faces all land animals is only somewhat lessened in the arthropods by the waterproofing epicuticular coating over the exoskeleton. As noted above, insects and arachnids are relatively small animals and their high surface:mass ratio would, all other things being equal, tend to increase their liability for passive loss of water to a dry aerial environment. Throughout land arthropods, we find resorption of water as a characteristic hindgut function. Thus, apart from a few plant-sucking arthropod forms like aphids, whose problem is *too much* water in the diet, the vast majority of terrestrial arthropods produce consolidated, dry faeces. Similar processes for the conservation of water characterize the structure and functioning of the organs of respiration and excretion.

Mechanistically, respiratory organs in land animals are constructed to allow appropriate gas exchanges with the environment (uptake of oxygen, release of carbon dioxide), while minimizing the concurrent water losses. If one excludes minute animals, only three patterns of air-breathing apparatus have been evolved in all terrestrial animals. Lungs have been independently evolved in pulmonate snails and in the higher vertebrates. Tracheal systems, involving air-tubes leading into the tissues, are the only other successful respiratory organs and are found in the majority of land arthropods, including all insects. A third pattern of air-breathing apparatus—sets of lung-books—is found in some arachnids.

The structure and functioning of lung-books can best be studied in scorpions where there are four pairs. Each pair is an invaginated pit, opening through a narrow slit to the exterior and containing a large number of leaf-like lamellae (like the pages of a book). Respiratory exchange depends on diffusion, but the arrangement of the respiratory surfaces (which are the "pages") results in a relatively small loss of water to the atmosphere. Lung-books have evolved from the gill-books, similarly structured but externally exposed, found in certain primitive aquatic arthropods like *Limulus*. The occurrence of lung-books in other arachnids will be discussed below.

Systems of tracheae are the successful respiratory organs of the majority of successful land arthropods. It is important to realize that tracheal systems represent the best functional adaptation of animals with an arthropod exoskeleton to terrestrial respiration. They have no narrow phyletic significance. Tracheae have been evolved independently by insects, isopod crustaceans, and spiders, by probably at least three other arthropodan stocks, and by onychophorans. The detailed structure and functioning of tracheal systems has been most thoroughly investigated in insects. Tracheae are essentially air-tubes branching from trunks into the tissues of the arthropod and involving extensions inward of the exoskeleton. Their cuticular lining is thickened in ridges to form a series of annular rings or a continuous spiral. These serve mechanically to hold the air-tubes open even if pressure within them is somewhat reduced. As extensions of the integument, the entire lining of the tracheal system has to be molted at each ecdysis. A few of the larger tracheate arthropods, including some insects, have muscular pumping of air resulting in rhythmic ventilation of the major tracheal trunks. Many, including the majority of insects, do not pump but can control the rates of diffusion of gases through the tracheal system by adjustment of the external openings or spiracles. Like ventilation rates in higher vertebrates, the control of spiracle opening and closing in some insects is based on detection of increased concentrations of carbon dioxide. Closure of spiracles is used by many arthropods to cut down losses of water vapor while resting, that is, while oxygen requirements are low.

The tracheal tubes branch into ever-finer subdivisions to supply all parts of the body and end in fine air capillaries, termed tracheoles, which are not lined with cuticle. These minute tubes pass intimately among the tissues, usually ending blindly *within* cells including muscle fibers. The walls of the tracheoles are, of course, not molted. As was first discovered in insects, a delicate balance within the tracheoles between capillary and osmotic forces allows a sort of "tidal" movement in their fluid contents which can draw air from the tracheal trunks deeper into the tissues at times of increased metabolic demands for oxygen. The elucidation in insects of this tissue mechanism formed an early triumph of insect physiology and will be discussed in Chapter 17. In general terms, tracheal respiration gains oxygen for the tissues of land arthropods while minimizing their water losses by an efficient *internal* exploitation of the properties of the exoskeleton.

A similar water-conserving physiology characterizes the different excretory organs found in terrestrial arthropods. In several groups of land arthropods other than insects, there are organs which correspond to the segmentally arranged coelomoducts of annelid worms. Coxal glands, which occur in several groups of arachnids and of other arthropods, are homologous with the excretory organs in crustaceans (including the "green gland" of the higher Crustacea). Coxal glands are made up of a blind sac leading through a convoluted tube or labyrinth by way of a storage sac or bladder to an excretory pore. These glands are always paired, but vary in position in different land arthropods. In many spiders there are two pairs of coxal glands opening at the bases of the first and third walking limbs. Scorpions have a single pair opening on the third legs and other spiders and ticks have one pair on the first limbs, while a few mites have four pairs of coxal glands. Functionally, coxal glands excrete guanine or uric acid in a semisolid form after resorption of water. An exception similar to that of the aphid diet, noted above, involves argasid ticks, the so-called soft ticks, which take in huge blood meals during a short period of attachment to the host. This can involve a fourfold weight increase and involves a copious secretion of very dilute urine by the coxal glands during and immediately after feeding. From an evolutionary viewpoint, it would seem that the segmental excretory organs of primitive marine arthropods, which became reduced in numbers during the evolution of the higher groups of crustaceans, were similarly present in the ancestors of land arthropods and persist somewhat variably today, particularly among some groups of the arachnids.

However, the important functional excretory organs of the best-adapted land arthropods are not such segmental structures, but diverticula of the gut, called Malpighian tubules. These are the organs of nitrogenous excretion not only in the insects but also in millipedes, centipedes, and several groups of arachnids, including spiders. As was the case with respiratory tracheae, independent evolution of these excretory structures in several stocks of land arthropods is most likely. The nitrogenous excreta produced in Malpighian tubules are mainly in the form of uric acid crystals. Once again, tubule functioning has been most extensively studied in insects, so that the chemistry of the process will be summarized in Chapter 17. It is worth noting that a solid or semisolid urine consisting mainly of uric acid is also characteristic of the water-conserving vertebrates: reptiles and birds. In almost all terrestrial arthropods possessing them, the Malpighian tubules open into the hindgut so that faeces and nitrogenous waste are discharged together. The functional parallel with reptiles and birds is again striking, must have been evolved independently, and emphasizes the common need to conserve water in all stocks of animals which have colonized land.

Land and Aerial Locomotion

The efficiency of the interacting lever systems of the arthropodan machine— the jointed limbs—is nowhere so manifest as in locomotion on land. The limbs of insects, arachnids, and the myriapodous groups are all—to use crustacean terminology—stenopodous and uniramous. These elongate systems of rigid, tubular sections and articular membranes have their condyles (and thus also their axes of articulation; see Chapter 13) arranged in specifically precise orientation. The structural arrangements permit certain mechanical efficiencies.

The slender limb with its pointed tip allows the propulsive force to be applied through a single point on the substrate (through the point d'appui) and, in the stepping action of a series of limbs, allows each successive limb to take over almost the same point d'appui in turn. A limb length greater than the diameter of the trunk, combined with the specific condylar axes (Figure 16-1B), not only allows the body to be carried "high"—clear of all contact with the substrate in most land arthropods—and slung between the limbs in a stable fashion, but also allows for controlled changes in the limb's effective length.

This last factor is of paramount importance in mechanically efficient stepping, allowing the distance between the midline of the arthropod and the point d'appui of the limb (M–P in Figure 16-1A, B) to remain constant while the trunk moves forward during each step. (This ensures that no propulsive energy is wasted in lateral swayings of the trunk.) The actual propulsive force can be generated either by the contraction of muscles largely extrinsic to the limb itself, acting across its proximal articulations (the case in the majority, and probably also in the most primitive terrestrial limbs), or by contraction of intrinsic muscles acting across more distal joints. In both cases, the intrinsic musculature—flexors and extensors—is responsible for the continuous changing of the angles between podomeres during the propulsive backstroke of the step. Such continuous adjustment in the effective length of the limb allows the M–P distance to remain the same for the duration of the backstroke, so that effective propulsion is maintained in the direction of the midline. To put it another way, the distance between the limb tip and its basal articulation on the trunk must be shortest at the mid-backstroke. In each limb, the propulsive backstroke is followed by a recovery movement forward with the distal parts raised clear above the substrate, which recovery stroke ends with the limb tip at a new point d'appui. In almost all terrestrial arthropods, the limb is overall less flexed (i.e., has more extensor muscles contracted) during the recovery stroke and so the limb tip swings forward in an arc, much further from the trunk (M–R on Figure 16-1A, B) than the line of points d'appui (the "footprint" line). Again in a generalization, the muscles, both intrinsic and extrinsic, which are contracted during the "power" backstroke of the limb, will anatomically be more massive

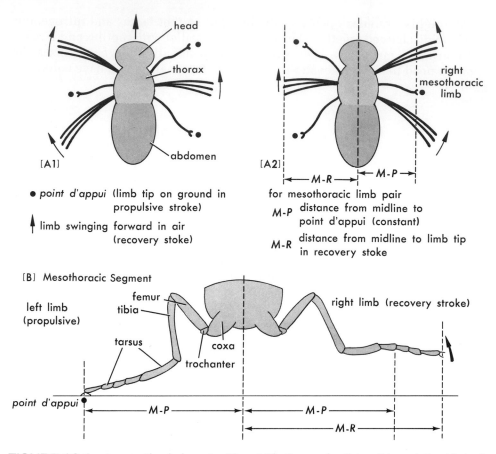

FIGURE 16-1. Locomotion in insects. **A1** and **A2:** Successive "steps" in a stylized insect (basically dipteran) which is walking slowly. The leg pairs are in opposite phase and so the insect is always stably supported as a tripod upon three *points d'appui*. All limb tips are further from the midline of the body during recovery strokes (*M–R*) than while they are exerting propulsive force (*M–P*), and the mesothoracic limbs work on wider spaced lines of *points d'appui* (i.e., leave wider footprint lines). **B:** Left and right mesothoracic limbs in a stylized insect (basically dictyopteran) diagrammed during a "step" in which the left limb is propulsive.

than their antagonists. Likewise, the muscles which are contracted to bring about the forward swing of the recovery stroke will be slighter and much less powerful.

Essentially this mechanical pattern is utilized by land arthropods in a wide variety of locomotory activities: slow crawling, fast running, digging, jumping, climbing silk threads or vertical surfaces, and even running while hanging upside down from ceilings. This variety is achieved not so much by change in the functional pattern of power and recovery strokes (except for jumping) as by anatomical differences in the proportions of the podomeres of limbs, in the distances between articulations, and in the pattern of extrinsic limb muscles. Mechanically, what differ are the proportions of each lever arrangement around its fulcrum.

The locomotory mechanisms of land arthropods probably evolved from those of an annelid-like ancestor with large parapodia used in crawling (as in present-day *Nereis* involving power and recovery strokes in a metachronal rhythm with the opposite sides of each segment exactly out of phase—see Chapter 11. An intermediate stage would be represented by the undifferentiated locomotion of presently living walking-worms (Onychophora—see Chapter 18), with soft worm-like bodies carried upon a series of paired limbs. The locomotory sequences in walking-worms like *Peripatus* can be related to those in the myriapodous land arthropods. Finally, reduction in limb numbers brought about the efficient eight- and six-legged patterns of arachnids and the six-legged pattern of all insects. This sequence—errant-polychaetes to walking-worms to myriapods to insects—certainly represents the progressive specialization in land locomotion which must have occurred—though, of course, these four groups of living animals cannot represent the actual evolutionary lineage and may not reflect phylogeny at all closely. Present understanding of the probable course of locomotory specialization in land arthropods is almost entirely based on a beautiful and extensive series of studies on many types of myriapods and on *Peripatus* by Sidnie M. Manton, which studies combine, in the best traditions of functional morphology, superbly detailed accounts of the microanatomy of muscles and of exoskeleton with a detailed mechanical analysis of the locomotory sequences in the healthy living organism. Her papers, mostly published in the British *Journal of the Linnean Society of London* (*Zoology*), should be consulted by interested students not only for a detailed account of structure and function in the locomotion of myriapods which cannot even be adequately summarized in a book of this scope but also as an example of how relatively simple observational work (using analysis of flash photographs and motion-picture film and measurement of tracks made on smoked paper) can lead to an understanding of the mechanics of complex muscle systems and to hypotheses on their evolution.

As a relatively simple example of the mechanical possibilities of "many-legged" systems in land arthropods, the differences between a typical millipede (class Diplopoda) and a typical centipede (class Chilopoda) can be considered (Figure 16-2). Millipedes are somewhat slow-moving, requiring relatively powerful propulsion as they move continuously through soil and decaying vegetable matter and ingest some of it. Centipedes are fast-moving, free-running animals, which are predaceous carnivores searching for and running down their prey. The type of gait shown by the former has been analyzed by Manton and can be termed "low-geared," while the gait of centipedes, as first elucidated by E. Ray Lankester early this century and reinvestigated by Manton, is "high-geared." For equivalent motor size (assessed as volume of locomotory muscles), millipedes are arranged with gearing like bulldozers, centipedes like racing cars. Figure 16-2 illustrates some of the ways these differences of gait are achieved in a millipede like *Spirostreptus* and a centipede like *Scolopendra* (or *Cryptops*). The most significant functional difference is that the relative durations of the propulsive backstroke and the recovery stroke are 7 to 3 for *Spirostreptus* and less than 3 to 7 for *Cryptops*: longer propulsive strokes for power, shorter propulsive strokes for speed. This correlates with differences in the metachronal succession of limb movements (Figure 16-2), the phase differ-

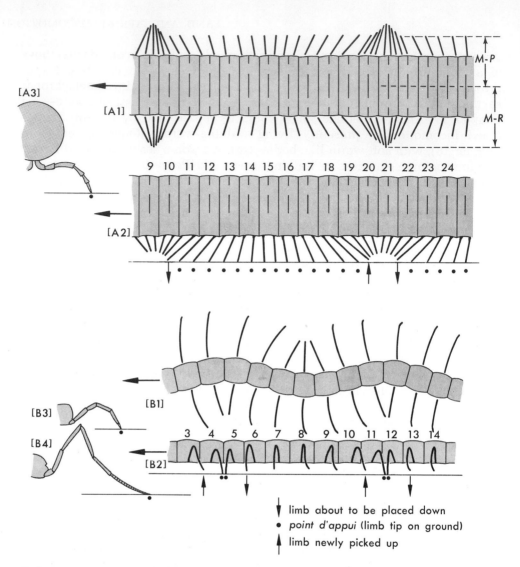

FIGURE 16-2. Locomotion in millipedes and centipedes. **A1:** Dorsal view of a typical millipede (such as *Spirostreptus* or *Gymnostreptus*) in motion, showing sixteen of the short, wide diplosegments with 32 pairs of limbs, left and right sets being exactly in phase with each other. **A2:** Lateral view of the same millipede, showing that the majority of the limb tips are on the ground at once, giving the *longer* strokes of propulsion characteristically "low-geared like bulldozers." **A3:** The millipede limb, relatively short and attached nearly midventrally to the short, wide, cylindrical displosegment. **B1:** Dorsal view of a typical centipede (such as *Scolopendra* or *Cryptops*) in motion, showing twelve of the long, narrow segments, each with a pair of limbs in opposite phase, and the undulations of the body which occur at speed and are mechanically inefficient. **B2:** Lateral view of the same centipede, showing the larger phase difference between segments with less than a third of the limb tips on the ground at once, giving the short swift strokes of propulsion characteristically "high-geared like racing cars." **B3:** The centipede limb (in *Scolopendra*), relatively long and attached laterally to the longer, thinner segment. **B4:** The much longer limb of the common house centipede *Scutigera*. For further discussion, see text. [Partly adapted from E. Ray Lankester, *Quart. J. Micros. Sci.*, **47**:523–582, 1904, and from S. M. Manton, *J. Linn. Soc. (Zool.)*, **42**:93–166, 1952, and 299–368, 1954.]

ence between successive legs being very small for the millipede and relatively large for the centipede. Structurally, the millipedes have many short, wide segments and many relatively short legs, while centipedes have fewer long, narrow segments with (in *Cryptops*) somewhat longer legs (and, in other centipedes, much longer legs). One result is that at any one time in the millipede, the majority of the limb tips are on the ground, while in the centipede, less than one-third of the limbs are simultaneously supporting (and propelling) the trunk. Thus, while it is possible for the millipede to have its pairs of legs in phase with each other, in the centipede the gait almost demands that the paired legs be in opposite phase: to ensure stability and support of the trunk without sagging. Thus we see that the powerful bottom gear of millipedes involves long backstrokes with many legs pushing simultaneously (the metachronal waves involving large numbers of pairs of legs). In contrast, the faster gait of centipedes involves few legs propelling at once by backstrokes of short duration. They show some sacrifice of stability to speed and a tendency to reintroduce the mechanical inefficiency of lateral undulations of the trunk. Swaying of the trunk segments alternately to each side involves energy waste and results from the propulsive limbs being widely separated and alternately placed.

Turning to the more successful land arthropods, spiders and insects, we have a significantly higher degree of tagmatization, with a smaller number of limbs, each relatively longer than those in myriapods. Significantly, the articulations of the three or four pairs of legs with the trunk are always close together near the center of gravity of the body. For example, in the insects the central tagma, the thorax, consists of only three segments—prothorax, mesothorax, and metathorax, usually fused but each bearing a pair of limbs. Thus, mechanically, in spiders and insects the propulsive force of the few long legs is applied to the body near a single center, avoiding any undulatory tendency. In general, insects stand or walk with the leg pairs moved in opposite phase (Figure 16-1A), and thus these hexapodous organisms are stably supported as tripods, the left mesothoracic leg being on the ground with the right prothoracic and metathoracic limbs. A few spiders move habitually on six legs like insects, while in others, using eight legs, there is a phase difference of a quarter of a cycle between one leg of each pair and its partner.

For obvious reasons of both static posture and locomotion, in arthropods with a reduced number of longer legs attached close to a single center, the legs must fan out anteriorly and posteriorly. This is as true of decapod crustaceans as it is of insects and spiders. Locomotor efficiency is increased by separating the *points d'appui* of the limbs, but this also leads to a functional differentiation not often discussed. If we take a characteristic insect, for example, and ignore for the moment the middle (mesothoracic) pair of legs, the forelegs and the hindlegs execute their propulsive backstrokes by strikingly different muscle contractions. Apart from the extrinsic limb muscles (which are also differentiated), most of the intrinsic flexors of a prothoracic leg will be contracting during the propulsive backstroke (in other words, a front leg ends its power stroke and begins its recovery stroke more flexed than at the beginning of its step). In contrast, contraction of the intrinsic extensors is involved in the propulsive stroke of a metathoracic leg (that is, the leg ends its power stroke more extended). If the earlier generalizations are accepted, then the attentive reader can deduce the size differences to be found in the muscles of the fore- and hind-limbs of insects.

Arachnid Diversity

The second largest class of the arthropods, the Arachnida, are structurally rather uniform. They all have 21 segments in two tagmata: the prosoma and opisthosoma. The prosomatic segments, completely fused in most forms, bear the rather stereotyped six pairs of appendages. These are, from the anterior end: the preoral chelicerae—usually prehensile and concerned in food capture; next, the pedipalpi, which may be tactile sensory organs or relatively enormous chelipeds like those of crabs; and then four pairs of walking legs. The opisthosomatic segments show some variation, being clearly separated and divided into two series in primitive forms like scorpions, or completely suppressed by fusion as in the big specialized group, the Acarina, or ticks and mites. Respiration is by either lung-books or tracheae, or both, and the excretory organs are either coxal glands of coelomoduct type found usually as a single pair in various middle segments of the prosoma or Malpighian tubules (see page 317). Most modern classifications of arachnids involve ten orders (see Table 16-1), of which two are of enormous extent—the Araneae (spiders) and the Acarina (mites)—and five are certainly of minor importance.

An intermediate-sized group, the order Scorpionida, are clearly the most primitive living arachnids and the oldest (Silurian) known terrestrial arthropods in the fossil record (see Chapter 34). Scorpions are cryptic and nocturnal

TABLE 16-1
Outline Classification of the Arachnida

Phylum ARTHROPODA
  Class ARACHNIDA
    Subclass PECTINIFERA
      Order Scorpionida
    Subclass EPECTINATA
      Order Chelonethida (= Pseudoscorpionida)
      Order Opiliones (= Phalangida)
      Order Solifugae[a]
      Order Uropygida[a,b]
      Order Amblypygida[a,b]
      Order Palpigrada[a]
      Order Ricinulei (= Podogona)[a]
      Order Araneae[c]
      Order Acarina[c]

Notes:
  [a] Minor groups of less than one hundred species.
  [b] These two orders were formerly grouped together as the order Pedipalpi.
  [c] The dominant arachnid orders: Spiders (Araneae) number about 34,000 species, while mites and ticks (Acarina) must number much more than 9000 species.

animals of the warmer latitudes. Characteristically fearsome animals to man, they have the pedipalps as powerful grasping chelipeds, and a long flexible abdomen bears the stinging apparatus with its sharp, hollow, curved barb for venom injection. In fact, scorpions are predaceous carnivores feeding mainly on insects but also on all other suitably sized small invertebrates. For example, a cockroach caught by a scorpion of about its own length will be held in the pedipalps while the abdomen is flexed over the back of the body and the venomous barb stabbed into it. The venom is a mixture of substances but apparently contains a paralyzing neurotoxin. The form of the scorpion opisthosoma is very primitive by arachnid standards, and is made up of a seven-segmented preabdomen bearing a pair of sensory appendages called the pectens and containing the four pairs of lung-books, and a five-segmented postabdomen of elongate annular segments with the last segment bearing both the anus and the stinging apparatus (Figure 16-3; see also Figure 1-4).

Two further orders of intermediate size are the Chelonethida (Pseudoscorpionida) or false-scorpions and the Opiliones or harvestmen. Chelonethids are minute arachnids living as predaceous carnivores on small arthropods in the soil and in leaf litter, although a few species live in the upper zones of the marine littoral, and one (Figure 16-4) lives in old buildings all around the world. The pedipalps resemble those of scorpions but have associated poison glands, while the abdomen is relatively wide and rounded posteriorly. The more familiar long-legged harvestmen, order Opiliones, number about 1800 species. The legs are always extremely long and slender, and in some tropical forms reach lengths of 16 centimeters. There is no constriction between the prosoma and opisthosoma in harvestmen, and the latter shows visible segmentation. The tracheal system of Opiliones differs from that in other arthropods

FIGURE 16-3. A typical scorpion. The sting is actually a postanal telson, and the holding claws are chelate pedipalps. There are also four pairs of walking legs and a pair of small chelicera at the anterior end (between the bases of the large pedipalps). Photo courtesy Carolina Biological Supply Company.]

FIGURE 16-4. A minor arachnid order: pseudoscorpions. Less than 5 millimeters in length, these small predaceous arachnids live in soil and in old buildings. They might be termed "librarians' friends" since they often feed on the tiny insects called book-lice (Psocoptera) which eat the bindings of old books. [Photo by Frank A. Romano III, of a specimen prepared and mounted by the author.]

and seems to have been independently evolved, and there are peculiar odoriferous glands borne in the prosoma. Harvestmen are omnivorous animals living mainly in humid habitats. The five minor orders, none of which shows consistently archetypic features, include: the Solifugae or sunspiders, the Uropygida or whipscorpions, the Amblypygida or scorpion-spiders, the Palpigradi, and the Ricinulei. Each of these five groups numbers less than one hundred species.

There remain the two huge and successful orders of the arachnids. The Araneae or spiders number about 34,000 species and are abundant predaceous carnivores in many types of vegetation in all parts of the world. The prosoma is joined by a waist-like pedicle to the usually unsegmented opisthosoma. The chelicerae are of moderate size and bear poison glands. Spider venoms vary, but most of them, although very effective against the usual prey of insects and other small arthropods, are not dangerous to man. However, recluse-spiders (*Loscosceles*) have a haemolytic venom which, in man, can cause a dangerously spreading necrotic ulceration from the site of a bite; while the notorious black-widows (*Latrodectus*) have a powerful neurotoxin which can cause muscular convulsions, general nausea with severe "pressure"-pain, and (in rare cases) death from respiratory paralysis (Figure 16-5).

The pedipalps are small in female spiders but have enlarged knobby ends in males where they form the copulatory organs, basically sperm reservoirs with injection apparatus of varying levels of anatomical complexity (Figure 16-6B). Before mating, a male deposits his sperm in a tiny silk web from which he then fills his two pedipalpal organs, before seeking a female and courting and copulating. In the courtships of some species, female pheromones are important in species recognition, in others specific musical vibrations are plucked on the web by males, and in others (especially in jumping spiders) limb-waving and male dancing are involved in elaborate patterns of precopulatory behavior.

Respiration is by lung-books and associated tracheal tubes. There are usually eight pairs of eyes on the prosoma, which although they are simple ocelli, are

FIGURE 16-5. A notorious spider. The female black-widow spider (*Latrodectus*) has a neurotoxic venom somewhat dangerous to man. [Photo courtesy Carolina Biological Supply Company.]

more highly developed than those of other arachnids. There are usually three pairs of spinnerets on the opisthosoma and these secrete the characteristic silk threads. The silk is a complex albuminoid protein, which when hardened has great tensile strength and elasticity. All spiders continually lay a dragline for their own safety. Many spiders build webs to capture prey, and other uses for silk include lining burrows, enclosing eggs, encasing sperms at copulation, constructing special copulatory chambers, constructing similar chambers for molting or for going into a diapause within. Certain chelonethids and mites also spin silk threads.

The other major order, the Acarina, comprises the much more than nine thousand species of mites and ticks. Even apart from the numerous parasitic forms, these are the arachnids of greatest economic or ecological importance to man. Mites can occur in fantastic numbers as destructive pests of crop plants, stored foods, and other natural products. The mite body is completely fused, even the division between prosoma and opisthosoma being imperceptible (Figure 16-6A). Except for the four pairs of legs, the paired appendages are usually minute and associated with the mouth-parts, which can be arranged functionally for either biting or piercing and sucking. Larval stages have six pairs of legs. All but the smallest mites have tracheae as respiratory organs. Most mites are minute, and the group includes the smallest adult arthropods where the mature size is less than 0.1 millimeter. One superfamily of the Acarina has secondarily returned to aquatic life in both the sea and fresh waters. Many free-living mites are predaceous carnivores, while others are omnivorous scavengers.

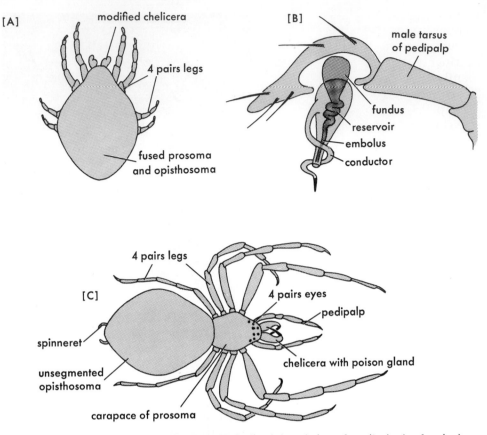

FIGURE 16-6. Arachnid organization. **A:** Stylized dorsal view of a mite (order Acarina), with the completely fused body and the four pairs of legs of the adult condition. **B:** Simplified diagram (from a cleared whole mount) of the modified pedipalp from a male spider (order Araneae) used as a copulatory organ after being "loaded" with sperm earlier deposited in a miniweb. **C:** Stylized dorsal view of a female spider (order Araneae), with a single carapace over the prosomal segments and an unsegmented opisthosoma, and with four pairs of walking legs, and one pair each of sensory pedipalps and poisonous chelicerae.

Ticks are essentially similar in body form to mites, though somewhat larger. A significant number of diseases in agricultural and domestic animals (and a small number in man) are transmitted by ticks. The causative organisms (see Chapter 6) of Texas fever in cattle and relapsing-fever and Rocky Mountain spotted fever in man are transmitted by ticks. Toxic salivary secretions of the ticks themselves can cause a general systemic reaction—called "tick paralysis" —in heavy tick infestations of cattle (and, more rarely, of man).

Minor Arthropods

The four minor classes of arthropods (see Table 10-1) include two which are closely related to the major myriapodous groups, millipedes and centipedes (Figure 16-7). Myriapod locomotion has already been discussed, but it is worthwhile recapitulating their general features here. The major class Diplopoda encompasses the millipedes, predominantly herbivorous animals which move through deposits of dead leaves and loose soils and have diplosomites, each bearing two pairs of legs. The major class Chilopoda is the centipedes, predominantly carnivorous with relatively long legs and a capacity for fast locomotion. The minor class Pauropoda includes about sixty species of minute, soft-bodied myriapods, probably closely related to the millipedes, which live in forest litter. The minor class Symphyla also has about sixty known soil-dwelling species. They superficially resemble centipedes and adults have twelve pairs of legs, insect-like mouth-parts, and a pair of spinnerets on the penultimate segment. In some structural features, symphylans resemble the centipedes and in others the apterygote insects.

The remaining two minor classes are probably distantly allied with the arachnids. The class Merostomata comprises two distinct and unequal groups: the extensive fossil subclass Eurypterida, including the giant water scorpions, and the subclass Xiphosura, encompassing the five species of horseshoe crabs like Limulus (Figure 16-8). The horseshoe crabs are marine arachnid-like animals with the body in three tagmata. There are a prosoma with a large semicircular carapace bearing two median and two lateral eyes and then a hexagonal "abdomen" consisting of the mesosoma bearing six pairs of appendages, including a genital operculum and a series of leaf-like lamellae forming the gill-books, fused to the metasoma without paired appendages but bearing the elongate spine of the telson. The paired limbs of the prosoma comprise a pair of small chelicerae and five pairs of walking legs—the first four of which are chelate, bear gnathobases, and closely resemble the single pair of pedipalps in scorpions. Horseshoe crabs are scavengers of the littoral and sublittoral on the Atlantic coast of North America and along the coasts of Southeast Asia. They are the only surviving marine arachnid-like animals, and fossil xiphosurans are known from the Ordovician period onward, but, unlike other surviving primitive animals, horseshoe crabs are neither rare nor retiring animals. There are few possible predators on Limulus except man, and, teleologically, horseshoe crabs seem to regard the rest of the littoral faunas of the regions where they occur with supreme indifference.

FIGURE 16-7. A large centipede. A typical subtropical scolopendrid, about 12 centimeters long. Visible are the head with paired antennae and poison jaws (or "maxillipeds"), the wide trunk segments each with a single pair of limbs, and the last pair of limbs on either side of the "telson" modified as sensory "posterior antennae." Compare with Figure 17-2. [Photo by the author.]

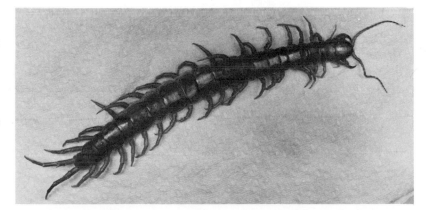

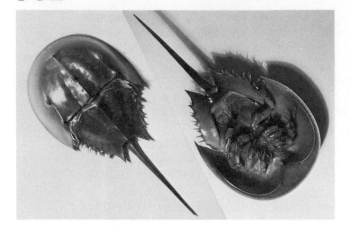

FIGURE 16-8. A living archetype, the horseshoe crab *Limulus:* dorsal (left) and ventral (right) views of a young adult. Note the three tagmata, prosoma, mesosoma, and metasoma with telson spine. Of the six pairs of limbs on the prosoma, five are chelate and all have gnathobases, while the limbs of the mesosoma form an operculum and gill-books. [Photos by the author.]

The final minor arthropod class is the Pycnogonida, the sea-spiders or no-body-crabs (Figure 16-9). These marine arthropods are widely distributed but never abundant. Although pycnogonids have been found in all the world's oceans and at all depths, the greatest diversity of species and genera occurs around the Antarctic, and both boreal and temperate faunas of the Pacific are much more diverse than those of the North Atlantic. In all pycnogonids, the trunk is greatly reduced, and the four pairs of walking legs usually enormously enlarged. The head bears a proboscis, a pair of chelicerae, and a pair of palps. The abdomen is a tiny vestige, and the alimentary and reproductive systems are largely enclosed in the legs. Detailed study of their segments and appendages suggests some homologies with both arachnids and crustaceans, all much modified by the peculiar body plan. Pycnogonids are dioecious animals, and in many the male broods the eggs until they are hatched. In the majority, the larva which hatches from the egg, termed a protonymphon, has only three pairs of appendages, though these are quite distinct from the corresponding structures in the nauplius larva of crustaceans. Pycnogonids are browsing carnivores, feeding on sessile sponges, coelenterates, and ectoprocts. Without further evidence of their ancestry, the Pycnogonida must necessarily be placed in a separate class of the arthropods.

Arthropod Phylogeny

As already discussed, the great bulk of available evidence suggests that the arthropods evolved from annelid-like animals. Consideration of the relationships between the annelid-arthropod stocks and the rest of the many-celled animal phyla will be postponed to Chapter 35. As we have seen when discussing tracheal respiration and other mechanisms, convergent evolution of structures and functions has occurred to a considerable extent in arthropods. Thus, although there is a unique structural and functional basic plan common to all arthropods, several authorities feel that the group may be of multiple origin

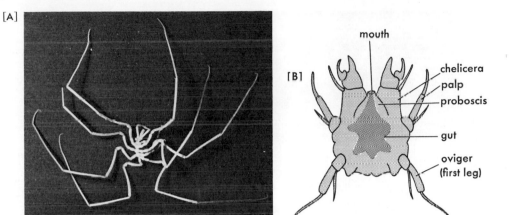

FIGURE 16-9. Class Pycnogonida (a minor arthropod group). **A:** Dorsal view of an adult "sea-spider" or "nobody-crab", *Nymphon* (phylum Arthropoda, class Pycnogonida). **B:** The earliest larval stage, or protonymphon, of a pycnogonid. Compare with the crustacean nauplius larva (Figure 10-3A). [Photo by the author.]

from protoannelid stocks. A diphyletic origin is favored by a number of investigators including Manton and O. W. Teigs. Certain interrelationships of the major classes are more universally accepted. The old association of the Crustacea with the insects and myriapodous groups in the subphylum Mandibulata is almost certainly erroneous and should be discarded, since the mandibles of crustaceans and insects are not homologous either in development or in fossil evidence of their phylogeny. Most authorities agree on a close relationship between the insects and myriapodous groups. The insects themselves are relatively stereotyped, but their probable evolution and the interrelationships between the major orders will be discussed at the end of Chapter 17. There is also a general agreement that eurypterid-like ancestors probably gave rise to the arachnid groups. The evolutionary process probably involved intermediate forms resembling the modern scorpions, and in turn, the eurypterids were almost certainly derived from modified trilobite stocks. The probable close relationship between the Crustacea and the trilobites (see Figures 14-1E and 16-10) has already been documented. The proponents of a diphyletic origin of the Arthropoda, including Manton, strongly maintained the hypothesis that the myriapods and insects evolved independently from soft-bodied segmented animals like the Onychophora (see Chapter 18). However, this would assume that the evolution of a spacious haemocoel and reduction of the true coelom antedated (as in *Peripatus*) the evolution of the arthropod exoskeleton. As already discussed, there are a number of functional aspects which make this sequence unlikely and also much embryological evidence which suggests that the primitive arthropodan coelom was metamerically divided and probably much more extensive. Those zoologists who prefer the alternative hypothesis that the arthropods form a monophyletic group, and I am among them, must admit that, on presently available data, the insects and myriapodous classes cannot readily

FIGURE 16-10. A fossil trilobite. A specimen of *Phacops* sp., about 33 millimeters long, in a fine-grained black shale of Silurian age (see also Figure 34-3). [Photo by the author. The fossil specimens in Figures 16-10, 24-9, 34-3, 34-4 and 34-5 were all loaned from Syracuse University teaching collections, courtesy of Dr. James C. Brower.]

be related to the probable stem-stock of both arachnids and crustaceans—the trilobites. In spite of this, we feel that the extensive homologies of structure and of function (discussed in Chapters 10, 13, and elsewhere) unite the diverse stocks of arthropods as animals with a uniquely integrated pattern of organs and physiological processes—in other words, as a single phylum.

17 INSECT SUCCESS

WE have been assessing "success" in the major phyla of animals in terms of both species numbers and ecological energetics. In terms of number of species (750,000), insects are the most numerous class of living animals. In terms of numbers of individuals, insects are the most numerous land animals, being outnumbered on the planet only by the planktonic copepods of the world's oceans, and just possibly by free-living nematodes. In bioenergetic terms it is clear that the vast majority of the organic carbon molecules synthesized by the green plants of land and freshwater environments are consumed by insects.

Although no other group of organisms (plants, protozoans, molluscs, or vertebrates) can compare with them in terms of species diversity, insects are remarkably stereotyped in structure and function. In terms of systematics, the great diversity of insects occurs below the taxonomic level of orders. There is nothing in insect diversity to compare with the contrasts in anatomy, physiology, and ecology between a bivalved clam and an octopus (in the phylum Mollusca) or between a sea-squirt and a human (in the phylum Chordata). In spite of this, it is probable that much more than half the species of eucaryotic organisms living in the world today belong to this one class, the Insecta. To all the other features that have made the arthropodan body plan so successful in terrestrial environments, including epicuticular protection against water loss and microorganism invasion and the superbly efficient lever systems of

305

the jointed limbs, the insect plan has added the power of flight. Insects are the only invertebrates that can fly. Insects were undoubtedly the earliest animals on this planet to fly, and they remain the only group of flying animals whose wings were not evolved secondarily from paired walking legs. The extinct flying reptiles, all birds, and the several small groups of flying mammals were all evolved much later than insects (see Chapter 34).

There is one obvious limitation to insect success: there are no large insects. Almost all living insect species lie in a size range between 2 millimeters and 4 centimeters in maximum dimension, with weights usually in the milligram range. This restriction of size is related to some of the features wherein insects show their greatest physicochemical and mechanical efficiencies, including flight and respiration by tracheae. Another environmental limitation is striking. Insects have had no success in the sea. They are primarily a terrestrial air-breathing group which has had moderate success in readapting to aquatic environments in fresh waters. Except for tiny numbers of littoral species, they are absent from marine habitats, where not only the copepods but many other groups of crustaceans are dominant members of the animal communities.

The Winning Plan

As we have already noted, the insects are remarkably stereotyped in structure and function. The differences used in the ordinal classification of insects concern the pattern of wings, the extent of metamorphosis during the series of molts leading to adulthood, and the type and functioning of mouth appendages.

As with the other major groups of the arthropods, the insects have a nearly diagnostic pattern of tagmatization of the segments in the adult body form. There are three tagmata: head, thorax, and abdomen; and the numbers of segments in these and their paired appendages conform to a fixed pattern. The insect head consists embryonically of six segments, the thorax consists of three segments bearing typically two pairs of wings and three pairs of walking legs, and the abdomen has normally eleven segments without any locomotory appendages. It should be noted that this threefold tagmatization does not correspond to the segments involved in the similar pattern of tagmata given the same names in the class Crustacea. The alternative class name for the insects, the Hexapoda, refers to the six legs of the adult insect, a pair attached to the pleura of each of the three segments of the middle tagma: the prothorax, the mesothorax, and the metathorax. The two pairs of wings are carried on the mesothorax and metathorax, the second and third segments of the thorax, and may have been evolved from rigid stabilizing fins first developed as horizontal extensions of the terga of these two segments and used in landing from jumps. Relatively few species of insects are wingless today, and only about half of these are primitive forms in which the two wing pairs have never evolved (subclass Apterygota—see the next section). The other wingless insects reflect some secondary adaptation, for example, to life as ectoparasites (as in fleas and lice). But the vast majority of insects have wings, though in many orders the wings are functionally reduced to a single pair. For example, in the enormous order Diptera, the true flies, only the mesothoracic wings are functional, and the posterior pair are converted into tiny halteres, which are sense-organs controlling posture, like the statocysts of aquatic animals or our own semicircular canals of the ear. In the other enormous order, Coleoptera (beetles), the metathoracic wings are the functional pair, while the anterior wings have become protective elytra, which completely enclose and protect the functional hindwings when the beetle is not flying. In the order Hymenoptera (ants, bees, wasps, and their allies) flight is accomplished by the muscles attached only to the mesothoracic wings. Although two pairs of membranous wings are developed, they are interlocked by means of chitinous hooklets, and the posterior pair is driven by being coupled to the actively flapped frontwings. Apart from the three pairs of legs, and the two pairs of wings, the only important appendages of the insect are on the head.

On the insect head there are normally one large pair of compound eyes and three ocelli (or simple light-sensitive organs). There are also one pair of antennae and three paired appendages surrounding the mouth. An unpaired head process, the labrum, forms an upper lip and another unpaired process, the hy-

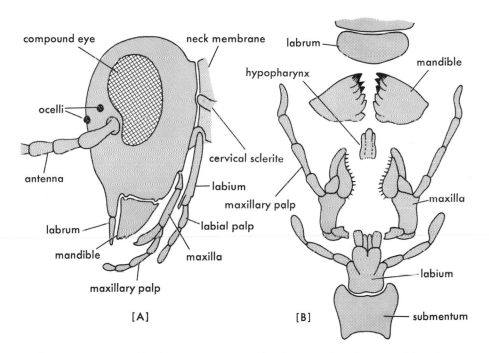

FIGURE 17-1. Archetypic insect mouth-parts (basically stylized from conditions in cockroaches, order Dictyoptera). **A:** Side view of head, showing the single labrum in front of the mandibles (lateral jaws) and the paired maxillary and labial structures behind them. **B:** "Exploded" ventral view of the mouth-parts. The labium (lower lip) can be regarded as made up of a fused pair of second maxillae, and the hypopharynx is the sclerotized "tongue-like" floor of the mouth. The bulk of biting and chewing is done by the mandibles, the other mouth-parts serving to gather in pieces of food and hold them in position to be chewed by the mandibles. It is important to note that the paired structures called mandibles in insects and those in crustaceans are not homologous.

popharynx, forms a floor or posterior wall to the mouthparts. The paired elements consist of mandibles anteriorly (although these are the major lateral "jaws" as in the crustaceans, their morphological origins are different), with behind them a pair of separate first maxillae, and behind that a labium formed of a fused pair of second maxillae. In the more generalized condition of these mouth-parts (perhaps corresponding to more primitive insects—Figure 17-1), both the free maxillae and the fused labium bear palps, although the mandibles do not. In many groups of more advanced or specialized insects, this basic pattern of biting and chewing mouth-parts can be modified for piercing and sucking, as in fleas, aphids, and mosquitoes; or for lapping fluids and sponging them up, as in many dipterans including house-flies; or into a long coiled siphon for collecting nectars from flowers, as in moths and butterflies. These will be more specifically listed in the systematic section which follows.

Only primitive insects and some larval stages show the full number of eleven segments in the abdomen. In most adult insects, only nine can be counted, and in some, there are further reductions. The only appendages are a pair of sensory cerci, like backwardly pointing antennae which are borne on the last segment of the abdomen. These are best developed in more primitive insects, and in the juvenile stages of a few higher groups. A possibly archetypic insect pattern, the "campodeiform" pattern, is illustrated in Figure 17-2; it is typified by such anterior and posterior sense-organs. A few insect groups have secondary sexual structures on the last segments of the abdomen, which may have been derived from other paired abdominal appendages.

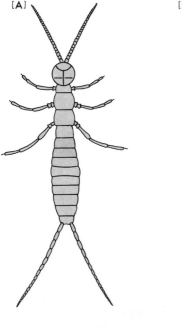

[A]

[B]

FIGURE 17-2. A possibly archetypic insect pattern. **A:** An adult primitive wingless insect, a bristle-tail, *Campodea,* about 4 millimeters long. **B:** A late aquatic nymphal stage of the plecopteran stone-fly *Perla,* about 3 centimeters long. Note the external wing buds in this preadult stage (the Exopterygota of earlier classifications; see text) and the generally "campodeiform" pattern.

The molt stages of insects, called instars, may change gradually with growth toward the adult structures, without any stage involving a major change of form during a particular molt. This is the pattern in the more primitive insects, where the successive series of only gradually changing young stages are called nymphs. In the more highly organized insects there are complex metamorphic molts, with the young hatching initially as larvae and then going through an inactive pupal stage preceding the emergence of the adult. The result of each kind of development is an adult insect, called an imago.

Insect Diversity

The 750,000 species of insects, though morphologically stereotyped, can be classified on such bases as the type of wings, the pattern of life-cycle and metamorphosis, and the type of mouth-parts. Such characteristics *do* have considerable functional and ecological significance, and they allow 33 orders of insects to be distinguished (see Table 17-1).

TABLE 17-1
Synopsis of a Classification of Insects

Phylum ARTHROPODA: Class INSECTA

Subclass **APTERYGOTA**
    Order COLLEMBOLA (springtails)
    Order PROTURA
    Order DIPLURA
    Order THYSANURA (silverfish and bristle-tails)
Subclass **PTERYGOTA**
  Division **PALAEOPTERA**
    Order EPHEMEROPTERA (may-flies)
    Order ODONATA (dragon-flies)
  Division **POLYNEOPTERA**
    Order DICTYOPTERA (cockroaches and mantises)
    Order ISOPTERA (termites)
    Order PLECOPTERA (stone-flies)
    Order ORTHOPTERA (grasshoppers and crickets)
    Order DERMAPTERA (earwigs)
    (and four more minor orders)
  Division **PARANEOPTERA**
    Order ANOPLURA (true sucking-lice)
    Order HEMIPTERA–HETEROPTERA (true bugs)
    Order HEMIPTERA–HOMOPTERA (aphids and leafhoppers)
    (and four other orders)
  Division **OLIGONEOPTERA**
    Order NEUROPTERA (lacewings)
    Order COLEOPTERA (beetles)
    Order TRICHOPTERA (caddis-flies)
    Order LEPIDOPTERA (moths and butterflies)
    Order DIPTERA (true flies)
    Order SIPHONAPTERA (fleas)
    Order HYMENOPTERA (wasps, ants, and bees)
    (and four more minor orders)

These orders are grouped into two very unequal subclasses, Apterygota and Pterygota. The former contains only four minor orders of fundamentally wing-less insects. Apart from the primitive apterous condition, they retain paired abdominal appendages and perhaps the most unspecialized mouth-parts found in insects. The four orders listed in Table 17-1 include the relatively abundant springtails and bristle-tails, such as *Lepisma* and *Campodea* (Figure 17-2). There are about 1500 species of apterygote insects as against the 700,000 or so of the subclass Pterygota or fundamentally winged insects (among which there are some orders where the wings are *secondarily* reduced or even absent).

The subclass Pterygota can be divided into four divisions, the first of which, the Palaeoptera, comprises the more primitive winged insects in which the wings cannot be folded and are held permanently at right angles to the body. Several extinct orders of insects, including the earliest flying insects, would be included in this division, along with the two living orders, Ephemeroptera (may-flies) and Odonata (dragon-flies—Figure 17-3A). Both orders have their young stages living in fresh waters as nymphs which undergo only simple metamorphosis as they grow and mature to the adult. Both show rather unspecialized chewing mouth-parts and two pairs of primitive membranous wings which cannot be folded. The aquatic nymphs of all may-flies and of some Odonata are structurally "thysanuriform" or "campodeiform" (see Figure 17-2, in which a similar stone-fly nymph is compared to an adult apterygote insect).

The remaining three divisions are neopterous, being able to fold their wings

FIGURE 17-3. Dragon-fly and mantis. **A:** An adult dragon-fly (order Odonata), a surviving representative from a group of mostly extinct insect orders which included the earliest animals ever to fly. Nymphal stages of dragon-flies live in fresh waters. **B:** Close-up of a mantis (order Dictyoptera) eating a grasshopper, the prey being grasped in the subchelate prothoracic limbs. [Photos courtesy Carolina Biological Supply Company.]

in various ways. The divisions Polyneoptera and Paraneoptera encompass the winged insects which hatch as nymphs and undergo a series of molts accompanied by only gradual and simple metamorphosis. Their wing buds develop externally in preadult stages, and the groups were included in earlier classifications in the Exopterygota or Hemimetabola. The remaining division, Oligoneoptera (see Table 17-1, encompasses the ten important insect orders where the young hatch as larvae, do not show external wing buds, and undergo a special complex metamorphic molt or molts, associated with an inactive pupal stage preceding the adult emergence. In earlier classifications, the Oligoneoptera were, for these features, termed the Endopterygota, or Holometabola. The division Paraneoptera includes certain forms which are in some ways intermediate between the other two neopterous divisions, and an incipient pupal stage is found in some bugs and other representative families of the division.

The Polyneoptera comprises nine orders, of which the five likely to be familiar to the reader are set out in Table 17-1. The Dictyoptera and Orthoptera, cockroaches and grasshoppers and their allies, are typical of the division with nymphal stages slowly leading to the adult form, with biting and chewing mouth-parts, and with relatively great powers of running and jumping but poor powers of flight (Figure 17-3B). They include the largest presently living insects. In New Zealand, the historical lack of mammals has allowed some of the niches normally occupied by mice and voles to be exploited by "invertebrate mice" or wetas, which are giant wingless orthopterans living as nocturnal herbivores. The adult of the largest species has an individual biomass equivalent to that of a small rat, and different species of wetas correspond to woodland, grassland, coastal, and alpine species of mice in other parts of the world. The stone-flies, order Plecoptera, have aquatic nymphs which are again campodeiform (Figure 17-2).

The division Paraneoptera encompasses seven orders, including the Anoplura, or true sucking-lice, where the wings have been secondarily lost and the adults are parasites on birds and mammals, and the two large hemipterous orders. (Figure 17-4A). All three orders have piercing and sucking mouth-parts and a gradual metamorphosis toward the adult. Both the Hemiptera-Heteroptera and the Hemiptera-Homoptera are large diversified groups containing large numbers of species, and some forms, such as aphids, are often extremely abundant as individuals, as a result of fantastic fecundity.

Finally we have the eleven orders of the division Oligoneoptera, or Holometabola, of which seven more familiar orders are set out in Table 17-1. Two of the smaller orders are the Trichoptera or caddis-flies, where both larval and pupal stages are aquatic, building characteristic cases of stones, shells, or plant material, and the Siphonaptera or fleas, where the wings are secondarily absent and the mouth-parts are adapted for piercing and sucking and their mode of life as ectoparasites on birds and mammals (Figure 17-6A). The order Lepidoptera comprises the familiar moths and butterflies, with two pairs of membranous wings covered with scales, in some cases brilliantly pigmented, and with the adult mouth-parts modified to form a long coiled tube for siphoning off free fluids like nectars, while the larvae have biting and chewing mouth-parts (Figure 17-5A). Approximately half the known species of insects fall in the order

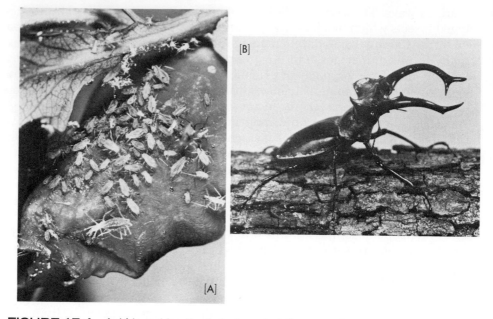

FIGURE 17-4. Aphids and beetle. **A:** A characteristic mass of young (nymphal) aphids (order Hemiptera-Homoptera) resulting from the parthenogenetic reproduction of wingless females which allows rapid exploitation of the infested plants. **B:** A male stag-beetle (order Coleoptera) with the hypertrophied mandibles as a secondary sexual characteristic. They are neither much used in feeding nor of value in defense, but they play a part in copulation. [Photos courtesy Carolina Biological Supply Company.]

Coleoptera or beetles, where the forewings are modified as elytra which can completely enclose the membranous functional hindwings when at rest (Figure 17-4B). Several groups of beetles have campodeiform larvae early in development, and several show hypermetamorphosis with additional major changes in structure and way of life occurring within the larval stages.

The order Diptera or true flies is another of the largest orders of insects, with the forewings functional and membranous, and the hindwings modified into balancing organs or halteres (Figure 17-5B). The larvae have biting and chewing mouth-parts, while the adults have either piercing and sucking mouth-parts or a peculiar proboscis adapted for lapping fluids or sponging them up. The blood-sucking forms include mosquitoes, horse-flies, black-flies, and biting midges. Many are important as vectors of tropical diseases, including malaria, sleeping sickness, yellow fever, and elephantiasis. Some species of house-flies, of the genus *Musca* or another, may be the most numerous overt land animals. Finally, another enormous order is the Hymenoptera, comprising ants, bees, wasps, and ichneumon-flies, where the two pairs of membranous wings are interlocked by means of chitinous hooklets (Figure 17-6B). Many are social insects, and a large proportion of these show division of labor among several specialized castes. The most elaborate pattern behavior, and most sophisticated methods of communication between individuals of a species, so far discovered in any invertebrates occur in hymenopterous insects.

FIGURE 17-5. Butterfly and true flies. **A:** A typical member of the order Lepidoptera is the monarch butterfly, here collecting nectar from a flower head. **B:** A female culicine mosquito (order Diptera, or true flies). Only females are blood-suckers and thus are potentially transmitters of yellow fever and the filarial (see Chapter 9) diseases. In the other major tribe of mosquitoes (the anophelines), females are the potential vectors of malaria (see Chapter 6). **C:** A cleared mount of the head of a house-fly (order Diptera) showing the fleshy labella at the tip of the mouth-parts which has many tiny perforations and is used as a "lapping-sponge" to take up liquid food, partially digested externally by enzymes in the saliva which has been "spat" onto the foods surface. This feeding method in house-flies and blow-flies is responsible for the "accidental" transmission of pathogenic microorganisms from faeces to food, for example, in epidemics of bacterial dysentery. [Photos courtesy Carolina Biological Supply Company.]

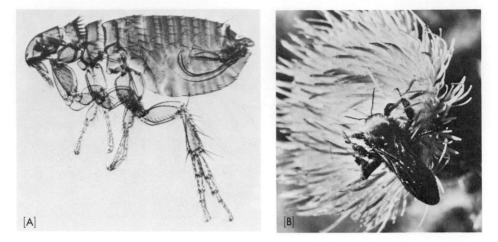

FIGURE 17-6.   Flea and bee. **A:** A cleared whole mount of a male dog-flea (order Siphon-aptera, a group of secondarily wingless insects), with piercing and sucking mouth-parts which are well-adapted for the ectoparasitic way of life. **B:** A bumble-bee collecting nectar from a composite flower. The order Hymenoptera (or ants, bees, and wasps) encompasses those insects with the most elaborate social behavior (see text). [Photos courtesy Carolina Biological Supply Company.]

Representatives of all the other major arthropod groups appear in the geological record earlier than the insects. The earliest fossil insects are from the Carboniferous (Pennsylvanian) period and include pterygote forms resembling present-day dragon-flies and cockroaches (see Chapter 34). The campodeiform body may reflect an ancient apterygote insect body form, and has some resemblances to the myriapodous minor class Symphyla (see Chapter 16 and Table 10-1).

Insect Respiration

The only kind of respiratory structures found in insects are tracheae. These are anatomically identical with those of the other terrestrial arthropods already described (Chapter 16). External openings, termed spiracles, lead into invaginations of the exoskeleton to form armored air-tubes, or tracheae, which branch from trunk systems throughout the insect's organs and tissues. The finest branches of this tracheal system form the minute air capillaries (tracheoles), lying intimately between or even within, the individual insect cells. The walls of the tracheoles (unlike the rest of the tracheal system) are not lined with cuticle, and thus they alone of the air-tube system are not molted. Tissue gas exchange takes place through the walls of the tracheoles. As V. B. Wigglesworth first discovered, the extent of gas-filled tracheole can vary with functional demands. In a resting insect muscle, much of the tracheole is fluid-filled. In active muscles and, significantly, in fatigued muscles, there is fluid only at the tips of tracheoles, and air can be seen to extend right into the muscle fibers. The physical mechanism of this, as elucidated by Wigglesworth, involves the osmotic pressure in the tissues surrounding the tracheoles acting in opposition to the capillarity forces within the fine tubes. When there has been increased muscle contraction, the tracheole is surrounded by hypertonic fluids (metabolites such as lactic acid having accumulated locally), and fluid is withdrawn from the tracheole into the tissues, air moving in to replace it (Figure 17-7B). In contrast, in a resting piece of muscle tissue, there are hypotonic fluids around the tracheole and fluids diffuse back into the tracheole lumen to fill most of the terminal tubules (Figure 17-7A). In both the fluid-filled tracheole and the gas-filled tracheole, oxygen diffuses to an extent determined by concentration gradients, but the rate of diffusion in gas-filled tracheoles will be much higher. Thus the extension of the gas-filled part of the tracheole deeper into the tissues in the active state will improve the supply of oxygen to the cells. As several physiologists have commented, the process is analogous to the dilation of blood capillaries in active vertebrate muscles.

The physiology and ecology of those insects and insect larval stages which live in fresh waters have many interesting aspects. The terrestrial tracheal system is readapted for aquatic life in many ways, as in the similarly evolved aquatic pulmonate snails (see Chapter 21). Some aquatic insects are merely divers, surfacing at regular intervals to breathe air. Others use an exposed gas bubble as a "physical gill" to extract oxygen from the water by diffusion and pass it thence into the tracheal system. In several insects, this process has evolved into plastron respiration, in which a thin film of gas is carried between minute hydrofuge hairs to form a physical gill which does not require renewal at intervals. Other insects have become fully aquatic by developing tracheal gills in which fine ramifying tracheae lie immediately below an extremely thin cuticle.

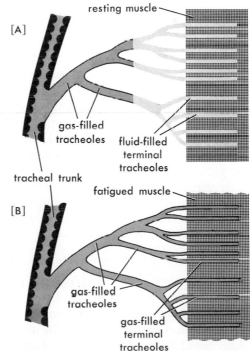

FIGURE 17-7. Respiration in land arthropods: the functioning of tracheoles. **A:** Fluids fill most of the terminal branches in resting muscle tissue, where there are hypotonic fluids surrounding the tracheoles. **B:** The terminal tracheoles are gas-filled for more efficient respiratory exchange in active or fatigued muscle tissue, where the accumulation of metabolites has rendered the fluids surrounding the tracheoles hypertonic, bringing about withdrawal of fluid from each tracheole. For further discussion, see text. [Modified from various figures by V. B. Wigglesworth.]

Two final points about tracheal systems concern the evolution of arthropods. Since they depend on gas diffusion, active arthropods using tracheal respiratory systems are probably limited in size. It is almost certainly this, rather than the mechanics of the exoskeleton, which limits the maximum size of insects. The other matter which must be recapitulated is that, in possibly seven or eight lines of arthropods, tracheal systems have been independently evolved in the course of colonization of land.

Insect Excretion

As noted in Chapter 16, the coxal glands found in several groups of arachnids and other arthropods are probably homologous with the excretory organs in crustaceans. However, the important functional excretory organs of the best-adapted land arthropods are not such segmental structures but diverticula of the gut, called Malpighian tubules. These are the organs of nitrogenous excretion in insects, and are also found in millipedes, centipedes, and several groups of arachnids. Once again, independent evolution of these structures is most likely.

In such land arthropods as insects with Malpighian tubules, nitrogenous excreta are mainly in the form of uric acid crystals. A solid or semisolid urine of this composition is, of course, also characteristic of the water-conserving vertebrates: reptiles and birds. Once again, the functioning of Malpighian tubules has been most studied in insects, and best elucidated by V. B. Wigglesworth. There are two distinct histological regions in each tubule: a proximal portion where the lining cells are differentiated with a brush border and a distal region where they are less differentiated and are said to have a honey-comb border. The processes leading to expulsion of uric acid crystals into the gut, and thence to the exterior with the faeces, will be best understood with reference to Figure 17-8. Uric acid can be combined with potassium bicarbonate and water to form the relatively soluble potassium acid urate, which is actively transported from the haemocoel into the lumen of the tubule in its distal section. The contents here are neutral or faintly alkaline, while in the proximal part they are acid. In the lumen of the proximal part, potassium bicarbonate and water are reabsorbed, resulting in the precipitation of uric acid crystals.

In combination with this mechanism, there is usually resorption of water in the gut, and in many insects the blind distal ends of the Malpighian tubules are closely applied to the water-resorbing section of the hindgut. In other cases, the Malpighian tubules lie within blood sinuses. Once again, each aspect of the integrated concert of functions in a well-adapted terrestrial animal must involve conservation of water.

Insect Hormones

In discussing the control of molting and growth in crustaceans (see Chapter 13), we noted that it had been in insects that some of the most sophisticated experimental studies on hormonal coordination in invertebrates have been car-

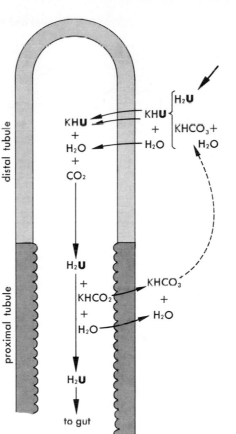

FIGURE 17-8. Excretion in land anthropods: the functioning of the excretory organs of insects, the Malpighian tubules. The processes within the two histologically distinct regions of a single tubule are illustrated. They lead to the expulsion of the insoluble crystals of uric acid into the gut, following the transport of the relatively soluble potassium acid urate into the lumen of the distal part of the tubule. For further discussion, see text. [Modified from various figures by V. B. Wigglesworth.]

ried out. The investigations of Carroll M. Williams and his associates in the United States and of V. B. Wigglesworth, J. Harker, and others in Britain are of particular significance.

As previously noted in Chapter 13, it seems that the molting hormones throughout the arthropods are closely related chemical substances, and those which are active in one group like the crustaceans will show some activity in all other groups such as insects, arachnids, and xiphosurans. The molting hormone characteristic of insects is α-ecdysone. It was first isolated as a pure crystalline hormone by Butenandt and Karlson in 1953–54, when they obtained 25 milligrams of it from a series of extractions that started with half a metric ton of silkworm pupae. It proved to be an unusually water-soluble steroid, related to cholesterol which occurs in all many-celled organisms. Skeleton formulae are shown in Figure 17-9, including that for the 20-hydroxyecdysone known as β-ecdysone (or ecdysterone or crustecdysone), the characteristic molting hormone of crustaceans. Closely similar steroids are manufactured by certain plants including the yew tree and the prolific pest-fern bracken. These analogs of molting hormones have probably been developed by the plants as a protection against insect attack. We noted that in the crustaceans control of molting and of other processes, including maturation , is brought about by the interplay

[A] Ecdysones and Related Compounds

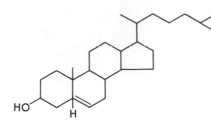

[A1] cholesterol

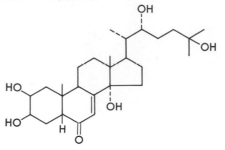

[A2] α-ecdysone

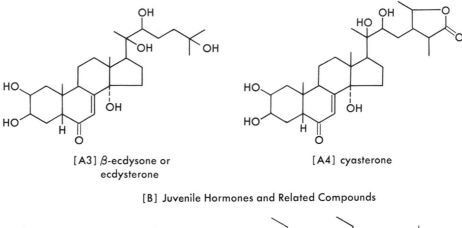

[A3] β-ecdysone or
ecdysterone

[A4] cyasterone

[B] Juvenile Hormones and Related Compounds

[B1] farnesol

[B2] juvenile hormone A

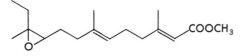

[B3] juvenile hormone B

[B4] methyl todomatuate

FIGURE 17-9. Structural formulas of molting and juvenile hormones. Molting hormones are steroids related to cholesterol (**A1**). The normal molting hormone of insects is α-ecdysone (**A2**), and that of crustaceans is β-ecdysone or ecdysterone (**A3**). Cyasterone (**A4**) is a potent molt-evoking agent stored by amaranthacean plants in their root tissues. Juvenile hormones are related to farnesol (**B1**), and there are several naturally occurring forms (including **B2** and **B3**) used by insects, often in mixtures. The methyl ester of todomatuic acid (**B4**) is the notorious active agent found in timber of the balsam fir, and present in papers manufactured from balsam fir pulp.

of a pair of antagonistic hormonal systems. In insects, the prothoracic glands secrete α-ecdysone, and certain gland tissues attached to the ventral side of the insect brain and forming the corpus allatum secrete an "antagonist" hormone called juvenile hormone (JH). This, too, has now been chemically isolated and identified. There are actually a number of compounds with juvenile hormone activity occurring as differently proportioned mixtures in different insects. Once again, similar substances have been found in a variety of plants. A notorious example is a substance of intense juvenile hormone activity, which occurs in the wood of the balsam fir, and is present in paper made from pulp of that tree species. In the laboratory, culturing of certain hemipteran bugs on such papers causes them to develop into giant immature nymphs, while culturing them on papers made from other tree pulps allows normal development to adults. Related compounds have now been synthesized that are far more active than the natural juvenile hormones, and most of these are related to farnesol (see Figure 17-9), which is widely distributed as an intermediary metabolite in the biosynthesis of sterols in many organisms.

In 1977, evidence was beginning to come in of such synthetic compounds with JH-activity affecting differentiation processes in arthropods other than insects. For example, G. G. Payen and John D. Costlow showed that on xanthid crustaceans one JH-mimic had effects similar to those naturally controlled by the crab's androgenic glands. Other evidence suggests that complex metamorphic processes in crustaceans, such as those in the life-cycle of barnacles, may be affected. As with the ecdysones, it may prove that the other hormonal controls first investigated in insects are, in fact, characteristic of the arthropods as a whole.

The anatomical basis of the control system in insects is illustrated in Figure 17-10. Apart from the prothoracic glands, which secrete the molting hormone α-ecdysone, various parts of the insect brain or cerebral ganglia are concerned in the control of molting and indeed of all the serial differentiations implicit in the development of the insect larval and pupal stages. Perhaps the most important features (Figure 17-10) are the neurosecretory cells of the *pars intercerebralis* and, on the ventral side of the paired ganglia, double-paired bodies—on each side, a corpus cardiacum and a corpus allatum. The origins of both are fantastically similar to those of the parts of the vertebrate pituitary. The corpus cardiacum is neural in origin and consists of both neuroendocrine and more normal nerve cells, while the corpus allatum is of epithelial origin and consists of nonneural gland tissue. The corpus allatum is the source of juvenile hormone, which in the insect system is the antagonist of α-ecdysone. It not only acts as a feedback inhibitor of the prothoracic glands, preventing or slowing the secretion of α-ecdysone, but it also acts directly on the same targets as does the molt hormone. This pair of hormones clearly have direct effects on enzyme mechanisms, and on those membrane permeabilities that are particularly associated with the secretion and hardening of a new exoskeleton at each molt; they also appear to have important effects on the chromosomal material in the nuclei of *all* cells in the developing insect. It is important to remember that all the genetic information necessary to make all stages of the insect is present at all times and through all stages of the insect development. In other words, the cell nuclei in the tissues of the larval caterpillar also contain all the information

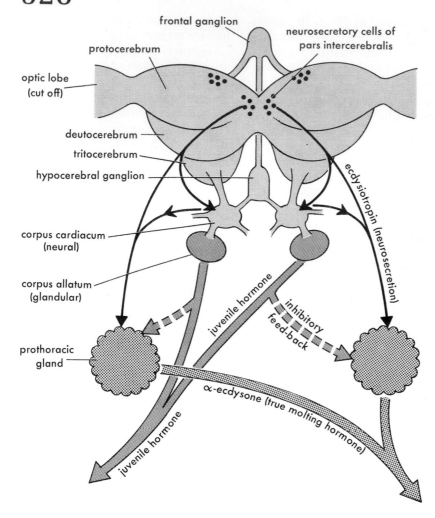

FIGURE 17-10. Brain and endocrine organs in insects. This stylized diagram shows the interactions among the neurohormone ecdysiotropin, produced by the neurosecretory cells of the *pars intercerebralis;* the molting hormone α-ecdysone, produced by the prothoracic glands; and juvenile hormone (JH), produced by the corpora allata. Note that secretion of α-ecdysone by the prothoracic glands is stimulated by ecdysiotropin and inhibited by JH. Both α-ecdysone and JH circulate generally in the insect's bloodstream and can have as targets any cells (see Figure 17-11). In crustaceans β-ecdysone (ecdysterone) produced by the Y-glands and the neurosecretory material produced in the X-organs (and stored in the sinus glands) show similar interactions (see Figure 13-6).

necessary to make the structures (and carry out the functions) of the pupal stage, as well as of the imaginal insect or butterfly. Similarly, the cells of the butterfly still contain all the necessary genetic information to make both larvae and pupae. A simplified version of the combined effects of the two hormones is illustrated in Figure 17-11. If the chromosomes of the insect cells are exposed to a relatively high proportion of juvenile hormone along with a low proportion of α-ecdysone, then the mRNA (messenger RNA) which leaves the nucleus to cause differentiation in the cell and the synthesis of appropriate structural proteins is *larval* mRNA. Similarly, with more equal proportions of juvenile hormone and ecdysone as the "hormonal cocktail" stimulating the DNA of the insect cells, it will be *pupal* mRNA which is produced. Finally, at the time of the molt to the adult, there will be no juvenile hormone and the effect of pure α-ecdysone will be to generate *adult* mRNA, which will in turn produce the imaginal or adult structures. It is possible that some of this selective "turning on" of the insect genetic material can be spatially visualized by the "puffing" of specific loci in the chromosomes under certain conditions of tissue culture. It is just possible that cytogeneticists will be able to map the positions of larval, pupal, and imaginal genetic material on such things as the giant chromosomes of *Chironomus* in much the same way—by careful and laborious genetic work —as linkage maps were constructed some decades ago for *Drosophila*. In the meantime, knowledge of the action of juvenile hormone in association with α-ecdysone is proving a powerful research tool in the hands of developmental biologists working on insect systems. In addition to all this, there is clear evidence in some insects that the other groups of neurosecretory cells within the brain (Figure 17-10) are concerned in the stimulation of both the corpora cardiaca and the corpora allata. Thus within the system there are several possibilities for negative feedback systems which, of course, allow this control system to be largely self-regulating. In other words, one could classify this (along with the many similar systems in vertebrate physiology) as basically a system for efficient homeostasis in development and growth.

Despite this, it is worth noting that, like all systems based upon central ner-

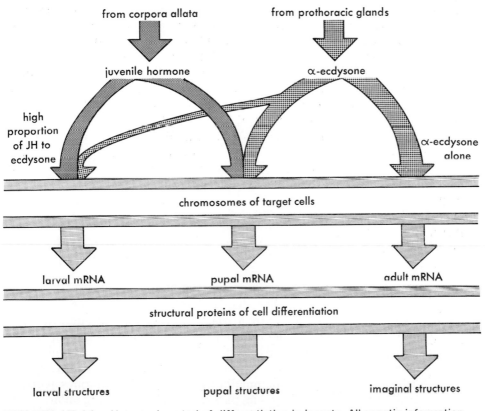

FIGURE 17-11. Hormonal control of differentiation in insects. All genetic information necessary to make larval, pupal, and adult structures coexists in all cell nuclei. The simplified diagram shows how the composition of the "hormonal cocktail" circulating in the blood of the developing insect changes from a low proportion of α-ecdysone with a high proportion of JH ("turning on" the production of *larval* mRNA) to α-ecdysone alone without any JH ("turning on" the production of *adult* mRNA).

vous control, it is possible for this hormonal system to respond in an almost "reflex" fashion to certain neural inputs, either from sense-organs sampling the environment or from sense-organs monitoring some internal condition of the insect's bodily organization. For example, one of the earliest discoveries important to our knowledge of the insect molt-cycle dates from the early work of V. B. Wigglesworth. He and his associates discovered that the bug *Rhodnius* goes into a molt immediately after it has had a large blood meal from a suitable host. This can now be documented in some detail. The initial stimulus comes from proprioceptors or stretch receptors in the gut wall of *Rhodnius*. These sensillae detect when the gut wall is stretched after the bug has engorged itself with blood. Their signal passes to the central nervous system where it stimulates the neurosecretory cells of the *pars intercerebralis*. These in turn produce a neurosecretion called ecdysiotropin which circulates in the blood of the *Rhodnius* as a short-lived chemical messenger stimulating the prothoracic glands to production of an appropriate dose of α-ecdysone. This causes the bug to move into proecdysis, the initial stages of a new molt. The external environment can affect the molt-cycle in a similar fashion. As Carroll Williams and his associates have so beautifully demonstrated in silkworm pupae, these seemingly unresponsive and inert stages of insect development have light-sensitive cells which carefully monitor light in their immediate environment. When, after a suitable temperature regime, they are exposed to light/dark phases of an appropriate day length, they will signal to the central nervous system, and the neurosecretory cells of the *pars intercerebralis* in the pupal brain will respond by neurosecretion, and here also trigger the secretion of α-ecdysone by the prothoracic glands, thus sending the pupa into the first stages of the final molt, which will result in the emergence of the adult moth.

Other hormones and neurohormones undoubtedly occur in insects and are beginning to be investigated. In some blood-sucking bugs, after a meal of blood, a neurohormone is produced which is a diuretic and promotes rapid water excretion and thus concentration of the meal. Other hormones are responsible for color changes in some insects, and still others for behavioral changes, such as the shift into the migratory phase in certain locust species. In a later section of this chapter on insect behavior, we will consider the occurrence of pheromones, similar substances secreted externally and used in communication.

Hormones produced by vertebrates may affect the life-cycles of insects. A truly remarkable linking of the reproductive cycles of rabbit-fleas with their hosts has recently been elucidated by Miriam Rothschild. Although they can and do jump from host to host, rabbit-fleas live for long periods of time on one rabbit, feeding mainly at the relatively hairless surfaces of the rabbit's ears. On adult rabbits, rabbit-fleas never copulate or reproduce. However, if a female rabbit becomes pregnant, mammalian reproductive hormones (corticosteroids) circulating in her bloodstream, pass into the fleas causing enlargement of their reproductive structures and certain behavioral changes (the feeding rate of fleas increases and they tend to remain permanently on the rabbit's ears). When the doe gives birth to a litter, the fleas respond to a further hormonal change in her blood. They then leave the ears and cluster round her mouth where they transfer to the newly born conies as they are licked by the doe.

Some factor present in the blood of rabbits only when they are less than one-week old now triggers copulation and egg-laying by the rabbit-fleas. In this way, the life-cycles of the rabbit and its rabbit-fleas are hormonally bound together.

Insect Flight

The capacity for flight is central to the dominance of insects. Other flying animals—pterodactyls, birds, and bats—adapted walking legs for purposes of flight relatively late in their evolutionary histories. Insects have been flying for about 340 million years, using lateral extensions of the thoracic segments initially in gliding, and then in flapping flight. Archetypically, the wings are arranged in two pairs on the meso- and metathoracic segments, but reduced to a single functional pair in several distinct orders. The mechanics of insect flight are still in part controversial, and there are several variant patterns. Only a partial synopsis can be attempted in a book of this scope. The simplest case, mechanically, can be illustrated in a stylized model aphid (Figure 17-12), in which the articulation of the wing is greatly simplified. In this, the flapping of the wing results from the antagonistic contractions of the indirect flight muscles which can deform the mesothoracic and metathoracic segments, mainly by moving the fused terga (dorsal roof plates) of the thorax. As shown in the aphid mesothorax (Figure 17-12) contraction of the dorsoventral muscles elevates the wings by pulling down the tergum and at the same time slightly elongates and laterally extends the thoracic box, thus stretching the longitudinal muscles. Then contraction of the longitudinal muscles shortens the thoracic box, the tergum moves upward, depressing the wings and stretching the now-relaxed dorsoventral muscles. As these two sets of indirect flight muscles work in rhythmic alternation, the so-called direct flight muscles (attached directly to the wing bases at their articulation) alter the "angle of attack" of the wings so that qualitatively they can be acting as lifting airfoils throughout the cycle. To achieve this lift by having appropriate small angles of attack during both downstroke and upstroke, the direct flight muscles twist the wing so that its leading

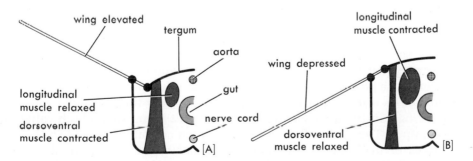

FIGURE 17-12. Insect flight. **A** and **B**: Successive positions of the indirect flight muscles in a stylized insect (basically hemipteran), showing the dorsoventral muscles of the thorax as antagonists of the longitudinal muscles. This is an oversimplification of the mechanics of wing movements in aphids: see text for further discussion.

edge is twisted downward during the downstroke and upward during the upstroke. This cycle of change in the angle of attack seems to be true for all medium and larger insects in flapping flight forward at moderate rates of wing beat and is accompanied in many by changes in wing camber also brought about by the indirect flight muscles. (It is obviously different in hovering flight, at faster beat frequencies, and in very small insects.) Insect flapping flight has received detailed experimental and kinematic studies only in the desert locust (by T. Weis-Fogh and Martin Jensen), and therein the principles of maintained airfoil action outlined above hold true (although the muscle systems are different from the aphid), and analysis can follow the principles of simple propeller theory. Crudely summarized, during the downstroke all of each wing's length contributes to lift, and the outer parts also give an excess of thrust; while during the upstroke, much of the wing's length still contributes to lift, but there is no thrust. Thus, under many circumstances, much less power is required for the upstroke than for the downstroke, and in certain cases the wing is actually raised by the wind (the resultant of the forward motion), and so there can be a *negative* power requirement for the upstroke. Such is the case in the desert locust where calculated figures of the power required for aerodynamic work in the downstroke and in the upstroke were found by Jensen and Weis-Fogh to be in reasonably close accordance with experimental measurements of work done. However, this basic pattern of aerodynamics is modifed in many insects.

One difference of some evolutionary significance is that in the more primitive groups of flying insects such as the dragon-flies (Odonata), contractions of the direct flight muscles supply the forces for the downstroke and upstroke. Depressor and elevator muscles for both pairs of wings can be distinguished anatomically. The neuromuscular arrangements of such muscles are also simple: each contraction resulting from a single motor nerve impulse, the wing-beat rate being about 20–28 per second, and being termed synchronous. In contrast, in the faster small insects belonging to groups usually thought less primitive, such as Hymenoptera and Diptera, neuromuscular function is asynchronous, each nerve impulse setting up an active state in the muscle, during which a variable number of contractions (perhaps 40) can occur. The muscles exhibiting this myogenic mechanism are termed fibrillar. Frequencies of wing beat can thus be high: 250 per second for hive-bees; 200 per second for *Musca*, the house-fly; and even about 1000 per second in some ceratopogonid midges.

Perhaps surprisingly, the flight muscles of insects (even including these fibrillar muscles) are neither unusually strong nor do they show unusually rapid shortening when compared to those of other animals including vertebrates. However, since power is measured as work per unit time, the high frequency of twitches in the synchronous muscles and even more the high frequency produced myogenically in the fibrillar muscles result in unusually high values for power output. These are reflected in the oxygen consumption and in the very high rates of utilization of carbohydrates and of fats as metabolic energy sources in flying insects. The flight muscles all work aerobically, and there is never any fatigue (or accumulation of intermediary metabolic products) in them as can occur in active locomotory muscles of other animals. Adaptively, this aspect of insect muscle physiology reflects the need for flapping flight to be sustained. A flying insect could not stop in midflight to allow oxidative recovery from an accumulation of lactic acid! A continuous supply of oxygen is ensured by the branches of the tracheal system which penetrate folds in the cell membrane of each muscle cell until they lie deep among the mitochondria and the fibrils of the flight muscles. Both at the level of the blocks of flight muscles, and at the level of the individual muscle cells, the whole insect flying mechanism is geared for very high rates of continuous power output, in fact the power:mass ratios are closely comparable to those for small aircraft engines. Efficiencies run as high as 50%, but there is still "wasted" energy produced as heat. One can think of the muscle machinery of a flying insect as air-cooled since most of this excess heat is removed in flight through the tracheal air-tubes. On the other hand, most insects fly well only at relatively high temperatures (37–40°C). Many insects have to have a preliminary warming-up period for their muscles before they can take off. In some large beetles, warming up produces a detectable buzzing sound, and in some larger moths, a slight vibration of the wings can be seen. In 1968, A. E. Kammer was able to demonstrate a neurally and mechanically efficient pattern of warming up in certain Lepidoptera, including the sphinx moth. This species has two synergistic paired sets of flight muscles which, in normal flying, work together and produce a flapping frequency of about 50 per second. During the warm-up period the signals from the central nervous system to the two sets of muscles fire alternately, causing the muscles to contract in antiphase to each other. When the flight engines are sufficiently warm, the central nervous system then switches its pattern of motor signals to synchronous firing so that the muscles work together moving the wings in flapping flight.

Finally, in many insects — perhaps most—power is saved in flight by various mechanisms of elastic recoil. E. G. Boettiger has provided a detailed analysis of the wing-hinge system in the dipterous fly *Sarcophaga*, where the skeletal arrangement is such that when the wing is in any intermediate position between greatest elevation and greatest depression, the curved surfaces of the skeletal elements are strained, involving storage of energy. Their recoil can "click" the wing to the up position, and this "click" or "snap" mechanism like that of a child's metal cricket ensures a magnification of each muscle contraction and a rapid movement to each extreme position in turn. Thus the minimum change of length required of the fibrillar flight muscles (with an amplification of about 500 times at the wing tips) allows them to operate nearly isometrically. A more obvious example of intrinsic elasticity concerns the cuticle of the hinge in some insects (first studied in locusts by Weis-Fogh). This hinge material is largely a peculiar rubber-like protein called resilin with cross-link bonding of a type unique in structural proteins. Mechanically, this intrinsically elastic hinge allows the energy of the "negative work" of the upstroke already mentioned to be stored, and to contribute to the following downstroke of the wing. Finally, the entire flight muscles have been shown to possess a strong passive-elastic component themselves. All such mechanisms of elastic recoil enable the insects possessing them to minimize the dimensions of their flight muscles. In summary, the locomotory mechanisms of land arthropods include the most sophisticated patterns found in animals, and these are all functionally based on the unique properties of the arthropod exoskeleton.

Other Insect Functions

As we have seen, terrestrial adaptation and the nature of the exoskeleton determine the patterns of respiration, excretion, growth, and locomotion in insects. A brief consideration of the processes of nutrition, circulation, and reproduction seems appropriate here.

The insect alimentary canal is a simple tube, though often with some convolutions, running from the mouth to the anus. Structurally and functionally, it is divisible into a cuticle-lined foregut, an endodermal midgut, and a hindgut—again with a cuticular lining. In the majority of insects, valvular arrangements separate the three regions, and, if diverticula occur other than the Malpighian tubules, they form gastric caeca attached to the anterior end of the midgut or a blind storage crop attached to the foregut (as in many Diptera, butterflies, and moths).

The form of the foregut varies with the diet of insects. Various forms of mouth-parts are, of course, associated with each of the major orders of insects, but functionally they fall into three groups. Most primitive are chewing and biting mouth-parts (suitable for predaceous carnivores, foliage-eaters, and other macrophages); while more specialized are those for collecting fluids (products of bacterial decomposition, or angiosperm nectars); and those adapted for piercing and sucking (not only animal blood but also tissue fluids of plants). Functionally, the foregut is primarily a storage organ, with a crop formed as a dilation or a diverticulum, and a capacity appropriate to dietary habits. There can also be a "gizzard-like" region for internal trituration. Some digestion may take place here, the enzymes originating elsewhere, but there is no absorption. Functionally, the main site of production of digestive enzymes, of digestive breakdown, and of absorption is the midgut. Being lined, as are its caeca, with endodermal epithelia, it lacks the locally secreted epicuticle and procuticle of the other two gut regions. A peculiar loose lining—the peritrophic membrane—is secreted in various ways, is functionally semipermeable, and, although extremely thin, is probably of similar composition (of chitin and protein) to the deeper layers of the procuticle in the exoskeleton. It permits passage of liquids and solutes in both directions but apparently prevents solid particles of food from coming into contact with the gastric epidermis. Thus, unusually clearly demonstrated for invertebrates, digestion is entirely extracellular. The hindgut is again lined with invaginated ectoderm and therefore with an exoskeleton-like cuticle. Its major function appears to be uptake of water from both the faecal remains from the midgut and the nitrogenous excreta from the Malpighian tubules. It is worth noting that while the peritrophic membrane is continuously secreted and renewed (in many cases forming a sheath around the faecal pellets on their expulsion), the cuticular linings of foregut and hindgut have to be reformed and shed along with the exoskeleton at every molt.

As in all arthropods, the circulatory system is an "open" one, the heart being tubular with a series of ostia and the only true blood vessel being the anterior aorta leading toward the head from the heart. Blood is released anteriorly and returns through a series of sinuses to the pericardial space and thence through the ostial openings into the heart. In view of the nature of the respiratory structures, insect blood does not need to be effectively involved in the transport of oxygen to the tissues. A large part of the osmotic pressure of insect blood is maintained by amino acids, and the ionic ratios found in herbivorous insects involve unusually high potassium to sodium values. This latter fact creates difficulties in applying the usual physiological explanations to some aspects of neuromuscular function in plant-eating insects. There are large numbers of nucleated blood cells, and about thirty types in six classes have been described. Most are phagocytic. The ratios of blood-cell types change with the molt cycle in most insects but the functional significance of many of the differentiated cells remains obscure.

The great majority of insects are dioecious, although parthenogenesis has been described in several groups. The paired testes in the male usually consist of a series of follicles, within each of which can be found a series of groups of developing spermatozoa contained in packets. The several stages of sperm maturation are to be found in a serial arrangement. There are usually storage sacs, termed seminal vesicles, and in some cases accessory glands. The female system is similarly arranged for the serial production of gametes. Each ovary consists of a series of finger-like ovarioles, each of which is an epithelial tube with the oögonial germ cells at the free end and a contained series of eggs in various stages of development. The number of eggs which can be laid at one time obviously corresponds to the number of ovarioles. In many cases each laying is enclosed in an oötheca or other egg capsule or mass secreted by the lower part of the female genital system (oviducts, vestibulum, or vagina) or by various kinds of accessory glands. Hormonal mechanisms are undoubtedly responsible for determining the onset of reproduction in insects, and probably for controlling fecundity (see page 319). However, the secondary sex characters do not ever seem to be controlled by hormones originating in the insect gonads. The contrast with crustacean endocrine physiology has already been noted.

Insect "Brains" and Insect Behavior

The central nervous system of insects is essentially like that of the annelid worms and of the other arthropods in that it consists of a ventral chain of segmentally arranged ganglia linked by ladder-like connectives. Just as in the crustaceans, least fusion of ganglia is found in some primitive insects and in larval stages of others, while the most extensive fusion of ganglia into "brain" masses occurs in the adults of higher forms. There are some difficulties in allocating particular parts of the fused system to individual head and thoracic segments, either developmentally or in terms of their evolution from the basic annelid plan. In the adults of higher insects, the anterior "brain" consists of a protocerebrum, a deuterocerebrum, and a tritocerebrum, which last is probably an anterior thoracic ganglion in origin. Behind the mouth there is the suboesophageal ganglionic mass (or "hindbrain") consisting also of three pairs of fused ganglia, and behind this there are the abdominal ganglia, primitively eight in number, but showing varying degrees of fusion in the higher insects. We have already discussed the importance of certain neurosecretory cells in these "brains" in relation to the hormonal controls of the insect life-cycle. In terms of normal neural activity, it might be said that the anterior "brain" in front of or above the mouth receives most of the sensory input, being connected to eyes, antennae, chemoreceptors, and so on, while the suboesophageal brain behind or below the mouth has many motor neuron cells and is the origin of the motor nerves running to the mouth-parts directly and, via the more posterior segmental ganglia, to the insect's organs of locomotion, the walking legs and the wing muscles. Although the anterior group of ganglia seem to have no important motor nerves running directly from them, they are clearly a center for some motor coordination, and for considerable correlation or association of various inputs from sense-organs. The function as a coordinating center seems to be achieved largely by inhibitory processes, by limiting or interrupting the occurrence of certain reflexes whose main "wiring" involves other parts of the central nervous system. In general, damage to the anterior ganglionic mass results in an insect performing sets of movements and responses, often in quite elaborate series, in a repetitive and uncoordinated and, to the human observer, "frenzied" manner.

In a consideration of the functioning of the insect central nervous system, it is important to recognize that, in contrast to other higher invertebrates and to vertebrate animals, there are relatively few nerve cells involved. The neurons of the central nervous systems of all animals are, by and large, of similar dimensions, and this means that there are very many more of them in the brain of an octopus, or even of a small mammal, than in a typical insect. In functional terms, the behavior of insects which is based on this nervous system may appear to be elaborate, but is always relatively stereotyped and predictable. Insects are capable of only limited learning, and in this they contrast markedly with the behavioral potentialities of either the octopus or the rat. The more elaborate forms of insect behavior seem to be genetically innate, programmed into the central nervous system, and unmodifiable by training or by sequential spontaneous experience. It seems that the relatively low number of "association" neurons (see Chapter 25 for a discussion of the contrasting condition in the octopus and its allies) results in the insect's responses to its environment being those of a well-designed minicomputer with a moderately large repertoire of algorithmic programs, which result in apparently elaborate behavior, but not those demanding both the storage capacities and the ability to modify and combine programs of a larger computing center. Perhaps because of this low learning ability, insects have been valued by students of animal behavior from naturalists like J. H. Fabre to modern ethologists.

Insects display abundantly the simpler kinds of orientation to their environments that were first described by Jacques Loeb as taxes and kineses. These are based on the still simpler reflex arcs referred to earlier in worm locomotion, and in the discussion of insect hormonal controls. As outlined in Chapters 11 and 16, all annelid and arthropodan locomotion is based on a series of reflexes, and its apparent coordination owes as much to the proper sequence of signals received from the environment as to any extensive integration in the central nervous system. Locomotory movements which are generalized and nondirectional in response to an external stimulus are called kineses. As Gottfried Fraenkel showed more than forty years ago, the wing movements of flight in Diptera, and in certain other insects, result as a generalized kinesis whenever the tactile receptors on the tarsi of the walking legs are *not* stimulated by contact with a surface. Many cryptic and nocturnal insects, whose preferred environmental conditions include high humidity and darkness, demonstrate kineses which involve increased locomotion (and increased rates of turning during that locomotion) while they are exposed either to drying conditions or to brighter illumination. Anyone can observe this sort of behavior in insects and other cryptic arthropods simply by turning over (particularly in temperate woodland or garden situations) a suitable large stone or fallen log. The intensity of activity (and particularly of turning) in the disturbed arthropods remains high, until they get back into their preferred conditions of high humidity and shade. It is important to note that, for this kind of behavior to be effective, there need not be any directionality at all. A kinesis is a nondirectional increase in activity upon stimulus.

In contrast, many kinds of insect behavior depend upon the simplest kind of directed movements, or taxes. These can involve an input through appropriate sense-organs from a chemical, visual, or tactile difference in the immediate environment. An important morphological feature here is that insects, like all segmented animals, and even flatworms, have their sense-organs arranged in bilateral symmetry. Some of the simplest insect taxes involving movements toward preferred conditions along gradients, or toward a food source as a stimulus, may involve insects in "scanning" comparisons of the intensity of the stimulus using alternately the sense-organs of the left and right sides of the body. Somewhat more complicated are the telotaxes, particularly light-compass orientation, whereby insects can move in a straight line at a constant angle to a stimulus source. Use of the sun in this way is demonstrably of importance in the "navigation" of ants, outward to a food source and then back to their nest. With ants traversing suitable ground, an experimenter can produce a regular pattern of disorientation in the ants (heading, for example, at right angles to the intended direction) by a cunning use of mirrors.

In many groups of insects, individuals can "communicate" with other individuals of the same species, and to us, anthropomorphically, this would seem a distinctly "higher" form of behavior. In fact, much insect communication does not involve any more complex neural organization than was required for the kineses and taxes in response to physical conditions in the environment, but adds to this only a source of chemical, light, or sound stimulus, which is (in most cases) species-specific. We will postpone discussing the tactile communication used by social bees for the moment. Chemical means of communication in insects are typified by the sex attractant odors used, for example, by the silkworm moths. A hormone-like substance used externally, and thus termed a pheromone, is used by a virgin female to attract males of the same species over long distances (even several miles). The female releases the pheromone into the air under suitable weather conditions and a spreading plume of the odor flows downwind. Males in the track of this, even receiving as little as 40 molecules per second of the pheromone scent, respond by being stimulated to fly upwind. This positive anemotaxis, or flight directly into the wind's source, results in the assembly of sometimes few, sometimes many, males in the close vicinity of the female putting out the "call." In the last decade, a number of the sex pheromones of insects have been chemically synthesized. Certain economically important species, such as the gypsy moth pest, can be trapped if the appropriate pheromones are used; and this kind of control has less deleterious effects for the environment and for humanity than a corresponding widespread use of insecticides could have. Chemical communication is also used extensively by ants: in some cases, the proper community spirit in these social insects is achieved by the possession of a nest odor peculiar not to the species, but to the individual community of ants. In such cases, if ants from another nest, even if of the same species, intrude into a colony they are immediately detected as having the wrong smell and attacked, expelled, or killed. Some ant species can lay chemical trails to be followed by other foragers from the same nest.

Insects also use the emission of light in visual communication between individuals of certain species. Among the beetles are the nocturnal fireflies, which have light-producing organs used to produce species-specific "codes" involving pattern and timing of flashes. The male fireflies cruise while the females remain stationary on low vegetation; the males emit the specific flash pattern, and if a female replies exactly two seconds after each flash made by the male, he orients toward her, finds her, and can subsequently mate. A human experimenter with a flashlight and a good sense of timing (particularly of a two-second "rest") can decoy males toward a trap. Some species of fireflies form huge assemblages which fall into rhythmic synchronous flashing. Drs. John B. and Elizabeth M. Buck have shown how, in some of these cases, individual flashes recruit nearby fireflies and bring them into synchrony, much as additional humans are added to the rhythms of community singing or of a marching parade. In some parts of Southeast Asia, there are arboreal fireflies where the males gather in thousands on trees and then flash twice a second in perfect unison, lighting up the whole tree like a Christmas display.

Several different kinds of insects use sound production, essentially species-specific songs, for mating purposes, and these include mosquitoes, certain beetles, some bees, and the well-known cicadas (Hemiptera-Homoptera) and katydids (Orthoptera). In temperate North America, large black crickets belonging to the genus *Gryllus* are common, not only in old fields and shrubby roadsides, but also inside older houses. In late summer and fall, the mature male has a calling song, put out when he has a spermatophore of ripe sperm ready for transference to a female, consisting of an almost continuous series of chirps recurring about a rate of about three per second. Crickets like *Gryllus* cannot fly, but use their reduced forewings in a scraper and file arrangement to produce a series of sound pulses which make up the chirps. (On acoustic analysis, the sound pulses are each about 20–30 milliseconds in duration, with "rests" of about 50 milliseconds between them. Both frequency and intensity can be modified and even the human ear can detect the increased volume and apparently increased pitch in the cricket's chirping.) Each male when he begins his calling song is occupying a territory "on the hearth" or among the roadside grass-stems and herbs, which he defends against all other males. In crickets, the two sound receptors (or "ears") have their tympani or drum-heads each placed on a major podomere of a foreleg. They are thus paired symmetrically and can be used directionally. If a female cricket of the correct species hears the calling song she is attracted to the male's territory and is there "courted" by a patterned sequence of movements and by a characteristic courting song, wherein louder single chirps are separated by a succession of lower sounds of reduced intensity. In contrast, when another male cricket enters the defined territory, a rivalry song ensues with both males stimulating each other to increase intensity and frequency ("I can sing better and higher than you"), until the aggressiveness leads to a rather formal fight, usually without real damage, involving kicking, forelimb wrestling, and biting movements. One male then retires from the territory. Human investigators have found that all the different sequences— of mating in females and of singing-back and aggression in males—can be produced (in the absence of potential mates or potential opponents) by the playing of appropriate recordings to crickets. It is possible to show experimentally that the nerve cells which produce the appropriate songs in crickets are themselves situated in the mushroom bodies of the forebrain.

Sound reception and sound production are used in other ways by insects. Many larger moths fly at night and can fall prey to insect-hunting bats. The bats locate their prey by sonar means, emitting ultrasonic chirps and locating both food insects and obstacles by the echo patterns which return to their ears. At least four families of moths have evolved auditory organs, in this case, on the side of the metathoracic segments. Although with rather simple ears in terms of numbers of sensory cells, a noctuid moth can detect the chirps made by a bat at a distance of about 30 meters; although the bat itself can use the echo return effectively only at about 3 meters. Once again, the ears being paired on either side of the body allow the moth to determine the direction of the sound; a faint bat chirp will cause the moth to turn away, a loud (that is, close) bat chirp causes it to dive toward the ground. A few moths of one family have developed not only substances distasteful to bats within their body tissues but also a system of producing ultrasonic clicks in response to bat chirps which effectively "warn off" the bats.

All of the insect behavior described so far is "pattern" behavior, innately set for the species and not modified by experience during the individual insect's lifetime. In contrast to this, insects do show some simpler capacities for learning. In the early 1960's, Adrian G. Horridge and Graham Hoyle were able to study simple learning in thoracic ganglia of the cockroach. Their experimental preparations involved animals from which the heads had been removed, and they were able in a series of elegantly simple experiments to train specific legs of roaches to avoid an electric shock by withdrawal of the limb (Figure 17-13). In some of the most straightforward experiments, training took about 30 minutes, and the trained response was retained for a number of hours. More recently, Hoyle and his associates have tried to map out the nerve cells involved in learning in these thoracic ganglia. The neural maps involved in central programming can be explored in this system because the total number of neurons is only a few hundred. On the level of whole animal behavior, many habituation changes have been observed in ants, bees, and wasps, which reflect simpler forms of behavioral modification or learning. However, in the great majority of insects, including even these more highly evolved and social forms, almost all behavior, however elaborate, follows set, programmed patterns which appear to be part of the genetic inheritance of each species.

Even in cases of apparently complex (and in anthropomorphic interpretation, apparently motivated) behavior, we are observing fixed patterns, always

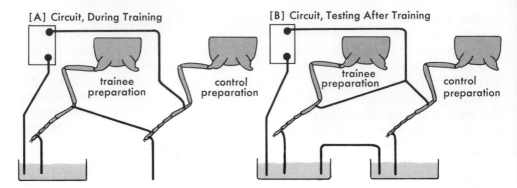

[A] Circuit, During Training

trainee
preparation

control
preparation

[B] Circuit, Testing After Training

trainee
preparation

control
preparation

FIGURE 17-13. Simple neural learning in cockroaches. The preparations are decapitated cockroaches used in pairs. During training (**A**) they are wired in series and suspended so that if the leg of the animal being trained is extended it receives a tarsal shock by contact with a dish of saline. The control animal receives the same number of shocks, but randomly and unrelated to its degree of extension. After a training period of 30–45 minutes, the preparations are wired in parallel (**B**) for the testing situation. Now both trainee and control can receive a shock after extension, but the trained preparation "remembers" to keep its leg raised much more of the time and receives significantly fewer shocks than the control preparation. In the most straightforward experiments (such as that shown here), training extended over 30 minutes and the trained response was then retained for several hours. [Partly modified from Horridge, 1962, and Hoyle, 1965.]

played out in a certain way by individuals of an insect species without any modification to meet local circumstances or to respond to previous events. It is probably for this reason that some of the more elaborate sets of courtship behavior are found among predatory insects, in which the object is not only successful mating, but the prevention of the partner providing an easy "kill" for feeding. Some of the "balloon-flies" or empidids, which are predaceous on other insects, have patterns which involve the presentation of a food insect to the female by the male immediately before copulation. The name originates with those empidid species where the male, after killing a suitable insect, will wrap it in a ball or balloon of silk which he has secreted. This is then presented to the female, and copulation ensues while she is occupied in the unwrapping of the silken package and the subsequent ingestion of her meal. Apparently in a few, certainly more chauvinistic and perhaps more highly evolved, species the female is presented with an empty ball of silk which she unravels vainly during copulation. More than a century ago, the great French naturalist, Fabre, had observed that in the praying mantis, the smaller male was often consumed by the female after, or sometimes during, copulation. More recently, K. D. Roeder was able to demonstrate that this sexual cannibalism could promote, rather than hinder, successful reproduction and the survival of the species. A receptive *and* hungry female will usually strike at an approaching male as she would at any prey insect, catching his head in the jacknife-like subchelae of the forelegs and eating it first. Destruction of the male's brain removes an inhibitory control over the ventral neural ganglia which then signal for continuous copulatory movements of the abdomen, along with some strange rotatory locomotory movements of the limbs, as long as these parts remain uneaten. It can be demonstrated neurophysiologically that the stimulus patterns responsible for the copulatory movements are produced endogenously in these abdominal ganglia, and that they can be released in an attacked male by contact with the female, but are more effectively released and can become continuous if he is decapitated. Thus the male "loses his head" in order to carry out his reproductive purpose more effectively.

The behavior patterns of the social insects—no matter how much human moralists may use them as examples of the altruistic, and political or counter-culture evangelists as patterns of the proper communal organization—are fixed patterns without "choice." Termites feed largely on woody materials with cellulose as a main component. Like cattle, they depend upon a culture of micro-organisms in their guts to digest cellulose. Social organization in the Isoptera (termites) probably evolved because, for nutritional health, termites are required to eat one another's faecal pellets and cast exoskeletons after each molt in order to conserve stocks of the essential microorganisms. We have already noted how, in the true ants (Hymenoptera), social cohesion in a particular colony is achieved by distribution of specific pheromones, and this may be aided by frequent regurgitation of food for exchange between members in the nest to help maintain the common nest identity of taste and smell. Many ant species are nearly blind, and as already noted, taste-smell is used in trail-blazing as well as for establishing colony loyalty. More elaborate "social" behavior patterns, including the marching columns of army ants, which can attack and destroy much larger animals, and the fungus-farming activities of the leaf-cutting ants, and even the "rescuing" of strayed or separated nest mates in many forms, are all dependent upon programmed movements released in response to specific stimuli combined with the appropriate smell-signal of that particular colony.

Tactile communication is also important in social insects. Certain ants perform tandem running with the antennae of a follower in contact with the abdomen of the "leader." The "soldier" caste of termites have massive mandibles which they use in a synchronized drumming against the nest walls to alarm all the termites of a colony to the presence of intruders. The other termites detect the drumming tactilely through the tarsal joints of their feet. Communication by hive bees in the total darkness of the hive is achieved by a most elaborate system of dancing which is perceived by the other workers crowded around feeling the direction of the dance. Our knowledge of the dance language of honey bees was first elucidated by Karl von Frisch, for which work he shared a recent Nobel award. Von Frisch demonstrated that a scout bee which has located a new food source is able to communicate (back in the hive): the distance, the direction, and the nature of the food resource. This can then be exploited by an exponential buildup of other food-gathering workers flying directly to and fro between the hive and the source, without scouting or stray movements in other directions. The nature of the food source is conveyed to the other bees in the darkness of the hive by regurgitation of nectar, but the direction and the distance are communicated in the tactile language of the dance. In the case of more distant groups of newly blooming flowers, for example, the successful scout bee performs a waggle dance on the vertical surface of the comb. This consists of a series of figure-eight loops with a waggling run between the two loops (Figure 17-14), the distance, and perhaps the richness of the food source,

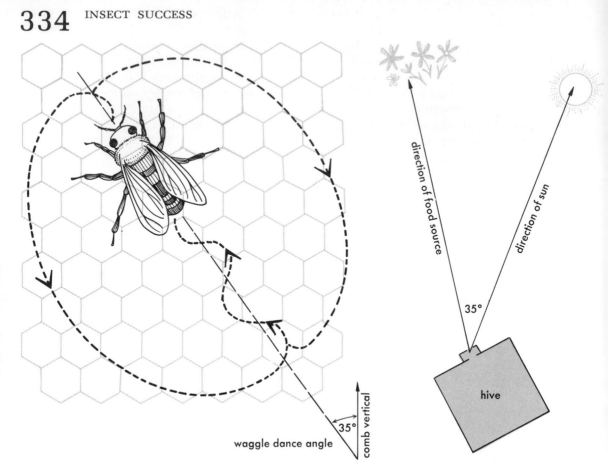

FIGURE 17-14. Hive bee dance language. In the darkness of the hive, on a vertical comb surface, a successful scout bee communicates (in the tactile bee dance language) the distance of a new food source by the frequency of the body waggles on the central part of her figure-eight dance. The direction is indicated by the angle the waggle run makes with the comb vertical (in this case 35° counterclockwise). This means that the food lies at 35° counterclockwise (at the hive entrance) from the incident angle of the sun at that time of day. If the dance is continued for any length of time, the change in apparent solar angle is compensated for. Angles can be indicated through 360°; for example, if the food source had been found in the opposite direction from the hive (away from the sun), then the waggle run would have been vertically downward on the comb surface. [Modified from von Frisch, and others.]

is related to the number and the vigor of the body waggles. Perhaps the most remarkable aspect is that the direction is indicated by the angle made by the waggle-run with respect to gravity on the vertical surface. This angle, if averaged over several runs, corresponds to the angle between the sun's direction at the entrance to the hive at that time, and the direction in which the food source lies. Scout workers can compensate for the apparent movement of the sun through the sky by changing the waggle-run angle, if the dance persists for more than a few minutes. Bees can determine the sun's direction even when it is obscured by broken clouds, as a result of their capacity to discriminate polarization patterns formed by sunlight in the blue sky, and there is also evidence that they are able to detect the direction of the earth's magnetic field.

Most of the other complexities of behavior in hive bees involve no "learning" as such. The elaborate systems of work interactions between different ages of nonreproductive females or workers, the production of a few males or drones in season, and even the production of a new queen by a colony left queenless, are all matters of pattern behavior involving the production of species-specific signals and appropriate pheromones. The human temptation to regard the succession of activities performed by a worker bee as in some way involving a series of training processes must be resisted. The average worker bee spends her first five days or so as a nursemaid tending the developing young, and as a caretaker cleaning around the hive; and then, increasingly over the next ten to fifteen days, the worker bee becomes concerned in building more wax cells and capping them in comb construction; and then (at an adult age of about twenty days) she reduces all these in-hive activities, becoming a field worker and making several trips each day to collect pollen and nectar. Eventually, if she survives to the old age of about a month or so, she will become a scout bee, searching for new food resources. For obvious reasons of liability to environmental accidents, but also because they become worn out by their intensive activity, there is a high mortality rate among scout bees. None of this is learned; it can be demonstrated that all the patterns of behavior involved in comb-building, in food-foraging, and in scouting and the dance-language, are genetically fixed and present in all of the young nursemaid workers.

Insect Ecology and Man

Just as the systematic description of insect diversity can fill extensive textbooks of entomology, so any inclusive coverage of the ecology of insects lies far beyond the scope of this volume. Indeed, it has been justly claimed that the community ecology of *all* terrestrial environments and of most freshwater ones consists mainly of the ecology of the insects inhabiting them. Our attempts to control and exploit terrestrial environments—through agriculture, by forestry, and even in creating human dwellings—bring us into direct conflict with the long successful and ecologically dominant insect groups.

In our discussion of the importance of insects in the transmission of human diseases, it is important to distinguish between facultative or almost accidental transmission and the part played by true intermediate hosts or vectors, where an essential part of the life-cycle of the disease-causing organism takes place during the transmission from human to human by the insect. An example of this latter case already discussed is malaria, where the anopheline mosquito acts as a vector, in which (see Chapter 6) essential stages of sexual reproduction and sporozoite production must occur before a new human host can be infected. In contrast to this, we have the transmission of such diseases as bacterial dysentery by blow-flies. Here the transmission results from the blow-fly's habit of walking over and licking carrion and faeces, as well as fresh foodstuffs which may be subsequently consumed by humans. Blow-flies and house-flies have taste receptors on their feet as well as on the extrusable proboscis, and as V. G. Dethier and his associates have shown, the flies can taste and select for a wide variety of amino acids and other organic compounds by a complex system of coded signals generated by a very small number of sensitive "taste" cells. Thus the programmed tasting activities of such flies, as they explore all potential food resources in their environment, can result in accidental transmission of pathogenic microorganisms (for example, from faeces to food). In some vectors, there is no important life-cycle transformation within the insect host but merely some multiplication of the disease organisms. This is the case in two groups of diseases transmitted by fleas (Siphonaptera): plague, and certain forms of typhus. A rat-flea Xenopsylla cheopis, transmits plague bacilli hypodermically from rat to rat, and from rat to human. If the flea feeds on an infected rat, the plague bacilli multiply in its gut and cause a blockage which forces the flea to feed more frequently, and to regurgitate some of its gut contents (including the plague bacilli) at each subsequent meal. It has been estimated that, during one epidemic in the fourteenth century in Western Europe, between 25 and 40% of the entire population was killed by plague in one year. In more recent times, plague and, more frequently, typhus have resulted from wartime conditions producing undesirably close association between people and rats, and unsanitary multiplication of the rats themselves. It is important to note that plague is still an important disease of many rodents, being continuously present in wild rodent populations, including some in the Western United States.

Human epidemic typhus is caused by a rickettsia (a miniature intracellular bacterium with some culture characteristics like a virus), which is transmitted by sucking lice (order Anoplura). Once again it is a disease of wartime; louse-borne diseases spread as epidemics only when unwashed people are crowded together. Typhus and other infections, including the now almost forgotten European relapsing fever and "trench" fever, which are all carried by lice, can sometimes be injected hypodermically into the new human host through the louse's sucking mouth-parts. More often, the pathogens are introduced through skin abrasions caused by scratching and by lice being crushed. There are two species of human lice, Pediculus humanus (with its two varieties on head and on torso), and Phthirus pubis (the so-called crab-louse of the pubic region). Both became much rarer in the second half of the twentieth century, after the use of synthetic organic insecticides had become widespread. Now in the late 1970's, when we have become acutely aware of the ominous implications to the environment of the accumulation of residues from such biocides as DDT, it is salutary to remember the humane purposes which these insecticides typified when first used. DDT was developed as an insecticide in 1939–40 by a Swiss chemist, Paul Müller, who received a Nobel award in 1948. It was manufactured in wartime in Britain and in the United States, initially in secret. In 1944, typhus broke out in the overcrowded civilian population of Naples, but was brought under control within a few weeks after a mass delousing program using DDT. In World War I, epidemic typhus had killed several millions of Russians, Serbians, and Austrians; and, at Naples in 1944, for the first time in history, a wartime epidemic of typhus was halted. Müller's Nobel prize was merited. In more recent years, lice have made something of a comeback in certain western societies. The head and body louse (Pediculus humanus) has returned with longer hair and beards; particularly in underwashed communities. The crab-louse (Phthirus pubis), whose "claws" on the mesothoracic and metathoracic limbs are actually subchelate and adaptively fitted for anchoring on the stouter kinds of human hair, can be an irritating but relatively harmless venereal pest and, like the pinworm *Enterobius* (see Chapter 9), has increased in the "permissive societies" of certain social and age groups—its occurrence on eyebrow hair as well as pubic hair indicating the same human behavior patterns as involved in transmission of *Enterobius*.

Pathogenic organisms going through essential stages of their life-cycle in the insect vector are sometimes called cyclopropagative or cyclodevelopmental, and they include the protozoan causative organisms of malaria and of sleeping sickness (see Chapter 6), transmitted by mosquitoes and tsetse flies, respectively; and various nematode worms causing filarial diseases, including "river-blindness" (see Chapter 9), transmitted by other mosquitoes and black-flies. Control of the insect vector is a necessary part of any attempt to reduce the incidence of these endemoepidemic tropical diseases, and this may require widespread use of insecticides.

"Pest" insects, which can infest human crops or stored foods, may represent a greater threat to the survival of man on this planet than do the insect disease vectors. It is important to realize that as regards the interests of humanity, the vast majority of insect species are neutral, less than 5% can be considered pests, and less than 1% (but still some thousands of species) are beneficial to man. Pest species exist because of the alterations from natural communities brought about by man's activities. Obviously, crop agriculture produces excellent concentrations of host plants for many insect species, whose preferred food sources are thus provided in uninterrupted abundance. The storage of crops already harvested again provides an unnatural concentration of resources available to appropriate insect species; and to a lesser extent, the large-scale culture of single species of stock animals by man provides increased and concentrated opportunities for *their* insect parasites. Large-scale crop monoculture by man can also produce a local modification of microclimate, with conditions of humidity, temperature, and light particularly suitable for the rapid development and multiplication of a pest insect, in conditions which may be unfavorable to the development and multiplication of its natural enemies. In other ways, man's activities have increased the incidence of insect pests of food crops. These include the accidental introduction of a pest insect species into favorable new areas, where its natural predators, parasites, and pathogens do not occur. Even more significant has been overextensive use of insecticides, resulting in killing off of the natural enemies of a pest species as well as of competing forms and, in some instances, creating conditions for the selective survival of insecticide-resistant forms of the pest species. Again, it may be salutary to remember the great economic benefits conferred by the use of insecticides over the last few decades. As recently as the years 1945–1948, the relief teams of the United Nations working in central and southeastern Europe were losing about 23% of food stores brought in for relief purposes to rodents and insects. In 1976, the chemical industry in the United States estimated that use of insecticides against pest insects of both crops and stored foods resulted in a national saving of twenty billion dollars each year. Even those ecologists who are most opposed to continued widespread use of certain insecticides have to admit that that estimate could only be exaggerated by about 30%. Just as Dwight D. Eisenhower, no antiwar liberal, could state that "the cost of one modern bomber is this: a modern brick school in more than thirty cities''; so this author, no chemical industry lobbyist, can note that reducing insecticide uses in agriculture and food distribution by 50% (without taking other control measures) could result in a 30% increase in basic food costs to the consumer accompanying a proportional loss of real income to *all* farmers and farm workers.

Today, the whole field of economic entomology is concerned with a much more ecologically sophisticated approach to the problems of management of the insect pests of agricultural crops and of stored food products. Particularly in some cases, methods of biological control by the deliberate introduction and multiplication of predators upon, or parasites of, the pest species can be very successful. Chemical control with insecticides can be limited to critical stages in the life-cycle of the pest species, once these have been identified. Other human weapons in this war include the breeding of resistant plant varieties, modifications in planting time or harvesting time of crops, and more elaborate forms of crop rotation, as well as use of the specific insect attractants discussed above (see page 329). It may well be that the worst period of somewhat indiscriminate use of long-lasting insecticides, such as DDT, has already passed.

To biologists of all kinds, the relatively mechanistic responses and simpler neural arrangements (along with the short life-cycles and high reproductive rates) of insects have made them ideal animals in many laboratory studies. Modern genetics is unimaginable without the fruit-fly, *Drosophila,* and modern cytology without it and some of its dipterous relatives. Population dynamics owe much to studies on beetles such as *Tribolium,* economically important as pests of stored grains. Neurophysiological studies on sense-organs and control of muscles, endocrine and neuroendocrine studies on reproductive and molt-cycles, and behavioral studies at all levels have been facilitated by being conducted on suitable insects.

Insect Archetypes and Phylogeny

At the end of Chapter 16, we briefly considered the probable phylogeny of the arthropods as a whole and, while we prefer the hypothesis that the arthropods form a monophyletic group, we admitted that there was some difficulty in linking the insects and myriapodous classes considered as one group to the similar stem-group which would serve for both arachnids and crustaceans among living classes, and to the extinct trilobites. When we consider the insects alone, it is important to note that they appear in the geological record much later than the representatives of all the other major arthropod groups. The earliest fossil insects are from the Carboniferous (Pennsylvanian) period and include, besides wingless insects, pterygote forms resembling present-day dragon-flies and cockroaches (see Chapter 34). It is most likely that the earliest hexapods, or true insects, were evolved somewhat earlier, in the Devonian period, from many-legged arthropods which had already become terrestrial. It is particularly important to realize that the origins of insects are clearly linked to the earliest extensive development of a land vegetation, and that the later evolution of certain specific insect groups was similarly dependent upon the evolution of particular kinds of plants. For example, both the Lepidoptera and the bees appear in the fossil record contemporaneously with plants bearing more complex flowers. There has clearly been much co-evolution of plants and insects involving adaptations of the plants to ensure pollination paralleling evolution of the insect's mechanisms to handle nectar and pollen. Similarly, insects such as lice, fleas, and blood-sucking Diptera and hemipteran bugs could only begin their adaptive radiation in the late Mesozoic and early Tertiary periods with the evolution of their hosts, the warm-blooded vertebrates: birds and mammals.

Even the earliest wingless insects clearly showed a high degree of tagmatization (see Chapter 10). As with the arachnids, great success as terrestrial animals seems to have depended upon the specialization of a few limbs and segments, loss of all extensive limb series, including the abdominal limbs, and the development of the head as a group of segments fused into a strong box bearing the organs both of food intake and of sensory scanning of the environment. The parallel reduction of the thorax to three segments bearing six legs had a mechanical basis (see Chapter 16) in allowing tripod stability while alternate legs move. Along with this, the abdomen became limbless, though still articulated into individual segments, and evolved principally as a container for the digestive, reproductive, and excretory organs. The campodeiform body of some aquatic nymphs of present-day stone-flies (Plecoptera) and other groups may reflect an ancient apterygote insect body form (see Figure 17-2). One of the surviving apterygote insects, the bristle-tail, *Campodea*, is built according to this plan, and it has some resemblances to that of the myriapodous minor class Symphyla (see Table 10-1 and page 301). This basic insect body form, with complete hexapodous tagmatization, but without wings and without further specialization or reduction of segments, and with characteristic paired sensory cerci extending *both* anteriorly and posteriorly, may be as close as we can come to an archetype for the insects: to a first pattern of six-legged success as a land arthropod. Initially, there was competition from the two-tagmata, eight-legged arachnids, which became the successful spiders and mites; but the development of wings in insects clearly must have contributed to their dominance over these immediate rivals (a twentyfold superiority in number of species, and probably a many-hundredfold dominance in ecological energetics in favor of the insects). For small animals in a terrestrial environment, where food resources are much more "patchy" than they are in the sea, the evolution of wings conferred a tremendous advantage. Wings permitted relatively fast avoidance of adverse physical conditions and of predators, but also allowed rapid movement from one group of specific food plants to another. As Knut Schmidt-Nielsen has pointed out, flight is more efficient than walking in terms of energy expenditure (and thus of food intake) per kilometer for most insects.

Evidence from the fossil record (see also Chapter 34) suggests that early insects before the Carboniferous were probably wingless. Up to the beginning of the Permian, it is probable that all the insect groups had hemimetabolous development, with the molts accompanied by only gradual and simple metamorphosis. The more advanced holometabolous insects such as the true flies, the beetles, and the Hymenoptera evolved in the Permian. Thus the modern insect types were mostly all established by early in the Mesozoic, except for such forms as butterflies, which had to await flowers, and for the fleas and so on of birds and mammals. It is clear that insects still continue to evolve. For example, major changes in the geographic distribution of genera, and adaptive radiation of new species, have accompanied the climatic changes of the last million years or so. Insects long antedated vertebrates as successful land animals, and their obviously continuing capacity for adaptive change may give them more reasonable gambling odds (as compared with mammals like man) in the future evolutionary sweepstakes of terrestrial life on this finite planet.

18 MINOR METAMERIC PHYLA

In a relatively natural classification, the largest phylum of animals, the Arthropoda, can be arranged to include ten distinct lines of animals with a chitinous exoskeleton and jointed limbs, as we have seen. There remain four very small groups of animals, metamerically segmented and clearly but distantly related to the annelid-arthropod stock. These are sufficiently distinct from one another and from the two major metameric phyla that each must be considered as a distinct minor phylum. As is the case with the minor pseudocoelomate phyla (see Chapter 8), the case for combining such minor groups into superphyla is not a good one. The composite groups formed would be far looser in their relationships than the major phyla and, given our ignorance of their true relationships, it is best to treat them as separate minor groups.

The Walking-worms

From the point of view of broad phylogenies, the phylum Onychophora is certainly the most interesting of the minor metameric phyla. The so-called walking-worms, of which there are about eighty living species and a few fossils, are claimed in many textbooks as the missing links between the annelids and the arthropods (Figure 18-1; see also Figure 1-15A). They exhibit a mixture of the characters of the two groups, and although in no way really intermediate between modern annelid

341

[A]

[B]

FIGURE 18-1.   Neither annelid nor arthropod. Dorsal (**A**) and lateral (**B**) views of a "walking worm," *Peripatus* (phylum Onychophora). [Photo **A** courtesy General Biological Supply House; photo **B** by Dr. Joseph L. Simon of a living specimen collected by Dr. Donald J. Zinn.]

worms and the Arthropoda, they probably preserve some of the structural and functional patterns of the ancient stocks from which the true arthropods arose. They are found in tropical and south temperate regions and are restricted to highly humid habitats. Like other less well-adapted terrestrial animals, they are cryptic forms, with all the appropriate behavioral reactions.

Although the body surface is covered with a chitinous cuticle, there is no real exoskeleton. Chemically, the cuticle of onychophorans is very permeable and, mechanically, it is thin, flexible, and not divided into a series of articulating plates. Below this are the typical layers of muscles and connective tissues of a worm-like body. However, these muscles surround an enlarged haemocoel, and as Sidnie M. Manton has elucidated, the body-wall muscles are concerned only in changes of length and of shape of the body, the main processes of locomotion being carried out by the movements of the appendages alone. The speed of locomotion can be modified by changes of gait, and this involves changes in the extension of the body: faster gaits being associated with an elongated body, slower gaits with a shortened one. The paired limbs themselves vary from 14 to 43 pairs in different species. Each leg is a stumpy protuberance ending in two claws and with, on the ventral side, from three to six pads which serve as the walking soles which contact the substratum. The body is actually lifted off the ground by the legs which move in a series of steps. In each, the antagonistic contraction of muscles arranged both within the limb and attached to its base, perform in a pattern resembling that of the limbs of primitive myriapods. The head has a series of paired appendages vaguely like those of arthropods but exhibiting no clearly homologous structures. Associated with each pair of walking limbs is a pair of coxal glands, similar to the coelomoducts in annelids, true nephridia (by the restricted definition based on *embryonic* development which was elaborated in Chapter 12) being absent. On the other hand, the respiratory organs are tracheae, again independently evolved in this group. Onychophoran tracheae arise from small pores, which cannot close like arthropodan spiracles, and are scattered all over the surface of the body. The tracheae form a series of tufts of minute tracheal tubules, which do not branch but run as simple tubules to each tissue. The nervous system is essentially annelid in form with the addition of a large bilobed brain, located supraoesophageally, to which are linked the sensory tentacles and eyes. The moderately well-developed sense-organs are obviously important in the cryptic and nocturnal behavior so important to their survival as poorly homeostatic animals on land. The reproductive organs and the relatively large eggs are essentially arthropodan in their structure and functioning. Some genera are viviparous.

Several zoologists believe that the arthropodan features of the walking-worms are so marked that the group ought to be considered as a subphylum of the phylum Arthropoda. While it seems likely, as Manton and others propose, that forms like *Peripatus* are representatives of the primitive stock from which myriapods and some other arthropods evolved, it seems better to retain a separate phylum for the group. Pragmatically, the diagnosis of the phylum Arthropoda, with its significant functional features, would lose much if stretched to accommodate the onychophorans.

Water-bears or Tardigrades

The phylum Tardigrada consists of minute animals (50 microns to 1 millimeter) with a cuticular exoskeleton divided (in some) into segmental plates and molted during growth in all. There are four body segments, each with a pair of stumpy appendages ending in cuticular claws (Figure 18-2). There are a few marine species, but most are semiaquatic, freshwater animals—living in soil water, under or on mosses, lichens, and liverworts. Tardigrades or water-bears are cosmopolitan and ubiquitous, and it seems to me that they can always be found in the moss which grows on the roofs of old buildings. They have a complex buccal apparatus involving sharp stylets, which are used by all tardigrades to pierce plant cells and suck out their contents. Physiological specialization allows tardigrades to survive in their peculiar habitats.

Although at first sight arthropod-like, tardigrade anatomy involves a jumble of similarities to other phyla, which in most places cannot reflect real relationships. Although there is a molt-cycle, the cuticular exoskeleton is not chitinous and does not show the characteristic articulations of arthropods. Development is direct and there are no larval stages. Although the cleavage pattern is holoblastic, it does not conform to any of the patterns found in the major animal phyla. While protostomous in early development, tardigrades form five coelomic spaces by enterocoelous means (like echinoderms and chordates!). Only one reproductive coelomic pouch persists, the adult body-cavity being derived from the segmentation cavity and being therefore pseudocoelous. In most

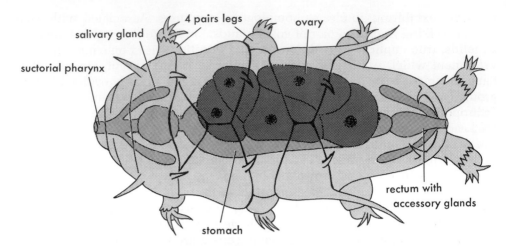

FIGURE 18-2. A typical tardigrade (adult female). The internal organs of digestion and reproduction show cell number constancy. The kind of tardigrade illustrated has partially retractile legs. Adults are about 0.4 millimeter long. They can be found in isolated patches of damp moss on rock faces of mountain peaks and on older roofs in cities.

cases, subadult growth seems not to involve cell division, and therefore cell number constancy seems to occur. (This last feature, along with the pseudocoel, involves similarities to rotifers, Nematomorpha, and the other pseudocoelomate minor phyla (see Chapter 8). Three tubular glands attached to the hindgut have been claimed as excretory organs, but there is little or no evidence for this. There are no circulatory or respiratory organs. The nervous system is obviously metameric, and the muscles show some segmental repetition.

The most important feature of tardigrade physiology is the capacity for diapause. Most of the life of each individual tardigrade seems to be spent in a desiccated, shriveled state. Metabolism continues but is reduced to a very low level in this diapause, which is ended as soon as free water is again available. The animal then swells to four to five times its diapausing volume, and becomes active and feeds on any available plant cells. While in the desiccated state, tardigrades are resistant to fantastic environmental extremes. There are authentic records of tardigrades emerging after dry laboratory storage for seven years or surviving exposure to temperatures of $-272°C$. Individual tardigrades in diapause can also be submerged in saturated salt solutions, in absolute alcohol, and even in ether, and survive. The nature of the cuticle obviously requires further investigation, as does the molt-cycle, which, however, is clearly different in many respects from that in arthropods.

Tardigrades have separate sexes, internal fertilization, and moderately large eggs. It seems that tardigrades, like rotifers and gastrotrichs, can produce eggs of two sorts: those for immediate development, which are thin-shelled, and those resistant to desiccation, which diapause. It is possible that eggs can also be parthenogenetic, as in rotifers and cladocerans, and males are relatively rare in some populations. Although most textbooks give figures of from 300 to 400 described species for the phylum Tardigrada, and postulate that many more remain to be discovered, this seems to result from excessive "splitting"; there are probably only about 170 extant species. It is characteristic of cosmopolitan freshwater groups relying on passive dispersal that there is a very high degree of infraspecific interpopulation variation and, conversely, that species and genera are relatively widespread and ubiquitous.

One last feature of tardigrades deserves to be mentioned. They have an alluring charm for certain biologists, out of all proportion to their importance. Many invertebrate physiologists have revealed to me their dreams of a profitable series of researches on tardigrade functions—a vision I share. Perhaps an hour spent at a microscope watching some tardigrades come out of diapause would enroll any student of biology in this company of dreamers.

Phylum Linguatulida

The pentastomids or linguatulids form a minor phylum of about sixty species of worm-like, coelomate, parasitic animals. Once again, there seems to be some relation to a prearthropod stock. The body is covered by a thick chitinous cuticle and there is a regular molt-cycle during larval development (Figure 18-3A,B). They are found in the lungs and respiratory passages of certain reptiles, birds, and mammals. The host of the adult is always carnivorous, and in most linguatulids the life-cycle requires an intermediate host, usually a herbivorous vertebrate, for the larval development. In some forms the larva resembles a tardigrade. There are no respiratory, circulatory, or excretory organs, but the nervous system is metameric, with the annelid-arthropod form. Like the onychophorans, the linguatulids seem to derive from an unknown prearthropod stock.

Phylum Echiuroidea

The echiurids form a minor phylum of marine animals which have obvious resemblances to annelid worms, but which show no trace of metamerism as adults, and which are characterized by the possession of an enormous nonretractable grooved proboscis which is used in feeding (Figure 18-3). There are more than eighty species, and the majority live in shallow water of the sublittoral.

The body wall is annelid-like and there are setae which are chitinous and formed in sacs as in earthworms. Development involves spiral cleavage and a typical trochosphere larva leading to a metamerically segmented embryo. Almost all traces of metamerism are then lost during later development. The excretory system and circulatory system are like those in primitive annelids, although the central nervous system is much simpler.

On the other hand, some features are distinctly unlike the annelid worms. The coelom is an open cavity without septa and the gut is a very long, convoluted tube. The diagnostic feature of the group is the large, extensible grooved proboscis—actually a cephalic lobe corresponding to the prostomium of annelid worms—which contains the brain. The edges of the proboscis are rolled to form a gutter which is ciliated and this is used in obtaining food. Most genera are detritus-feeders, catching food particles on mucus secreted by the proboscis

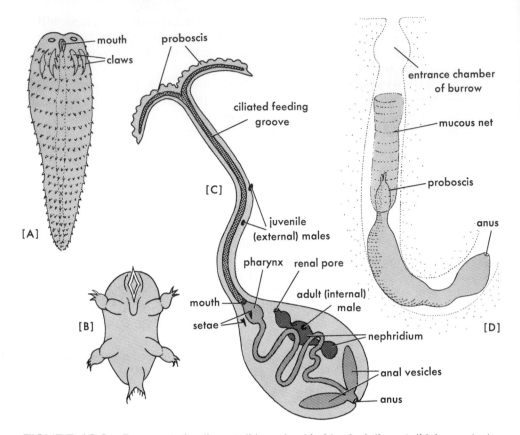

FIGURE 18-3. Representative linguatulids and echiurids. **A:** A linguatulid (or pentasto-mid) parasite from the lungs of a carnivorous vertebrate. **B:** A larval linguatulid from an inter-mediate host (a prey species for the primary host, such as a small rodent or rabbit). If such larval linguatulids get into the "wrong" host (such as man), they may become inactive and persist as the contents of calcified cysts. **C:** The echiurid worm *Bonellia*. The large female lives in marine sediments, feeding on detritus by the extensile bifurcate proboscis (anatomi-cally similar to the prostomial segment of an annelid worm); the minute male, at maturity, be-comes an internal parasite in the female's nephridium. **D:** Diagram of another echiurid worm, *Urechis*, in its burrow. Filter-feeding involves use of a mucous net which is periodically in-gested.

and carrying it back to the mouth along the ciliated groove. The Californian genus *Urechis* contructs U-shaped burrows and spins mucous nets to catch par-ticulate food. These mucous nets can apparently be spun with a pore size fine enough to stop protein molecules of molecular weight about 500,000, while passing those of molecular weight around 100,000. This filtration of protein molecules and other small particles does not seem to involve adsorption onto the mucus. Several of the detritus-feeding species can extend the proboscis to ten times the length of the trunk, that is, for distances of about 1 meter.

This minor phylum seems to have arisen from a group ancestral to the pres-ent-day Annelida, in which metameric segmentation had already been evolved.

19 GENERAL CHARACTERISTICS OF MOLLUSCS

AMONG the relatively few patterns of animal construction which have been highly successful is the molluscan plan. This is true both in terms of the number of individuals and the number of species, and also in the more significant ecological sense which was discussed at the beginning of this book. Clearly, the Mollusca constitute a major phylum; there are probably about 110,000 living species of molluscs, mostly belonging to the three major classes: Gastropoda, Bivalvia, and Cephalopoda. That is, in terms of numbers of species, the Mollusca must rank along with the Nematoda, just after the arthropods. (Surprisingly, there are more than twice as many molluscan species as there are vertebrate species.) The phylum name is derived from *mollis,* meaning soft, referring to the soft body within a hard calcareous shell, which is usually diagnostic. Thus most molluscs are readily recognizable as such. The more obvious functional homologies in the group arise from the extensive use of ciliary and mucous mechanisms in feeding, locomotion, and reproduction. The basic molluscan plan of structure and function is remarkably uniform throughout the group, but there is no single standard molluscan shape. An extreme diversity in external body form has been based upon this plan—oyster, chiton, snail, and octopus are all molluscs.

Three major groups, and a number of smaller ones, are set out in the table of classification (Table 19-1). The Gastropoda constitute a large, diverse group with the shell usually in one piece.

347

TABLE 19-1
Outline Classification of the Mollusca

Class A MONOPLACOPHORA (mainly fossil; but one living genus, *Neopilina*)
Class B APLACOPHORA
Class C POLYPLACOPHORA
Class D SCAPHOPODA
Class E ROSTROCONCHIA (fossil only)
Class F GASTROPODA
  Subclass I Prosobranchia
    Order Archaeogastropoda
    Order Neritacea
    Order Mesogastropoda
    Order Neogastropoda
  Subclass II Opisthobranchia[a]
    Order Bullomorpha (or Cephalaspidea)
    Order Aplysiomorpha (or Anaspidea)
    Order Thecosomata[b]
    Order Gymnosomata[b]
    Order Pleurobranchomorpha (or Notaspidea)
    Order Acochlidiacea
    Order Sacoglossa
    Order Nudibranchia (or Acoela)
  Subclass III Pulmonata
    Order Systellommatophora
    Order Basommatophora
    Order Stylommatophora
Class G BIVALVIA (or PELECYPODA)
  Subclass I Protobranchia
  Subclass II Lamellibranchia[c]
    Order Taxodonta[d]
    Order Anisomyaria[d]
    Order Heterodonta[d]
    Order Schizodonta[d]
    Order Adapedonta[d]
    Order Anomalodesmata[d]
  Subclass III Septibranchia
Class H CEPHALOPODA (or SIPHONOPODA)
  Subclass I Nautiloidea (Tetrabranchia)
  Subclass II Ammonoidea (fossil only)
  Subclass III Coleoidea (Dibranchia)
    Order Teuthoidea (squids)
    Order Sepioidea (cuttlefish)
    Order Polypoidea ("normal" octopods)
    Order Cirroteuthoidea (deep-sea octopods)
    Order Vampyromorpha (only *Vampyroteuthis*)

Notes:
  [a] This subclass is sometimes divided into two superorders: Tectibranchia (the first seven orders) and Nudibranchia.
  [b] These two orders of planktonic opisthobranchs are sometimes united in the order Pteropoda.
  [c] No longer used to designate whole class.
  [d] There are several alternative ordinal arrangements for the many well-established superfamilies of lamellibranch bivalves. An attractive one on functional grounds (but unacceptable phyletically) is to divide the subclass into two: Filibranchia and Eulamellibranchia (see text).

This shell may be coiled as in typical snails—that is, helicoid or turbinate—or it may form a flattened spiral, or a short cone as in the limpets, or it may be secondarily absent as in the "slugs." Most gastropods are marine, but many are found in fresh waters and on land—in fact, they are the only successful non-marine molluscs. The Bivalvia are a more uniform group with the shell in the form of two calcareous valves united by an elastic hinge ligament. Mussels, clams, and oysters are familiar bivalves. The group is mainly marine with a few genera in estuaries and in fresh waters. There can be no land bivalves since their basic functional organization is as filter-feeders. The third major group, the Cephalopoda, includes the most active and most specialized molluscs. There is a chambered, coiled shell in *Nautilus* and in many fossil forms; this becomes an internal structure in cuttlefish and squids, and is usually entirely absent in octopods. There are only about forty genera of chitons, or "coat-of-mail shells" with eight plates, placed in the Polyplacophora. The other living minor groups number only a few species each: the "elephant's-tusk-shells" or Scaphopoda, the worm-like Aplacophora, and the Monoplacophora encompassing several fossil families and the one living genus *Neopilina*. The only completely extinct molluscan class, the Rostroconchia, was limited to the Palaeozoic.

More than 99% of living molluscan species are snails and bivalves, and, ecologically, these two classes can make up an important part of the animal biomass in many natural communities. Certain snails of the seashore, such as *Littorina, Hydrobia,* and *Nassarius,* are known to occur in several parts of the world at densities of hundreds per square meter (Figure 19-1), and one species of *Nassarius* on Cape Cod occurs in patches with up to 23,000 individuals per square meter. Several investigators have recorded densities for the minute estuarine snail *Hydrobia ulvae* in almost "pure cultures" of double that figure, and we estimated that one area of less than 2.5 square miles (just over 6 square kilometers) in the Clyde Estuary in Scotland supported a population of 3×10^{10} (or 30 billion) individual snails. This vast molluscan population was potentially panmictic (potentially all interbreeding) as a result of tidal activity patterns, including temporary mucous rafting.

Even less obvious molluscs—nocturnal or cryptic ones like the slugs shown in Figure 19-1—can be abundant. One English zoologist, with a garden of one-quarter acre in extent, removed four hundred slugs from it each night for many years without any obvious reduction in the population. Thus, even on land, molluscs can be of ecological importance.

Archetypic Functioning

The unique features of molluscs depend on the modes of growth and of functioning of the three distinct regions of the molluscan body. These are (Figure 19-2) first, the head-foot with some nerve concentrations, most of the sense-organs, and all the locomotory organs; secondly, the visceral mass (or hump) containing organs of digestion, reproduction, and excretion; and thirdly, the mantle (or pallium) hanging from the visceral mass and enfolding it and secreting the shell. In looking at any mollusc it is important to realize that whatever

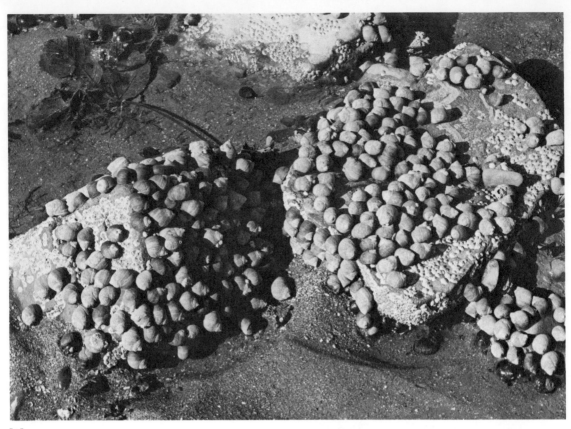
[A]

[B]

FIGURE 19-1. The abundance of some molluscs. **A:** Periwinkles, *Littorina littorea,* exposed on boulders on the seashore at low tide. **B:** Night-crawling slugs, *Agriolimax* and *Arion,* in a rough garden in upstate New York. [Photo **A** © Douglas P. Wilson; photo **B** by the author.]

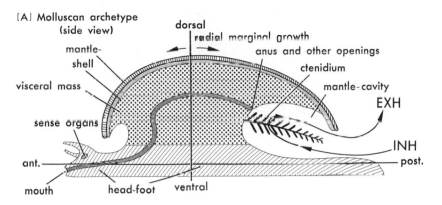

FIGURE 19-2. The molluscan archetype and its pallial complex. Note the three parts of the molluscan body: head-foot, visceral mass, and mantle-shell. Water circulation through the mantle-cavity (and pallial complex) is from ventral INHALANT to dorsal EXHALANT. Compare with Figures 19-3 and 20-4.

the shape of shell it is always underlain by this fleshy fold of tissues, the mantle. In its development and growth, the head-foot shows a bilateral symmetry with an anterioposterior axis of growth. Over and around the visceral mass, however, the mantle-shell shows a biradial symmetry, and always grows by marginal increment around a dorsoventral axis (Figure 19-3). It is of considerable functional importance that a space is left between the mantle-shell and the visceral mass forming a semi-internal cavity: the mantle-cavity or pallial chamber within which the typical gills of the mollusc, the ctenidia, develop. This mantle-cavity is almost diagnostic of the phylum: primarily a respiratory chamber housing the ctenidia, but with alimentary, excretory, and genital systems all discharging into it. Although its basic functional plan is always recognizable, it has undergone remarkable modifications of structure and of function in different groups of molluscs. In addition to its respiratory function, it provides the feeding chamber of bivalves and of some gastropods, a marsupial broodpouch in some forms, and an organ of locomotion in a few bivalves and in the most highly organized molluscs—the Cephalopoda. The head-foot functions largely by muscles which show many of the usual fast reflexes of bilaterally symmetrical animals, while the visceral mass and the mantle-cavity function slowly and continuously using mucus and cilia. For conceptual purposes, it is possible to regard an unspecialized mollusc as being made up of two animals: a muscular animal responsible for locomotion and retraction into the shell, which carries about a ciliary animal responsible for respiration, excretion, and feeding, i.e., for all general visceral activities. In different forms of molluscs these two functional aspects—and their corresponding and clearly distinct growth patterns—assume different levels of relative importance. The ultimate adult form of a mollusc may result from dominance of the mantle-shell over the head-foot in growth, as in clams, or *vice versa*, as in slugs.

In the more primitive living snails and bivalves, there are paired ctenidia of characteristic basic form (Figure 19-4). The filaments or plates always alternate on either side of an axis. The axis always contains a dorsal afferent blood vessel

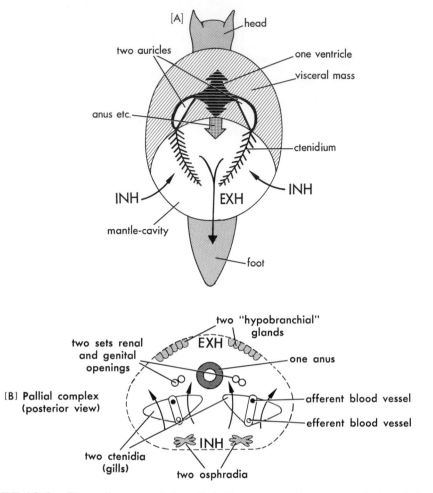

FIGURE 19-3. The molluscan archetype II. **A:** Dorsal view of the symmetrical posterior mantle-cavity, with a pair of feather gills (aspidobranch condition) and two auricles in the heart (diotocardiac condition). **B:** Cross section through the mantle-cavity of the archetype showing all the elements of the pallial complex in a bilaterally symmetrical array. Compare with Figures 19-2 and 20-4 through 20-7.

bringing deoxygenated blood to the gill, and a ventral efferent blood vessel carrying the oxygenated blood back to the heart and thence to the rest of the molluscan body. The flow of blood through the filaments (or lamellae or plates) is therefore from dorsal to ventral. The cilia covering the exposed surfaces of the filaments are so arranged as to produce a flow of water through the gill—and through the mantle-cavity—in the opposite direction to the blood; that is, the water flow between adjacent filaments is from ventral to dorsal. This counter-flow system in the gills is—with few exceptions—structurally and functionally homologous in its arrangement throughout the molluscs. The water flow is actually created by the lateral cilia on the faces of the filaments. The other cilia, which fringe the filaments, are termed the frontal on the ventral edge and the abfrontal on the dorsal edge and are concerned with cleansing. Another characteristic feature is the presence in most molluscan gill filaments of a supporting skeletal rod which lies immediately under the frontal cilia (that is, facing the incoming water current in the leading edge of the lamella).

Morphological and functional constancy is again seen in the arrangement of the ctenidia in the mantle-cavity. In almost all molluscs, the mantle-cavity is

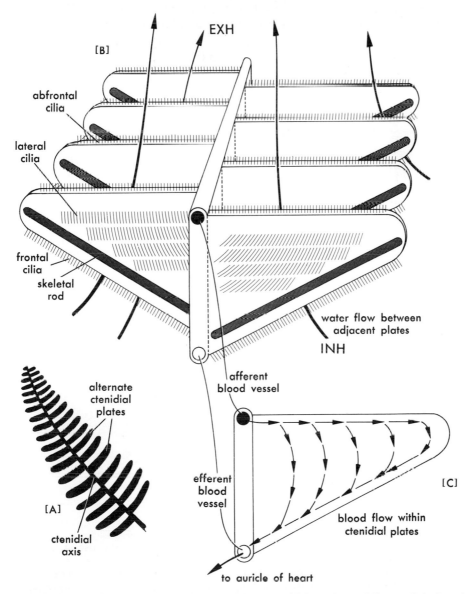

FIGURE 19-4. The archetypic molluscan gill: the aspidobranch ctenidium as it is found in primitive gastropods and other groups. **A:** An aspidobranch or feather- or plume-gill. **B:** Stereogram to show the water current from ventral inhalant to dorsal exhalant between adjacent gill plates—a current created by the *lateral* cilia. **C:** A single gill plate to show the countercurrent flow of blood within the plate, which results in physiological efficiency of oxygen exchange.

functionally divided by the gill curtain into an inhalant part which is usually ventral and an exhalant portion usually lying dorsal to the gills. Significantly, the anus, and the openings of the kidney and genital ducts, all discharge into the exhalant current; that is, they discharge dorsal to the bases of the ctenidia. More primitive forms of both snails and bivalves have the anus in the center of the roof of the mantle-cavity with two genital openings and two kidney ducts symmetrically placed. Further apart and below are two ctenidia (Figures 19-2 and 19-3). Below these, again in the path of the inhalant current, lies an osphradium, or pair of osphradia, pallial sense-organs which sample the incoming water current. Finally, above the ctenidia in the exhalant region adjacent to the

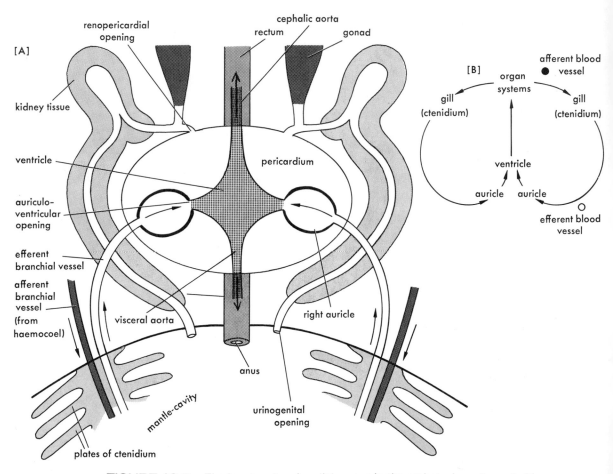

FIGURE 19-5.   The heart and pericardial organs in the archetypic mollusc. A: Diagrammatic dorsal view of the heart and associated organs in the molluscan archetype. The mantle-cavity is posterior (compare with Figures 20-4 and 20-8), and the heart and other structures are bilaterally symmetric. Oxygenated blood from the paired ctenidia flows by way of paired efferent branchial vessels into the paired auricles of the heart, to be pumped out both anteriorly and posteriorly by the single central ventricle. The paired coelomoducts open from the pericardium by renopericardial openings, and the gonads discharge by way of these ducts. The coelomoduct walls are enlarged to form the adult kidney tissue, and both genital and renal products are discharged through openings alongside the anus in the exhalant (dorsal, central) part of the mantle-cavity. B: Pattern of blood circulation in the same archetypic mollusc.

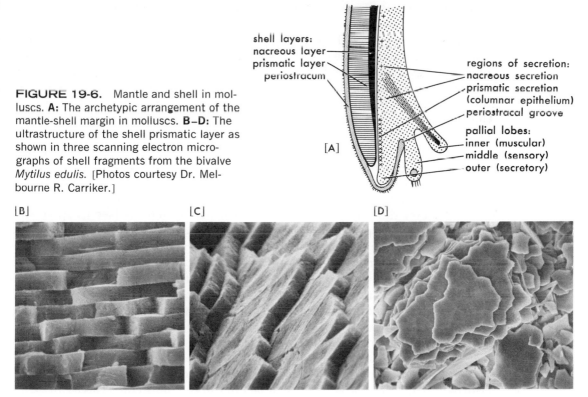

FIGURE 19-6. Mantle and shell in molluscs. A: The archetypic arrangement of the mantle-shell margin in molluscs. B–D: The ultrastructure of the shell prismatic layer as shown in three scanning electron micrographs of shell fragments from the bivalve *Mytilus edulis.* [Photos courtesy Dr. Melbourne R. Carriker.]

shell layers:
nacreous layer
prismatic layer
periostracum

[A]

regions of secretion:
nacreous secretion
prismatic secretion
(columnar epithelium)
periostracal groove

pallial lobes:
inner (muscular)
middle (sensory)
outer (secretory)

[B]   [C]   [D]

anus lie the misnamed hypobranchial glands. All these structures form what can be defined as the pallial complex, a functionally integrated group of structures which are consistent in their arrangement throughout most of the Mollusca (Figures 19-2, 19-3, and 19-5).

The detailed structure of the shell and the mantle edge itself are also consistent in the molluscs. There are three parts to the mantle margin: an outer secretory lobe which produces the shell, a middle sensory lobe bearing tentacles and sometimes eyes (see Figure 22-10), and an inner or muscular lobe capable of considerable cross-fusion or extension into siphons (Figure 19-6). The shell which is secreted is basically made up of calcium carbonate crystals enclosed in a meshwork of protein. The crystals may be of calcite or of aragonite, and the proteins, which vary considerably in both nature and extent, are said to be tanned. The flexibility or brittleness of molluscan shells corresponds to differing relative amounts of protein and of crystalline calcium carbonate. Structurally, the shell itself is always in three layers: the outer periostracum, which is highly organic; the prismatic layer, which is the most massive part (Figure 19-6); and the innermost or nacreous layer, which has least organic content. The periostracum is secreted in a characteristic groove between the outer and middle lobes of the mantle. The main mass of the shell is secreted by a group of columnar cells on the outside of the outer mantle lobe, and scattered cells all over the outer face of the mantle secrete the nacreous layer. These last cells can also secrete around any intrusive particle or organism and wall it off to form a pearl.

In primitive gastropods (see Chapter 20) and bivalves with two gills, the efferent blood vessels from the ctenidia lead into two auricles which in turn lead into a single ventricle (Figure 19-5). This central muscular chamber then pumps out the blood to the body through a closed system of arteries which lead into a spongy system of sinuses and lacunae (the so-called haemocoel) surrounding the viscera and muscles. It should be noted that the true coelom is limited to the pericardial space and the connected cavities around the gonads. The gametes are carried to the exterior through coelomoducts which run from the pericardium to the exhalant part of the mantle-cavity. These are true mesodermal coelomoducts (by the definitions set up by E. S. Goodrich—see Chapter 12), and their glandular walls form the adult excretory organs of all molluscs. Protonephridia (again by Goodrich's definition) do occur in some molluscan larvae and in many embryonic molluscs. In subsequent gastropod evolution, the asymmetry developed in pallial and cardiac structures (see Chapter 20) allows functional separation of the reproductive function (passage of gametes to the exterior) from the excretory function (ionic and osmotic regulation and disposal of nitrogenous residues). As shown in Figure 20-8, in the majority of advanced snails the left coelomoduct becomes the functional renal organ, while the right one is the gonoduct for the solitary gonad.

Throughout the molluscs, the cardiac structures are also closely linked to the pallial complex: if there are a symmetrical pair of ctenidia, there will be a symmetrical pair of auricles on either side of the muscular ventricle of the heart; if one ctenidium, one auricle; if four, four. There are auriculoventricular valves to prevent backflow, and the ventricle drives blood both anteriorly and posteriorly into the major arterial systems, again through nonreturn valves. In turn, these serve the haemocoelic spaces with a relatively low-pressure circulation of relatively large volumes of blood. Note that body fluids in molluscs are almost all blood—just as body cavities are almost all haemocoel. The respiratory pigment is usually hemocyanin in solution, so that neither circulatory efficiency nor blood oxygen-carrying capacity is high (compared to even lower vertebrates like fishes). However, molluscs are mostly "sluggish" animals with low metabolic (and hence respiratory) rates. As we shall note later (Chapter 24), the most active molluscs—cephalopods like squid and octopods—have modified the circulatory system for greater efficiency. The blood in the haemal meshwork of all molluscs has another functional importance, since it is used as a hydraulic skeleton to transmit forces generated by *distant* muscle contraction. A few minutes spent carefully watching the movements of any living snail or clam should convince an observer on this point. The characteristically extensible soft parts of molluscs such as tentacles, the foot, the siphons, and so on, can all be rapidly withdrawn by muscular contraction, but are only slowly extended again by blood pressure, by blood being shifted into them from another part of the molluscan body. Strictly speaking, it is correct to refer to this as a *hydraulic* skeleton since forces are transmitted by *movements* of fluid. As we discussed in Chapter 11, the fluid-filled coelom of annelid worms usually functions (for example, in the locomotion of earthworms) as a *hydrostatic* skeleton, although in a few annelids, hydraulic transmission of forces in this molluscan fashion became important. In molluscs, the underlying anatomical pattern of obvious retractor muscles within each structure, without obvious antagonists locally placed, is characteristic. This reliance on distant antagonists, along with the unchanging total blood volume in the haemocoelic hydraulic skeleton, together are responsible for many of the peculiar features in the mechanics of molluscan locomotion and other movements. For example, limitations arise in the number of extensile structures which can be dilated and protruded at one time. If a pulmonate snail is observed in copulation, the sensory and locomotory organs of the head and foot are flaccid and crumpled since a large proportion of the blood volume is involved in dilation of the genital structures. Further, in most gastropods and bivalves, the total blood volume is limited to that which can be withdrawn along with the tissues into the closed shell. To a limited extent, these difficulties can be bypassed by compartmentalization of the haemocoel. Certain mechanically efficient molluscan organs in which this is the case have been investigated. In these organs, the capacity to seal off a potentially variable amount of blood internally allows the use of intrinsic muscles in local functional antagonism and thus bypasses some of the limitations of a constant volume hydraulic system. Examples of such localized hydrostatic compartments involve the use of radial muscles to extend open siphons (by "thinning" the siphonal walls) in tellinid clams (see Chapter 22), and the superbly efficient jet-locomotion of cephalopods (see Chapter 24). Another exceptional case is the use by naticid snails of an *internal* system of pedal water-sinuses as a hydraulic skeleton of variable volume (see Chapter 20). Other bivalves use the seawater in a temporarily closed-off mantle-cavity to transmit forces generated by one set of muscles to bring another set to precontraction length in simple antagonism. It should be emphasized that all these are somewhat exceptional, and the rule of distant antagonists working through a haemocoelic, hydraulic skeleton of fixed volume applies to the vast majority of mechanical systems in molluscs.

Cilia are used to create water currents for respiration and for feeding; to move materials through the gut, the kidneys, and the genitalia; to cleanse exposed surfaces; and, in some forms, for locomotion. Uniquely molluscan is their use in "sorting surfaces," which can segregate particles into different size categories and send these to be disposed of in different ways in several parts of the organism. One of the simpler types of sorting surface is illustrated in Figure 19-7: the epithelium is thrown into a series of ridges and grooves, the cilia in the grooves beating *along* them and the cilia on the crests of the ridges beating *across* them. Thus fine particles impinging on the surface can be carried in the direction of the grooves, while larger particles are carried at right angles. Such sorting surfaces occur both externally on the feeding organs and internally in the gut of many molluscs. For example, on the labial palps of bivalves, they are used to separate the larger sand grains (which are rejected) from the smaller microorganisms which pass thence to the mouth. It should be noted that, although such sorting is by size alone, it can result in a separation of valuable food from inedible particles. Some molluscan sorting surfaces can achieve segregation into four or more size categories (Figure 19-7). It should be noted that many molluscs have the capacity to alter the thresholds of discrimination on these surfaces, since such organs as labial palps can be expanded to a varying extent by blood pressure. In our simple example, the maximum size of the particles which pass along the grooves (to be accepted as food) and *not* across the

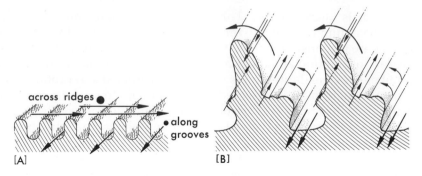

FIGURE 19-7. Ciliated sorting surfaces, which are used externally and internally in molluscs for the mechanical separation of particles of different sizes. On a simple sorting surface (**A**), large particles are carried *across* the ridges, fine particles *along* the grooves. On more complex sorting surfaces such as (**B**), five categories of particles can be sorted in different directions.

ridges (to be rejected) would be raised by an increased expansion of the surface.

In all molluscs the alimentary canal is extensively ciliated, with many internal surfaces organized as sorting areas. Functionally, digestion is mixed: in most, both extracellular and intracellular breakdown is carried out. Although in a few highly specialized molluscs most of the digestion is extracellular, the typical molluscan gut is organized to deal slowly but continuously with a steady stream of finely divided plant material passing in from the feeding organs, whether these be filter mechanisms or the characteristic rasping tongue (or radula—Figure 19-8) of the grazing molluscs. Molluscs of all groups (with the sole exception of the bivalves) have this radula (see scanning electron micrographs, Figures 19-9 and 19-10), a chitinous surface bearing teeth and covering a tongue supported by an odontophore (Figure 19-8). It is capable of being moved in a variety of rhythmic strokes by the group of muscles forming the buccal mass. The radular apparatus has a twofold function: it serves both for rasping off food material (mechanically like an inverted version of the upper incisor teeth of a beaver) and for transporting the food back into the gut like a conveyor belt.

A to-and-fro movement of the chitinous ribbon (at rates of about forty strokes per minute) is achieved by alternate contractions of the antagonistic muscles attached to it, which radular muscles are shown in Figure 19-8 lying above and below the odontophore cartilage. At the same time the whole odontophore (or "tongue") is protruded through the mouth. The teeth of the radula are mostly directed backward (so that the inward stroke is the effective one in rasping) and are arranged in characteristic rows *across* the ribbon. There is continuous secretion of new rows of teeth in the radula-sac and, at the other end of the ribbon, worn teeth are continually breaking off. All rows of teeth in the ribbon are exactly alike, and the pattern of teeth is species-specific. Because of this, radular patterns (or "dental formulae") have been much used in molluscan systematics, and older conventional texts on molluscs devote many pages to accounts of the detailed structure of representative examples. It is worth noting that the rhipidoglossan type, with most numerous teeth in each row (Figures 19-8B and

19-9A), is found in the most primitive snails, and more advanced carnivorous snails (once called "stenoglossans") show reduction to the rachiglossan or three toothed pattern (Figures 19-8B and 19-9B) or to the toxoglossan type, with a specialized single tooth per radular row (Figure 19-10B and page 378). Unfortunately, little is known about the adaptive significance of even the major types of pattern, since careful studies of function have been carried out only for about six species. This is still an open field, where relatively simple observational work (using analysis of motion picture film and measurement of the

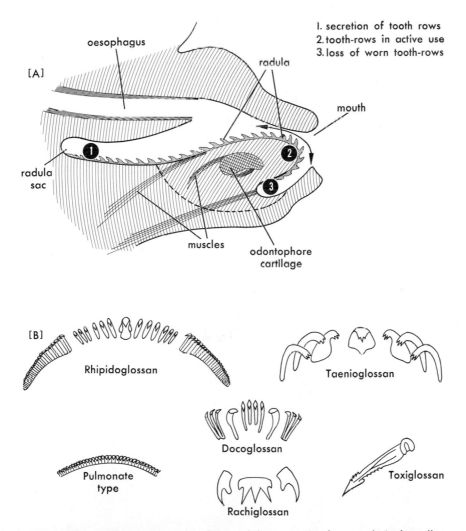

FIGURE 19-8. A: The arrangement of the radular apparatus in an archetypic mollusc, as shown in a vertical section through the midline of the head. **B:** Various types of radular tooth rows as found in gastropod molluscs. The rhipidoglossan pattern is probably the most primitive; the rachiglossan is found in many carnivorous neogastropods; and the toxiglossan involves a highly specialized single tooth (one per radular row), which is used, in combination with a gland secreting a neurotoxin, as a poisonous harpoon by *Conus* and its allies. See also Figures 19-9 and 19-10.

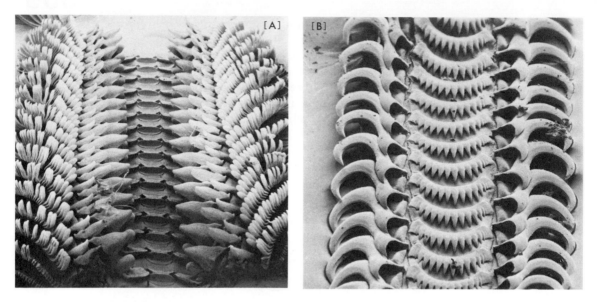

FIGURE 19-9.   Radular teeth rows, in scanning electron micrographs. **A:** The primitive many-toothed rhipidoglossan pattern, as shown by the archaeogastropod *Haliotis ruber*. **B:** The carnivorous rachiglossan (stenoglossan) pattern with only three "tooth-units" per row, as shown by the neogastropod *Nassarius particeps*. [Photos © Dr. T. E. Thompson.]

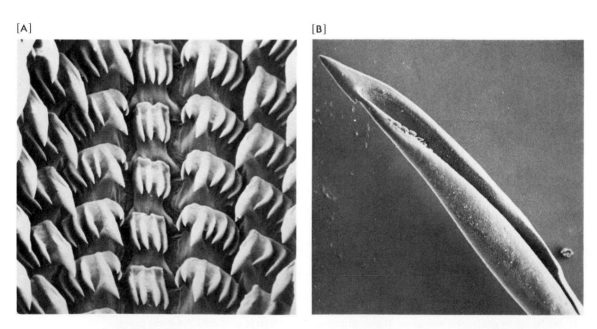

FIGURE 19-10.   More radular teeth, in scanning electron micrographs. **A:** An opistho-branch radula for grazing on sponges and other sessile animals, as shown by the nudibranch *Cadlina modesta*. **B:** One specialized tooth represents a radular row: the neurotoxin-bearing harpoon of the predaceous carnivore *Conus mediterraneous*. [Photos © Dr. T. E. Thompson.]

rasping traces left by snails and chitons) could readily lead to an understanding of the mechanics of the muscle systems involved and perhaps to detailed ecological correlation. Even the most elaborate grazing movements of mouth, odontophore, and radula are rhythmically repeated in stereotyped sequences, and are clearly examples of fixed action patterns (like some elaborate insect behavior briefly discussed in Chapter 17). In the pulmonate snail *Lymnaea*, the 46 muscles of the whole buccal complex (including odontophore and radula) are controlled by nine pairs of motor neurons with coupled input connections. Electrophysiological studies on these buccal action patterns may yield data of more general biological significance on such complex movement sequences coordinated by small numbers of nerve cells.

There is a wide range in the degree of complexity of the molluscan nervous system, and in this, the phylum resembles the chordates. The nervous system of a chiton is not dissimilar to that of a turbellarian flatworm, whereas the nervous system and sense-organs of a cephalopod like an octopus are equalled and exceeded only by those of the most highly organized vertebrates. In most typical molluscs, the nervous system is in an intermediate condition. This state involves five groups of ganglia associated with the radula, with the head organs, with the foot, with the visceral mass, and with the mantle. These five "local brains" are linked by long commissures. The relative simplicity is probably related to the fact that, in molluscs other than the cephalopods, the main effectors controlled by the nervous system are cilia and mucous glands. In fact, apart from the muscles which withdraw it into its shell, the typical mollusc is a slow-working animal with little fast nervous control or quick reflexes.

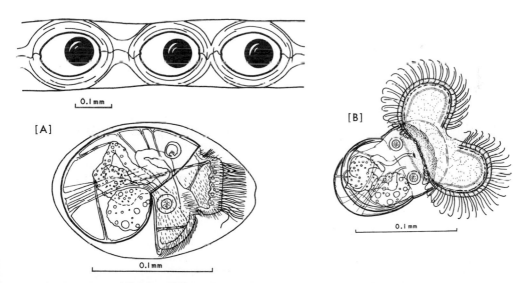

0.1 mm

[A]

[B]

0.1 mm

0.1 mm

FIGURE 19-11. Egg and veliger of a marine pulmonate snail. **A:** Egg from egg-mass of the salt-marsh snail *Melampus bidentatus* shortly before hatching. In this case all earlier larval stages (including trochophore) occur *within* the egg-membrane, and a fully developed veliger emerges at hatching. **B:** The free-swimming veliger larva of *Melampus,* which lives in the plankton for about 14 days (between two successive sets of spring high tides); see text, Chapter 21. [Figures from Russell-Hunter, Apley, and Hunter, *Biol. Bull.,* **143:**623–656, 1972.]

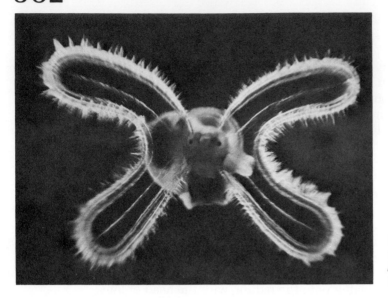

FIGURE 19-12. The planktonic veliger larva of the snail *Nassarius incrassatus*. [Photo © Douglas P. Wilson.]

In all primitive molluscs, the sexes are separate and fertilization is external after the spawning of eggs and sperms into the sea. The zygote undergoes spiral cleavage and eventually becomes a trochophore larva, with a characteristic ring of locomotory cilia (see Chapter 27). This later develops into a typical wheel-shaped velum bearing long cilia, and then this veliger larva develops some of the characteristics of the group of molluscs to which is belongs (Figures 19-11B and 19-12). A mantle rudiment appears and secretes a characteristic shell, while a visceral mass and the head-foot have also differentiated before the ciliated larva settles to the bottom and metamorphoses into a miniature of the adult mollusc. Some examples of molluscan larvae in the marine zoöplankton are discussed in Chapter 33. In general terms, molluscan larvae first become recognizable as gastropods, as bivalves, or as chitons when their mantle-shell rudiments take on the growth patterns characteristic of each class. In more advanced molluscs, eggs are larger (and fewer), fertilization may become internal (with complex courtship and copulatory procedures), and larval stages may be suppressed. This suppression of larval stages is sequential: trochophore stages may be retained within the egg and the young molluscs hatch as veliger larvae, or both trochophore and veliger stages are passed within the egg membranes and miniature adults (or "spat" stages) hatch out.

20 THE EVOLUTION OF GASTROPODS

THE class Gastropoda is the largest and the most varied group of the molluscs. It encompasses a range from certain marine forms which can be numbered among the most primitive of living molluscs to the highly evolved terrestrial air-breathing slugs and snails. In relation to the generalized primitive mollusc already discussed, all gastropods have undergone torsion. This implies that the visceral mass and mantle-shell have become twisted through 180 degrees so that the mantle-cavity has become anterior and placed behind the head. In spite of confused statements in many textbooks, this torsion has nothing directly to do with shell coiling: some gastropods are not coiled at all, but all gastropods at some time in their phylogeny, and at some time in their development, have undergone torsion. With the mantle-cavity facing forward, the vast majority of gastropods became asymmetrical with successive reductions of the originally paired structures of the pallial complex. The most typical shell is turbinate (growing along a "screw spiral"), and this, along with the pallial simplifications, means that almost all snails do *not* have similar left and right halves (Figure 20-1). Even in a few forms in which the shell is lost and a secondary return made to bilateral symmetry of external features, there are usually marked asymmetries of internal anatomy. The hydrodynamic, sanitary, circulatory, locomotory and other problems associated with torsion and with asymmetries of the pallial

363

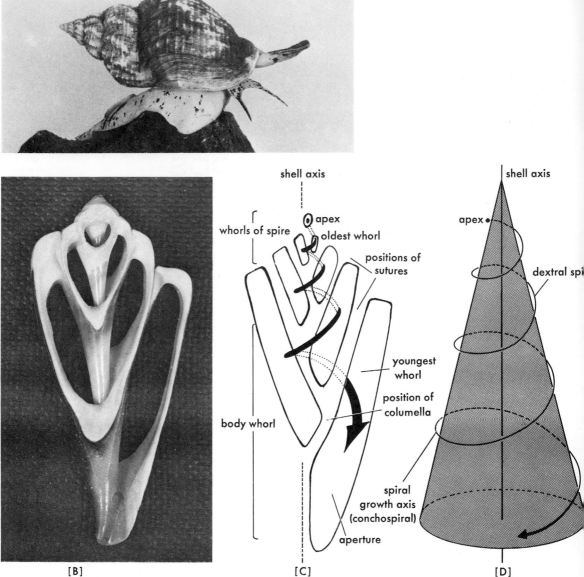

FIGURE 20-1. Gastropods. **A:** A typical carnivorous gastropod, *Buccinum undatum,* shown crawling with the inhalant siphon to the mantle-cavity held extended in advance. **B:** Longitudinal section ground through the shell of a specimen of *Conus spurius* from Florida, to reveal the central columella and spiral of whorls expanding to the aperature. (Whole shells similar to this one are shown in Figure 20-14, and a similar X-ray photograph in Figure 20-13.) **C:** Diagram of the shell in **B** to show the relationship of the whorls sectioned. **D:** Stylized representation of the conchospiral (or central spiral axis of growth) in a turbinate gastropod shell, as viewed with the apex tilted toward the observer. The conchospiral is represented as a logarithmic or equiangular spiral line surrounding a hypothetical cone, but it should be remembered that the spiral line is merely the center line of a steadily expanding *conic* tube (the helicone or conchospiral tube, approximately circular in successive transverse sections in a few "globose" gastropod shells, but appearing as a succession of flattened rectangles in **B** and **C.**) it should also be remembered that this conic tube of shell is actually secreted by the tissues of the mantle edge, differential growth of the cells of which generate the ongoing spiral; see also Figure 20-3C,F. [Photo **A** © Douglas P. Wilson; photo **B** by the author.]

complex are solved in a variety of ways, and gastropods show a diversity of functional morphology unequalled by any comparable group in the entire animal kingdom.

As set out in Table 19-1, the usual systematic arrangement of the gastropods involves three somewhat unequal subclasses. The first, largest, and most diverse is the Prosobranchia, which is made up largely of marine snails (but see Chapter 21) with an anterior mantle-cavity containing two complete ctenidia, or one complete ctenidium, or a "half-ctenidium" (that is, a one-sided, comb-shaped or pectinibranch ctenidium). Adult prosobranchs always retain clear effects of torsion including the streptoneurous (twisted commissure) condition of the nervous system. Prosobranchs are divided into three orders. First, we have the order Archaeogastropoda, with relatively primitive gills of the aspidobranch type (feather-shaped as in the archetype of Chapter 19—see Figure 19-4), and with the gonads opening to the exterior through the coelomoduct of the functional kidney. All two-gilled snails belong to this order, as do all those formerly called Diotocardia (with two auricles) where cardiac and pallial structures retain vestiges of their primitive symmetrically paired state. Secondly, we have the Mesogastropoda, the largest order, with simplified pallial and cardiac structures, with a functional separation of genital from renal ducts, and with a pectinibranch ctenidium in all but a few forms. Thirdly, the Neogastropoda (or Stenoglossa) are the most highly specialized prosobranchs. All are marine carnivores with appropriate radular apparatus (of few teeth per row, hence the alternative ordinal name) and a long extensible proboscis, all are pectinibranch, and all have separate renogenital arrangements.

The other two subclasses (Opisthobranchia and Pulmonata) are each considerably more uniform than the subclass Prosobranchia and, in both, the effects of torsion are reduced or obscured by secondary processes of development and growth. The marine subclass Opisthobranchia consists largely of "sea-slugs" in which the shell and mantle-cavity are reduced or lost, and there is a bilaterally symmetrical adult with a variety of secondary gill conditions. The final subclass, Pulmonata, consists of gastropods with the mantle-cavity modified into an air-breathing lung and no ctenidia. There are a few littoral marine forms (see Chapter 21), but the order Basommatophora is mainly made up of the freshwater lungsnails, and the order Stylommatophora consists of the successful land snails like *Helix* plus a few shell-less families of land slugs. In the development of both opisthobranchs and pulmonates there is a tendency for suppression of free-living early larval stages, as well as a reduction of torsion to about 90 degrees in the egg-contained embryonic stages. Thus, in many of those species in which a mantle-cavity is retained, it opens to the right side in the adult snail. Despite the detorsion exhibited in the two subclasses of "higher" snails, all gastropods are derived from ancestors which had undergone a full torsion through 180 degrees, bringing the mantle-cavity to face forward (above and just behind the head). The *anterior* mantle-cavity, as opposed to the posterior mantle-cavity of all other major molluscan stocks, is diagnostic of the class Gastropoda.

Torsion

As can be seen from Figure 20-2, torsion involves the twisting of all internal organs into a loop. Since the earlier anatomists found that such twisting occurred in the commissures of the nervous system, they set up the group name Streptoneura, which used to be applied to the basic gastropod groups. As was first elucidated by Garstang about fifty years ago, torsion appears to have selective value mainly to the larval gastropod. It can be regarded as the result of a relatively simple mutation, or recombination, of immediate selective advantage to the larva. As can be seen in Figure 20-2 in the pretorsion condition, retraction of the larva into the developing mantle-shell involves withdrawal of the foot first and of the delicate velum and head last. However, after torsion, both velum and head are first withdrawn affording them fastest protection, and the foot only thereafter. The posterior part of the foot may develop a protective plate or operculum. In fact, during development, torsion is a fairly rapid process arising from a simple asymmetry of the withdrawing musculature. The majority of workers on the molluscs have long regarded torsion as bringing the adult gastropod only disadvantages as a result of the twisting of the internal organs. However, J. E. Morton has pointed out that swinging the mantle-cavity anteriorly has certain advantages for gill-breathers, allowing them to take the inhalant respiratory current from undisturbed water ahead of the snail. Further,

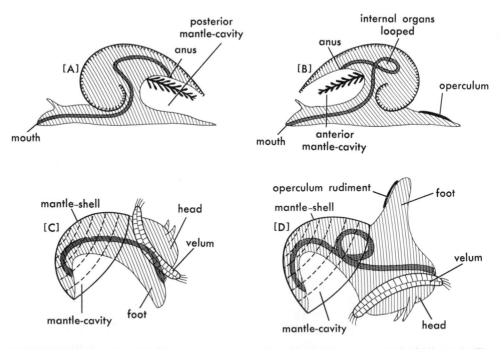

FIGURE 20-2.   **A** and **B:** The hypothetical relationship between an archetypic "untorted" pregastropod mollusc (**A**), and a primitive gastropod after torsion with its anterior mantle-cavity (**B**). **C** and **D:** The actual process of torsion in the development of a gastropod larva or veliger. In the veliger before torsion (**C**), the head and velum are last to be retracted; in the veliger after torsion (**D**), the foot is last to be retracted.

it should be noted that torsion adds the mantle sense-organs to those of the head region, allowing the osphradium along with the head tentacles, eyes, and so on, to continuously monitor the environment into which the snail is moving. An older theory postulating functional advantages to adult gastropods (best set out by, though not originating with, A. Naef), now revived, has a satisfying simplicity. It suggests that torsion allows snails to pull along coiled (or elongate) shells more easily. However, as originally postulated, the evolutionary sequence would involve not "functionally integrated archetypes" (as discussed earlier) but hypothetical animals with considerable incompatibilities in their organization. Torsion obviously occurs early in larval development, and it may be, as M. T. Ghiselin has suggested, that the adaptive advantages (of the torted condition in locomotion) that are significant are not those affecting adult snails, but those affecting newly settled spat.

Intellectual exercises in phylogeny, and discussions of relative adaptive significance, are always seductive, though they are concerned with matters ultimately unprovable (see Chapters 27 and 34). While they continue (and their continuation is necessary to the peculiar science which is biology), each hypothetical organism invoked needs to have been *possible* (see Chapter 2), with an integrated concert of structures and functions that *could* have worked together. It is also important that biologists be periodically reminded of basic logical principles and of the pitfalls of semantics. Theoretical arguments on gastropod torsion have long been numerous—and their proponents not always careful of the functionally possible and of the exclusion of sophistry. Tragically, the accumulation of empirical evidence from studies on larval development has been much slighter: the details of the morphogenesis of the muscles involved in torsion are known for very few gastropods.

Although some stocks of primitive gastropods have the shell in the form of a simple cone—that is, in the limpet form—and others have secondarily acquired the limpet shape, the great majority are coiled. Most coiled gastropod shells are not planospiral but turbinate (or helicoid). The ubiquity of coiled shells obviously involves the best utilization of the competitive patterns of growth of the different parts of the molluscan body discussed earlier, and cannot be explained simply. However, one obviously important part of the explanation is that after torsion the anus is in front and thus the animal cannot easily increase its internal organs by elongation unless this involves extension of the visceral hump dorsally. Probably the most convenient disposition of an extended visceral mass is in the compact form of a spiral coil. If it is remembered that the mantle-shell grows by marginal increment, one can see that it involves only fairly simple growth gradients along the mantle edge for its growth to continue to cover the growing mollusc and maintain this characteristic shell shape. As can be seen in Figures 20-1B, C and 20-3F, yet another simple change in the growth gradient of the mantle edge will allow the development of a turbinate rather than a planospiral coiling and thus accommodate greater bulk of visceral mass in a shell with smaller linear dimensions. The turbinate shell, however, if its aperture were symmetrically placed over the head-foot, would be mechanically unbalanced. As seen in Figure 20-3E, the readjustment for balance of a dextrally coiled shell involves a shift of the shell to the animal's left and its carrying the apex of the spire more posteriorly. Since the majority of

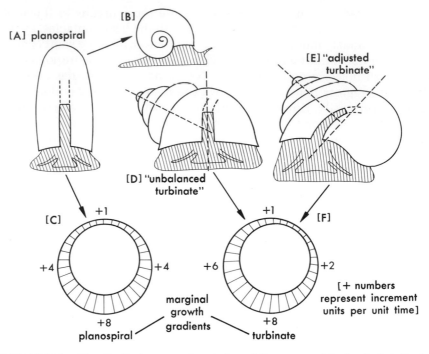

FIGURE 20-3.   Hypothetical course of evolution of the asymmetrical coiled shell of most gastropods. The symmetrical planospiral shell (**A–C**) is first modified to turbinate coiling (**D**) by a simple change in growth gradients (**F**), thus allowing greater bulk of visceral mass to be carried in a smaller shell. This arrangement (**D**) is mechanically unstable, and a shift of shell position to the left with the apex carried more posteriorly (**E**) achieves a readjustment for balance. For further discussion, see text.

turbinate-shelled gastropods show dextral rather than sinistral coiling, there is a tendency in those gastropods for the right side to become reduced—in fact, in the majority of living gastropods there is considerable asymmetry, particularly of the organs of the mantle-cavity, with the right side of each pair being reduced or absent.

Mantle-cavity Evolution

The story of the evolution of the gastropods, as elucidated by C. M. Yonge and others over the last forty years, is essentially one of the modification with increasing asymmetry of the functioning system of the mantle-cavity and its contained organs, and of the many internal structures such as the heart, the excretory organs, and gonads which are intimately connected with the pallial complex. In all, there is a tendency for the right-hand member of each pair of anatomical structures to be reduced. The most primitive living gastropods are called Zygobranchs, because they have two ctenidia, or Diotocardia, because there are two auricles in the heart. In one of the most primitive genera, such as *Pleurotomaria*, or in "keyhole limpets" such as *Fissurella*, the pallial organs are organized exactly as in our generalized primitive mollusc, except that the mantle-cavity as a whole is at the front of the visceral mass above the head. As shown in Figure 20-4, the inhalant water current comes in to the mantle-cavity from both sides ventrally and the exhalant current passes out centrally and dorsally. Various forms of zygobranch gastropods show a variety of adaptations of mantle and shell to avoid the problem of sanitation which arises. In other words, such adaptations as the various shell slits or complex openings found, including those developed in many fossil forms, all accommodate the exhalant current, and are so arranged to ensure that the current bearing with it faeces and genital and kidney products does not discharge directly over the head. As can be seen (Figure 20-4D,E), perhaps the most efficient of these adaptations is that found in the keyhole limpets. However, relatively few living gastropods are zygobranch with two ctenidia.

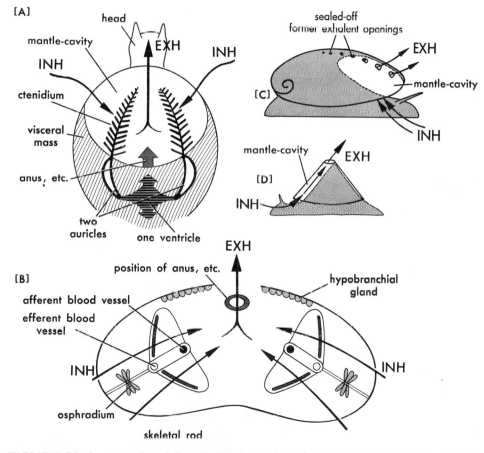

FIGURE 20-4. Evolution of the gastropod mantle-cavity I. **A:** Dorsal view of the symmetric mantle-cavity in a primitive gastropod, which can be termed zygobranch (with two feather-gills) or diotocardiac (with two auricles in its heart). **B:** Cross section through the mantle-cavity of the same primitive gastropod to show elements of the pallial complex. **C:** The arrangement of this type of mantle-cavity in an abalone. **D:** The arrangement of this type of mantle-cavity in a keyhole limpet.

In most gastropods the right ctenidium is completely lost and in these we find no dorsal shell slits or exhalant openings, since the inhalant water current passes from the left side of the mantle-cavity through the gill and the exhalant current passes out on the right side of the snail instead of over its head. A minority of those asymmetric one-gilled gastropods retain a feather-shaped gill (usually referred to as the aspidobranch condition), illustrated in Figure 20-5A, B. The snails which show this condition of the pallial complex include such genera as *Acmaea*, the common tortoiseshell limpet of the Atlantic coast, *Valvata* in fresh waters, the big worldwide group of top-shells (*Trochus* and its allies), and the tropical littoral snails such as *Nerita*. As can be seen in Figure 20-5B, because of the way the aspidobranch gill is placed on an axis diagonally disposed across the mantle-cavity with filaments on either side, there is a tendency for any particulate material being swept through the mantle-cavity to accumulate on the dorsal side of the afferent membrane and cause dangerous clogging of the pallial structures. It is because of this that such genera as *Acmaea*, *Nerita*, and *Trochus* are ecologically limited to relatively hard substrata and are unable to invade areas of the sea bottom or seashore covered with mud or silt.

By far the most successful marine gastropods are those which show a further reduction of the pallial structures—involving both further anatomical asymmetry and greater hydrodynamic efficiency. These are the so-called pectinibranch gastropods, where, as the name implies, the gill is comb-shaped (Figure 20-6A). This one-sided ctenidium, with only one set of filaments on the axis and the other side of the axis fused to the wall of the mantle-cavity, can generate a much more efficient water current than any other arrangement of gills in gastropods. This pallial pattern is illustrated in Figure 20-6B. It should be noted that the individual gill filaments are themselves essentially unchanged: the functional relationships of their cilia and blood vessels unmodified from that in primitive snails with two "feather-gills." A corollary of the asymmetry is that the rectum and the kidney and genital ducts are elongated so that they

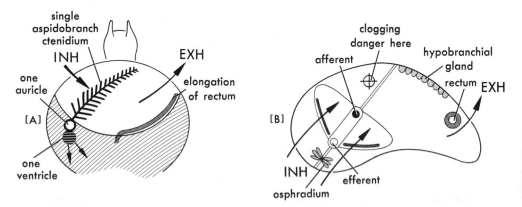

FIGURE 20-5. Evolution of the gastropod mantle-cavity II. The mantle-cavity and pallial complex in an asymmetric snail with a single aspidobranch ctenidium (or feather-gill), in dorsal view (**A**) and in cross section (**B**). Although monotocardiac (with a single auricle heart corresponding to the single gill's efferent circulation), such aspidobranch snails are limited to clean waters over hard substrates because of a danger of clogging in the dorsal part of the mantle-cavity.

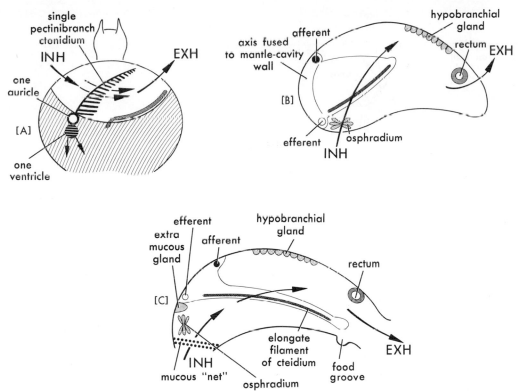

FIGURE 20-6. Evolution of the gastropod mantle-cavity III. The mantle-cavity and pallial complex in an asymmetric snail (monotocardiac) with a single pectinibranch ctenidium (or comb-gill) with the axis fused to the mantle wall, in dorsal view (**A**) and in cross section (**B**). The majority of prosobranch gastropods (including all the successful neogastropods) have this pallial organization and are found in a wide variety of habitats. **C:** A cross section, similar to **B**, in a form like *Crepidula*, where elongation of individual ctenidial filaments allows the gill to be used in filter-feeding.

do not discharge in the midline but along the right-hand side of the mantle-cavity, that is, in the new exhalant region. Most familiar snails of the seashore, including such genera as *Busycon*, *Nassarius*, *Littorina*, and *Polinices* and very many others, have pectinibranch gills and this arrangement of the mantle-cavity. A slight modification is found in such forms as *Crepidula* where considerable elongation of the individual filaments of the gill has occurred (Figure 20-6C), resulting in increased efficiency of the ctenidium as an organ of particle filtration and therefore as a food-collecting organ, but possibly decreased efficiency as a set of respiratory surfaces. A final reduction of pallial structures is found in the subclass Pulmonata (Figure 20-7), in which the mantle-cavity is an air-breathing lung, and there are no ctenidia.

The evolution of increasing asymmetry in the pallial structures, including osphradia and hypobranchial glands as well as the ctenidia, is paralleled in the structures of the heart and of the coelom. As we noted earlier, an alternative name for most of the Archeogastropoda was the Diotocardia (with two auricles delivering blood to the single central ventricle of the heart). The top-shells such as *Trochus* present an intermediate condition with a single aspidobranch

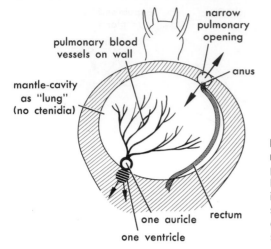

pulmonary blood
vessels on wall

narrow
pulmonary
opening

anus

mantle-cavity
as "lung"
(no ctenidia)

one auricle

one ventricle

rectum

FIGURE 20-7. Evolution of the gastropod mantle-cavity IV. Dorsal view of the mantle-cavity in a pulmonate land snail. There is now no gill and the mantle has become vascularized as a lung wall, but the remaining structures of the pallial complex *and* the heart are still orientated as in asymmetric prosobranch snails (including the monotocardiac condition and the pallial extension of the rectum—here reaching the lung opening).

gill, one osphradium, and a single gonad (right), but with two auricles and two coelomoduct kidneys (left and right). Other archaeogastropods like *Fissurella* and other "keyhole limpets" have two ctenidia (and a secondarily symmetrical pallial complex), but with a single functional kidney and single gonad (both right). The Neritacea are peculiar in having a single left aspidobranch ctenidium (with a vestigial right ctenidium in some forms), two auricles, and a greatly enlarged (paired) renopericardial coelom, but a single kidney (left) and a single genital coelomoduct with its gonad (right). Finally, all snails with pectinibranch gills (both Mesogastropoda and Neogastropoda) are monotocardiac with a single kidney (left) on the opposite side from the gonadal coelomoduct. Figure 20-8 shows the interrelationships of the coelomic structures with the heart and pericardium in the major groups of molluscs. In all the gastropods,

FIGURE 20-8. (opposite) Dorsal diagrams of the pericardial cavity and its associated organs (heart, kidneys, gonads, rectum, and various ducts) in a series of typical molluscs. Molluscan (**A**) and gastropod (**B**) archetypes are illustrated first, but all the other conditions shown occur as actual states of these organ systems in presently living forms: in bivalves (**C**), in chitons (**D**), in three groups of aspidobranch gastropods (**E–G**), and in "modern" pectinibranch gastropods (**H**). In all molluscs, urinogenital ducts are derived from pericardial coelomoducts (see Chapter 12, and especially Figure 12-5). In bivalves and chitons, separation of genital from renal functions has been achieved despite the retention of paired kidneys and gonads. In the more successful groups of gastropods (pectinibranch prosobranchs and "higher" forms), urinogenital separation results from the asymmetry of a single (right) gonad and a single (left) kidney combined with a loss of pericardial connections to their ducts (**H**). In more primitive aspidobranch gastropods (**E–G**), a greater variety of arrangements of urinogenital organs is found. Note that **E**, **F**, and **G** are all ranked as archaeogastropods and are forms with aspidobranch gills, while the great majority of living gastropods (mesogastropods, neogastropods, and "higher gastropods") have the pericardial organ system arranged as in **H**. Note the enlargement of the pericardial spaces to reform a coelomic body cavity in neritaceans, and the fact that a genitopericardial canal persists in a few of them (**F**) and in a few pectinibranch prosobranchs (**H**). It should also be noted that such forms as fissurellid limpets, with secondarily symmetrical pallial structures including a pair of aspidobranch gills (and two equal auricles), can still have markedly asymmetric urinogenital organs. Compare with Figures 19-5, 20-4 through 20-7, and 23-1.

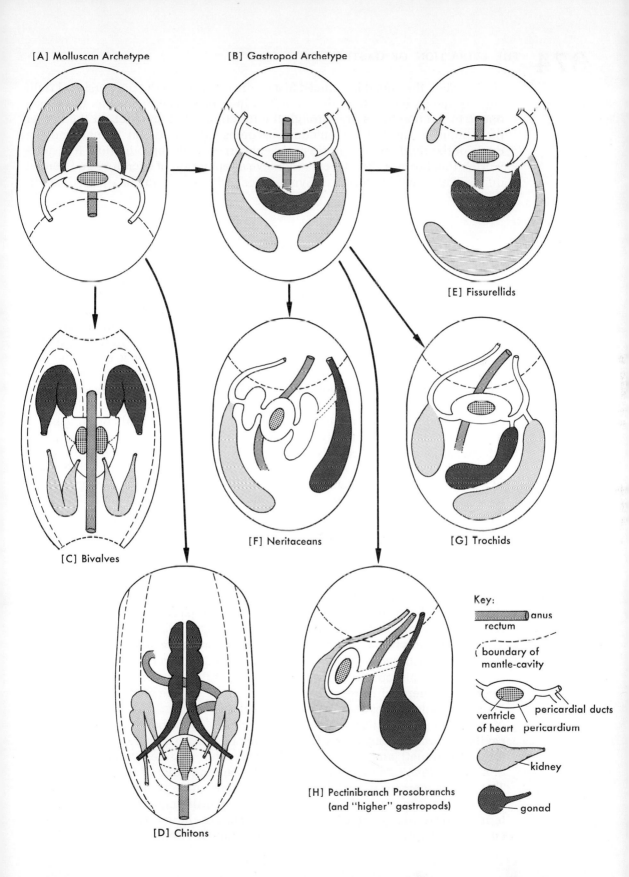

[A] Molluscan Archetype

[B] Gastropod Archetype

[E] Fissurellids

[C] Bivalves

[F] Neritaceans

[G] Trochids

[D] Chitons

[H] Pectinibranch Prosobranchs (and "higher" gastropods)

Key:

anus
rectum

boundary of mantle-cavity

ventricle of heart
pericardial ducts
pericardium

kidney

gonad

after torsion, the gonad is reduced to a single opening on the posttorsional right side. In the archeogastropods, the left kidney is reduced and thus the gonad opens to the mantle-cavity through the functional kidney duct. In the higher prosobranchs, the left coelomoduct becomes the functional kidney and the ducts of the right are now exclusively concerned with reproduction. It is noteworthy that the class Bivalvia also achieve a separation of excretory and genital duct systems, while remaining symmetrical with all organs in a paired condition.

Prosobranch Diversity

In many different groups of gastropods, limpets have been evolved. They have a conical shell with a wide opening and are attached to rock surfaces by the enlarged and flattened foot which is rounded and muscularized to form an efficient sucker. All limpets move relatively slowly over rock or other hard surfaces, protecting themselves by clamping down the shell opening against the substrate by contraction of the enlarged shell muscles. There is never an operculum on the foot, and limpets are defenseless once detached. Apparently the coiled gastropod shell has evolved into the limpet form on many occasions by a rapid enlargement of the mantle-shell edge so that the snail is completely accommodated in one open half-whorl. In many cases the initial coiling (shown in the larval stages) is entirely lost in the adult which is a wide cone. In a few others, initial coiling is readily recognizable on the upper surface of the shell. Among the more primitive groups of gastropods there are zygobranch limpets like *Fissurella* and aspidobranch one-gilled forms like *Acmaea*. Among the mesogastropods there are forms like *Crepidula* (see Figures 20-6C and 23-3) and *Capulus*. There are also opisthobranch limpets, and at least four separate lines of pulmonate limpets have been evolved in fresh waters and on the littoral (see Chapter 21).

Pedal attachment of limpets to rocks and one type of snail locomotion, by retrograde pedal waves, show common features in their muscle arrangements and mechanics. As work by E. R. Trueman and his associates has made clear, this retrograde locomotion with muscular waves passing in the opposite direction to the movement of the animal is characteristic not only of primitive gastropods and chitons but also of free-living flatworms and nemertines. The process is illustrated in Figure 20-9, which is a diagram of a longitudinal section through the foot in a form like the limpet *Patella*. Only relatively small forces are generated in respect of each pedal wave. The contraction of dorsoventral muscles raises the sole off the substrate, compressing isolated vesicles of the haemocoel against a thick layer of muscle fibers above. The mucus-filled cavity so formed beneath the foot is under negative pressure (the attachment force of a limpet), and this allows the dorsoventral muscles at the leading edge of a pedal wave to antagonize those of the lagging edge of the wave in front. Thus the movement of pedal waves in this retrograde fashion can be brought about without major hydraulic movements, by utilizing both internal and external hydrostatic systems for the transmission of these forces. As will be discussed in Chapter 21, the locomotion of pulmonates on land is by direct rather

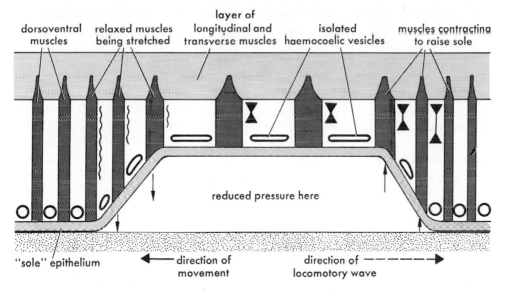

dorsoventral   relaxed muscles   layer of   isolated   muscles contracting
muscles   being stretched   longitudinal and   haemocoelic vesicles   to raise sole
                             transverse muscles

reduced pressure here

"sole" epithelium   ◀——— direction of   direction of —————▶
                       movement   locomotory wave

FIGURE 20-9. Diagram to show the muscular basis for a single retrograde locomotory wave in the foot of a limpet like *Patella*. The relatively small forces involved are all generated by the dorsoventral muscles, contraction of which lifts the pedal "sole" off the substrate distorting the isolated haemocoelic vesicles between the muscle columns. The negative pressure formed under the sucker-foot helps restretch the relaxed dorsoventral muscles anteriorly (that is, to the left of the diagram). Slow pedal waves passing in the direction opposite to the locomotion of the animal are used in similar fashion by certain nemertines as well as by chitons, and by limpets and other primitive aquatic gastropods. Compare with Figure 21-3. [Modified from Trueman, 1969, and Chapman, 1975.]

than retrograde locomotory waves and these involve a rather different mechanism with more specialized muscles. It is of some evolutionary interest to reflect that many small molluscs move like the smallest flatworms by cilia beating in a mucous trail. Larger primitive molluscs moving by muscular means seem still to have employed mechanisms essentially similar to those of flatworms.

In general, the archaeogastropod prosobranchs are limited to locomotion on hard substrata by the clogging danger associated with the aspidobranch gill (Figure 20-5D). Again the Neritacea (Figure 20-8) are somewhat exceptional: basically a group of tropical intertidal snails (some becoming almost amphibious in an evolutionary parallel to the mesogastropod littorinids—see Chapter 21), as are the many successful species of the genus *Nerita* itself, they have independently evolved many terrestrial and a few freshwater genera, although no marine and gill-bearing neritacean has been evolved for life on muddy or even sandy seashores. In all their paralleling of evolutionary trends in the major lines of the Mesogastropoda and of higher snails, the various stocks of archaeogastropod Neritacea all retain their peculiar pericardial and coelomic arrangements (Figure 20-8), many have recognizable degrees of diotocardia, and some even show a vestigial organ corresponding to the right ctenidium. In this book, they are considered as a rather separate suborder *within* the order. Achaeogastropoda, but their completely independent evolution from the rest of the gastropods could well be expressed systematically by considering the Neritacea as a separate order (within the subclass Prosobranchia) or even as a separate subclass.

Of course, marine gastropods are not limited to crawling over rocks and hard substrata. As we already noted, solution of the problems of sanitation and of pallial hydrodynamics by the evolution of the pectinibranch ctenidium has freed the mesogastropods and neogastropods with that comb-gill to invade those areas of the seashore and of the sea bottom which are covered with sand, silt, or mud. In fact, these two groups of prosobranchs seem to be found in every habitat in the marine environment. In the next chapter, we will discuss how one group of mesogastropods, *Littorina* and its allies, are among the most successful of seashore animals, some species invading the highest levels of the littoral and becoming physiologically almost terrestrial in the process.

We have already noted how the mesogastropod slipper-limpet, *Crepidula* (see Figures 20-6C and 23-3), has become a filter-feeder. Other high-spired mesogastropods including *Turritella* live in depressions on muddy bottoms and feed by ciliary mechanisms on suspended organic material. The related pelican's foot shell, *Aporrhais pes-pelecani* (Figure 20-10), lives *within* the substrate and is a deposit-feeder, collecting organic detritus with an extended proboscis. An even more sedentary habit is adopted by the worm-shells [including species of the genera *Vermetus, Siliquaria* (Figure 20-11), and *Vermicularia*], which use extruded mucus threads to entangle the plankton organisms on which they feed. In worm-shells, there is a typical turbinate snail shell only during early life, but this becomes an "uncoiled" (Figure 20-11) and irregular tube (like that of a serpulid worm—see Chapter 12) during continued growth as a sessile adult. Other mesogastropods living in sand and mud are more rapidly moving carnivores. Prominent among these are the naticids or moon-snails. In life, these moon-snails, such as *Polinices*, seem too large for their shells with an enormous foot, shaped like a wedge or plowshare (see Figure 20-12). Uniquely among molluscs, these naticids use a series of pedal water-sinuses for inflation

FIGURE 20-10. *Aporrhais pes-pelecani.* Living specimens of this detritus-feeding gastropod, long known by its descriptive species name, "the pelican's foot shell." Found in muddy gravels on the continental shelves, it is a north Atlantic representative of a largely Indo-Pacific group of mesogastropods, the Strombacea, many of which are much sought after by shell collectors. [Photo © Douglas P. Wilson.]

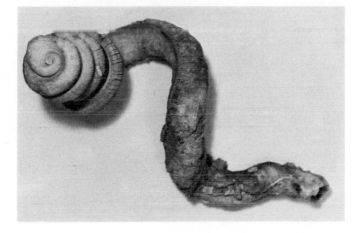

FIGURE 20-11. A "vermetid" gastropod. Three probably unrelated families of mesogastropods have become totally sessile and are either ciliary filter-feeders or plankton-feeders using long mucous threads spun from the foot to entangle their food. In all three groups, the lengthened shell grows "uncoiled" in an irregular tube (here about 9 centimeters long) resembling that of a serpulid annelid worm, with only the youngest part of the shell (top left) resembling a "normal" gastropod. [Photo by the author.]

and extension of the foot, and considerable intake of seawater is involved. A fully expanded specimen of *Polinices* can weigh 3.5 times its own contracted weight; retraction, with expulsion of seawater, takes about 3 seconds; and expansion about 6 minutes. By allowing moon-snails to expand in seawater labelled with inulin, we were able to demonstrate that about 92% of the uptake of seawater enters the pedal water-sinus system (the rest going into the mantle-cavity), and that moon-snails can remain expanded for three or four days with relatively little exchange of seawater between these pedal water-sinuses and the external medium. The hypertrophied foot, made possible only by the pedal water-sinuses being added to the normal molluscan haemocoelic hydraulic skeleton (itself of necessarily constant total blood volume), is the basis of naticid success as rapidly moving carnivores. This huge foot is not only of adaptive value in moving through sand, but also serves in predation, when it can be wrapped around the clams of various sorts which form the preferred prey of moon-snails. There is a protrusible proboscis which can be used to bore through a bivalve shell, with the chemical aid of an accessory boring organ.

It is among the neogastropods (or stenoglossans) that we find the greatest degree of specialization as carnivores and carrion-feeders. Internally, the gut is much simpler than the pattern to be described in Chapter 23, which is typical of primitive gastropods and the majority of bivalves with elaborate ciliary sorting areas associated with the continuous slow feeding of herbivores and filter-feeders. In the carnivorous neogastropods, digestion is wholly extracellular and the anterior gut is developed as a protrusible proboscis. The short radula with three strong teeth, or sometimes only one, to the row (see Figure 19-10B) is carried at the tip of the proboscis, which in some genera can be extruded for four or five times the length of the rest of the snail. In neogastropods, the osphradium, which in more primitive molluscs was involved in mechanical detection of sediment for the mantle-cavity, has become a chemoreceptor used in hunting down suitable prey by water-borne "scent." The osphradium is placed at the inner end of an inhalant siphon, clearly seen in the photograph of *Buccinum* (Figure 20-1), and neogastropod shells all show a siphonal notch

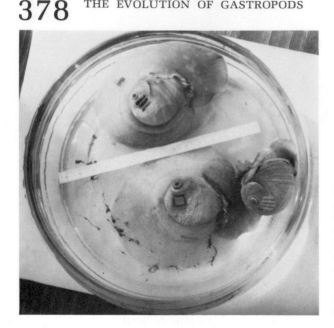

FIGURE 20-12. Expanded moon-snails, *Polinices duplicatus.* Experimental moon-snails during investigations of uptake into the pedal water-sinus system (see text). In two snails the hypertrophied pedal tissues are almost fully expanded; the third snail (on the right) is of the same shell and tissue size but nearly completely withdrawn into the shell after massive expulsion of pedal water. (The scale rule is 15 centimeters in length.) [Photo by the author.]

(Figure 20-13). Common names for various genera of these carnivorous snails are whelks, drills, and tingles. The oyster drill, *Urosalpinx,* is one of the most destructive pests of commercial oyster beds. The whelks *Nucella* and *Thais* prey on mussels and other bivalves but most extensively on barnacles where the proboscis is used to force open the opercular plates rather than to bore an access hole as with bivalve prey. The most specialized neogastropods are the tropical toxoglossids belonging to the families Conidae or cone shells (Figure 20-14) and Terebridae or auger shells. Both groups have been prized by shell

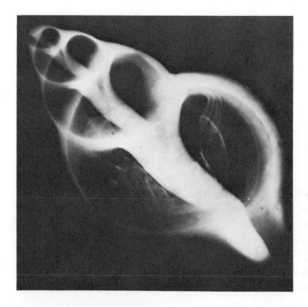

FIGURE 20-13. X-ray photograph of a typical neogastropod shell. Note the expanding series of whorls around the central columella. Compare with Figure 20-1**B**, **C**. The siphonal notch at the bottom left is characteristic of the neogastropods. This shell was occupied by a hermit-crab, whose limbs and carapace (less X-ray opaque than the shell) show faintly in the last whorl. [X-ray original and print courtesy of Frank A. Romano III.]

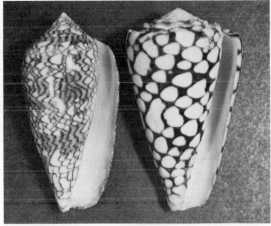

FIGURE 20-14. Two common Indo-Pacific cones. *Conus textile* (left, 5.6 centimeters long) and *Conus marmoreus* (right, 6.1 centimeters) are relatively common species of this specialized family of neogastropods, which are all predaceous carnivores using a single enlarged radular tooth envenomed with neurotoxin (see Figures 19-8 and 19-10, and text). The rarer cones are sought after by shell collectors, and the most valuable shells in the world are the thirty or so undamaged specimens of *Conus gloria-maris*. [Photo by the author.]

collectors for centuries. In them, the radula is reduced to a series of separate darts (see Figures 19-8B and 19-10B), which are used in association with a modified salivary gland secreting a neurotoxin for the capture of large prey animals, both invertebrate and vertebrate. Many species of cones respond only to particular species of prey by aiming the proboscis, striking with a single everted tooth, injecting toxin, and then feeding upon the paralyzed prey. There are three main categories of cones, each with appropriately proportioned harpoon-like teeth and appropriate toxins, which feed respectively on other snails, on annelid worms, and on fishes. A number of human fatalities have resulted from two species of fish-eating cones in the East Indies.

"Higher" Snails

In the two more specialized subclasses, Opisthobranchia and Pulmonata, the gastropod pattern of evolution by progressive reduction of the original paired complex of structures has been taken still further. Older texts refer to these two groups as the Euthyneura, since in the case of the Opisthobranchs, there is detorsion in the shell-less forms, and since in pulmonates the nerve ganglia have moved forward to form a compact ring, so that the twisting of the visceral loop can no longer be clearly detected.

The most typical Opisthobranch snails show complete loss of the shell and modification to the sea-slug or nudibranch condition. It is almost certain that these stocks of often extremely beautiful animals have been independently evolved—that is, that nudibranchs are polyphyletic. Numerous stages in the evolution of the secondarily symmetrical slug type coexist today and the larger ones exhibit various secondary gill conditions. Structurally, the most primitive living opisthobranchs are burrowing forms that have retained shells, and it seems likely that shell reduction in the group followed the evolution of a hypertrophied head-foot for plowing through soft substrata like the prosobranch moon-snails. In the barrel-shell, *Acteon* (Figure 20-15A), the shell is still large

FIGURE 20-15. Living opisthobranch snails. **A:** One of the more primitive opisthobranch "canoe-bubbles," *Scaphander lignarius,* which retains a thin external turbinate shell, shown here plowing its way through sand in search of its worm prey. **B:** In the sea-hare, *Aplysia punctata,* the shell is reduced to a flat paper-thin internal vestige, and the foot is expanded into "parapodial" flaps which can be used in swimming. There are two individuals, each about 8 centimeters long, on the small rock. Some species of *Aplysia* are among the largest gastropods (0.6 meter long), and several have giant bag-neurones of research value to physiologists. [Photos © Douglas P. Wilson.]

enough to allow nearly complete retraction of the head-foot, and the mantle-cavity is still relatively anterior. In such bubble-shells as *Haminoea* and *Retusa* the relatively fragile shell can no longer contain the snail and the mantle-cavity lies on the right side and is reduced in extent. Forms like *Aplysia,* the sea-hare (Figure 20-15B), retain a tiny internal shell and a small lateral mantle-cavity, while in the true nudibranchs, disappearance of the shell and mantle-cavity is complete, the viscera being contained within the head-foot (and we have a bilaterally symmetrical sea-slug with no external vestiges of torsion). Two of the most beautiful groups of nudibranchs are the dorids and the aeolids (Figure 20-16). The dorids retain what might be regarded as a greatly modified pallial complex with a ring of pinnate secondary gills surrounding a posterior anus and kidney opening. The aeolids are characterized by the cerata which not only provide an increased surface area for respiration (again as secondary "gills"), but also are penetrated by diverticulae of the gut where nematocysts derived from the slug's coelenterate prey can be stored and used for protection. The brilliant colors of such nudibranchs as dorids and aeolids are often associated with the development of glands secreting distasteful or toxic substances for defence against predators, as well as the foreign stinging-capsules of aeolids. One group of tropical opisthobranchs, including the genus *Tumano-*

[A]

[B]

FIGURE 20-16. Representative living nudibranchs. **A:** A typical dorid nudibranch, *Archidoris pseudoargus,* crawling toward the left with two tentacles at the head end and the characteristic circlet of gill-plumes expanded round the anus (about 7 centimeters long). **B:** An aeolid nudibranch, *Aeolidia papillosa* (about 6 centimeters long), crawling between two anemones (*Actinia equina*). Aeolids all have rows of elongated finger-like cerata on their dorsal surfaces. These cerata contain blind lobes of the alimentary canal, and their tips may be armed with functional nematocysts gained (undischarged) from the nudibranch's feeding on hydroid coelenterates. Many species of nudibranchs, including both dorids and aeolids, are flamboyantly colored, and many have chemical secretions (including sulfuric acid!) as defensive mechanisms. [Photos © Douglas P. Wilson.]

valva, have secondarily evolved bivalve shells, with a ligament and single adductor muscle but with a tiny gastropod protoconch on *one* valve. It should be noted that such bivalved gastropods (and the limpet-like anomiid bivalves—see page 420) are certainly *not* intermediate forms between the class Gastropoda and the class Bivalvia (thus not of major evolutionary significance), but are merely examples of secondary convergence in discrete stocks which illustrate the great plasticity of molluscan form and function. At least three groups of smaller opisthobranchs have developed a swimming habit; this is best shown by the pteropods, which are entirely planktonic and ciliary-feeding (see Figure 20-17; and Chapter 33). The opisthobranchs must have arisen from a prosobranch stock and (like the pulmonates) have not evolved directly from those more ancient types of archeogastropods with symmetrical organization of their pallial and cardiac structures. The few stem-genera of opisthobranchs have pallial and cardiac structures which are already asymmetric.

Finally, the most highly evolved gastropods belong to the Pulmonata. This group includes not only *Helix* and other land snails and slugs (see Figures 19-1 and 20-18), but also several genera which have secondarily returned to fresh waters or to the sea. All of them—whatever their habitat—are characterized by having *no* ctenidia. Their mantle-cavity has been modified into an air-breathing lung with a relatively small opening to the exterior (the pulmonary opening), and with the outgoing ducts of the gut and the kidney extending right across the roof of the mantle-cavity to the pulmonary opening. The mantle-cavity is lined with a pavement epithelium and, in life, kept moist with mucus.

FIGURE 20-17. A planktonic pteropod snail. A living specimen of *Spiratella retrovera* a tiny (a little over 1 millimeter across) opisthobranch living throughout its life-cycle in the oceanic plankton and feeding on green algal cells by ciliary "fields" on the expanded foot. [Photo © Douglas P. Wilson.]

FIGURE 20-18. A typical land snail. The characteristics of the subclass Pulmonata (the shadowy angle of the shell-aperture edge with the reflected mantle tissue marks the pneumostome leading into the air-breathing lung) and of the order Stylommatophora (the eyes borne on the *ends* of the longer, more posterior pair of tentacles) are well shown in this living specimen of *Mesodon* sp. found crawling on a forest floor. [Photo courtesy Barry S. and Susan M. Payne.]

This lining is underlain by a vascular network which still drains to a single auricle and thence to a ventricle which pumps out to the system of arteries of the body (Figure 20-7). As can be readily seen, from such features as the single-auricled heart and the asymmetry of the pallial structures, it is quite certain that the Pulmonata evolved from a group of gastropods which had already undergone considerable evolutionary reduction of the paired pallial system, and were certainly not evolved directly from primitive zygobranch snails. Both physiologically and histologically, the parallels between the lung in a pulmonate snail and the lung in a primitive air-breathing vertebrate such as a newt are striking. The mechanics of gas exchange and the mechanisms to minimize water loss are very similar, and the microstructural arrangements of the blood capillaries immediately below a thin layer of wide squamous epithelial cells are closely parallel.

As "higher" snails, both the Pulmonata and the various opisthobranch stocks show considerable concentration and shortening of the nervous system, and great elaboration of the genital ducts. The general significance of this is worth emphasizing. The changes in the respiratory and the other pallial organs which we have followed from the Archaeogastropoda to these "higher" forms are paralleled by increasing complexity (or functional modification) in several other systems of organs. Thus, in gut structure, nervous system, and reproductive organs, the Archaeogastropoda are least advanced, the various pectinibranch groups intermediate, and the nudibranchs on one hand and the pulmonates on the other the most advanced. When one examines the internal anatomy of a land snail such as *Helix,* one finds the most centralized nervous system of any snail, with very short commissures and fused ganglia, the regions of the ganglia being associated with particular sense-organs and so on. As regards reproduction, all primitive gastropods have simple gonoducts and discharge their eggs and sperms directly into the sea. Many, perhaps most, marine prosobranchs have separate sexes and employ external fertilization, giving rise to zygotes and ciliated larval stages in the marine plankton. The simple gonoducts of such primitive gastropods involve, of course, a coelomic space with a direct connection through the pericardium to the exterior in the exhalant current of the mantle-cavity. On the other hand, advanced forms, particularly among the opisthobranchs and pulmonates, are more often hermaphroditic and always have more complex genital ducts. These secondarily evolved gonoducts have no obvious connections with either pericardium or pallial structures and lead to an external opening outside the mantle-cavity and usually on the right-hand side of the head-foot. Figure 20-19 shows an example of such a highly evolved genital system. This can be termed triaulic, because there are essentially three separate duct systems: one for outgoing sperm leading to the penis or other intromittent organ, one for incoming sperm leading from the vagina to a spermathecal organ, and a third, not only for the passage of eggs to the exterior, but also for the provision to these eggs of additional food and water supplies and of various protective membranes and shells. It is of fundamental importance that such more complex genital ducts are associated with the production of large eggs or with viviparity (the retention of developing young within the maternal ducts). Ecologically, the stocks which possess them are those which have managed to colonize nonmarine environments. Once again, there are parallels with the vertebrates. The largest land snails lay eggs

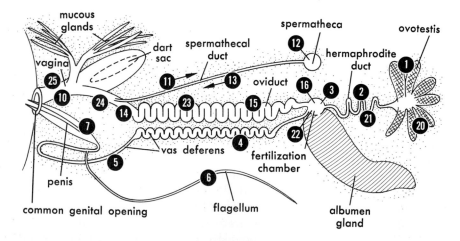

1 –7 = Route taken by sperm going out to fertilize partner
10 –16 = Route taken by partner's sperm after copulation
20 –25 = Route taken by eggs (fertilized at 22 by partner's sperm)

FIGURE 20-19. The reproductive system in a land snail such as *Helix.* There are functionally three duct systems.

with calcareous shells which in many aspects of the physiology of their development are closely similar to the eggs of reptiles and birds.

Some evolutionary, and much physiological and ecological, interest attaches to the fact that the molluscan group which is most successful in land and freshwater habitats is the subclass Pulmonata. Evolved first as an air-breathing group without ctenidia, some pulmonates have secondarily returned to aquatic life in fresh waters (Chapter 21). Thus, unlike the case of the higher vertebrates, in which the main path of evolution ran marine stocks → estuarine stocks → freshwater stocks → land stocks, in these snails the course was marine stocks → land stocks → freshwater stocks.

21
THE TRIALS OF
NONMARINE MOLLUSCS

MOLLUSCS are largely marine. Apart from a small number of bivalve genera living in brackish and fresh waters, all nonmarine molluscs are gastropods. Despite the soft, hydraulically moved bodies and relatively permeable skins typical of all molluscs, some snails are relatively successful as land animals, although they are largely limited to more humid habitats.

Land faunas, both past and present, have always been phyletically limited: the only continuously successful and "perfectly" adapted terrestrial animals belong to certain groups of vertebrates and of arthropods. The amniote vertebrates which are successful on land—reptiles, birds, and mammals—lie beyond the scope of this book; and we have already discussed the arachnids and insects and their immense ecological success as terrestrial animals (Chapters 16 and 17). We noted how important the arthropod body plan, particularly exoskeleton and limb functions, was to their terrestrial adaptations of air-breathing, of water conservation, and of locomotion on land. Much of the interest surrounding the evolution of nonmarine snails arises from the lack of such functional preadaptations as jointed limbs and skin impermeability in the molluscan body plan. Despite this, one group of gastropods, the pulmonate snails, brings the phylum Mollusca into third place (though well behind Arthropoda and Chordata) in terms of numbers of terrestrial species and places it close to the phylum Annelida (represented in terrestrial soils by the oligochaete earthworms) in significance to terrestrial ecology. All other phyla and classes in which land stocks have evolved are represented by small numbers of cryptic species of little ecological significance. They include turbellarians, rhynchocoels, leeches, and isopod crustaceans. The vast numbers of soil nematodes can be regarded as aquatic rather than terrestrial, and the earthworms remain somewhat amphibious in their physiology and restricted in their habitat. Seen in this framework, the success of the gastropod subclass Pulmonata as land animals is remarkable.

Leaving the Sea

In the evolution of all nonmarine animals, two "intermediate" environments, provided by the intertidal zone and by estuaries, respectively, have been important in the adaptation of originally marine stocks to life on land or in fresh waters. Swamps—particularly less transitory swamp areas in the tropics—provide a similar intermediate environment for evolutionary exchanges between fresh waters and land. First, primary physiological adaptations for life on land concern water control, conversion to air-breathing, and temperature regulation. Secondly, if fresh waters are entered through the brackish estuarine environment, while the animals remain essentially aquatic, their primary adaptations concern capacity for osmotic and ionic regulation. Thirdly, the primary physiological requirement for stocks entering fresh waters from land concerns the readaptation of an air-breathing respiratory system to aquatic life. In general, all nonmarine forms also possess appropriate reproductive and excretory adaptations. There is evidence that all of these "routes" out of the sea have been used by different molluscan stocks.

Terrestrial gastropods belong to two distinct groupings. The largest number are members of the order Stylommatophora of the subclass Pulmonata, but they are joined (particularly in the tropics) by certain genera from scattered families in the subclass Prosobranchia which, though each is independently evolved from marine relatives, are collectively termed the "land operculates." In brackish and fresh waters there are a small number of bivalve genera, in addition to both prosobranch and pulmonate gastropods. Throughout most of the world, the majority of freshwater snails belong to the order Basommatophora (Pulmonata), and they are clearly derived from air-breathers, showing various degrees of readaptation to aquatic life. There are some primitive pulmonates living today in salt marshes and in the intertidal zone, and there is some contradictory evidence regarding the initial nonmarine habitat occupied by the ancestors of the major subclass Pulmonata. However, it is clear that the freshwater pulmonates (the so-called higher limnic Basommatophora) derive from air-breathers and are like the freshwater insect stocks discussed in Chapter 17 in showing recognizable morphological and physiological series of differing degrees of readaptation to aquatic life. Thus the evolutionary trend in these pulmonate snails from land *into* fresh waters has been contrary to that which, as vertebrates ourselves, we regard as the more usual evolutionary route. Even although some modern vertebrate morphologists would play down the significance of the surviving lungfishes, all would agree that the higher vertebrates reached land by an evolutionary progression by way of estuaries, fresh waters, and swamps. There is no doubt that not only the bivalves, but also the prosobranch gastropods of fresh waters, evolved by way of estuarine habitats since they all retain typical gills and extensive ciliation on all upper body surfaces of head-foot and mantle. Similarly, it is clear that some of the land operculates are closely related to stocks of prosobranch snails still living in the intertidal zone, and they have (independently of the Pulmonata) turned the mantle-cavity into a lung for aerial respiration. Evolution of progressively more terrestrial forms in some prosobranch stocks is still going on in the littoral zone, as we noted it was for isopod crustaceans (Chapter 15). No one would claim that the contemporary shore-snails were descended from stocks ancestral to present-day land snails, but they undoubtedly face the same physiological problems and live in the same periodic and variable habitats as did the actual Jurassic ancestors of land forms. Study of contemporary littoral species, and of their physiological mechanisms, can help clarify our concepts of how such an "unlikely" animal machine as one built on the basic molluscan plan—involving elaborate ciliary, mucous, and hydraulic mechanisms of great efficiency in an aquatic environment—can maintain itself on land. Investigations of physiology and ecology in all "amphibious" stocks in the intermediate environments may help set limits to our hypotheses regarding the evolution of major nonmarine animal stocks.

One superfamily of mesogastropod prosobranch snails (Littorinacea) have representative species at all levels of the seashore, from the practically terrestrial conditions around the average high water mark of spring tides, down to the zone exposed only by the lowest spring tides where conditions are not dissimilar from those offshore in the shallow sea. In the intermediate zones, littorinids, like all larger intertidal animals, are alternately aerial and aquatic in habit (Figure 21-1). Most of these "middle-level" species are adapted for respiration both as air-breathers and by aquatic ventilation of ctenidia or other exchange "gill" surfaces. On both sides of the temperate North Atlantic, the commonest littorinid, and one of the most abundant animals of the seashore (see Figure 19-1), is the common periwinkle, Littorina littorea. At lower levels in the intertidal, usually associated with the common fucoid seaweeds, is the smooth winkle, Littorina obtusata (Figure 21-2); while on the higher levels of rocky beaches, one finds the rough winkle, Littorina saxatilis, and at the highest levels of the shore in some parts, its congener, Littorina neritoides. The vertical zonation of these littorinids is illustrated in Figure 21-1, along with certain details of their reproductive and respiratory scheses. Besides the four species of the genus Littorina, three other snails are shown. Lacuna vincta is a littorinid living sublittorally (that is, a truly marine form), and Acmaea is another sublittoral marine form, in this case a more primitive aspidobranch snail (see Chapter 20). At the other "land" end of the series is added Melampus bidentatus, a primitive amphibious form belonging to the subclass Pulmonata. As well as vertical scales for two specific localities, Figure 21-1 shows the percentage of time in each semilunar tidal cycle of about fourteen days when each particular zone is actually bathed by seawater. At the lower end of the series, Lacuna and Acmaea are submerged for 83–100% of their lives; while at the terrestrial end, Littorina neritoides and Melampus are bathed by the sea for from less than 2% to about 5% of each cycle. In general terms, the forms from the highest levels of the shore have markedly greater powers of resistance to desiccation, to high and varying temperatures, and to varying salinities than do the forms like *Lacuna* and *Acmaea*, which in many aspects of their physiology are indistinguishable from the great majority of marine snails living elsewhere in the sea. Likewise the forms at the highest levels are air-breathers, and those at the lowest respire under water using gills. However, when we look in detail at structural and functional modifications for increasingly terrestrial or aerial life, we find that these have not been acquired continuously in a smooth series of parallel adaptations in all physiological systems.

Taken as a whole, the general adaptational trends run from marine forms to land ones—from *Acmaea* to *Melampus* in Figure 21-1—with many irregularities intervening in each functional series. For example, the middle columns of the figure show certain aspects of reproduction and the "best-adapted" form in this respect for near-terrestrial life is *Littorina saxatilis*, which is usually viviparous bringing forth miniature adult snails. Unpredictably, *Littorina obtusata* comes "next best" in reproductive adaptation. The most marine species, *Acmaea*, spawns tiny individual eggs and has the primitive molluscan larval history by way of trochophore and veliger larvae as free-swimming forms. However, *Melampus*, though laying egg-masses which have some degree of protection against desiccation and destruction, still has a free-swimming veliger for a limited period of its life-history. The right-hand columns of the figure are concerned with respiratory structures and functions, and largely result from reinvestigation of the physiology of respiration in these littoral forms by Robert F. McMahon and me. The pallial organs range from an aspidobranch ctenidium (a gill suitable only for clean waters) in *Acmaea*, through typical pectinibranch ctenidia in the three lower littorinids, to lungs with a vestigial ctenidium in *Littorina saxatilis* and *L. neritoides*, and to a pulmonate lung with no trace of ctenidial structure in *Melampus*. Our assessments of ratios between rates of oxygen uptake under water and exposed to air at various temperatures varied from 5:2 (aquatic:aerial) for *Acmaea* to 1:10 for *Melampus*, with the three midlittoral littorinids being intermediate. In the temperature responses of their respiration, only *Littorina littorea* shows a considerable capacity, and *Melampus* a somewhat lesser one, for a zone of thermal regulation in oxygen uptake; and capacity for response to low oxygen conditions is less related to vertical zonation but more to the microhabitats of the individual species (for example, neither *Littorina saxatilis* at the top end nor *Acmaea testudinalis*, with its "clean water" gill, at the bottom ever enters reducing environments, and neither shows any capacity to cope with low oxygen stress). Similar series of physiological adaptations (not shown in the figure) would involve water control and nitrogenous excretion in these snails. Although the series are less clearcut than those for the evolution of terrestrial arthropod excretion discussed in Chapter 16, it seems established that the "water-saving" excretion of nitrogenous waste as uric acid is greater in *Littorina neritoides* and *L. saxatilis* than it is in *L. littorea* and *L. obtusata*.

Thus many aspects of structure and function in littoral snails are directly related to the pattern of vertical distribution shown by these species in the intertidal environment. However, as noted earlier, the evolution of more terrestrial adaptations of anatomy and physiology in these littoral snails has proceeded

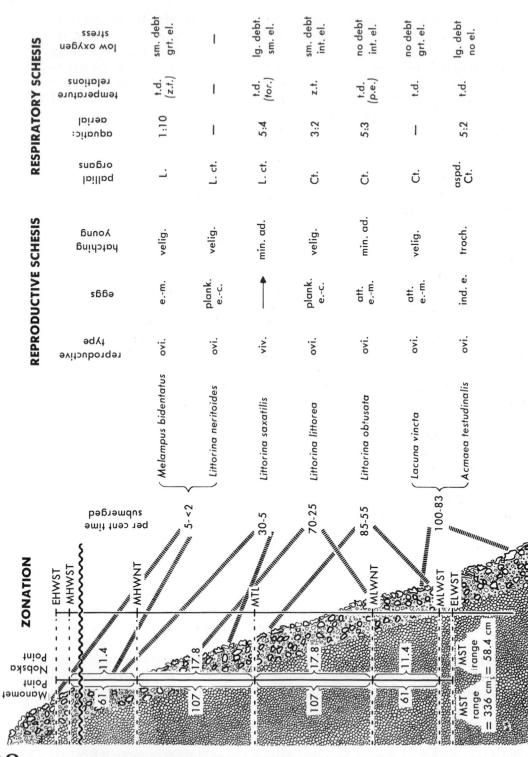

390

FIGURE 21-1.   (opposite) Evolutionary trends in seashore snails. Vertical zonation in the intertidal with reproductive and respiratory scheses of seven species of littoral snails. Average percentages of time spent submerged are indicated along with the actual mean spring tide (MST) ranges in centimeters at both Manomet Point and Nobska Point, Massachusetts. The seven species are *Melampus bidentatus, Littorina neritoides, L. saxatilis, L. littorea, L. obtusata, Lacuna vincta,* and *Acmaea testudinalis.* In the three columns concerned with reproduction, the basic reproductive type is indicated as oviparous (ovi.) or viviparous (viv.); the egg-laying pattern as individual eggs (ind. e.), egg-capsules (e.-c.), or egg-masses (e.-m.), which may be attached (att.) or planktonic (plank.); and the hatching young as trochophore larvae (troch.), veliger larvae (velig.), or with the body form of miniature adults (min. ad.). In the four columns to the right concerned with respiration, the pallial respiratory organs are designated as having the typical prosobranch gill or pectinibranch ctenidium (Ct.), the more primitive feather-shaped gill or aspidobranch ctenidium (aspd. Ct.), the mantle-cavity as an air-breathing lung (L.), or as a lung with a vestigial ctenidium (L. ct.); the average ratio of aquatic to aerial oxygen consumption (both assessed at about 22°C) is shown; temperature relations (similar for both aquatic and aerial respiration) are indicated as showing a zone of thermoregulation (z.t.) or as being largely temperature dependent (t.d.), the latter being modified by some thermoregulation at higher temperatures (t.d. (z.t.)), by torpor at higher temperatures (t.d. (*tor.*)), or by involving passive endurance (t.d. (*p.e.*)); and, lastly, the effects of low-oxygen stress involve payment of a large (lg.), a small (sm.), or no oxygen debt and great (grt.), intermediate (int.), small (sm.), or no elevation of post-stress rates of oxygen uptake at lower (10%) oxygen concentrations. Although the general adaptational trends from *Acmaea* to *Melampus* are from sea to land, it is clear that the evolution of near-terrestrial structures and functions has proceeded anacoluthically. [From McMahon and Russell-Hunter, *Biol. Bull.,* **152:**182–198. 1977.]

anacoluthically, or in discontinuous series not necessarily matched with each other. This pattern of evolution should not be surprising to any modern biologist. Processes of natural selection (including those of the evolution of terrestrial forms in the littoral) proceed by blind variation and recombination, followed by systematic elimination and selective retention. Except in fully terrestrial animals which have moved elsewhere on land, we should not expect the "best" land eggs and the "best" land lungs to have evolved contemporaneously in the same species.

FIGURE 21-2.   A living periwinkle. A specimen of *Littorina obtusata* crawling on a frond of the brown seaweed *Fucus vesiculosus* from the lower part of the intertidal zone. *L. obtusata* is principally aquatic and gill-breathing, but has largish eggs which hatch as miniature adults (see Figure 21-1). [Photo © Douglas P. Wilson.]

Prosobranchs and Pulmonates

As we noted in the last chapter, two subclasses of "higher" snails—opistho-branchs and pulmonates—show the most extensive reduction of paired pallial structures characteristic of progressive evolution in the gastropods, and also show processes of detorsion and considerable elaboration of the genital ducts appropriate to nonmarine environments. Of the two, the Opisthobranchia have not colonized either land or fresh waters. In contrast, the Pulmonata, except for a few intertidal genera which may be among the most primitive, are essentially land and freshwater molluscs. The other nonmarine snails belong to a variety of prosobranch superfamilies including the Neritacea, the Archaeotaenio-glossa, the Littorinacea, and the Rissoacea, but in terms of numbers of species and ecological importance, these land operculates and freshwater operculates must be regarded as less successful nonmarine snails. It can be postulated that they represent many independent invasions of nonmarine habitats by way of the littoral or estuarine waters. In a few cases, this direct independent coloniza-tion can be documented. For example, *Pomatias,* one of the two genera of land operculates in western Europe, is clearly related to and anatomically not much modified from the littorinids of the highest levels of the intertidal. Land oper-culates are relatively rare in temperate regions of the world, making up less than 2% of land-snail species. They are only abundant as species and as indi-viduals in the warm, damp forests of certain tropical areas including Southeast Asia, the East Indies, and Central America, but perhaps most prominently in the Greater Antilles. In Cuba, Hispaniola, and Jamaica, there are unusually large numbers of land operculates. In Jamaica, there are between 400 and 550 species of land snails, of which only about half are pulmonate, but nowhere else in the world are operculates so numerous.

It has already been noted that land prosobranchs are clearly polyphyletic, with diverse genera and families, each related to comparatively unspecialized marine forms, although showing convergent evolution in their adaptations for life on land. One of the stocks, the Neritacea, derives independently from the archaeogastropods (see Chapter 20) and belongs to a group which has evolved completely independently of—though occasionally parallel to—the rest of the modern gastropods. Another prominent family of land prosobranchs is the Cy-clophoridae, which has developed from the most primitive stock of the Meso-gastropoda; yet another, the hydrobiids (also prominent in estuaries and fresh waters), derives from the Rissoacea.

All of these share with the major group of land snails (the pulmonate su-perorder Stylommatophora) equivalent structural, functional, and behavioral modifications which allow them to function as land animals. Anatomically, the mantle-cavity is always a gill-less vascularized lung, and the reproductive sys-tem is adapted for internal fertilization and the production of relatively large eggs, or for viviparity. Functionally, they can all make physiological adjust-ments which allow them to alternate periods of diapause with periods of activ-ity under suitable microclimatic conditions. Hibernation and aestivation are merely names for the more lengthy periods of inactivity which can occur in drying or otherwise unsuitable conditions. Behaviorally, land snails—both prosobranch and pulmonate—are cryptic or nocturnal or both, and one seeks land snails in daylight hours by turning over rotting logs and stones in the shade of damper woodlands or undergrowth. Land prosobranchs live only in these sorts of conditions in the tropics in environments of very high humidity and relatively high and constant temperatures. In contrast, the true Pulmonata are much more widespread geographically, largely because they are better adapted for terrestrial life, and occur in all kinds of soils and vegetation types from the edge of the snow line to the fringes of deserts.

The origins of the subclass Pulmonata lie in some mesogastropod group in which the reduction of paired pallial structures to a single ctenidium and the corresponding changes of the heart to a monotocardiac condition had already occurred, but the details of the interrelationship (if any) between the opisthobranchs and pulmonates and the specific origin of both in a more primitive marine snail stock remain somewhat controversial. The minor superorder of pulmonates, the Systellommatophora, including both aberrant terrestrial veronicellid slugs (also common in Jamaica) and the intertidal onchidiids, air-breathing sea-slugs with some resemblances to opisthobranchs, lies apart from the mainstream of pulmonate evolution, may not be a natural group, and need not be considered further here. The successful land snails are all relatively uniform in anatomy and belong to the superorder Stylommatophora, while the successful snails of fresh waters form the "higher limnic" families of the superorder Basommatophora. Certain more primitive forms in each group, such as *Succinea* from the first and *Melampus* from the second, have considerable resemblances both in their functional anatomy and in their life-style as amphibious snails, though air-breathing. The fossil record is not particularly helpful in this case (see Chapter 34), although some fossil pulmonates are known from the Carboniferous. The most primitive pulmonate gastropods alive today are two distinct marine stocks of "lower" Basommatophora. One of these is the family Ellobiidae, of which *Melampus*, living in salt marshes at the highest levels of the littoral, is typical. As already noted, *Melampus* retains a veliger larva (see Figure 19-11) and, although in other ways a well-adapted air-breathing form, has to have elaborate semilunar synchrony of reproductive phases and of larval stages to allow it to live where it does, and yet retain the primitive marine pattern of producing large numbers of small eggs.

A number of workers on pulmonate morphology and physiology, including J. E. Morton and me, would regard these ellobiids as being not far removed from the ancestral stem-group of both modern land snails and freshwater pulmonates. However, another primitive pulmonate group comprises intertidal limpets like *Siphonaria*. They also have a larval stage, though it is postveliger; some workers, including Drs. Eveline and Ernst Marcus of São Paulo, regard this and their possession of "an opisthobranch-like gill" as making these pulmonate limpets a truly marine stem-group for the pulmonates. In contrast, C. M. Yonge regards *Siphonaria* as a form returned to the littoral zone from a more terrestrial stock, with a readaptation to aquatic respiration by the acquisition of new pallial gills, the true ctenidium having been lost at some earlier evolutionary stage. It can be argued that *neither* the ellobiid snails like *Melampus* nor the marine pulmonate limpets like *Siphonaria* represent stocks which have ever been further onto land than the supralittoral fringe. The retention of planktonic and free-swimming veliger larvae in the first, and of "ambivalent swimming-crawling" postveligers in the second, strongly suggests that both have always remained amphibious stocks, in touch with the sea for reproduction. As with the prosobranch stocks, the acquisition of an air-breathing lung in the higher levels of the littoral probably preceded the invasion of land by stocks of the Pulmonata. A somewhat anomalous position (but one consistent with the view that the pulmonate invasion of nonmarine environments was actually polyphyletic) is occupied by the primitive freshwater pulmonate *Chilina*. This genus shares many primitive features with the Ellobiidae and has a much more "archaic" nervous system than *Siphonaria*, with retention of a streptoneurous condition (or, more correctly, of a chiastoneury of the visceral connective loop in a figure-8 shape, as in all more primitive gastropods retaining the full effects of torsion). *Chilina* obviously represents a separate invasion of fresh waters (perhaps by the estuarine route) totally independent of the "higher limnic" families of Basommatophora. Unfortunately, there is a lack of functional studies on *Chilina*: for example, knowledge of its excretory mechanisms would help. The possibility that *Siphonaria* has returned to midlittoral life from a higher zone on the shore could explain the apparent "incomplete readaptation to aquatic life." It is worth noting that, within the archaic prosobranch group Neritacea (mentioned above), one otherwise entirely terrestrial family living in tropical habitats has a single genus, *Smaragdia*, which has secondarily returned to marine life (like a prosobranch analogue of dolphin evolution). Once again, the fossil record does not help: the earliest identifiable siphonariids are Cretaceous. It is important to note that this controversy does not affect our conclusions regarding the evolution of the freshwater pulmonates. No matter whether ellobiids, or siphonariids, or some other marine group of protopulmonates lies at the origin of the Stylommatophora and the "higher" Basommatophora, and no matter whether the more distant origin was in a mesogastropod stock or a more ancient, but monotocardian, one, it remains clear that the successful pulmonate snails of fresh waters derive from stocks which once were more completely terrestrial.

The freshwater species can be ranked as a series showing progressively greater degrees of adaptation to aquatic life. At one end of the series are marsh-dwelling forms, primarily air-breathing and nonaquatic whose habits result in their occasional submergence in water. At the other end are the aquatic species including the ancylid limpets, where the pulmonate mantle-cavity is completely absent and neomorphic gill-lobes have been developed. It is important to realize that the amphibious pulmonates of fresh water are structurally unspecialized in terms of renal and genital organs and even of the nervous system, when compared to those more completely adapted for aquatic life in fresh waters, such as the ancylid limpets. The opposite would be the case if the group had followed the more usual evolutionary pattern (that of the vertebrates) of preadaptation in fresh waters leading to terrestrial life.

Land Snails and Slugs

Compared to all the other major groups of gastropods, the subclass Stylommatophora, comprising the land snails and slugs, is unusually stereotyped in anatomy and relatively uniform in physiology and behavior. All show adaptations for maintenance of internal water in aerial environments of variable humidity, respiration of gaseous oxygen, and resistance to (or avoidance of) a range of temperatures wider than any in aquatic environments. In addition, their modes of locomotion, reproduction, and nitrogenous excretion are suitably terrestrial. The pulmonate slugs, which have secondarily lost their shells, differ only in the need to cope with more extreme water losses, and in the fine details of the behavioral mechanisms which preserve them from desiccation and death.

Unlike land arthropods and amniote vertebrates, land pulmonates have skins readily permeable to water. In dry air, deprived of access to water, a typical slug of the genus *Limax* will lose about 3% of its weight per hour if motionless, or up to 16% per hour if stimulated to move continually. Thus, if they are kept in drying conditions, death will result within a few hours for all slugs and for most snails. Starting from fully hydrated tissues, several slugs and *Melampus* can survive the loss of more than 60% of their body water. All land pulmonates have a capacity for rapid rehydration on their return to more humid circumstances, but some doubt surrounds the fraction of the uptake due to active drinking. Of course, shelled snails can greatly cut down on water losses by periods of withdrawal into their shells. Except in a few wet marshy habitats in temperate regions and in the lower levels of tropical rain-forests, land pulmonates survive only by strictly regulating their activity according to changes in environmental temperatures and humidities. Many of the species are nocturnal, taking advantage of cooler nighttime temperatures and the presence of free water on plant surfaces in the form of dew, and they may have regular diurnal rhythms of activity which control each evening's arousal. In other cases, perhaps in all slugs, activity is induced by falling temperatures and suppressed by rising temperatures. These stimuli account for occasional daytime activity of slugs and snails after rainshowers, as well as the regular nocturnal rhythm. It is a highly significant fact that daytime slug activity is never obvious if rain is continuous: in such conditions, although humidity levels are good, the temperature-change stimulus is lacking. Similarly, most slugs and snails turn away from strong air currents and move downwind; also, if exposed to light gradients, they become active when illuminated and seek more shaded areas. The ecological significance of these and of other responses is that they move pulmonate snails and slugs away from desiccating conditions. On the other hand, land pulmonates must avoid pools of free water to prevent dangerous overhydration. Even shallow water can cause entrapment because efficient crawling locomotion is hindered by swelling, while deeper water will drown most slugs and land snails.

Continued evaporation from the moist skin of a slug or snail effects a degree of temperature regulation in all but fully saturated air. By use of thermistors, it can be clearly demonstrated that most larger slugs maintain body temperatures lower than the surrounding air. In fact, they are rather like wet-bulb thermometers. In an experiment at an external temperature of 33.7°C and at 24% relative humidity, the body temperature of a slug was found to be 21°C, that is, well below its thermal death point. Measurements on shelled snails like *Helix* showed similar reduction of internal temperatures in fully extended snails and in living snails from which the shells had been removed. In contrast, snails, when fully withdrawn into the shell, showed internal temperatures far above corresponding "wet-bulb" temperatures, but still appreciably lower than those of the air around.

In relation to survival in different environments, therefore, a snail has a choice which the slug has not. Except in saturated air, both can protect themselves from overheating by evaporation, but this cannot continue indefinitely without access to replacement water. The slug is compelled to lose water by evaporation from the body surface when external temperatures are low, that is, even in circumstances which confer no adaptive advantage. In contrast, the shelled snail can protect itself against overheating in excessively warm but not too humid surroundings by expansion, and alternatively, can cut the rate of water loss by withdrawal into its shell in other circumstances.

Most pulmonate shelled snails can withdraw into the shell for days or longer periods of time and then use a layered secretion of calcareous material and hardened mucus (termed the epiphragm) to seal off the mouth of the shell. Oxygen consumption is lowered, water loss is cut, and the snail is in a long-term diapause (often termed hibernation or aestivation without much reference to seasons).

Land locomotion is another somewhat unexpected molluscan ability. The land pulmonates, like all other gastropods, use the muscular foot for adhesion to and crawling over relatively hard substrata. Considerable quantities of mucus are secreted, and fixed points are provided by those parts of the large, flat "sole" of the foot in contact with the mucous trail. In land snails and slugs the locomotory waves are direct (in contrast to the retrograde waves employed by some worms, discussed in Chapter 11, and also used by many marine gastropods—see Chapter 20). The system is one of a hydraulic skeleton, with in this case pedal waves moving in the same direction as the animal. The musculature involved in each pedal wave (shown in Figure 21-3) is remarkably similar to that used in the locomotion of turbellarian flatworms (see Chapter 7). Oblique muscle fibers are particularly important. Those directed inward and rearward contract between waves and pull the body of the slug forward while exerting a backward thrust on the substratum. The oppositely directed oblique muscle fibers (that is, running inwardly and *anteriorly* from the sole) serve to pull the epithelium of the sole forward and form each pedal wave. Compared to that of most marine molluscs, the blood pressure of land pulmonates is relatively high, particularly during land locomotion. Little is known about mechanisms involved in the relatively efficient and fast burrowing carried out by many slugs. Moving over surfaces, the locomotion is surprisingly efficient: even small snails and slugs can achieve speeds of 20–30 centimeters per minute, and it has been recorded that one species of *Helix* can drag fifty times its own weight horizontally, or nine times its own weight up vertical surfaces.

As noted in the last chapter, the complexity of the triaulic hermaphrodite reproductive system of pulmonates is impressive. The eggs produced are all relatively large and hatch as miniature adults. Courtship preceding copulation is often elaborate, and in some slugs, it can last for up to an hour and involve circular crawling patterns with oral stimulation and transfer of mucus preceding a copulation which is effected in about thirty seconds. Fertilization is

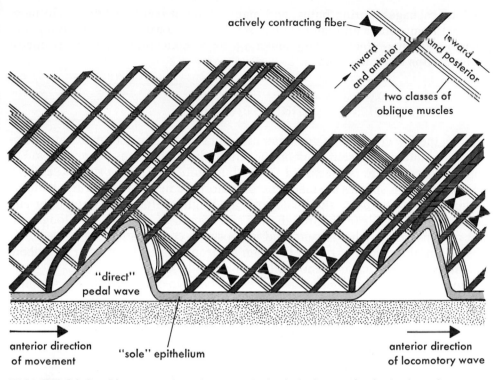

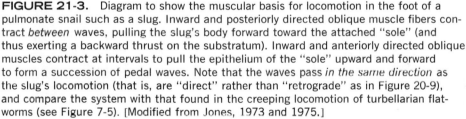

FIGURE 21-3. Diagram to show the muscular basis for locomotion in the foot of a pulmonate snail such as a slug. Inward and posteriorly directed oblique muscle fibers contract *between* waves, pulling the slug's body forward toward the attached "sole" (and thus exerting a backward thrust on the substratum). Inward and anteriorly directed oblique muscles contract at intervals to pull the epithelium of the "sole" upward and forward to form a succession of pedal waves. Note that the waves pass *in the same direction* as the slug's locomotion (that is, are "direct" rather than "retrograde" as in Figure 20-9), and compare the system with that found in the creeping locomotion of turbellarian flatworms (see Figure 7-5). [Modified from Jones, 1973 and 1975.]

usually reciprocal. The elaborate and species-specific behavior in slugs may correspond to the use of stimulatory darts in land snails like *Helix* and serve as an important factor in the reproductive separation of closely related species.

The surprising and unwelcome abundance of slugs in temperate gardens was alluded to in Chapter 19. It is perhaps unexpected to consider land molluscs as agricultural pests, but they do damage a wide variety of crops and are of economic importance in many parts of the world. They are particularly important as pests of root crops such as potatoes or sugar-beets, but they can also destroy considerable acreage of spring wheat in milder, temperate regions. In the case of potato crops, slugs constitute the third most important pest, after viruses carried by aphids and the nematode species called collectively potato-cyst-eelworm. It is difficult to get estimates for the amount of slug damage to horticultural crops, but it is possible that this exceeds (in monetary and labor terms) the agricultural losses. They can cause widespread damage to crops like lettuce, various cabbages, spring bulbs, and many kinds of bedding flowers. In more general terms, slugs and land snails prefer either to eat already-decaying and possibly fungus-infected plant materials or to concentrate on the youngest growing tips of tender plants.

Snail Selection

Among the few convincing demonstrations of selection in action on natural animal populations is recent work on the land pulmonate genus *Cepaea*. Two European species are widespread and abundant: *Cepaea hortensis* and *C. nemoralis*. Both species are markedly polymorphic in shell colors and patterns (Figure 21-4). The basic shell color may be bright yellow, dull brown, or any shade from very pale fawn through pink and orange to red. Up to five longitudinal dark bands may be present, and all possible combinations of presence or absence of them, and of fusion between adjacent bands, have been described. All of these variables are genetically determined.

Most natural populations of *Cepaea* consist of two or more different varieties

FIGURE 21-4.   Shell banding patterns in *Cepaea*. Part of a sample of shells collected by the author from rough limestone moorland on the Isle of Lismore, off the west coast of Scotland. The shells are all from the same interbreeding local population, are in the range 1.5–2.2 centimeters, and include both five-banded forms (three upper row) and fused-banded (two- and three-banded, lower row). The polymorphy would be even more obvious in a color photograph; the light areas between the bands are bright yellow in some shells, pink in others, and pale orange-fawn in others. All such variables of background color and banding pattern are genetically determined, and the effects of different selection pressures in different habitats have been demonstrated (see text). [Photo by the author.]

living together. The proportions of varieties in each colony correlate with the general class of background in the habitat where the snails are living. In woods where the ground is brown with decaying leaves and fairly uniform in appearance, unbanded browns and single-banded pink shells are more common. In hedgerows and on rough green herbage, the yellow five-banded form tends to be commoner. The less green the background, the lower the proportion of yellow shells, and the more uniform the background, the more common unbanded or one-banded forms are. Important predators include the European songthrush, which collects Cepaea and similar snails and cracks them by banging them on stone "anvils," leaving the broken empty shells after extracting the snails. Thus it is possible to compare the proportions of each variety eaten by thrushes to the proportions present in the snail colony. Thus direct evidence of selection pressures can be gained. In typical rough herbage, unbanded shells were subject to significantly greater thrush predation than banded shells. Other studies showed that the proportions of differently patterned snails killed change with seasonal changes in the background vegetation. In a wood in early spring when the background was brown, proportionately more yellows (43%) were taken than in late spring (14%) when the background was green. This would seem to be a particularly clear demonstration of a shift in selection pressure. Interesting differences occur on sloping downland with coarse grass where rabbits were the principal predators rather than thrushes. Such localities have unexpectedly high proportions of pinks and low proportions of yellows. Experiments showed that rabbits were likely to be exerting visual selection by tone alone and not by color and tone as do thrushes and the other bird predators. Rabbits are colorblind, and to them a rich pink shell could appear more like green grass than a yellow one. Direct evidence of predation rates was obtained from only a few localities, but it is amply supported by indirect evidence from many populations which show clear correlations between the proportions of shell morphs and the habitat backgrounds. Of course, if selection of these polymorphic snails was entirely due to predators, the variation would be reduced to a few pure stocks, one suitable for rough herbage, one for woodland, and so on. This does not occur. Other selective agents must tend to maintain the complex polymorphism, and considerable efforts of population geneticists are directed to assess these.

A. J. Cain has pointed out that, in comparably sized land snails more cryptic than Cepaea spp. (that is, limited to shady habitats), a dull brown color with transparency of shell and mantle is the general rule. As noted more than 35 years ago by the great American conchologist H. A. Pilsbry, light shell colors (or whiteness) with greater shell opacity are found in more exposed snail species living in the open (including bush-climbing species and forms dwelling on mountain rocks or in desert fringes). This has survival value in relation to exposure to strong sunlight. Since Cepaea nemoralis is of intermediate habits and occurs in a considerable variety of vegetation, Cain now believes that the remarkable polymorphism is itself appropriate to such a wide-ranging species. Banding can be regarded as valuable disruptive camouflage for light color morphs of a largish snail like Cepaea. In the course of a recent (1977) discussion of selection pressures, he noted that, in a locality where Cepaea nemoralis is known to be common, it is usually easy to find a few quickly, but "only when one gets down on hands and knees and searches, does one find how many and of what colors and [banding] patterns have been missed." This field assessment of the human collector as a differentially efficient predator has significance to discussions of the selection involved in such cases of "balanced" polymorphism.

Shell polymorphism is undeniably stable. Evidence from subfossil and fossil populations of *Cepaea* shows that the range of shell banding has persisted for tens of thousands of years, and it is probable that the genes and the polymorphism are older than the presently separated species pair *Cepaea nemoralis* and *C. hortensis*. In other words, the ancestral stock which gave rise to these species was already polymorphic on the same allelic bases.

Molluscan Gills in Brackish and Fresh Waters

The bivalve genera and the prosobranch gastropods, living in brackish and fresh waters, are both directly derived from, and in some cases closely related to, marine forms. Thus, all are gill-bearing, retaining the typical molluscan ctenidia, and are essentially aquatic animals which have acquired some capacity for osmoregulation. This necessary adaptation involves more drastic physiological changes in bivalves with their enormous gill surfaces used in feeding (see Chapter 22).

The salt concentration in the blood of the typical freshwater clam, *Anodonta*, is only equivalent to about 4% seawater. In such molluscs, as in several other invertebrate groups, drastic reduction of the blood concentration is probably essential for colonization of brackish and of fresh waters. For reasons discussed in earlier chapters, the production of urine hypotonic to the blood is necessary to offset the passive influx of water, but the lower blood concentration allows higher efficiency in osmoregulation and reduced osmotic work. For example it has been estimated that only 1.2% of the total metabolic energy of *Anodonta* is used in the work of pumping out water as hypotonic urine. Some estuarine bivalves can, by closure of the shell valves, be effectively sealed off from short-term changes in the external medium, but this mechanism cannot be used by totally freshwater forms.

The four major freshwater bivalve families are remarkably uniform in structure, showing little of the adaptive radiation found in marine bivalves (see Chapter 22). The larger freshwater mussels are all forms essentially similar to *Anodonta* and *Unio*, and the smaller bivalves include the cosmopolitan and ubiquitous genera *Pisidium* and *Sphaerium*. In some freshwater habitats a single species of bivalve may be abundant, and, particularly in hard-water rivers in many parts of the world, there are areas of stream bottom paved with larger unionids or smaller sphaeriids (the latter at densities reaching 1400 per square meter).

In the case of the freshwater prosobranchs there is clear evidence that brackish and fresh waters have been colonized from the sea many times by different stocks. Several genera of freshwater prosobranchs are closely related to estuarine or to littoral marine forms, and in a few cases, freshwater species have marine congeners. They range from species of the hydrobiid genus *Potamo-*

pyrgus, which have become immigrants to fresh water within recent (and recorded) historical time, through forms like *Theodoxus fluviatilis*, which colonized fresh waters during the Gunz-Mindel interglacial (that is, relatively recently in the geological time-scale), to well-established freshwater genera such as *Bithynia*, *Valvata*, and *Viviparus*. In general the freshwater prosobranch snails, as gill-breathers, are more limited to running or at least well-oxygenated waters than are freshwater pulmonates. In total about ten unrelated groups of the Prosobranchia are found in the world's fresh waters, but they are all alike in having complex genital ducts associated with internal fertilization and with viviparty or with the production of relatively large eggs. A number of them, including species of *Bithynia* and *Viviparus*, are filter-feeders, using the gill in a similar fashion to that of *Crepidula* (see Chapter 20). This again bespeaks a continuously aquatic evolution for these forms, and it is significant that none of the freshwater pulmonates has a comparable feeding mechanism.

Peculiar freshwater prosobranchs, with more elaborate secondary adaptations for particular habitats or particular modes of feeding are found in the few large lakes known to be relatively ancient. These include Lake Tanganyika in east central Africa, Lake Baikal in Russia, and Lake Ochrida in the Balkans. The best-known of these is Lake Tanganyika, the prosobranchs of which show radiation into a variety of forms (probably more than 84 species, 66 endemic) which have been derived by adaptive radiation of a single melaniid stock. Early naturalists, impressed by the diversity and the relatively massive sculptured shells of the prosobranchs (Figure 21-5), deduced erroneously that the Tangan-

FIGURE 21-5. Snails from an ancient lake. Three species of the "thalassoid" (or "marine-appearing") freshwater prosobranch snails from Lake Tanganyika. **A:** *Melanella nassa;* **B:** *Limnotrochus thomsoni;* **C:** *Neothauma tanganyicense.* These, and many other, peculiar species are endemic to Lake Tanganyika. They are not, in fact, of marine relict origin, but have evolved from freshwater stocks in this relatively ancient lake. [Photo by the author.]

yika snails represented a marine relict fauna "cut off" by catastrophic earth movements from the sea. In fact, 58 of the 66 endemic species belong to a single freshwater family, the Melaniidae, and, despite the variety of shells (some like neogastropod carnivores), all species are herbivorous (as deposit-feeders or radula-grazers) with a gastric crystalline style (see Chapter 23). This speciation is rather like that of Darwin's finches in the Galápagos Islands, and occurs because these few lakes have been in existence for 100,000 to 1 million years, as compared to the majority of freshwater habitats in the world, which are much more transitory, with ages from 100 to 10,000 years in almost all cases. To reiterate, with the exception of these prosobranchs in ancient lakes, the world's freshwater snail faunas are predominantly pulmonate.

Molluscan Lungs in Fresh Waters

The dominant snails of fresh waters are the so-called higher limnic families of the subclass Basommatophora. Although they are freshwater snails, their most important feature is the absence of ctenidia, the characteristic gills, present and structurally homologous in representatives of all other aquatic molluscan groups (see Chapter 19). The gill-less mantle-cavity has a roof richly vascularized with thin-walled vessels, its opening to the exterior is narrow and muscular, and the rectum and kidney duct do *not* open into it; the mantle-cavity is a lung. As with the secondarily aquatic insects and insect larval stages living in fresh waters (discussed in Chapter 17), where the tracheal system of air breathing had to be readapted to aquatic life, so in the pulmonate snails we have an evolution from temporarily diving forms to truly aquatic ones which can remain submerged. Such swamp-dwelling snails as *Lymnaea truncatula* and *L. palustris* are amphibious, able to withstand occasional submergence but capable of being drowned. Closely related forms such as *Lymnaea stagnalis* and *L. peregra* live mainly as aquatic animals—feeding and reproducing under water—but are still merely "divers," surfacing at regular intervals to take in air. When such species occur in large lakes, they are limited to the margins and to shallow water within easy reach of the air interface. Many natural populations of these species are involved in seasonal or in short-term migrations, where the respiratory needs of adult snails may result in their temporary starvation (by forcing migration onto near-sterile substrata after a rise of lake level), or may cause them to migrate away from young they have produced. Other populations of these and of other pulmonate species visit the surface less frequently and typically have a gas bubble in the lung of about 84% nitrogen which is used as a "physical gill" to extract oxygen from the water by diffusion. This ingenious method of gaining oxygen, employed much more extensively by aquatic insects (see Chapter 17), depends upon the maintenance of a bubble of relatively high nitrogen content from which oxygen is removed to the tissues of the snail as it diffuses into the bubble from the higher concentrations of dissolved oxygen in the water outside. Apart from microanalysis of the lung gases, there is an interesting experiment which can be carried out on animals suspected of using a gas bubble as a physical gill. It was first used by August Krogh and R. Ege to demonstrate the mechanism in insects, and some years ago Andrew E. Henderson and I were able to apply it to freshwater pulmonate snails. If snails with this respiratory pattern are held in experimental tanks and allowed to surface into an atmosphere of pure nitrogen, they will then remain submerged in well aerated water for much longer periods with that lung-full of nitrogen than with an ordinary air bubble. Contrariwise, if allowed to surface and obtain a lung-full of pure oxygen, they can only remain submerged for a much shorter time, since that bubble cannot be used as a physical gill. In species such as *Lymnaea peregra* and *Physa fontinalis*, where some populations employ a physical gill, other populations, including some in relatively shallow water, have the lung remaining water-filled throughout life, and all respiratory exchange is cutaneous. In some other populations, the lung contains a gas bubble, but a relatively high content of both carbon dioxide and oxygen suggests that its use as a physical gill is unlikely and a hydrostatic function as an internal bubble giving buoyancy more likely.

In still other pulmonate species, particularly the ramshorn snails of the family Planorbidae, and the freshwater limpets like *Ancylus* and *Ferrissia*, secondary gills have developed. These gill-lobes are neomorphic—clearly not homologous with any part of the ctenidium of other molluscs—and lie outside the original mantle-cavity. Cilia are almost totally absent from the pallial organs of most freshwater pulmonates—reflecting an incomplete readaptation from life on land. In many cases, well-ciliated surfaces could obviously increase the efficiency of the secondary (neomorphic) gills. In fact, ciliated surfaces on some of these pulmonate gill-lobes are achieved by the expedient of everting (in development) part of the rectum. It should be noted that even in land snails where external cilia are limited to the sole of the foot, the lining of the gut remains ciliated. It is highly significant that the true ctenidium—lost in the evolution of terrestrial pulmonates—is never regained by those that return to an aquatic life in fresh water. This illustrates one aspect of the irrevocability of evolutionary loss, which used to be referred to as "Dollo's principle." Although much criticized, its application may still be valid in the case of complex structures or functions involving large bundles of integrated genetic material for their development and differentiation. In evolution generally, any complex structure is more easily lost than gained. Interestingly enough, the associated sense-organ of the mantle-cavity, the osphradium, is not totally lost in pulmonate snails as is the true ctenidium. It is present in the embryos of all land snails, and is redeveloped and functional in the more aquatic freshwater pulmonates.

Once again it must be emphasized that in the series of pulmonates considered to show progressive readaptation of an air-breathing stock to aquatic life, the amphibious forms have the least specialized genital and nervous organization, while the limpets with neomorphic gills, like *Ancylus* and *Ferrissia*, are also the most advanced in terms of the concentration of their nervous system (and thus degree of euthyneury, or reversal of the effects of torsion). This effectively disposes of the theory put forward by some authors that the lymnaeid and physid gill-less "divers" were derived from estuarine forms by way of the more aquatic freshwater limpets like *Ancylus*. *Chilina*, with its markedly less concentrated (and closer to streptoneurous) nervous system is the only genus of freshwater pulmonates which might derive directly from an estuarine proto-pulmonate stock. No matter whether pulmonates first became a nonmarine stock by way of the littoral zone or by way of estuaries, the pulmonate snails ran counter to the evolution of the higher vertebrates, and paralleled the case of the freshwater insects, in deriving their freshwater stocks from an air-breathing land one. Further, it is certain that the higher limnic forms do *not* represent a stage in the evolution of land snails from estuarine stocks.

In more general terms, it is remarkable enough that such an animal machine as one built on the basic molluscan plan, and so successful in marine environments, should be able to colonize fresh waters and land relatively successfully. We should not be surprised that no polyplacophoran, scaphopod, or cephalopod has ever colonized land or fresh waters, or that colonization by bivalves should be so restricted. It is of greater evolutionary interest that one of the subclasses of higher snails, the Pulmonata, has taken third place (admittedly well behind the arthropods and amniote vertebrates) in the race to exploit nonmarine habitats, both terrestrial and aquatic.

22 THE EVOLUTION OF FILTER-FEEDING BIVALVES

THE great majority of living bivalves are lamellibranchs; that is, they have enormously enlarged gills. These ctenidia—like those in a few genera of mesogastropods such as *Crepidula*—are enlarged by elongation of the filaments so that, together, adjacent filaments form a lamella. Each gill, although homologous both functionally and morphologically with the gills of more primitive gastropods in respect of its blood vessels and arrangement of cilia and so on, is far more extensive than is required for the respiratory needs of the animal. It is now the major organ of food collection in these filter-feeders. Briefly, a water current through the mantle-cavity is created by the lateral cilia. This flows through between the filaments of the ctenidium from the inhalant part of the mantle-cavity to the exhalant region. Any particulate matter remains on the inhalant face of the gill, and frontal cilia and mucus are used to make chains or boluses of material to pass to the mouth.

Bivalve Organization

At this point, it is convenient to consider the other diagnostic characters of the class Bivalvia. In the evolution of this group, extension of the mantle and its shell has occurred followed by its lateral compression, so that all parts of the body lie within the mantle. The visceral mass and foot both lie within the

405

mantle-cavity, and the head is lost. Of course, cephalic sense-organs would not be of much value within the cavity and out of contact with the environment. Significantly, it is in bivalves that the middle lobe of the mantle edge shows the fullest development of sense-organs. As discussed later, there is some controversy on the origins of the major molluscan groups, but on structural and functional grounds it is fairly certain that the bivalves were derived from an ancestor with a dome-shaped mantle-shell and with a posterior mantle-cavity with one pair of ctenidia. It is probable that this bivalve ancestor differed from most other molluscs in that the mantle was attached to the shell near its margin. The final stage of lateral compression was made possible by reduction of calcification along the dorsal midline, while near each end cross-fusion of mantle muscles produced the shell adductors. Many palaeontologists believe that both the major class Bivalvia and the minor one Scaphopoda have been derived from the totally extinct class Rostroconchia (whose significance was first recognized in 1972, and which have a bivalved adult shell but a single larval protoconch). Rostroconchs range from the lower Cambrian to the Permian (see Chapter 34), and are themselves believed to be derived from untorted pregastropods or from forms within the Monoplacophora.

In the bivalves—as in all other molluscs—the mantle and its secreted shell form a single structural entity. The description found in most textbooks of two discrete valves united by a ligament of different origin is totally erroneous. Developmentally, a single mantle rudiment appears early in the ciliated larva, and although growth patterns are such that anterior and posterior embayments appear in the originally dome-shaped rudiment, there always remains (Figure 22-1) a mantle isthmus. Usually, the material secreted by a mantle isthmus contains proportionately less crystalline calcium carbonate and proportionately more elastic tanned proteins, and forms the ligament of the bivalve shell. This elasticity is very important to the mechanical functioning of the bivalve. In all

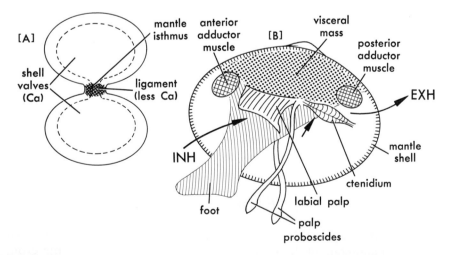

FIGURE 22-1. A: Later mantle-shell rudiment in a developing bivalve: the mantle isthmus secretes the less calcified ligament. B: An archetypic protobranchiate bivalve. The ctenidia are small and the labial palps relatively enormous with long extensible palp proboscides which are used in deposit-feeding.

bivalves the shell is closed by the action of adductor muscles, which run from one shell valve to the other. These, the largest muscles in any bivalve, have no single antagonist but can be stretched by several mechanisms, which include the elasticity of the horny hinge ligament and several kinds of hydraulic systems. The relative importance of each method varies in different types of bivalves. The position and functioning of the elastic ligament, however, are always similar. It connects the shell valves dorsally and is under strain when the valves are closed. This strain can be tensile in external hinge ligaments or can result from compression of internal ligaments. The force derived when the ligament is under strain tends to open the valves; that is, it acts against the adductor muscles. To put this another way, when the adductor muscles of a bivalve contract, closing the shell, they are also doing the work which will subsequently reopen the shell valves. This work involves compression or extension of the "springs" of the ligament.

Filter-feeding

Although most families of bivalves are structurally lamellibranch, and thus functionally filter-feeders, there are two groups of genera which are not so. One of these, the Septibranchia, consists of relatively rare, highly specialized bivalves which live in moderately deep water in the oceans. They have the typical gills secondarily replaced by a muscular septum, and they create a water current for feeding by pumping rather than by ciliary means. Accordingly, they are macrophages feeding on such things as copepod crustaceans of moderate size. Being a minor and highly specialized group, they need not be considered further here. The other group, the Protobranchia, are of considerably greater significance, since they are in many ways intermediate in form and function between the typical specialized filter-feeding bivalve and its more generalized molluscan ancestors. In such genera as *Nucula* and *Yoldia*, the gills are relatively small (Figure 22-1), more closely resemble the ctenidia of other molluscs, and are not extended in filter-feeding lamellae. As in all bivalves, the head structures are reduced, sense-organs and radula being totally absent. However, in the protobranchs, there are around the mouth mobile extensions of the lips forming palps and palp proboscides which are used in feeding. The protobranchs are essentially deposit-feeders, and each extensible proboscis is covered with cilia and is used to bring deposits from outside the clam in toward its mouth. The labial palps themselves form sorting surfaces of the kind already described. That is to say, they are capable of discriminating and separating particles of larger size from the fine organic particles which are valuable as food and which they pass on into the mouth. Recent work has shown that in many protobranchs the ctenidia, although not greatly enlarged, are capable of carrying out a certain amount of filtration and contributing to the food supply of the bivalve in this way. Figure 22-2A–C contrasts the mantle-cavity structures of typical protobranchs with those in a true lamellibranch. Feeding is largely by the palp proboscides in protobranch genera like *Nucula* (nuculids) and *Yoldia* (nuculanids), but not in a third group represented by *Solemya* (solenomyids), where the ctenidia are of greater importance. All protobranch bivalves have a

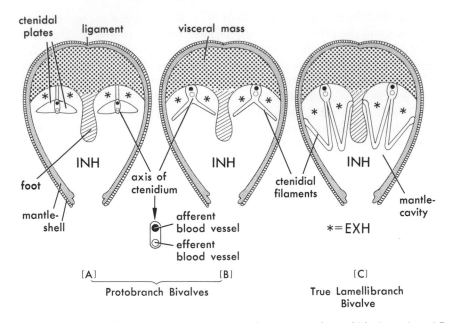

FIGURE 22-2.   Mantle-cavity structures as seen in cross sections of bivalves. **A** and **B:** In protobranchs. **C:** In a true lamellibranch with greatly enlarged ctenidia.

relatively large foot which can be protruded through the valve gape to expand into a "mushroom anchor" in surprisingly rapid locomotion. Nuculanids are of particular importance in the bivalve faunas of the depths of the oceans, and it has been noted that while protobranch forms usually make up somewhat less than 10% of the bivalves in the shallower inshore seas, they can comprise over 70% of the bivalves in deep-sea samples. This may be a true relict distribution, and it parallels the fact that seemingly protobranch bivalves were abundant in the fossil record well before the dominance of lamellibranch bivalves was established (see Chapter 34). Thus distributional ecology, as well as functional morphology and the fossil record, supports the status of the Protobranchia as a naturally separate subclass of the Bivalvia. The case for separating the subclass Septibranchia (as is done here—see above and Table 19-1) is much less valid. The subclass Lamellibranchia, made up of filter-feeding bivalves with greatly enlarged ctenidia, comprises by far the largest subdivision of the class Bivalvia and, in terms of numbers of individual and ecological bioenergetics, the most successful of all molluscan groups. Thus most recent workers on molluscs have come to regard the protobranch condition as relatively primitive and have agreed that it forms a necessary intermediate stage between the loss of direct feeding by a mouth which cannot now leave the shell and the acquisition of ciliary filter-feeding by enlarged ctenidia.

The majority of bivalves (perhaps 29,000 out of 31,000 living species) have essentially the same feeding process. The following description would apply to any of them, although Figure 22-3 is largely based on the structures in mussels such as *Mytilus.* In all lamellibranchs, the lateral cilia produce the water current between adjacent filaments. This water passes ventrally into the inhalant part of the mantle-cavity, and thence through the gills to the exhalant chamber above and within them. All food organisms and all suspended material are accumulated on the inhalant faces of the gill lamellae. Such material and food is then moved by the frontal cilia toward the ventral edges of the gills and accumulated in the food grooves with some mucus. As can be seen in Figure 22-3, the food grooves result from an infolding of the frontal surface of the gill filaments. In them the frontal cilia are functionally modified and beat anteriorly, so that the food material passes anteriorly along the ventral edges of the gills to between the labial palps. Here again, sorting is carried out on a size basis (see Figure 19-7). Fine material is carried by cilia into the mouth and thence into the oesophagus and stomach where it undergoes further sorting. Coarser particles

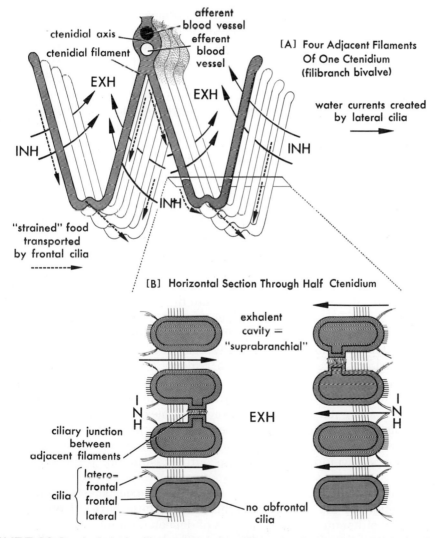

FIGURE 22-3. Archetypic gill structure in lamellibranchs. **A:** Stereogram of folded filaments in one ctenidium. **B:** Horizontal section through a demibranch (a half ctenidium).

accumulate at the edges of the palps and are periodically thrown off by muscular twitches onto the mantle wall. This material, which has been filtered off by the feeding structures but has never entered the gut, is usually called pseudofaeces; it is collected by the cleansing cilia of the inside of the mantle wall into ciliary vortices whose arrangement varies in different bivalves. In almost all, however, the pseudofaeces are finally expelled from the bivalve by spasmodic contractions of the adductor muscles which force water, together with the accumulated pseudofaeces, out through the normally inhalant openings to the mantle-cavity. All bivalves show these "spontaneous" spasmodic contractions of the adductor muscles, which thus have a cleansing function. It should be mentioned at this point that the anus and the renal and genital openings are in the exhalant part of the mantle-cavity in bivalves, as in all molluscs, and thus expulsion of the wastes or reproductive products is not accomplished by these spasmodic cleansing movements, but by the normal and continuous water flow of the feeding current. Two points should be noted in the diagram of the horizontal section through a half-gill (Figure 22-3). An additional group of cilia, the laterofrontals, have arisen and serve as a part of the filtering mechanism. In a classic series of research reports (1936–1943), Daphne Atkins reported beautiful studies by light microscopy on the variation of laterofrontal cilia and on the ciliary mechanisms of various lamellibranch groups. More recently, use of the scanning electron microscope has shown the laterofrontals to be compound cilia (compare with the "walking" ciliates in Chapter 6), with a finely pinnate structure which greatly increases their efficiency in the trapping of food particles and flicking them onto the frontal collecting tracts (Figure 22-4). On the other hand, there are no abfrontal cilia on the exhalant sides of the filaments. Functionally, this implies that there is no material which penetrates to the exhalant part of the mantle-cavity and has to be cleansed off the gill surfaces. A further point is that in such forms as *Mytilus* the adjacent filaments are held together only by occasional ciliary junctions, which function rather like modern dress-fastenings of Velcro. In certain other bivalves, such as the soft-shelled clam, *Mya*, these ciliary junctions are replaced in adults by tissue fusion between adjacent filaments. If living clams are dissected, it is immediately obvious which clams have tissue fusion between filaments and which have gills merely held together by ciliary junctions. This character of the nature of the interfilamentary junctions was formerly used in the classification of the bivalves. Recently, however, it has been realized that tissue fusion has been evolved independently in several lines of clams. Incidentally, there is evidence of a totally different sort that the significance of the vast size of the lamellibranch gill is alimentary and not respiratory. If measurements are made of the oxygen consumption of clams, it can be calculated that at the oxygen tensions of their environment, gills of approximately one-fiftieth of the surface area of those developed would suffice for the entire respiratory exchange of such clams.

From time to time claims have been made that mucous sheets are important in the filtration by the gills of lamellibranchs. These have all proved to be wrong, based either on misinterpretation of data on clearance rates or on direct observations on injured or unhealthy clams. The latter misunderstanding arises from the fact that many lamellibranchs will respond to the traumatic removal of

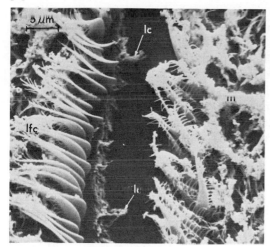

[A] [B]

[C]

FIGURE 22-4. Bivalve laterofrontal cilia as cirri. The great efficiency of the lamellibranch gill as a filter-feeding mechanism depends upon the "additional" group of cilia (the latero-frontals) on each ctenidial filament (see Figure 22-3B for orientation). **A** and **B:** Scanning electron micrographs of adjacent pairs of ctenidial filaments in face view. **C:** Interference photomicrograph of the edge of a living gill filament (all from the mussel *Mytilus*). In all pho-tographs the finely pinnate nature of these laterofrontals as compound cilia (or cirri) can be seen. On the right hand filament in **A**, the partially extended laterofrontals have been fixed as they cleansed from a small mass of mucus; while both filaments in **B** (a preparation which had been stimulated with serotonin or 5-hydroxytryptamine, at a concentration which is known to increase water flow through the gill while decreasing particle retention) have the laterofrontals (lfc) folded inward over the frontal cilia, thus "opening" the spaces between the filaments and increasing the efficiency of water propulsion by the lateral cilia (lc), which are seen to be organized in metachronal waves. In the living condition (**C**) the laterofrontals are shown extended and beating in metachronal rhythm, and thus this photomicrograph con-stitutes a "food-particle's-eye" view of the filtering apparatus of a typical lamellibranch bi-valve. [Photos courtesy of Dr. C. Barker Jørgensen of the University of Copenhagen.]

411

one shell valve and attached mantle by "shutting down" the gill by contraction of muscles in the ctenidial filaments, withdrawal of blood from the gill as a whole, and a massive secretion of mucus over its surface for protection. If the bivalve recovers, all of this mucous sheet is rolled up and moves through the rejection tracts to finish as pseudofaeces on the remaining mantle wall. In a fully extended and healthy gill there is only a little mucus on the frontal collecting tracts and water passes freely between the laterofrontal cilia into the exhalant chambers. Filtration of particulate material that will be "accepted" and passed over the labial palps into the mouth is carried out entirely by the spaced combs of the compound laterofrontals, which throw food particles onto the frontal collecting tracts without any mucus entanglement. This is in complete contrast not only to gastropod filter-feeders like *Crepidula*, which spin a mucous net as one of their feeding mechanisms, but also to such forms as *Urechis* (of the minor phylum Echiuroidea—see Chapter 18) and to all sea-squirts (Chapter 32). Those filter-feeders, using a mucous net or mucous sheets, are apparently able to retain large colloids, but there is no good evidence of such forms being involved in adsorption of organic molecules out of solution. In Chapter 12, we discussed older theories and recent experiments on direct uptake of dissolved organic materials from seawater by invertebrates. It is worth reemphasizing here that although bivalves such as mussels can be shown to take up dissolved organic molecules such as glucose and amino acids, detailed work on the net effects of the transport systems involved, along with assessments of the amount of dissolved organic matter available in unpolluted seawaters, demonstrates that the amounts that can be taken up could not constitute as much as 1% of the total energy requirements of these animals. Although some marine invertebrates, including certain polychaetes (Chapter 12), can cover a major part of their maintenance requirements by uptake of dissolved amino acids from sediments rich in organic material, this is clearly not the case in bivalves, despite their remarkably high surface:mass ratios. In summary, lamellibranch bivalves are true ciliary filter-feeders, relying only on the nice spacing of their laterofrontals (adaptively adjusted to each specific habitat, and capable of temporary shifts in "mesh size" by variation in the hydraulic turgor of their filaments), and do not need recourse to sheets of mucus, or to direct active transport of dissolved organic molecules for their nutrition.

The lamellibranch bivalve form has been enormously successful, incorporating as it does the most efficient mechanism ever evolved in any animal group to collect the largest plant crop in the world—the phytoplankton (see Chapter 33). Although the number of living bivalve species (possibly 30,000) is second to the class Gastropoda (over 74,000), the biomass of marine bivalves is much larger. Dense beds of mussels and of oysters are obvious features of the lower intertidal and the shallower seas in many parts of the world. Like the narrower bands of barnacles (Chapter 14), mussel-beds show up clearly not only in many aerial photographs of coasts but also even in some pictures from satellites. Less visible burrowing clams can be almost as abundant. One estimate gives densities of mactrids (small "surf-clams") reaching 8000 per square meter and covering about 2500 square kilometers of the Dogger Bank in the North Sea. Lamellibranchs are a major food of all bottom-living fishes and, in the economy of the sea, are second only to planktonic copepods like *Calanus* in annual turnover of calorific value for animal tissues in food-chains (Chapters 14 and 33).

Ecological Diversity

In spite of a relatively constant pattern in feeding mechanisms, bivalve species vary greatly in their actual body form. Much of this variation results from differential growth of different parts of the mantle-shell, though the adductor muscles and the foot also vary. The edges of the mantle may show different degrees of fusion and form various kinds of siphons—both inhalant and exhalant. Perhaps it is easiest to understand the adaptive value of these modifications by attempting to relate them to the ecology of the different forms involved. Indeed, any attempt to explain these modifications phyletically would require several books of the size of this one. One of the reasons for this is that recent work on functional morphology of bivalves has elucidated that the structural and functional adaptations associated with certain specific habits in bivalves are often clearly polyphyletic. The anatomical and physiological peculiarities shown by deep-burrowing bivalves, for example, have been independently evolved in several stocks and thus reflect convergence rather than close relationship.

The least specialized bivalves are active shallow-burrowers with a relatively large muscular foot and short siphons, a symmetrical globose shell, and the dimyarian condition of more or less equal anterior and posterior adductor muscles. Cockles and the quahog (*Mercenaria*—Figure 22-6) are typical of these forms. Once again we owe to E. R. Trueman and his associates our knowledge of burrowing locomotion in such unspecialized bivalves. The stages of a burrowing cycle (illustrated in Figure 22-5) involve two pairs of pedal retractor muscles as well as the anterior and posterior adductor muscles (and their antagonist, the elastic ligament), with the shell valves and the foot providing alternate fixed anchorages against which movement can take place. The point in the cycle (Figure 22-5A) when the two shell valves are being pushed against the substratum by the ligament while the adductor muscles are relaxed is the shell or penetration anchorage. With this as a basis, contraction of the circular and transverse muscles within the foot causes it to probe downward, the siphons becoming closed as the foot is extended to its maximum length. At the end of this stage the tip of the foot begins to dilate and partial dilation is immediately followed by a relatively sudden contraction of the adductor muscles (with the siphons closed). This contraction frees the shell valves from their anchorage in the substratum and at the same time forces some water out of the mantle-cavity around the foot, and drives much more blood into the foot, thus completing its terminal dilation and forming a new pedal anchorage. Contraction of the pedal retractor muscles then pulls the whole body of the clam down toward the foot, the shell valves remaining closed by the adductors until the bivalve is poised for the beginning of another burrowing cycle. In the globose, unspecialized clams which we are considering, there can be a further refinement of the process of pulling down on the pedal anchorage. Contraction of the

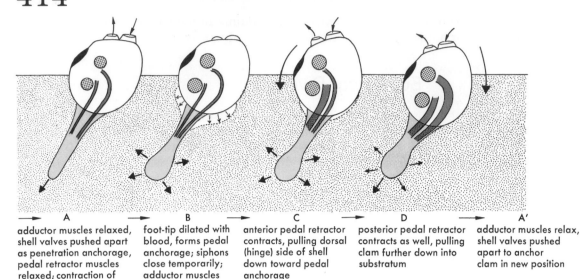

A	B	C	D	A′
adductor muscles relaxed, shell valves pushed apart as penetration anchorage, pedal retractor muscles relaxed; contraction of circular and transverse muscles in foot causes probing extension of foot	foot-tip dilated with blood, forms pedal anchorage; siphons close temporarily; adductor muscles contract, freeing shell valves and expelling water between valves around foot	anterior pedal retractor contracts, pulling dorsal (hinge) side of shell down toward pedal anchorage	posterior pedal retractor contracts as well, pulling clam further down into substratum	adductor muscles relax, shell valves pushed apart to anchor clam in new position

FIGURE 22-5. The stages of burrowing locomotion in a stylized bivalve. The cycle can be repeated many times, with alternate *points d'appui* being provided by the opening gape of the shell valves (the shell or penetration anchor) and by the dilated tip of the foot (the pedal or retraction anchor). For further discussion, see text. [Modified from Trueman, 1966, 1968, and other publications.]

anterior pedal retractor muscles can occur a little in advance of the other movements, and will cause the clam to be pulled forward obliquely on the pedal anchor (with the dorsal or ligament side of the shell being pulled into the substratum somewhat faster). At this point the siphons will reopen, and contraction of the posterior retractors will cause the shell to rotate in the opposite direction (the valve gape being now preferentially pulled down), and the rotation will work the clam further into the substrate before the adductor muscles relax and the valves are separated by the elastic action of the ligament to form a new shell or penetration anchorage for yet another cycle. In many clams each complete cycle will take a few seconds, but in the razor-clam, *Ensis,* there can be ninety cycles of probe and pull-down per minute. Razor-clams (Figure 22-6) in suitable soft sands can burrow to a depth of about 60 centimeters in less than ten seconds. In the slower-moving, more globose bivalves, such as quahogs and cockles (Figures 1-3B and 22-6), the shell valves bear a surface sculpture of radial ribs, concentric ridges, or short spikes. All of these serve to make the penetration anchorage (when the valves separate under the action of the ligament) more efficient. As already noted, protobranch bivalves are somewhat more active than the majority of true lamellibranchs, and in their locomotion they are aided by a foot with somewhat more elaborate musculature capable of allowing a distinctive centrifugal spread of the tip into a mushroom-anchor.

The adductor muscles of bivalves (and, to some extent, the pedal retractor muscles also) have long been of particular interest to physiologists and are now also important to biophysicists concerned with muscle function at the molecular level. Functionally, bivalve adductors are involved in two distinct kinds of mechanical abilities. They must be able to clap the shell valves shut quickly in an emergency (against a predator) or in the spasmodic cleansing movements which send the pseudofaeces out through the inhalant openings. They must also be able to maintain a prolonged closure, for example, to prevent tissue exposure at low tide. Accordingly, adductor muscles usually contain two types of muscle fibers. These are always histologically distinct, and although in some bivalves they are intermingled, in many scallops and in some protobranchs like *Nucula* they are segregated and anatomically "quick" and "catch" muscles can be distinguished. The quick muscles, which have some resemblances to vertebrate striated muscle, are able to contract rapidly in a twitch but are unable to sustain contraction for a long period. In contrast, the catch muscles (which were once thought to correspond to vertebrate "smooth" muscles) can only contract relatively slowly but when contracted, are able to stay in their shortened state for long periods with minimal expenditure of energy. At the level of whole-animal physiology, this means that some of the problems of fatigue and buildup of oxygen debts need not occur in these catch-muscle tissues during

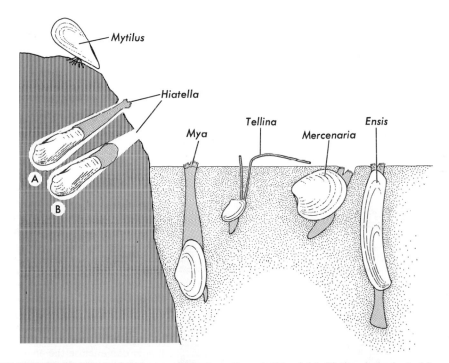

FIGURE 22-6. Ecological types of bivalves. From left to right: *Mytilus* is typical of byssally attached mussels, *Hiatella* of rock-borers in permanent burrows, *Mya* of sedentary deep-burrowing clams with long fused siphons, *Tellina* of active deep-burrowers using long extensible separate siphons for deposit-feeding. *Mercenaria* (formerly *Venus*) of active shallow-burrowers with globose shells and short siphons, and *Ensis* of the razor-clams which can withdraw rapidly using the foot as a mushroom anchor. (In culinary terms, *Mya* is the soft-shelled or long-neck clam, and *Mercenaria* is the quahog, hard-shelled, or little-neck clam.) [Partly after Russell Hunter and Brown, 1964, and Russell Hunter, 1949.]

prolonged closure of the clam's shells. At the molecular level, the known differences of ultrastructure and of electrical properties merit, and are receiving, continued investigation. As already noted, levels of activity in bivalves vary greatly: some can move relatively rapidly (and a few even swim), while others are cemented down and live as completely sessile animals like corals.

There are several obvious ecological types (each of which is almost certainly polyphyletic), and some of these types are illustrated in Figure 22-6. The active shallow-burrowers with short siphons, which never go deep into the sandy substratum, often have globose shells, and include such forms as the cockle, *Cardium*, and the quahog, *Mercenaria* (formerly *Venus*). These contrast with the active deeper-burrowers bearing long extensible separate siphons, such as *Tellina* and *Macoma*, which use their siphons like vacuum cleaners sweeping over the surface of the sea bottom. Then there are more sedentary deep-burrowers, such as the soft-shelled clam, *Mya*, which do not move from place to place, but have long siphons which can be almost completely withdrawn into the posterior shell gape. In *Mya* and its allies, these are *fused* siphons (Figure 22-6): both inhalant and exhalant channels are enclosed within a common sheath made up of massive bands of circular and longitudinal muscles. An extension of the flexible outermost shell layer or periostracum (see Figure 19-6) forms a coarse and much-wrinkled brown "skin" over this muscular double-tube. One classic French monograph (by F. Vlès in 1907) describes this periostracal skin as giving the massive fused siphons, "l'aspect du pénis d'un vieux

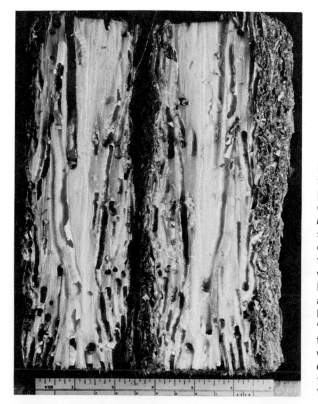

FIGURE 22-7. Wood from the submerged part of an old dock piling, split open to show the work of the bivalve "shipworm," *Teredo navalis.* Living bivalves can be seen (for example, at the inner end of the top straight burrow in the left-hand piece). The shell valves are reduced so that they no longer can enclose the animal and are used by it as the "auger-blades" in boring, while the "older" walls of its burrow (extending back to the outside of the wood) have been given a smooth calcareous lining of "shelly" material. [Photo © Douglas P. Wilson.]

cheval." In the course of a recent study on predation energetics in *Polinices* (see Figure 20-12), D. Craig Edwards notes that in attacking *Mya*, this carnivorous snail always drills through the shell valves, thus suggesting that the periostracum covering of the siphonal area in *Mya* is repellent to predators like *Polinices*. A more specialized way of life is that shown by the razor-clams, such as *Ensis*, which are fast-moving, the whole animal being withdrawn by pulling down on the foot, which functions like a mushroom-anchor.

Several bivalves bore in clays and peaty deposits and even in rock and wood. Various mechanisms are used, and these borers have mostly evolved from deep-burrowing stocks. Some borers, such as *Pholas* in rock and *Xylophaga* in wood, show obvious structural modifications including a sucker-foot and a gaping shell sculptured with tiny teeth like a rasp or file (Figure 22-7). These excavate their burrows (which are permanent) mechanically. Other rock-borers, including the date-mussels, *Lithophaga* and its allies, are relatively unmodified except for the possession of mantle-edge glands which secrete acid. They bore chemically and are found only in limestones and similar rocks (including corals). Another type of rock-borer is represented by *Hiatella* (Figures 22-6 and 22-8), adults of which excavate mechanically, though they lack any obvious anatomical specializations. The burrow in the rock is shaped like an elongated cone, the opening being narrower than the part occupied by the animal. This proves it to be a permanent burrow—a lifetime home for the bivalve. In 1949, I was able to demonstrate that the boring method employed by *Hiatella* depends on the use of a hydrostatic skeleton of variable volume, involving seawater within the mantle-cavity rather than blood within the haemocoel. In *Hiatella*, as in forms like *Mya*, the siphons are disproportionately large. In withdrawal, closure of the siphonal tips *precedes* contraction of the longitudinal siphonal retractor muscles. The increased water pressure which results forces the shell valves apart (the hinge ligament being relatively weak and contributing little mechanically). In this way, the pressure of the shell valves against the

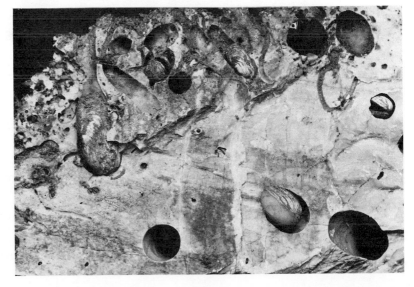

FIGURE 22-8. Limestone rock split open to show the boring bivalve, *Hiatella*. Those burrows near the bottom of the photograph probably result from about eight years' excavation by each bivalve. A boring sponge and a polychaete are responsible for the finer burrows at the top left. [Photo © Douglas P. Wilson.]

burrow is used in abrasion. The shortening of the siphons is also accompanied (as another result of the increased water pressure) by expansion of their basal parts (Figure 22-6), and this distension provides a fixed point about which the abrading movements can take place. The mechanism is effective but slow: the larger perforations in Figure 22-8 have been excavated over about eight years, in a hard limestone rock. In colder waters, small specimens of *Hiatella* probably abrade less than a cubic centimeter of rock each year. Finally, it is of some interest that the actual sequences of muscular movements employed for boring in this genus have evolved with little modification from the protective reactions used by closely related forms which do not bore.

As subsequently elucidated by Garth Chapman, the siphonal movements in *Hiatella* correspond to one of two principal patterns of siphonal extension in bivalves. In forms with considerable fusion of the mantle margins and relatively massive siphons like *Mya*, closure of the mantle openings allows the siphonal musculature to act as the antagonist of the shell adductor muscles around a temporarily constant volume of seawater. This is why, as Chapman (with G. E. Newell) showed, siphons are extended in such clams in a stepwise progression with the tips of the siphons *closed* during each extension. In com-

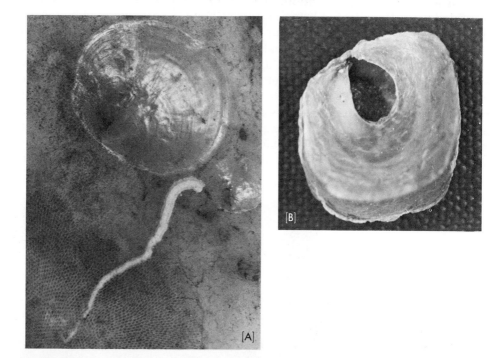

FIGURE 22-9.　Saddle-oysters or jingle-shells. **A:** Two living specimens of the bivalve saddle-oyster *Anomia simplex* on an intertidal rock surface near Woods Hole; the shiny domes are actually the left (or free) valves of the bivalves, and are known as "jingle-shells" or "mermaids' toenails" when detached after death. (In addition, a colony of the ectoproct *Electra* is growing from the bottom left corner, and serpulid tubeworm, *Hydroides,* is part of this "fouling community" all cemented down to the rock surface.) **B:** A detached specimen of *Anomia,* viewed from the right (or under) valve, showing the characteristic notch or embayment which surrounds the calcified byssal attachment "pillar." [Photos by the author.]

plete contrast with this is the extension of the long separate siphons in deposit-feeding forms like tellinid clams, where the relatively delicate siphons have their ends *open* during the process of protrusion. Chapman and Newell were able to demonstrate that this process involves the use of radial muscles within the siphonal walls working on a fluid content in the wall which has been isolated from the rest of the bivalve's haemocoel, and extends the siphons by "thinning" their walls. As we will discuss in Chapter 24, the efficient jet-propulsion of squids and cuttlefish depends upon a much more rapid, but mechanically essentially similar, contraction of radial muscles in the pallial wall.

Both sedentary deep-burrowers and the more active deposit-feeders with separable siphons consist of a number of stocks, probably independently evolved from active shallow-burrowers with globose shells. A totally separate

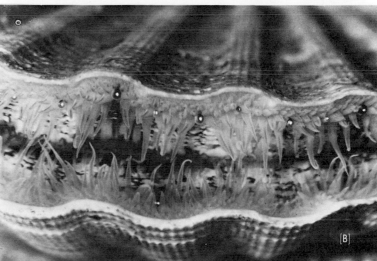

FIGURE 22-10. Scallops, actively swimming bivalves. **A:** Live specimen of *Pecten maximus*. Contraction of the single powerful adductor muscle can jet-propel the scallop for an escape jump of a meter or more, and slower flapping swimming is also possible. **B:** The mantle margin of *Pecten maximus* showing the sensory tentacles and shining blue pallial eyes (all on the *middle* pallial lobe—see Figure 19-6A), and behind them the more muscular veil (functioning like the rubber apron of a hovercraft) of the *inner* pallial lobe, which shows here as mottled white-and-black pigmented tissue. [Photos © Douglas P. Wilson.]

group of bivalves is the mussels, which are usually heteromyarian (having uneven adductor muscles) and live attached by byssus threads to hard substrata such as rocks or the pilings of docks. *Modiolus* and *Mytilus* are examples of these. Other bivalves are even more permanently fixed to hard substrata, either by a calcified byssus, like a miniature limestone pillar passing through an aperture in the lower shell valve, or by the lower valve itself becoming cemented down (by a modification of the secretion of the outer shell layer or periostracum). The former method is that of the Anomiacea or saddle-oysters, the latter that of the Ostreacea or true oysters and of at least four other bivalve stocks in which permanent cementation has been independently evolved. As "headless" filter-feeders, bivalves might be considered to be "preadapted" for sessile life, and several of these cemented stocks have become monomyarian (with a single adductor muscle centrally placed) with an almost radial symmetry. Oysters and their allies have been highly successful in their life-style and can form extensive reef-deposits like corals. Native oysters were "over-fished" early in human history and much human effort and ever-more elaborate techniques have been employed in the earliest true aquiculture—oyster-farming. Saddle-oysters like *Anomia* (also known as jingle-shells or "mermaids' toenails") are not uncommon on rocks of the lower littoral (Figure 22-9) throughout the world. One anomiid presents an extreme case of convergence and reversal in molluscan evolution. Appropriately named *Enigmonia*, it has resumed a crawling habit on an enlarged foot (though still lying on one side) and taken up life as a bivalve turned "limpet." Finally, there are a few bivalves which can swim more or less actively. These include some scallops like *Pecten* (Figure 22-10), and more specialized forms such as *Lima*. In these forms the swimming is by valve flapping, and this ability has arisen, interestingly enough, from the cleansing movements shown by almost all bivalves to expel pseudofaeces.

23 FUNCTIONAL ASPECTS IN SNAILS AND BIVALVES

THE uniformity of the plan of structure and function which underlies molluscan diversity has already been stressed. Basic homologies are clear in the structures of the mantle and in all the functioning systems connected with the pallial cavity: including, of course, the ctenidia, the heart and circulatory structures, and the excretory organs. We have already summarized the evolutionary trends involving the pericardial structures and their renal and genital derivatives (see Figure 20-8). As a result of their progressive asymmetries, the more advanced gastropod groups have kidney and genital structures which have moved further away from the archetypic molluscan pattern than have the corresponding structures of bivalves and of the smaller molluscan groups. The same is true in the evolution of the ctenidia, which is now summarized in Figure 23-1. To recapitulate, the gastropods show reduction from a pair of aspidobranch ctenidia to a single one, and from that to a one-sided pectinibranch ctenidium (or comb-gill), and then to no gill at all in the pulmonates. The bivalves show enlargement of gill leaflets to longer filaments and their subsequent folding into the true lamellibranch condition. Ctenidia are replicated in the chitons, and in the higher cephalopods, while still structurally homologous, are modified to resist the stresses of water currents produced by muscular pumping rather than by lateral cilia.

Apart from the pallial and cardiac structures, similar sets of

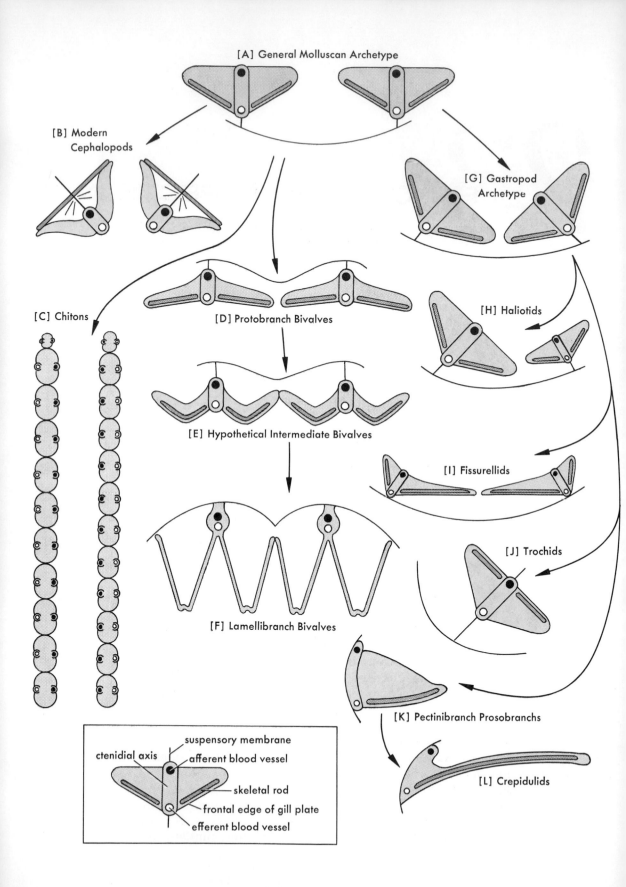

[A] General Molluscan Archetype

[B] Modern Cephalopods

[G] Gastropod Archetype

[C] Chitons

[D] Protobranch Bivalves

[H] Haliotids

[E] Hypothetical Intermediate Bivalves

[I] Fissurellids

[J] Trochids

[F] Lamellibranch Bivalves

[K] Pectinibranch Prosobranchs

[L] Crepidulids

suspensory membrane
ctenidial axis
afferent blood vessel
skeletal rod
frontal edge of gill plate
efferent blood vessel

structural and functional homologies can be discerned in the other internal organ systems, particularly in the gastropods and bivalves. Before we pass to a consideration of the most highly organized molluscs, the cephalopods, we will consider the digestive and reproductive systems in snails and bivalves. Particularly as regards the complex ciliary mechanics of the primitive gut pattern and certain peculiarities of gamete maturation and reproduction, these systems can be most instructive in any consideration of evolutionary origins and trends in the basic molluscan pattern.

The Primitive Molluscan Gut

The archetypic arrangement of the molluscan gut must have been closely similar to the pattern which is found today in the very many molluscs which are continuously feeding microherbivores. The many thousands of species of lamellibranch bivalves, those few snails which are filter-feeders, and the majority of more primitive gastropods which feed continuously by radula-grazing: all these have similarly organized guts and digestive processes.

Functionally, the organization of this primitive gut pattern is dictated by the need to deal with quantities of finely particulate food, embedded in a mucous strand with some inorganic material, which strand passes more or less continuously into the gut. Thus the physiology of primitive molluscan guts contrasts markedly with the intermittent feeding followed by cyclical processes of digestion so characteristic of vertebrates and the more highly organized invertebrates, including the more specialized carnivorous molluscs. Mechanisms for handling a continuous but slow stream of finely divided food material have been evolved in other filter-feeding animals, but nowhere have ciliated surfaces become so highly organized for continuous processing of material as within the molluscan alimentary canal. Most of the gut is ciliated, and the stomach and its associated diverticula have especially complex ciliary sorting tracts. Another characteristic feature of these primitive molluscan guts is the possession of a peculiar secreted structure: the crystalline style. In a few more primitive gastro-

FIGURE 23-1. (opposite) Molluscan gills (ctenidia). Cross sections, viewed from the posterior (but ventral in **C**), of the ctenidia in a series of typical molluscs. General molluscan (**A**) and gastropod (**G**) archetypes are illustrated, but the only other hypothetical pattern is that (**E**) shown as an intermediate bivalve condition. All the other gill patterns shown occur in presently living forms: in modern cephalopods (**B**), in chitons (**C**), in protobranch (**D**) and lamellibranch (**F**) bivalves, and in a series of gastropods (**H–L**). Compare with conditions in the pericardial and urinogenital organs in Figure 20-8 and note that different degrees of asymmetry shown here in the ctenidia of such archeogastropods as haliotids (**H**), fissurellids (**I**), and trochids (**J**) do not necessarily correspond to asymmetries of kidneys and gonads. The great majority of living prosobranchs (both mesogastropods and neogastropods) have the pectinibranch gill pattern shown at **K**, and **L** represents the elongation of the individual ctenidial filaments in forms like *Crepidula*, which lengthened filaments allow the gill to be used in filter-feeding. This crepidulid condition (**L**) has been evolved from a half ctenidium in a parallel fashion to the lamellibranch condition (**F**) in bivalves from a complete pair of ctenidia. Note also that the serial ctenidia of chitons (**C**), which continue to be replicated anteriorly with growth, are *not* developed in pairs, so that lateral asymmetry of numbers is common. [Derived, in large part, from various publications by C. M. Yonge.]

pods and in the protobranchiate bivalves, the style is secreted in, and occupies most of, the lumen of the first part of the intestine immediately after the stomach cavity. In the great majority of bivalves and filter-feeding snails the style-sac is completely separate from this anterior intestine (Figure 23-2). In a few bivalves there are anatomical connections between the style-sac and the typhlosole side of the anterior intestine, but the lumen of the sac is always functionally separated from the lumen of the intestine, except in some primitive snails. In the typical gut (Figure 23-2) the mouth opens into a short ciliated oesophagus—the cilia of which may be arranged in spiral rows which can impart a twist to the rope of mucus with contained food particles coming in from the feeding organs. This opens into a subglobular stomach from which there leave, ventrally, the openings of the anterior intestine and of the style-sac and, posteriorly, the paired basal ducts of the digestive diverticula. The anterior intestine has a fold, which is often elaborate, in its wall. This typhlosole usually extends along the floor of the stomach and around the openings of the digestive diverticula, sometimes in a complicated fashion (Figure 23-2). After the typhlosole ends in the anterior intestine, the rest of the gut-tube (posterior intestine and rectum) is anatomically undifferentiated. It ends in the anus which, as already stressed, discharges in the exhalant part of the mantle-cavity in all forms.

The hyaline rod of hardened muciprotein which is the style protrudes from

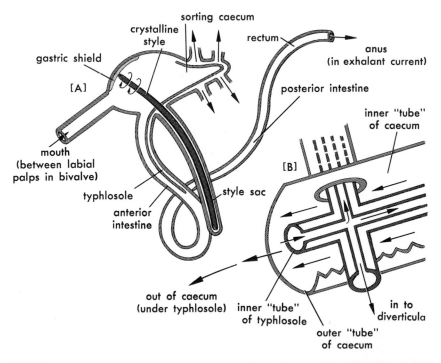

FIGURE 23-2. **A:** The arrangement of the alimentary canal in a filter-feeding mollusc. This gut is organized for the slow but continuous processing of a stream of finely divided food material. **B:** Part of the caecum in a lamellibranch bivalve. The typhlosole forms a tube within each tubular root of digestive diverticula, allowing further sorting. [Modified from Morton, 1958, and Owen, 1955.]

the style-sac into the stomach, and its free end bears on a cuticularized patch called the gastric shield. The style is formed in the lower part of the style-sac (although it may be secreted by cells of the typhlosole in some cases). It is continuously revolved by cilia lining the intermediate parts of the style-sac, and, in all lamellibranchs at least, the direction of revolution is clockwise if viewed from above the gastric shield. The hyaline material is impregnated with several secreted enzymes which always include amylases and glycogen-breaking enzymes (in a few cases lipases and cellulose-breaking enzymes have been detected). The style thus acts as an enzyme store allowing a slow but continuous release as the muciprotein softens in the higher pH of the stomach and the style is ground off against the gastric shield. It also acts as a stirring rod for the contents of the stomach lumen and in many molluscs can be seen to act as a windlass onto which the mucous rope of food material is being continuously wound or, in fact, pulled in from the oesophagus.

The rates of style revolution and of style wear are low: one revolution takes from 2 to 12 seconds, and about 6 millimeters of style per hour is worn off in an actively feeding, moderately large bivalve. In some deposit-feeding bivalves there is also trituration of food material by the relatively massive style acting against the gastric shield as a sort of millstone. In a few bivalves where there are connections between the anterior intestine and the style-sac, some particulate material may be recycled from the intestine, by passage into the style-sac and incorporation in the style material, and thus transported back to the stomach. Style-bearing molluscs all exhibit breakdown and resorption of the style under unfavorable conditions.

The pH patterns of the molluscan gut are characteristic: in a typical bivalve they can range from pH 4.4 in the style itself through about 5.7 in the stomach cavity up to pH 7.0 in the rectum. The higher pH of the stomach fluid helps bring about the disintegration of the style, although the style itself helps control the pH of the stomach contents and is believed also to exert some buffering action. The neutral pH of the posterior intestine and rectum is associated with a region of water reabsorption which serves to consolidate the faeces produced. It is obviously necessary to consolidate faecal material to avoid self-fouling of the feeding mechanism. Some bivalves are so successful in this that their faecal pellets or strands persist for long periods of time in marine bottom deposits, and in some cases the faeces of tellinacean clams can be identified to species by patterns impressed on them by the typhlosole and ciliary tracts in the more anterior parts of the gut, before consolidation.

The other important function of the stomach—sorting of particulate material —is carried out largely in the posterior caecum from which arise the major ducts of the digestive diverticula. The actual details of the ciliary sorting areas differ considerably in different molluscs, but they are all basically made up of surfaces covered with grooves and ridges functioning in essentially the same fashion as the labial palps in bivalves (see Figure 19-7). In general, their functioning directs larger and heavier particles toward the anterior part of the stomach cavity (and if not broken down on recirculation thence to the anterior intestine and rejection), while finer particles are carried across the ridges and grooves and recirculated again and again past the openings of the ducts to the digestive diverticula. Thus, after the extracellular digestion of the stomach lumen, dissolved materials and finely particulate food pass into the tubules of the digestive diverticula. In the walls of these tubules are phagocytic cells and, in most bivalves and the more primitive gastropods, these take up the particulate material and a further intracellular digestion takes place. Even the fine material which is not digested is phagocytized and formed into spherules which pass back in a rejection tract and thence under the typhlosole to the intestine. No material can pass into the hindgut of such a mollusc until it has circulated at least once across the sorting surfaces and more usually traveled around the stomach more than twice.

In some filter-feeding bivalves the complexity of the typhlosole folds at the entrance to the basal tubes of the digestive diverticula is impressive (Figure 23-2B), but it seems that such complexity may not really be necessary. In fact, those very bivalves which have elaborate gastric sorting and their typhlosole involved in the digestive diverticular ducts are the same bivalves which have the most elaborate sorting surfaces outside the gut in the labial palps of the mantle-cavity.

In general, in all molluscs (with the possible exception of a few advanced cephalopods), the initial processes of extracellular digestion have to be followed by phagocytosis and thereafter by intracellular digestion. The cells lining the finer tubules of the digestive diverticula are the site of this cellular ingestion and subsequent digestion, and it appears that in many molluscs the bulk of proteolytic enzymes come into play only within the "food vacuoles" of the cells of the diverticula. It should be noted that the application of the term "liver" to this mass of tubules which form the digestive diverticula in molluscs is totally erroneous, although common in textbooks. The many functions of the diverticular cells include absorption, phagocytosis, secretion, and possibly some excretion. In the bulk of the microphagous forms we have been describing, the absorptive and phagocytic functions are the only important ones.

Other Digestive Processes

The elaborations of the microphagous digestive system are replaced in carnivorous gastropods (and also in cephalopods) by what is, to our vertebrate eyes, a rather simpler digestive organization. In these forms feeding is intermittent and digestion is largely extracellular, glands such as salivary glands discharging into the anterior part of the gut being important sources of protein-breaking enzymes. In many of them, the digestive diverticula are also mainly secretory. In these forms a meal is taken and a cycle of secretion and absorption is followed in a pattern similar to that familiar in the vertebrates. Omnivores and carnivores, among the higher pulmonate snails, have an enormous repertoire of secreted enzymes. These include enzymes unusual in animals, for example, some which can break down sulfated polysaccharides, and even a chitinase. Once assumed to be a product of the alimentary flora, it is now known that in such snails as *Helix* the chitin-breaking enzyme is secreted by the cells of the digestive diverticula themselves.

In bivalves, some specializations of digestion can be correlated with specialized feeding habits. In wood-boring bivalves such as *Xylophaga, Bankia,* and

Teredo (see Figure 22-7), the normal diet filtered from the water passing through the mantle-cavity is supplemented, or replaced, by ingestion of wood shavings. In some species the wood particles ingested are stored in special diverticula. Analysis of faecal material shows that cellulose and hemicellulose have been digested and removed from the material during its passage through the gut, and there is no doubt that cellulose-breaking enzymes are present in the digestive tract of these bivalves. As has been pointed out by several workers, the styles of many bivalves, including these wood-ingesting forms, contain large spirochaetes. It has not yet been conclusively demonstrated whether the cellulytic enzymes originate in molluscan gut cells or in these microorganisms.

In the giant clams of the Indo-Pacific (the genus *Tridacna* and its allies), the bivalve body form is modified to allow farming of enormous numbers of symbiotic algae in modified pallial tissues. As in the corals (see Chapter 4), these protistans are called zooxanthellae and are probably modified dinoflagellates (see Chapter 6). The tridacnids have well-developed ctenidia and remain capable of normal filter-feeding, but much circumstantial evidence has accumulated, and a few direct demonstrations have shown, that the symbiotic zooxanthellae are used by the clams moving them internally and then digested as a supplementary food source. Many of the peculiar features of tridacnid anatomy, including the enormously hypertrophied tissues of the siphonal region of the mantle, are only explicable as adaptations for the better utilization of the zooxanthellae.

Sex Changes and Reproductive Methods

Primitive gastropods and all bivalves have simple gonoducts, actually coelomoducts, connecting with the pericardium. As already mentioned, the more advanced snails among the opisthobranchs and pulmonates have more complicated genital ducts associated with the production of large eggs or with viviparity. Ecologically, the snails which have colonized fresh waters and land environments almost invariably belong to stocks with complex genitalia.

As regards gastropods, we can make the generalization that many marine prosobranchs are dioecious, while opisthobranchs and pulmonates are more often hermaphroditic. However, ambisexual forms are found within a wide variety of molluscan groups. Indeed, the only two groups made up entirely of dioecious species are the Scaphopoda and the Cephalopoda. In other molluscan classes, there are many genera in which one species is dioecious and a closely related species is hermaphroditic. The ambisexual molluscs show a wide variety of types of hermaphroditism. A few forms are truly simultaneous hermaphrodites, in which eggs and sperm ripen at the same time. These include a number of pulmonate snails and one of the marine scallops, *Pecten irradians*. A surprisingly large number of forms show various kinds of consecutive sexuality where the ripening of the two sorts of gametes is asynchronous. Most usually, the male phase occurs first, and these species are said to show protandric consecutive sexuality.

One of the best examples of protandry is the slipper-limpet, *Crepidula fornicata*, which begins life as a male, passes through a hermaphrodite stage, and finishes as a female. In nature, slipper-limpets are found in stacks of from eight to twelve individuals (Figure 23-3), all oriented in the same direction, and each clinging by its sucker foot to the shell of the one underneath. The basal member of each chain is usually attached to a dead shell, is always a female, and in summer is usually brooding eggs. At the other end of the stack (Figure 23-3) are the smallest individuals, which are immatures about to become males. Each individual will complete the characteristic sequential sex change with size and position: egg → larva → immature → functional male → transitional → female → death. As well as a testis, the functional males near the top of the stack have an extensile muscular penis which can reach down to fertilize the larger females with functional ovarian tissue at the bottom of the stack. This strange life-style has long been known, and a series of studies by W. R. Coe and H. N. Gould demonstrated that the timing of the protandric sequence depends upon a complicated interaction of both physiological and environmental factors. The interplay of these factors is particularly interesting in the sex change and onset of femaleness. For example, an *isolated* small male will become a transitional and then a female at a much smaller size than is usual. On the other hand, males which are kept in the presence of several functional females will remain males after they have grown to a larger size than is normal. The availability of food and changes in water temperature, as well as the biotic influence of adja-

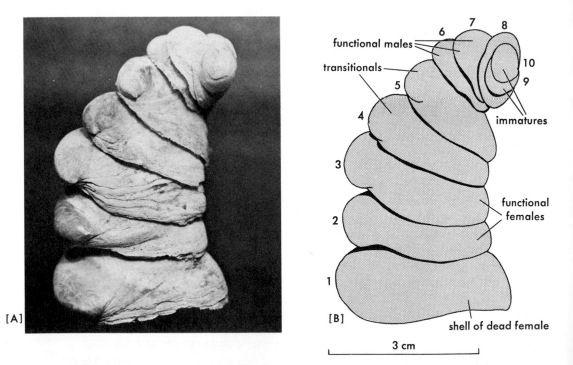

FIGURE 23-3.   Unit stack of *Crepidula fornicata*. The age sequence of the life-cycle (10, youngest; 1, oldest) is reflected in each unit stack "commune," and runs immature → functional male → transitional → functional female → death. In the early summer condition, eggs will be being "brooded" under individuals 2 and 3 (the functional females), and from each of the individuals 6 and 7 (the functional males) a large extensile penis can reach down the stack for fertilization. See also Figure 23-4, and discussion in text. [Photo by the author.]

cent individuals, affect the sex changes in Crepidula. It is certain that there is a species-specific pheromone which is liberated into the water by females and is responsible for maintaining the masculinity of the males. In Crepidula, as in most gastropods, there are neurosecretory cells in the cerebropleural-pedal complex of ganglia (or "brain"), and a few years ago Martyn L. Apley and I conducted some experiments involving the effects of brain implants in perturbing the normal sequence of sex change. Earlier experimental attempts to modify or to cause the sex change by injections of appropriate extracts from other limpets had failed. We had established a scale of penis condition which could be observed externally in living specimens (and afterward correlated with the condition of the gonads) which allowed us to establish distinctions between three successive development stages in immatures, two in fully functional males, and three in regressing stages in transitional animals that were becoming female (Figure 23-4). Crepidula fornicata proved to be an excellent subject for such operations, and we had only about 4% postoperational mortality in nearly two hundred implants. Most of our experiments involved implanting a brain (or brains) freshly removed from a functional male into the angle between the foot and visceral mass behind the head on the left side of a large male recipient. The operated animals were kept in isolation and the controls moved steadily through the transitional stage to become functional females. In a series of single implant experiments with male brains, we were able to delay the onset of sex change by an average of four days, and in a series of successive implants, were able to keep "giant" Crepidula as functional males for over seventy days. In other cases of successive implants we caused a gradual regression of the penis state to the immature condition. As regards experimental work on neuroendocrine controls of reproduction, it is clear that these molluscan studies are only at a level of sophistication reached by insect (see Chapter 17) and crustacean (see Chapter 13) physiologists about 25 years ago. Only in Octopus, as a by-product of "brain" research (see Chapter 25), do we have a clear example of molluscan endocrinology in the framework of a much simpler sexual function. M. J. Wells' work has clearly shown that the optic glands in Octopus are endocrine organs synthesizing a hormone which stimulates egg production and is specifically required for the protein synthesis of egg yolk. Blinding can bring about precocious sexual maturity in octopuses preceded by hypertrophy of these glands.

On the other hand, we are in a position to draw some conclusions regarding the adaptive value of protandry. A great deal of our molluscan growth studies, concerned with turnover of carbon and nitrogen in growth and reproduction, has emphasized the high energetic cost of femaleness. In our various molluscs, as well as in many other invertebrates, the energetic cost of egg production may amount to more than half of the energy available per year for growth and all other purposes in a female snail (or in a hermaphrodite acting as a female), while the "cost" of production of sperm may amount to only about 1% of the bioenergetic turnover of males of the same species. Although other theoretical biologists concerned with the evolution of sexual dimorphism have emphasized the advantages of having larger numbers of smaller, more mobile males for the "best" gene-flow in species-populations, it seems to us that protandry, like the adaptive value of sexual dimorphism in dioecious forms, is in-

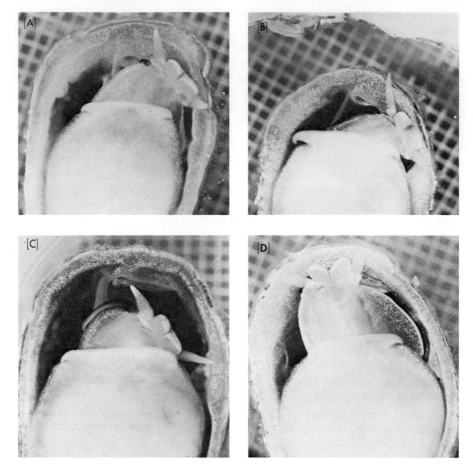

FIGURE 23-4.   Sex-change experiments on *Crepidula*. The condition of the penis (assessed on a scale **1** to **9**) proved to be a reliable external indicator of the state of the internal gonad during the natural shift immature → male → transitional → female. The penis is developed from the right-hand side of the slipper-limpet's head behind the right tentacle. These photographs, of experimental animals in isolation cells, are taken through the curved glass to which the sucker foot is attached and thus the penis appears to the left of the ventrally viewed head. **A:** Limpet with penis in condition **2**, an immature individual moving toward "maleness." **B:** Limpet with penis in condition **4**, a functional male individual. **C:** Limpet with penis in condition **5–6**, already starting into the transitional state. Viable sperm are no longer being produced internally, and the loss of black pigment from the penis occurs before it is much shortened by the resorption which follows. **D:** Limpet with penis in condition **9**. Resorption is almost entirely complete, and the individual is a functional female, already laying egg-masses. (Some experimental procedures can accelerate the sequence; others can "hold" limpets in condition **4** as functional males for unnaturally long periods by implants of neurosecretory material from other male limpets. The limpet in **A** had actually been a functional male, and was experimentally regressed by repeated implants.) For further discussion, see text. [Photos by the author.]

extricably linked to the higher bioenergetic cost of "femaleness." The two adaptive hypotheses are not necessarily mutually exclusive.

Mercenaria, the quahog of the Atlantic coast, is an example of a bivalve showing protandric consecutive sexuality. In this case, however, the position is a little more complicated. About 98% of the individuals in any population do show this simple protandry, being first males and then undergoing a single sex change and spending the remainder of their lives as females. However, a very small proportion of the population seems to consist of genuinely unisexual animals, both male and female.

A variation of this reproductive pattern is found in some forms where the change of sex is repeated, usually seasonally, with a more-or-less regular alternation from male to female. Such a pattern, termed rhythmical consecutive hermaphroditism, occurs in ptenoglossan prosobranch snails of the family Scalidae. It is perhaps best known in some oysters, particularly the European *Ostrea edulis* and the American species *O. lurida*, in which individuals first function as males, then as females, then as males again after each female spawning phase. The commoner American oyster, *Crassostrea virginica*, which (unlike the two species just mentioned) does not incubate its young, exhibits yet another pattern of sexuality which can be termed alternative sexuality. In it, the functional sex appears to be determined by water temperature and food supply, and the timing of the sex change is erratic. In *C. virginica*, as in *Mercenaria*, there are apparently a small proportion of true males and true females in some populations.

Some of the underlying causes of these extraordinary patterns of sex change are undoubtedly environmental. In these cases it is clear that all the individuals concerned are genetically hermaphroditic. In many cases it can be assumed that as regards their sexuality these forms have a relatively uniform genotype. However, in other forms, including the limpet *Patella* and perhaps such bivalves as *Mercenaria* and *Crassostrea*, there are populations with a variety of genotypes, most of which are essentially intersexual. This would imply the occurrence of multiple sets of sex-determining genes. Although some genetic analysis of this has been attempted, notably by G. Bacci of Naples, as yet there is no clear cytogenetic evidence for such unbalanced hermaphroditism involving a considerable variety of genotypes.

Hypotheses on the evolution of sexuality are worth discussing. In most many-celled animals it is generally accepted that dioecious forms are more primitive, and that hermaphroditic forms are secondarily derived. It is often obvious that hermaphroditism has a secondary adaptive significance, for example, in parasitic or freshwater animals. However, in the Mollusca, there is some evidence that the primitive state, in all the various groups of the phylum, was hermaphroditic. Whether the evidence for this is acceptable or not, it is clear that one of the more primitive patterns within many molluscan groups was that of hermaphroditism with a simple protandric sex change. Among longer-lived land and freshwater pulmonates, which we regard as showing true simultaneous hermaphroditism, a few cases are claimed of juvenile animals being ready to act as males at an earlier stage in their life-cycle. After further growth, they then turn to simultaneously ripening eggs and sperm, which combined activity continues for the rest of their lives. In our survey of conditions in more typical freshwater pulmonates, we have, with one exception, found truly simultaneous hermaphroditism. The exception is the freshwater limpet, *Laevapex*, in which Robert F. McMahon and I have shown functional protandry. However, in this species protandry has immediate adaptive advantages and probably does not represent any vestigial retention of a primitive pattern in this limpet, which is one of the more specialized of higher limnic Basommatophora (see Chapter 21). As I have stated elsewhere, protandric hermaphroditism in bivalves and gastropods is associated with variable size at sexual maturity, with capacity for degrowth in severe conditions, and with a lack of endogenous senescence. These features are common to the majority of species in the two largest molluscan classes and reflect a plasticity of growth patterns and of life-cycle controls totally unlike conditions in arthropods and higher vertebrates.

24 INVERTEBRATE ZENITH: THE CEPHALOPODS

THE third major group in the phylum Mollusca, class Cephalopoda or Siphonopoda, including such forms as cuttlefish, octopus, and squids, has an indisputable claim to encompass not only the most highly organized molluscs but also the most highly organized of all invertebrate animals. All marine and all dioecious, and almost all predaceous carnivores, the group includes many species which show as complicated patterns of innate behavior as some birds and a capacity to learn or be trained exceeded only by certain groups of higher vertebrates. Most swim by jet-propulsion at speeds in excess of those attained by any other invertebrate, and the largest living nonvertebrate animals are members of this class. The cephalopods—large, fast-moving, "brainy" animals—are notably nonmolluscan, or unsluggish, in their behavior. However, they are clearly constructed on the basic molluscan plan of structure and function. Among themselves they are much more stereotyped than the rest of the molluscs in their general anatomy: a successful pattern of animal machine successfully exploited.

By the chances of evolutionary history, the cephalopods are not as ecologically successful as are the more simply organized molluscs. There are probably only about 550 living species of cephalopods, as against over 74,000 snails and over 30,000 bivalves. However, there are at least 9000 known fossil species of cephalopods, some of which were among the commonest marine animals of their times.

433

Before we discuss structure and function in a "typical" cephalopod, it is worth introducing four generic "types" which characterize four ways of life: *Octopus*, *Loligo*, *Sepia*, and *Nautilus*. All four show the diagnostic characters of the class, having the head modified into a circle of tentacles and the upper and anterior parts of the foot modified during development into a funnel whence pass exhalant jets from the mantle-cavity for locomotion. Implicit in this description are a criticism of the usual group name, Cephalopoda, as a misnomer and a plea for the more correct but little-used group name Siphonopoda. Forms

[A]

[B]

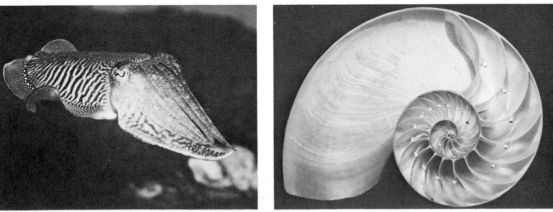

[C]

FIGURE 24-1. Characteristic cephalopod molluscs. **A:** *Octopus vulgaris*. **B:** A typical cuttlefish, *Sepia officinalis*, "resting" or "hovering" in midwater. **C:** A "half-shell" of *Nautilus*, ground to show the shell chambers with the "necks" for the siphuncular tube in every septum. The living animal occupies only the largest chamber and the rest are gas-filled and serve as a buoyancy mechanism. [Photos **A** and **B** © Douglas P. Wilson; photo **C** of specimen prepared and photographed by the author.]

like *Octopus* have eight arms equal in length, usually crawl on the sea bottom, and, although they can swim by jet-propulsion, rarely do so (Figure 24-1). The squids are animals like *Loligo* with ten arms, two of which are long and more retractile than the others, and are fast swimmers in the open waters of the sea. Cuttlefish such as *Sepia* have ten arms like squids but a shorter, less elongate body. They swim inshore, near the bottom, occasionally resting on it (Figure 24-1). The genus *Nautilus* comprises the only living species with an external chambered shell and numerous cephalic tentacles. Only the outer chamber of the coiled, divided shell is occupied by the animal, the rest being gas-filled and serving as a buoyancy organ.

Functional Organization

In the functional morphology of cephalopods, there are four principal differences from the rest of the molluscs. First, although completely bilaterally symmetrical, they have their orientation changed from the archetypic pattern considered earlier. The anterioposterior axis of the head-foot has swung into line with the dorsoventral axis of the visceral mass (Figure 24-2), so that the head lies in the same line with the elongate visceral mass, while the foot becomes the exhalant funnel of the mantle-cavity. Secondly, the importance of cilia is

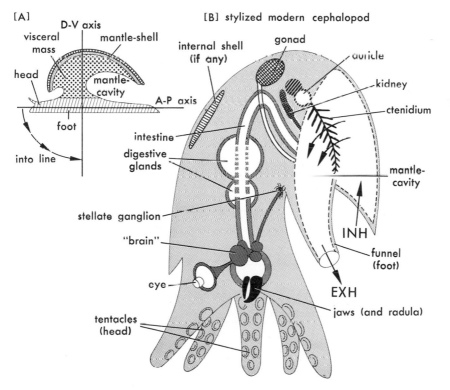

FIGURE 24-2. Functional organization in cephalopods. **A:** The reorientation of the axes of the archetypic mollusc in cephalopods. **B:** Stylized view of a "modern" cephalopod split in the midline. For further discussion, see text.

greatly reduced, the cilia of the mantle-cavity and ctenidia being functionally replaced by powerful muscles which carry out "pumping" for the respiratory current and for locomotion by jet-propulsion. Internally, the importance of cilia in the gut structures is also reduced. Thirdly, in all modern cephalopods except *Nautilus*, the shell is either an internal structure of little skeletal significance or, as in the octopods, entirely absent. Thus, unlike all other molluscs, including their Palaeozoic predecessors, modern cephalopods have no shell to retreat into. Their greatly concentrated and complexified nervous system is undoubtedly associated with a different and more active method of responding to changes in environmental circumstances. Fourthly, as can be seen from Figure 24-3A,B, the water circulation in the mantle-cavity is reversed from the usual molluscan pattern. It is, of course, brought about by muscular action, rhythmic contractions of the mantle-cavity wall creating the water current propelled through the siphon.

It is noteworthy that the ctenidium of a typical cephalopod (Figure 24-3B) is recognizably homologous with, and derived from, the aspidobranch type of ctenidium found in primitive members of the other molluscan groups. As a consequence, although skeletal elements are now provided at the new leading edges of the lamina, the blood circulation through the gill lamellae is still in the same direction. Thus the paradox arises: that in the most active molluscs the efficiency of a counterflow system in the respiratory structures is sacrificed to the replacement of ciliary propulsion by muscular pumping. On the other hand, a simple dissection of any cephalopod (or a glance at Figure 24-2B) will show that the anus and the openings of the excretory and reproductive systems, while anatomically at a position different from that in all other Mollusca, are functionally in the same relation to the ctenidium and to mantle circulation; that is, they are on the exhalant side of the gills. In spite of the loss of the counterflow system, there is a much greater rate of oxygen uptake by these ctenidia, and this, combined with the faster and more efficient water currents

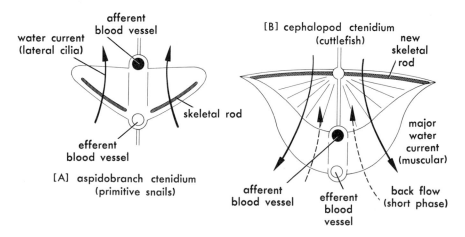

FIGURE 24-3. Ctenidial organization in cephalopods. Comparison of the ctenidial arrangement in primitive snails (**A**) with that in cephalopods (**B**). Note details in Figure 19-4; for further discussion, see text.

through the mantle, ensures that the high respiratory needs of an active cephalopod are met.

Much of the success of cephalopods as predaceous carnivores comes from the speed and efficiency of their locomotion. By jet-propulsion through their funnels, squids can move in any direction, and they are much more difficult to net at sea than are many fishes. There is some variation in squids and cuttlefish in the arrangements of valves which control the directions of the water current but most of them have connectors, like snap fasteners or poppers, which clip the sides of the mantle to the head. These domed fasteners hold the funnel and mantle together and an internal extension of the funnel forms a collar-valve and prevents the exit of water laterally during the exhalant phase of the pallial movements. Thus all the water is expelled through the funnel by contraction of the circular muscles of the mantle. Antagonists to these are provided by the radial muscle fibers in the pallial wall whose contraction causes that wall to thin out and therefore swell like a balloon, thus increasing the capacity of the mantle-cavity and pulling water in on either side and below the funnel (Figure 24-4). The initial contraction of the circular muscles in the exhalant phase brings about closure of the valve system again and so the main expulsion is through the jet of the funnel.

In all living cephalopods except *Nautilus*, there are two ctenidia similar to those in the most primitive zygobranch gastropods. The efferent branchial vessels from these two gills are connected through two auricles to a single, central ventricle of the heart (Figure 24-5). This sends blood through the arterial system to the organ systems of the body. In these modern cephalopods the blood returns through a better-defined system of veins which unite at the bases of the gills, whence accessory pumps (sometimes called branchial hearts) send the blood into the gills through the afferent branchial vessels (Figure 24-6). Cephalopods also have a truly closed circulatory system (which is lacking in the rest of the molluscs) with capillaries between the arteries and veins. The interposition of a branchial supplementary pump system and the existence of capillaries are the anatomical bases for a much more efficient and higher-pressure blood system than is found elsewhere in the molluscs. Despite this mechanical efficiency, the blood itself remains typically molluscan, with the relatively low oxygen-carrying capacity provided by the respiratory pigment haemocyanin in solution as a functional liability. In fact, the blood of fishes with the pigment haemoglobin in corpuscles can carry from four to five times as much oxygen per unit volume of blood as squid blood can carry (although squid oxygen transport is at least twice as efficient as that of any other invertebrate). Another liability is a complete lack of the oxygen storage in venous blood and in muscle tissues (the latter in myoglobins) which occurs in all vertebrates. Cephalopods are all extremely sensitive to oxygen lack, and cannot survive in those inshore or estuarine habitats which involve periodic reductions of oxygen tension to below saturation values. *Nautilus* is exceptional in both pallial and circulatory structures. It has four ctenidia which drain through four auricles to the central heart and has no accessory pumps. (This symmetry is maintained in other pallial structures of the complex, there being four osphradia and four excretory organs, although there are only two openings from the pericardium direct to the exterior.)

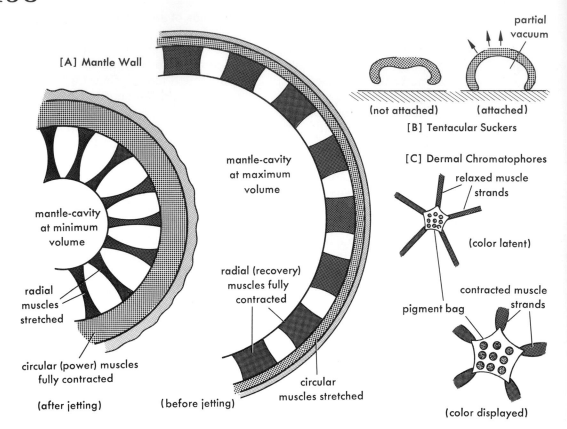

FIGURE 24-4. Muscle effectors in cephalopods. **A:** The propulsive jet is generated by contraction of the circular muscles of the mantle wall. Recovery, before the next jet, results from contraction of the radial elements, causing the wall to thin and "balloon" out, increasing the volume of the mantle-cavity, and drawing in more water. The diagram is oversimplified in showing the radials and circulars as separate muscle layers; regular bands of radials run *between* the major annuli of circulars, and the combined muscle layer is held together by a regular meshwork of nonelastic collagen fibers. **B:** Suckers (which may be sessile or stalked, and unsupported or bearing a chitinous ring at the opening) all function by muscles which enlarge the cup after attachment, thus creating a partial vacuum. **C:** The neurally controlled chromatophores of cephalopods are bags containing various colored pigments. Display of the color is accomplished by contraction of the chromatophore muscle strands, causing an "expansion" of the chromatophore bag.

As was probably the case in the great bulk of cephalopods known only as fossils, *Nautilus* uses its chambered shell as a buoyancy structure. It is only recently that the physiological mechanisms involved in this have been investigated. The chambers are formed initially fluid-filled. The fluid is absorbed by the epithelium of the siphuncular tissue, and gases diffuse in. Healthy animals achieve near to neutral buoyancy and, although the gas pressures are a little below atmospheric, the shell is strong enough to resist water pressures at depths of 600 meters (about 60 atmospheres).

It is not usually realized that the internal shell vestige found in cuttlefish such as *Sepia* is also used as an organ of buoyancy. This "cuttlebone" consists

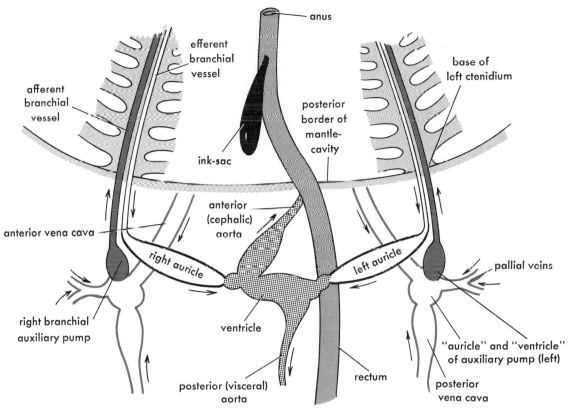

FIGURE 24-5. Diagrammatic ventral view of the heart and associated structures in a typical cephalopod (basically cuttlefish). The system is diotocardiac, the two auricles receiving oxygenated blood from the efferent branchial vessels of the pair of ctenidia (see Figure 24-6). Deoxygenated blood is pumped into the afferent branchial vessels of the gills by two auxiliary pumps (sometimes referred to as branchial hearts) which receive that blood from the body organs by way of a true venous system. The position of the rectum and its associated ink-sac is also shown.

of a number of thin chambers laid down one after another at intervals of a few days as the animal grows, forming a laminated structure, the spaces in which are little less than a millimeter in thickness. In life most of this is gas-filled, principally with nitrogen, and its average pressure is maintained at about 0.8 atmosphere by the animal balancing the hydrostatic pressure of the sea (which would tend to force water into these gas spaces) by a reduced osmotic pressure, creating an equalizing difference between the blood and the liquid in the ends of the interstices of the cuttlebone. Thus, in cuttlefish obtained at any depth in the sea, the cuttlebone liquid is markedly hypotonic to the rest of the animal's blood. The shell in squids is reduced to an elongate, light, chitinized "pen" which lies like an axial skeleton in the body, though it cannot have any great significance in muscle attachment. One small decapod cephalopod—the oceanic form *Spirula*—retains internally a miniature, coiled, chambered shell. Nothing is known of its physiology. Finally, the octopods have, as already mentioned, totally lost the shell.

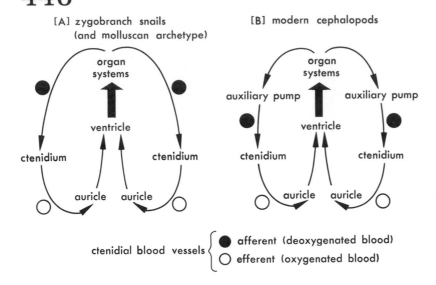

[A] zygobranch snails
(and molluscan archetype)

[B] modern cephalopods

FIGURE 24-6. Pattern of blood circulation in primitive snails (and molluscan archetype) (**A**) compared with that in "modern" cephalopods (**B**).

The suckers on the arms vary in form, but all function in the same way, a partial vacuum being created by muscular enlargement of the cup (Figure 24-4). In feeding, the prey is wrapped around by the arms and is pressed in toward the mouth which is armed both with a chitinous beak like a parrot and with a powerful radular apparatus. In some forms, including *Loligo*, the salivary glands secrete poison, and in all forms powerful proteases are produced which can be passed externally, so that in some cases the food is partly digested before actual ingestion. There are several internal glands secreting powerful enzymes, and the digestion is as efficient in terms of time as that in a higher carnivorous vertebrate such as a hawk. Absorption and utilization are also startlingly efficient and speedy. It is correct to talk about cephalopods as having a higher rate of metabolic turnover than any other invertebrates except some flying insects. Surprisingly, the gut is still based on the molluscan pattern, though muscles replace cilia as a means of moving food through it. However, the posterior part of the stomach still has ciliated leaflets which remove indigestible solid fragments and sort them toward the intestine for eventual expulsion.

Behavior

As a diverticulum of the posterior part of the gut most modern cephalopods have an ink-sac which opens to the exterior through, or beside, the anus in the exhalant part of the mantle-cavity. The true cuttlefish have the largest ink-sacs and can use them to lay down protective smokescreens as well as to produce smaller puffs. Nearly all modern cephalopods are able to produce compact puffs of ink which can act as a sort of *Doppelgänger* to disappoint potential predators. One small species of *Sepiola* reacts to danger by first darkening itself with its chromatophores, then emitting a compact puff of ink, then rapidly blanching its chromatophores and jetting away undetected by the predator, who gets a mouthful of ink. Note that this sequence of events can take place in less than a second.

The chromatophores of cephalopods are unusual in being bags of pigment which are expanded by muscles under nervous control (and contracted again by the elasticity of the bag—Figure 24-4). This means that rapid color changes are possible, unlike the cases of most other invertebrates where color cells are under hormonal control and can respond only relatively slowly to changed circumstances. Complete color changes are possible in less than a second and a delicacy of pattern control is achieved which is unequalled in the animal kingdom. It reaches its greatest development in bottom-dwelling forms like *Sepia*, which if placed on a submerged checkerboard is said to be able to give a creditable imitation of its background. Many other cuttlefish also display terrorizing, conspicuous patterns, including the temporary appearance of large black spots like huge eyes on the back. The direct nervous control also allows flickering patterns to be developed for purposes of concealment or, in some cases, for sexual display.

Many deep-sea species, and a few from inshore, have luminescent organs. In some the light is produced by glandular secretions of the animal itself, but in a few forms luminescent bacteria live as symbionts in special ducts or sacs. In some cases there are elaborate accessory structures (as in a few deep-sea fishes) with lenses, reflectors, and pigment screens which allow color changes in the emitted light under nervous control. Little or nothing is known of the adaptive significance of these elaborate light-producing organs—found especially in migrating profundal squids and cuttlefish—although various functions have been suggested, including species recognition, courtship display, food attraction, or predator repulsion.

All cephalopods reproduce by medium-sized eggs which contain sufficient food material for the young to hatch as miniature adults (Figure 24-8) with no intervening larval stages. In some, there are basic rituals of parental care of the eggs, though *never* of the young after hatching. (Some consequences of this are discussed in Chapter 25.) For example, some female octopods return to the egg-mass regularly to wash it with water from the funnel and to wipe each egg with the tips of the tentacles. Early-hatching young would simply be washed away. Otherwise, juvenile cephalopods are completely disregarded by their parents. The majority of species of cephalopods have elaborate courtship behavior, involving color changes and tactile caresses. In all, the male produces sperm in a packet called a spermatophore, and conveys it by an arm called the hectocotylus (which may be modified) from the male genital duct to the opening of the female duct. In *Loligo*, there are said to be a variety of alternative copulatory positions. In most cephalopods, the hectocotylus arm is thrust through the funnel of the female and the spermatophore pressed into the genital opening near the gill base. In a few species of octopods, the elaborate hectocotylus becomes detached from the male at a late stage of courtship, proceeds into the mantle-cavity of the female, moves about for some time by continued rhythmic contractions of its muscles, and then attaches by its suckers near the genital pore and later presses in the spermatophore.

Much of the behavior discussed above, and the learning capabilities to be surveyed in Chapter 25, is made possible only by the modified nervous system of all cephalopods. Ganglia of the basic molluscan categories are fused into a mass as a brain linked by two huge pallial nerves to the stellate ganglia, from which are innervated the circular and radial muscles of the pallial wall. Control of water-pumping and therefore of locomotion involves the giant fibers whose conducting rate can be as great as 20 meters per second, and whose diameters can reach 1 millimeter. We already met somewhat similar giant nerve axons in annelid worms (Chapter 11), again associated with fast "escape" responses. The first complete description of the cephalopod giant nerve fibers and the earliest electrophysiological work upon them were by J. Z. Young (whose investigations on brain function in *Octopus* we will discuss in Chapter 25). For the forty years since the initial discovery of their value, the largest of the motor axons in squid have provided superb material to biophysicists, biochemists, pharmacologists, and electrophysiologists; and much of our basic knowledge of the membrane chemistry and the electrical events associated with conduction in nerves has been derived from squid. It is the adaptational value to the cephalopod jet-propulsion mechanism that concerns us here, however. As already noted, exhalant water propulsion is created by a powerfull simultaneous contraction of the circular muscles of the mantle wall (antagonized mechanically by the smaller radial fibers which bring the circular muscles into their stretched precontraction state by thinning out the mantle wall and thus ballooning out the whole mantle-cavity in preparation for the "jet"). There are three sets of giant fibers in squid and similar forms (Figure 24-7). A pair of first-order giant cells are placed on either side of the brain in the posterior suboesophageal complex (corresponding to the pedal ganglia of other

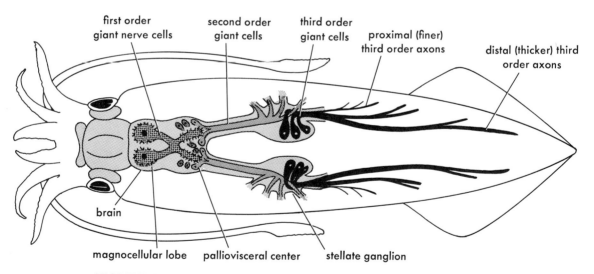

FIGURE 24-7. Stylized diagram of the giant axon complex in the squid. Many parts of the squid nervous system are omitted, and some are distorted and not to scale. The complex involves two sets of centers in the brain linked by the pallial nerves to the paired stellate ganglia, and then a radiating neural pattern which supplies motor nerves (some giant), not only to the mantle-wall muscles concerned in jet-propulsion, but also to the retractor and postural muscles of both the head and the siphonal funnel. Note that the first-order giant nerve cells show an unusual cross connection, and that the third-order axons are arranged so that motor impulses will reach all parts of the mantle-wall musculature simultaneously. Neural conduction speeds are proportional to the diameters of the axons involved, and here the fibers going from the stellate ganglia to the more distant parts of the mantle are considerably thicker than the fibers to closer muscles. [Modified from Young, 1938, 1939, and later publications.]

Mollusca—see Chapter 25). These respond to appropriate sensory inputs such as visual perception of danger by an output to their synapses with a number of second-order giant cells whose fibers not only innervate the muscles orientating the funnel but synapse with the third set of giant cell bodies within the stellate ganglia. From these motor control centers, the third-order giant fibers pass to all parts of the circular pallial muscles. These fibers are differentially graded in their diameters; the shortest running to the anterior parts of the mantle wall being finest, the longer fibers running to the posterior mantle having the greatest cross-sectional area. Conduction rate being proportional, the "firing command" from the stellate ganglia is received simultaneously by all parts of the circular muscles. The powerful propulsive jet through the funnel is caused by the resulting simultaneity of contraction. The finely adjusted angular orientation of the funnel allows as great a variety of movements in a cuttlefish (back, forward, up, down, horizontal spin, etc.) as do the small maneuvering nitrogen-jets of an Apollo command-service-module in space (Figure 24-8). Repeated fast contractions can take a squid up to speeds of 20 knots (usually backwards), or a slower downward-directed jet like that of a hovercraft can hold a squid nearly motionless in midwater.

Both high-speed and finely controlled maneuvering must depend upon a continuous input of information about the surrounding environment from the sense-organs to the large brain. In general, the brains of octopods and cuttlefish are even more massive than those of squids. J. Z. Young has calculated the number of cells in the whole nervous system of a medium-sized *Octopus* to be 300 million, with the number in the "higher" centers (see Chapter 25) of the brain at 168 million. As in higher vertebrates with large brains, the greatest numbers of cells are not those connected to motor fibers but rather those concerned in sensory integration and in memory functions. By invertebrate standards, cephalopods have enormous brains. The sense-organs are also impressively complex and magnificently efficient. The statocysts are capable of detecting direction and angle of acceleration as well as static posture (thus resembling the vertebrate labyrinth of the inner ear), and they are, structurally at

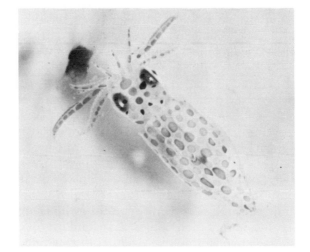

FIGURE 24-8. Newly hatched squid. This lively young specimen of the common Atlantic long-finned squid, *Loligo pealei,* has been hatched from the egg for about three hours, is trying out its chromatophores in response to a previous photographic flash, and has just ejected its first puff of ink. [Photo by the author.]

least, potentially sensitive to sounds. There are olfactory pits on the head, and chemotactile as well as mechanotactile sensillae on various parts of the arms and suckers. The eyes are, of course, the most striking sense-organs. They have often been compared to mammalian eyes, and cited as an outstanding example of convergent evolution. Except for *Nautilus*, all cephalopods have eyes with a cornea and iris diaphragm, a movable lens, and a retina. The last is presumably a little more efficient than that in the vertebrate eye, since it is not developmentally inverted, and thus the fibers of the optic nerve are collected together outside the eyeball. The lens of the eye is a complex structure which has a focal length of about two-and-a-half times its radius, although it is nearly spherical in form. This implies that the outer parts of the lens have a lower refractive index than the central regions, and in this cephalopods resemble fishes. The packing density of the retinal cells is similar to that in the higher vertebrates, and pigment movements allow light- and dark-adaptation. There are behavioral indications that cephalopods have color vision. The eyes are certainly the dominant sense-organs in most cephalopod behavior patterns. Experiments involving training have shown that they are capable of considerable image-formation and shape-discrimination. Octopus and cuttlefish are proving to be exceptionally useful experimental animals for neurophysiologists and physiological psychologists interested in localization of both innate and learned components in behavior. It may well be that the most significant advances in our future understanding of the cellular processes connected with "memory," and of the relation between optical discrimination of shape in the retina and the visual analyzing systems of the brain, will come from work presently being carried out on cephalopods (Chapter 25).

Survey of Types

The subclassification of the class Cephalopoda (Siphonopoda) is confused and a subject of controversy, not only between palaeontologists and zoologists but also among representatives within each group. However, a few main divisions must be mentioned. *Nautilus* is usually placed in a separate subclass, the Tetrabranchia, along with a number of fossil forms, which are assumed to be similar in having a four-fold arrangement of pallial structures and auricles. The subclass Ammonoidea is entirely fossil, but was an enormous and important group from the late Silurian to the end of the Mesozoic. They were animals with shell structures similar to the nautiloids (Figure 24-9), but the nature of their soft parts is unknown. The remaining cephalopods are usually placed in the subclass Dibranchia or Coleoidea, having two ctenida and two accessory circulatory pumps, an ink-sac, well-developed eyes, and a huge brain. There is no agreement among the authorities on the orders into which this subclass should be divided. However, apart from fossil forms, there are at least five main divisions of presently living forms. One of these encompasses the cuttlefish, not only inshore forms like *Sepia*, but also pelagic oceanic ones such as *Spirula*, and the small burrowing sepiolids which have large, rounded fins like wings on the sides of their bodies. A second group contains the squids, including besides *Loligo*, forms like *Architeuthis* which are probably the largest invertebrates. Specimens of over 18 meters in length have been recovered, and parts have been found which could be derived from specimens 42 meters long. These giant squids occur in association with toothed whales in a complex predator–prey relationship in which, although neither can be a major food resource for the other, each, while still growing, seems to be the *preferred* prey of the other as an adult. This is not without significance to the evolution of squids as a group and of some generic lines among cuttlefish. The outstanding efficiency reached by the modified mantle-cavity as an organ of jet-propulsion in all the pelagic cephalopods arose when that prime mover was in competition with the myotome muscle blocks (see Chapter 31) of those chordates which contemporaneously swam in the sea: various fishes, predaceous Mesozoic reptiles, and, finally, the dolphins and toothed whales. Size and speed ratios seem to have been kept in pace by selection pressures of alternate predation and competition in the evolution of pelagic cephalopods along with the swimming vertebrates. For example, both flying-fish and certain squid can briefly become aerial, leaving the water surface at speeds of about 20 knots in escape reactions. Without the pressure to "keep up with" the competitive vertebrates, the molluscan body plan would probably not have evolved the speed, the size, and the brains to become our "invertebrate zenith."

There are two very different orders of octopods (see Table 19-1), and they exhibit at least four very distinctive life-styles. First, forms like *Octopus* and *Eledone* are inshore bottom-dwelling—almost cave-dwelling—cephalopods in which the arms are locomotory and exploratory as well as organs of attack and feeding. Secondly, there are also octopods which float on the surface of the sea like *Argonauta* where the female makes a shell-like container for her eggs. Thirdly, there are deep-swimming bathypelagic octopods with transparent ge-

FIGURE 24-9. A typical ammonite. A nearly complete fossil specimen (11 centimeters across) of *Euagassiceras rhotomagensis* from Lower Jurassic (Lias) beds in the Teutoburger Wald, Germany. [Photo by the author, see caption for Figure 16-10.]

latinous mantles. Finally, there are some abyssal octopods living on the bottom at great depths, a few of which are blind. The fifth order of the Dibranchia is the order Vampyromorpha (see Table 19-1), the specialized, deep-water vampire-squids, black-pigmented, swimming like bats using a web between the arms, and having a number of characters intermediate between both octopod groups and the squids.

Although there are sound reasons for claiming the cephalopod molluscs as the most highly organized invertebrate animals, ecologically they are limited to the sea. The functional efficiencies of jet-propulsion and of massive brains have not been parallelled in other physiological systems, such as the excretory organs and the respiratory functions of the blood. These could not be readily adapted for life in estuaries or in fresh waters. Similarly, the lack of a rigid skeleton and the delicacy of cephalopod skin layers effectively prevent the evolution of successful terrestrial forms. Another significant feature is that the reproduction of most cephalopods involves large numbers of only moderate sized eggs. These are not truly cleidoic (closed-box) eggs, since the cephalopod embryos normally take up considerable quantities of inorganic salts from the sea in the course of their prehatching development. Another consequence of the numerical fecundity of cephalopods (a female *Octopus* can produce up to 150,000 eggs; a female squid of the genus *Loligo*, 70,000) is marked instability of their population dynamics. Natural populations of both octopods and squid are known to show exceptional abundance (to "plague" proportions from the viewpoint of clam- and crab-fishermen) in certain years, followed by periods of catastrophic decline. The fact that this fecundity has also prevented the development of parental care, with consequences to certain brain functions, will be noted in Chapter 25. In summary, the restriction of these "highest" invertebrates to the sea reflects real limitations imposed by cephalopod anatomy and physiology.

25 TOWARD PERCEPTION, MEMORY, AND LEARNING

THE last two decades have seen a flowering of molecular biology, following the elucidation of the molecular processes underlying genetic mechanisms. A similar fundamental advance in our understanding of memory and of "mind" could follow in this last quarter of the twentieth century, if appropriate discoveries are made about the physicochemical nature of complex processes in "higher" nervous systems. If this prediction comes true, the cephalopod molluscs will have played an important part. The sheer potential of this research area for the future of mankind justifies the attempt to summarize briefly here certain extensive but admittedly inconclusive investigations.

Even at present, experimental work by neurophysiologists and physiological psychologists using octopus and cuttlefish can already be correlated with certain deductions by neurosurgeons treating brain injuries in man, and with certain hypotheses developed from investigations on cybernetics and the development of "machine intelligence." Our future understanding of the cellular processes connected with learning will certainly be partially based on the cephalopod brain, and a further extension to a molecular level of explanation may involve neuropharmacological studies on other molluscan systems. There can be no more critical application of reductionist methodology than to fundamental questions on the working of brains. Molluscan studies already have a place of paramount importance in this scientific endeavor.

447

Biology, as part of modern science, involves the systematic organization of knowledge about living organisms by the development of explanatory hypotheses which are genuinely testable. Almost all modern science is reductionist in its interpretation of natural phenomena. It is thus based on only one of the dichotomous systems of thought which, in the Western world, we generally attribute to two Greek philosophers, Aristotle and Democritus. The latter's principles of atomism and causality lie at the basis of modern scientific thought. In contrast, Aristotle's holism and finalism as a system became a fundamental strand in the religious philosophy of the Middle Ages. Thus, despite the fact that Aristotle was himself a fine biologist and the founder of comparative zoology, the scientific revolution which began in the Renaissance (and continued in biology by way of Darwinian natural selection through molecular genetics) was essentially opposed to scholastic-medieval Aristotelianism. Unfortunately, this historical polarization has resulted in the views of Democritus and Aristotle on the phenomena of living organization being represented, not so much as a reductionist *versus* a holistic set of interpretations, as a conflict between a mechanistic and a vitalistic interpretation of these phenomena. Modern biology, adopting the sound scientific attitude, is reductionist in methodology and seeks mechanistic explanations.

As often discussed, there are at least three levels of the evolution of biological organization at which reductionist explanations are very difficult to apply. The first of these concerns the emergence of self-replicating living systems from previously inanimate macromolecules. Despite the recent advances of molecular biology there are still baffling problems in the origins of the genetic code and in the genesis of self-replicating life. A second set of problems is associated with the emergence of certain properties (which, for convenience, we will call consciousness) in the neural organization of certain higher animals. The third area of difficulty is, of course, the peculiarly human one of the completeness or noncompleteness of parallelism between mental states and the physical states of brain tissues or, crudely, mind-brain parallelism. For many humans, one or more of these "difficult" levels seem to necessitate a mystical (usually, but not always, religious) explanation rather than a mechanistic one. Despite these difficulties, scientific understanding of brain function can proceed only by the development of testable hypotheses from reductionist explanations.

The exceptional value of octopus and cuttlefish as experimental animals in brain research stems directly from their trainability, in a functional sense. In turn, this depends on the size of their brains, which are relatively large by invertebrate standards. When we discussed the functioning of the insect central nervous system in Chapter 17, we noted that insect brains were small and that the behavior of insects—although apparently elaborate—is relatively stereotyped and predictable. Much of the complex behavior of insects seems to be programmed into the central nervous system—that is, is genetically innate, and the learning abilities of insects are correspondingly limited. We noted that the central nervous systems of all animals are made up of neurons of similar dimensions. Thus there are very many more of them in the brain of an octopus or of a rat, or of a man, than in even the largest insect brain. It is a simple consequence of the changing surface:mass relationships with increasing size of animals that larger animals (even if they do not have proportionately larger brains) will have larger numbers of "association" neurons. It is fairly obvious that the number of sensory nerve cells connected to various receptors, and the number of motor nerve cells connected to effector organs such as muscles and glands, will increase in proportion to the surfaces of an animal, whereas the other nerve cells not connected to peripheral structures will increase with its volume. Thus in two similarly constructed animals, with exactly proportional brain size, if one species is twice the linear dimensions of the other, there will be approximately four times the number of sensory and motor neurons, but eight times the total number of neurons in the larger animal. The additional "spare" neurons (neither specifically sensory nor connected as motor) can form, with increasing brain size, more extensive "association" centers. More than 70% of the 300 million neurons in Octopus, and a higher percentage of the 12 billion in man, are potentially "association" neurons. As regards man, it may well be that possession of such huge numbers of nerve cells with their enormous numbers of dendritic branches (involving perhaps one thousand "contacts" per cell), and the mind-boggling numbers of potential interconnections between them, are necessary for the evolution of a mind capable of being boggled. This is not all that is required (see page 467).

The cephalopods are not only relatively large molluscs, but the brain:body ratio in such forms as cuttlefish and octopus is proportionately greater. Thus it is that, of all the diversity of invertebrates, these few cephalopod species have most to contribute to a Democritean, or better a Cartesian (for René Descartes, the great French philosopher of the seventeenth century), understanding of the processes by which we think about the invertebrates and about matters of lesser moment.

The Experimental Octopus

As noted in Chapter 24, it is a characteristic of cephalopod evolution that the lining up of the two axes of symmetry of the molluscan archetype has brought the head-foot and the pallial organs into close association at the front end of these mobile animals. In contrast to their condition in the rest of the Mollusca, the nervous control centers concerned with these three groups of organs must, even in the earliest cephalopods, have been closely placed. Even in Nautilus, we already have a "brain" as distinct from the widely separated ganglia found in primitive gastropods and in primitive bivalves. Nautilus still has visceral ganglia placed at some distance from the fused masses of the brain, which consist of a still recognizably molluscan assemblage with a fused cerebral mass above the oesophagus and an anterior pedal band and a posterior pleural one below. This neural concentration is rather imperfectly protected in Nautilus by annuli of cartilage forming a sort of set of "roll-bars."

In all other living cephalopods the brain not only is much more concentrated and "nonmolluscan" but also is more completely enclosed in a cartilaginous capsule or cranium (analogous to the cranial skull of vertebrates). Figures 25-1 and 25-2 show the main parts of the brain in Octopus. Connected to the supraoesophageal complex, itself made up of eight pairs of centers, are the enor-

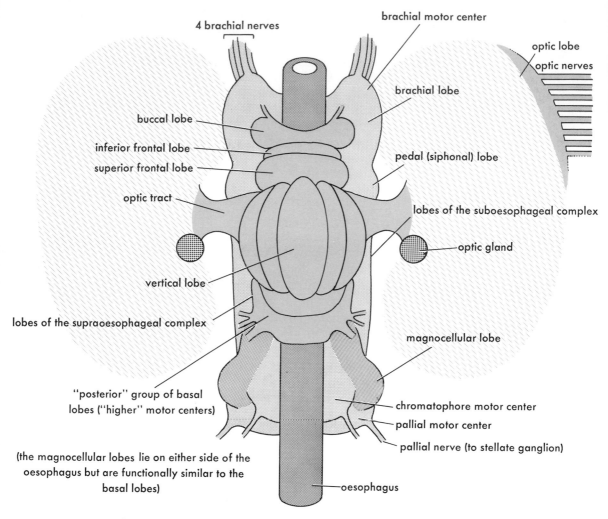

FIGURE 25-1. Diagrammatic dorsal view of the brain in *Octopus*. The upper lobes of the supraoesophageal complex correspond to the cerebral and buccal ganglia of other molluscs, while the suboesophageal complex is made up of (anteriorly) the pedal and (posteriorly) the fused pleurovisceral ganglia. There are now fifteen structurally and functionally distinct pairs of lobes in the brain of *Octopus*, comprising motor (output control and reflex) centers (including the brachial and pallial), higher centers concerned with the integration of motor patterns and responses (including the various basals and the magnocellular), and finally the so-called silent areas (superior frontal, vertical, and subvertical lobes) which interact with the optic and other lobes as the association centers concerned with memory (and thus with learning capacity). Functional aspects of these three "levels" of brain centers are discussed in the text.

mous optic lobes. Massive lateral lobes, no longer mere connective nerves, link this upper brain in saddle fashion over the oesophagus to an even larger and somewhat elongate suboesophageal complex, made up of two huge pairs of neural centers and some smaller ones. Before we consider this brain in more detail, structurally and functionally, it is worth noting for the last time those

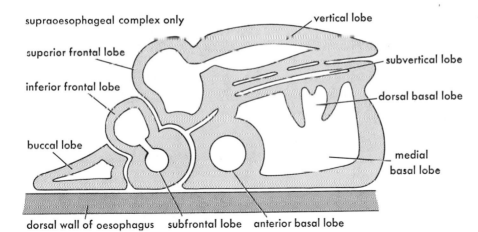

supraoesophageal complex only

vertical lobe

superior frontal lobe

subvertical lobe

inferior frontal lobe

dorsal basal lobe

buccal lobe

medial
basal lobe

dorsal wall of oesophagus   subfrontal lobe   anterior basal lobe

FIGURE 25-2. Diagrammatic sagittal section through the supraoesophageal complex of the brain of *Octopus*. Note that certain lateral lobes (including the magnocellular and optic tracts) which are part of this complex are not represented in this "midline" diagram. Compare with Figure 25-1.

basically molluscan features which still can be discerned. The upper, supraoesophageal region of the brain corresponds to the cerebral and parts of the buccal ganglia of the more primitive molluscan nervous system. The anterior part of the suboesophageal region corresponds to the pedal ganglia (and now supports the major motor and sensory connections both to the arms and to the siphon), while the posterior part corresponds to the fused pleurovisceral ganglia of other molluscs.

Each center or lobe of the brain in *Octopus*, and there are thirty such, shows a remarkable similarity to the cranial lobes of vertebrates, each being made up of a layer of nerve cell bodies or neurons with their nuclei, this layer enclosing a central mass of fibers, both dendritic and axonal. In many cases, each pair of structural lobes, distinguishable on anatomical and histological grounds, can be experimentally shown to be the functional center for a particular kind of central nervous activity. This degree of anatomical segregation of control functions, along with the fact that octopods (perhaps because they are inshore, bottom-living animals in nature) are remarkably tough experimental animals, has given rise to the fruitful and significant work of the last quarter-century, linking behavioral training to brain function in these animals. Originally developed at the famous marine Stazione Zoologica at Naples by J. Z. Young and his associates, notably B. B. Boycott, M. J. Wells, and N. S. Sutherland, this knowledge of the precise local site (in terms of individual nerve cells) of each kind of brain activity has grown until we know as much, if not more, about localized brain function in *Octopus* as we do in higher vertebrates including man.

The extensive and painstaking work of Young and his associates has shown that it is not entirely anthropocentric to regard the octopus brain as consisting of a hierarchical series of brain activities from the lowest semiautomatic reflex systems, which control the pumping of the mantle wall, situated in the posterior part of the suboesophageal complex, to the higher functions of memory and decision-making in the superior frontal and vertical lobes of the supraoe-sophageal complex. Apart from the more elaborate experiments involving be-havioral training which we will discuss later, the more basic work on brain function involves three techniques: direct electrical stimulation of specific parts, recording from specific parts by microelectrodes, and experimental sur-gical ablation of whole lobes or more specific lesions of neural tracts. The first of these three is a considerably cruder method of testing central nervous func-tion than are the other two, but a combination of the first and third can be very valuable in cases where recording from microelectrodes is impossible or hope-lessly complex in results. A fantastic amount of information about function can still be gained from detailed microanatomical studies, particularly selective staining of individual nerve cells and fibers by silver and similar techniques. We will now attempt a brief summary of the functional anatomy of the octopus brain, which cannot do justice to the vast extent of descriptive and experimen-tal work of Young and his followers. The book *A Model of the Brain* by J. Z. Young, along with his many research reports and his most penetrating Croon-ian lecture to the Royal Society, should be consulted by all interested students.

The posterior part of the suboesophageal complex is the origin of the two massive pallial trunks which run to the stellate ganglia on the inside of the mantle. As already noted, these contain the second-order giant neurons which in turn supply the third-order giant neurons to the effector muscles of the man-tle wall, both circular and radial. There are also, posteriorly, visceral lobes in which originate the nerves running to the gut and other internal organs. The anterior end of the suboesophageal complex sends four pairs of very large brachial nerves, of mixed sensory and motor function, one to each of the eight cephalic arms. (In the brains of cuttlefish and squid, of course, there are five pairs of such brachial nerves.) Also in the suboesophageal complex are lobes associated with the motor controls to the chromatophores on the mantle, and other centers concerned with the postural orientation of the funnel in jet-pro-pulsion, and with the effector action of the ink-sac. In an oversimplified gener-alization one could say that these suboesophageal centers, if locally stimulated, produce isolated sets of responses rather than the complete behavioral patterns of integrated muscular and other responses that characterize *Octopus* and the other higher cephalopods, which were exemplified in Chapter 24. Integration of these isolated control centers occurs in the next "higher" level of the octopus brain. In the control of more integrated behavior, the anterior and posterior basal lobes of the supraoesophageal complex become concerned. Here, crude direct stimuli from electrodes can produce coordinated responses. Stimulation of the posterior basal lobes produces coordinated sequences of swimming movements including those of the escape reactions involving the giant fiber systems. The sequence of events described for the small cuttlefish *Sepiola* in Chapter 24, with each event taking only a few milliseconds, would result from a similar overriding function of the posterior basal lobe in that species. In a similar fashion, stimuli applied to the anterior basal lobe can evoke the coor-dinated sequences of movements of head, tentacles, and those suckers used in prey capture and feeding. The paired buccal lobes are a third set of centers at this level of integration, being the origin of the motor nerves controlling the activities of the mouth musculature and thus the interacting movements of the beak and radula in the later flesh-tearing stages of feeding. A fourth set of centers at this level is the pair of lateral basal lobes, which are the higher motor centers integrating the motor responses for the particular areas of the lower suboesophageal complex which are concerned in turn with the control of the chromatophores. It should be noted that none of these higher motor centers has nerve cells whose motor axons run directly to any of the muscle systems involved: they are entirely integrating centers "driving" the appropriate suboesophageal motor centers from which the arm, siphonal, pallial wall, and chromatophore muscles are directly innervated. Surgical damage by ablation of, or lesions to, these basal integrating centers produces highly significant results. For example, isolation or loss of the left posterior basal lobe in a cuttlefish will cause the animal to swim in continuous circles to its right. Similarly, damage to the anterior basal lobe of one side of the brain in *Octopus* results in "circus" gropings of head and tentacles toward the undamaged side.

It is worth noting that two distinct kinds of neural integration in the lateral basal lobes are likely to be concerned in the more complex kinds of color changes in various cuttlefish which were mentioned in the last chapter. In the case of those species which could temporarily display dark patterns in the form of huge eyes on the back of the mantle as a conspicuous terrorizing pattern, the integration of this display probably involves a preset sequence of neural "wiring" in the lateral basal lobes. In other words, this would be a piece of innate behavior, genetically determined for the species, like some of the complex insect patterns discussed in Chapter 17. On the other hand, the imitation of a submerged checkerboard in forms like *Sepia* is not genetically predetermined, does not depend on an existing wiring pattern for its integration, but involves processes of neural learning. This will be more fully discussed in the next sections, but it is probable that in addition to the motor integration center for chromatophores provided by the lateral basal lobes, even "higher" centers of the cephalopod brain are involved in such imitative behavior.

Other lobes contain integrative centers at this intermediate level in the neural hierarchy, and these include the inferior frontal lobes which are concerned with the integration of the vast amounts of sensory information coming from tactile sensillae in the arms. The eight tentacles of a reasonably "alert" octopus move continually, rather like the fingers of a blind person sorting through the contents of a jumbled drawer. There are similar paired olfactory lobes attached to the optic stalks, which integrate the inputs from the chemoreceptors. Also at this level we have the enormous optic lobes themselves, which are involved in part in visual coordination, as we will discuss later, but which are also concerned both with the maintenance of general muscular tone and in the hormonal control of reproduction (see Chapter 23).

Finally we have a group of three brain regions which contain vast numbers of nerve cells but which have no direct sensory nerves as inputs nor do they originate any motor axons directly to effector organs. These are the superior frontal, vertical, and subvertical lobes, which have been termed the "silent areas." They are not even concerned with the integration of motor activities, since their removal causes no disturbance of the complicated activities of swimming and escaping or of seizing and ingesting prey. Extensive experiments on *Octopus* and *Sepia* have clearly shown that these three regions are the association centers concerned with memory and with learning. The vast number of neurons they contain, and the even vaster multiplicity of their interconnections, are responsible for the storage of sensory inputs from the past, or experience, which can be subsequently called upon to modify later behavior patterns. This is far removed from the patterned sequence of neural responses based on "wiring connections laid in" by the genetics of the species. It is not anthropocentric to refer to these as "higher" centers. They are, in fact, analogous to the cerebral cortex of higher vertebrates like humans. Like us and these other higher vertebrates, cuttlefish and octopods can learn from experience because they have information-storage capabilities, or memory, as a functional property of their brains.

Theories of Memory

A major difficulty for each of us humans in considering questions of memory and of learning is that each of us has conscious experience as a personal and uniquely private subjective experience: the so-called pure ego. This can create difficulties for a mechanistic and reductionist consideration of mnemic processes in animals and—even more obstructively—in man. Memory as we are discussing it here is not subjective, and we can avoid all philosophic dualism in considering experimental evidence about it and in attempting to develop hypotheses regarding its unit processes and physicochemical nature. Memory involves a physical system by means of which an organism records and retains the effects of stimulatory inputs to its nervous system in past time. In somewhat teleological terms, it involves the provision in the nervous system of a record to be consulted when future correlated and integrated actions (not as reflex responses invariably produced to a particular stimulus) have to be determined, and appropriate motor centers appropriately instructed. In this sense, precisely similar mnemic faculties are involved in the training of a flatworm to turn more often to the left in a Y-maze, in the training of a carp to respond to the sound of a bell by coming to pond-side for feeding, in the training of a chimpanzee to employ symbolic "language," and in the training of a human student to read and to reconsider memory in the light of experiments with invertebrates. In all these cases the mnemic record must be a physical system which has been modified by the learning experience.

Some changes—almost certainly unit changes—take place in a neural tissue mass when the process called learning occurs. Changes take place, although it may be very difficult to tell what these changes are at the cellular or molecular level, in the central nervous system of an animal when, by a training process, an obvious reflex action is prevented and a "conditioned reflex" substituted for it. Unfortunately, the enormous amount of work on experimental conditioning which has grown out of Pavlov's initial discoveries has not provided much data on the unit changes involved in mnemic records. Similarly, although there is much circumstantial evidence of electrophysiological changes detectable in mammals during training and also important recent conclusions from neurosurgery, neither of these yet allows us to define what Karl S. Lashley termed the engram or physical memory trace. Among many controversial possibilities, three more probable physical bases for memory may be considered. The first of these would propose that the process of learning actually establishes new patterns of interconnection between nerve cells in the central nervous system. There is considerable evidence against such establishment of a changed topography of dendritic and axonal connections between cells resulting from every learning experience in animals, and this theory has been largely abandoned. A second theory, essentially a functional and dynamic equivalent of the first, proposes that mnemic processes involve the establishment of new patterns of continual activity in nerve cells. By this hypothesis, there is no change at all in the neural circuits as such, but appropriately patterned changes in the action potentials flowing along them. This is obviously analogous to memory storage in electronic computers where physical wiring patterns are not changed, but functional properties are, by the "entering" of additional information. A variant of this second theory would involve the establishment of continuous activity patterns in groups of neurons in more than one place in the brain at once. A third hypothesis proposes that acquisition of a memory trace involves chemical changes in certain nerve cells, or groups of nerve cells. Two basically different kinds of chemical changes have been suggested: one involves modification of neurotransmitter secretion at synaptic junctions between cells, while the other involves changes in RNA and appropriate protein synthesis mechanisms in specific nerve cells during the learning process. The second and third theories may not be mutually exclusive.

One of the earliest unit-model systems for continuous activity as a basis for mnemic storage was proposed many years ago by J. Z. Young. Although in the light of current experimental data from *Octopus* and elsewhere, it is probably erroneous in certain oversimplifications, it remains a most useful explanatory system in setting out in the simplest terms how functional patterning by continuous activity could provide an effective mnemic neural mechanism. It is illustrated in Figure 25-3, and consists of an absolute minimum of six neurons. One of these is the motor neuron **M** and to it are connected two cells of input pathways **A** and **B**. There is also a chain of small neurons **X**, **Y**, and **Z**, wholly contained within this theoretical "learning center." Now a stimulus which produces a reaction without any previous learning is a reflex or unconditioned response and this would occur if an impulse in pathway **A** produced motor impulses in motor neuron **M** in an obligate fashion. On the other hand, a single impulse in path **B** occurring alone, although exciting neural pathways and synaptic connections to **M** is not enough alone to elicit a response in that efferent channel. We can then assume that a pair of stimuli arriving in **A** and **B** together is enough not only to stimulate **M** to react but also to excite activity in the cycle of neurons **X–Y–Z** (in both cases by an appropriate summation of simultaneous impulses arriving at the synapses to **M** and to **X**, respectively). Next we would theorize that the **X–Y–Z** cycle continues to work its sequence of impulses after the stimuli through **A** and **B** have ceased, and further that this continuous activity has the effect of so altering the threshold of **M** (by way of the synaptic connection from **Z**) that impulses in the sensory pathway **B** arriving alone are now sufficient to excite **M** and produce an efferent motor impulse from it. If we state this in terms of the whole organism, a conditioned response has now been produced to a stimulus in **B**, and that change produced by conditioning has been

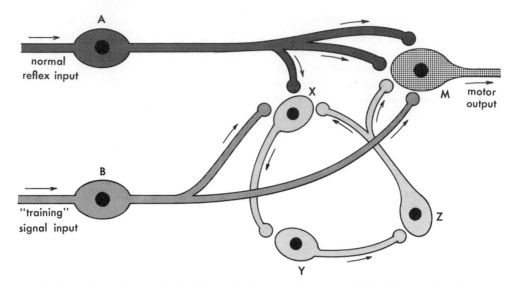

FIGURE 25-3.   An oversimplified model for mnemic storage. This minimum six-unit system, proposed as early as 1950 by J. Z. Young, involves a motor neuron, **M** and two input pathways, **A** and **B**. A chain of small neurons **X**, **Y**, and **Z** are contained within this "learning center." A stimulus in **A** can trigger a reflex response in **M** in an untrained animal, while a stimulus in **B** alone cannot. A training sequence of inputs in **A** and **B** together not only stimulates **M** to react but also can set up continuous activity in the **X–Y–Z** cycle. Thereafter a signal in **B** alone will cause a reaction in **M** (since the synaptic connection from **Z** has altered the threshold of **M**). Mechanistically, this theoretical minimum memory unit depends only upon the long-established property of synaptic summation at neural contacts. Thus, as long as continuous neural activity persists in **X–Y–Z,** the animal will "remember" to respond to the training signal **B** by the reaction controlled by **M**. For further discussion, see text.

the setting up of the **X–Y–Z** cycle and, so long as that continuous neural activity persists, so long will the animal "remember" to respond to a training signal input in **B** alone by the reaction controlled by the motor neuron **M**. If this theoretical minimum mnemic unit of six nerve cells were contained in the association centers of the brain of one of Pavlov's classic experimental dogs, then the sensory input in **A** would be the food presentation which would provoke a reflex response of salivation through its motor control **M**. In the naïve untrained dog, the ring of a bell would enter the unit through sensory pathway **B** and arriving alone would have no effect on other neural activity. However, a training period consisting of repeated simultaneous food presentation and bell-ringing would "condition" the trained dog to respond to the training signal of bell-ringing received alone through channel **B** by an impulse in **M** for salivation. The difference between the untrained and the trained condition in the dog's brain would be the continuous activity of the small association cells **X–Y–Z** in the trained animal.

It is worth noting that the only property required at the cellular level of functioning by this mnemic mechanism of continuous activity is the long-established one of synaptic summation. From some of the earliest work on neurophysiological mechanisms, conducted by C. S. Sherrington nearly a half-century ago, we know that summation occurs even in many classical reflex arcs. Thus in many vertebrates, even a simple stimulus with an arc involving only the "lower" centers of the spinal cord sets up impulses in several afferent fibers and each of these has endings at the synapses of a number of motor neurons in common. If one of these motor cells receives a single impulse it will probably not respond. It is only when several impulses arrive at almost the same time from a number of sensory fibers that the motor neuron will be activated and send out an impulse to the appropriate muscle or other effector cell. This is summation. More recently, J. Z. Young has developed the still relatively simple module which he terms a mnemon, and which is pictured in Figure 25-4. Experimental evidence that such simple modules each recording the consequence that followed stimulation of a particular "classifying cell" is the basis of the memory system in *Octopus* is very convincing, and there is also evidence from regular histology and from electron microscopy that the appropriate patterns of cells do exist in appropriate parts of the octopus brain. As discussed later, the likelihood of this, or something closely similar, being the unit of

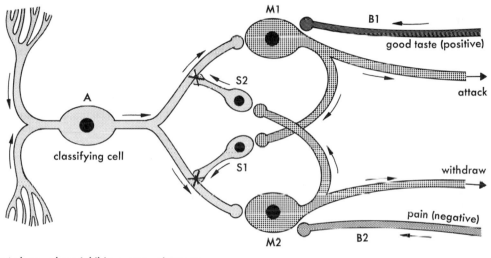

✕ places where inhibitory transmitter can
be produced (thus blocking pathway!)

FIGURE 25-4. A seven-unit memory module or "mnemon." This still relatively simple module was developed by J. Z. Young more recently than the six-unit system (Figure 25-3). There is histological evidence that closely similar patterns of cells do exist in those particular parts of the brain of *Octopus* which are electrophysiologically associated with the function of learning. As discussed in the text, octopuses can be trained to withdraw from, or to attack and feed upon, a crab presented at the same time as a visual signal. The mnemon has three inputs: **B1** and **B2** are the signals associated with positive (good taste) and negative (pain) training, respectively, while the classifying cell, **A**, derives in its many dendrites input from the "correlation" levels of visual input. The two small cells, **S1** and **S2**, have collateral connections from the opposite motor output cells, **M2** (withdraw) and **M1** (attack). There can be a bias in the system which results from training. This is thought to be brought about by activation of one of these small cells causing production of an inhibitory transmitter which blocks the opposite one of the alternative output pathways from the classifying cell. In this case, "memory function" depends upon contralateral inhibition, a feature of neural organization well known in the motor control of locomotion in bilaterally symmetrical animals (see Chapters 11 and 16). For further discussion of this memory module, see text.

memory is greatly strengthened not only by the fact that these workers have clearly demonstrated that two separate and distinct stores of mnemic records are made in the octopus brain during learning, but also by the fact that they can define anatomically and histologically precisely where these are situated. As shown in Figure 25-4, there are a minimum of seven components in each single memory unit or mnemon. There are two cells, **M1** and **M2**, concerned with output and driving successive higher and lower motor centers to produce the alternative actions of attack or withdrawal. There are three inputs, one being through the many dendrites of the classifying cell, **A**. These derive their inputs from secondary, or tertiary, sensory neurons, for example in the optic lobe "receiving" areas or in the deeper layers of the retina of the eye. The other two inputs, **B1** and **B2**, provide the "training" signals and are derived indirectly, through "higher" centers of integration and amplification, from sensillae which record, for example, good taste (**B1**) and pain (**B2**). Thus these inputs which can follow action signals will either be positive and reinforcing, as in the case of **B1**, or be negative with the opposite effect for **B2**. The system also includes two small cells, **S1** and **S2**, with collateral connections from the opposite output cells, **M2** and **M1**. Activation of these small cells is thought to produce an inhibitory transmitter blocking an unused output pathway for the classifying cell, **A**. This is the system which becomes biased toward one of the two alternative actions as a result of training. The unit is a basic memory store because its response is changed by this training process. Once again the functioning of this set of components is dependent upon a very well-established property of neural organization, that of contralateral inhibition. Our earlier discussions of locomotory mechanics in annelids (Chapter 11) and in land arthropods (Chapter 16) should have sufficiently emphasized the importance of this kind of control in the simpler neural arrangements of reflexes governing the effective action of pairs of muscles. Although the number of cell components for mnemons in the octopus brain is finite, it is still very large. J. Z. Young has estimated that there are possibly up to four million classifying cells in each of the optic lobes of the octopus. Thus, for those readers who find this oversimplification a bit too deterministic, we can restate the operation of the memory units in more stochastic terms. The operation of switching one mnemon, by training its classifying cell to the alternate response, reduces by half the uncertainty whether after the occurrence of a given input-output event the outcome, attack or withdrawal, will be good or bad for the animal.

There is even some evidence for the physical action of these small cells in creating presynaptic inhibition in the alternative outgoing pathways of the classifying cell. These small intrinsic or amacrine cells (of which there are many millions in the octopus higher lobes) can have fibers which when viewed by electron microscopy are packed with synaptic vesicles but which make contact with other fibers also filled with vesicles. This is of course anomalous in relation to the usual unidirectional arrangement of neurotransmitters at synaptic gaps between cells. The amacrine trunks are thus reservoirs of inhibitory transmitter which when stimulated appropriately can prevent signals from proceeding along the fibers which they contact. Not only have Young and his associates found this ultrastructural anomaly in *Octopus*, but workers on vertebrate spinal cord and retina have also shown it to be associated with the electro-physiological shifts of presynaptic inhibition.

The functional hypothesis is thus that the interaction of small cells and the classifying cells provides the basic memory unit for learning mechanisms. Young's work has produced an elaborate array of evidence for this, not only for classifying cells dealing with visual inputs but for those of the tactile system as well. Essentially the signals of results, positive or negative, may close un-wanted pathways through the sudden release of considerable amounts of a long-lasting inhibitory transmitter by the amacrine cells. If this also involves, as it could well do, the initiation of increased synthesis of the inhibitor in the cell, this in turn could involve changes in the RNA content in relation to the appropriate protein synthetic mechanism. Thus the claims by other workers of changes in the nucleic acid systems accompanying training would be ex-plained not as an involvement in carrying or recording specific information, but as providing, under unchanged gene control, mechanisms which are able to switch off unwanted channels. It is appropriate to note here that this is closely analogous to the functioning of the majority of electronic storage mech-anisms where, in multichannel systems, encoding is carried out by putting dis-tinct items into different channels (determined on a sequential either/or basis), rather than by passing different items along the same channel.

A further level of theoretical explanation for mnemic events in octopuses is provided by Young's introduction of re-exciting connections concerned with amplifying or optimizing signals as they arrive and redelivering them to the classifying cells. A possible layout of this is shown in Figure 25-5, and this process, which Young terms the "maintaining of the address," can best be un-derstood as a maintenance of input to the classifying cell which will, by a more complex version of summation, alter its threshold of response. The details can-not be spelled out here, but what we are considering in addition to the units of the mnemon are cellular connections in the vertical lobe system of the brain, divided into four: first and second upper subsystems and first and second lower subsystems (Figure 25-5). There is exceedingly good experimental evidence in *Octopus* that these upper sets of lobes, although not themselves the seat of the memory, are connected with activities which might be called reading-in or "addressing" the memory units themselves. The arrangement obviously also provides hierarchically for an integrated and optimized response. By having the centers thus arranged in series they provide, in Young's own words, an op-portunity to turn a "good" or "yes," and "bad" or "no" system, into what can be termed an "unless" system. In addition to this, the learning input channels can have both amplifiers and stabilizing negative feedback systems. Once again, evidence from experiments that both short-term and long-term mem-ories can be created strengthens the case for these particular patterns of neural circuitry.

Before we turn to behavior experiments with cephalopods, we can remark that this view of memory as an organ structure (in terms of a neural circuit) has a long conceptual history. At the end of the fifteenth century, Robert Hooke, whom we met in Chapter 6 as a founder of the microscopy of cells, and who was also an early urban redeveloper with the more famous Christopher Wren of

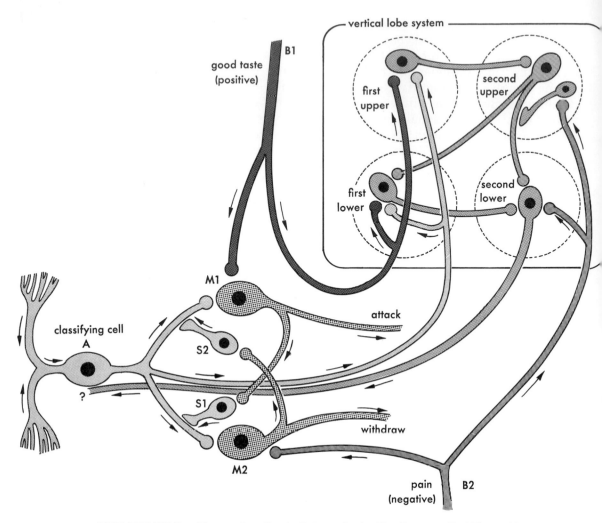

FIGURE 25-5. Memory function in *Octopus* brain. The "mnemon" of Figure 25-4 and a simplified wiring diagram (derived from the detailed studies of J. Z. Young and his associates) of the connections with the "higher" brain centers of the vertical lobe system. In this case, the classifying cell (**A**) and the small amacrine cells (**S1** and **S2**) of the mnemon would be contained within the optic lobe. These re-exciting connections are concerned with amplifying or integrating signals as they arrive and redelivering them to the classifying cells, a process which J. Z. Young has termed "maintaining the address" for an optimized response after learning. It is worth noting that studies by electron microscopy of the vertical lobe system in *Octopus* have revealed that, while the enormous number of synapses in the "first upper" subsystem (morphologically, the median superior frontal lobe) are conventional (that is, are apparently end-to-end one-way connections), the similarly large number of connections in the "second upper" subsystem (or true vertical lobe) are largely like those of the **S1** and **S2** amacrine cells (that is, are associated with presynaptic inhibition). With many such units and their parallel connections, the possibility is provided of a hierarchy of integrated responses. The bases of "levels" of memory, short-term and long-term, are discussed in the text.

a burnt-out London, regarded memory as constituting as much an organ system as the eye or ear and thought it to "have its situation somewhere near the place where the nerves from the other senses concur and meet."

Training, Recognition, and Split-brains

In the very earliest experiments in training *Octopus*, it was found that after merely six training exposures over the course of two days, an individual octopus could have "learned" to attack and feed on a crab presented alone, but to withdraw and avoid a crab presented with a white square. The training involves administering a mild electric shock if the octopus attacks the crab-plus-square. This is the basis for extensive experiments on the visual and tactile discrimination possible between shapes and surfaces respectively. The animal has marked powers of form discrimination, distinguishing between horizontal and vertical rectangles even when they approach "squareness." On the other hand, it has considerable difficulty in discriminating among oblique rectangles—a "sorting-out" which not only infant humans but also many small mammals can readily learn to employ (Figure 25-6). This limitation can be explained by the fact that the receptor cells of the cephalopod retina are arranged in groups of four subunits and these are oriented so that the microtubules within them are arranged in two cases in the horizontal plane and two in the vertical plane. Each set of four is related to a single optic sensory fiber. A further difference from the mobile scanning of the vertebrate eye is that the cephalopod eye is statically held with only an adjustment of the orientation of the pupil to hold the slit horizontal. As beautifully shown by M. J. Wells, this adjustment is achieved by an input from the bilaterally placed statocysts. Thus in terms of fine-functioning, the remarkable convergent evolution of the eyes of higher cephalopods with those of vertebrates, so impressive in terms of overall anatomy, is not paralleled either in ultrastructure or in basic physiology.

Young believes that in processes of visual learning in *Octopus*, cells in the optic lobes already preset to respond to particular dimensional ratios become conditioned to signal either attack or withdrawal according to the "result" signal ("good taste" or "pain") reaching them. At this stage the conditioning or learning process is very limited, and without generalization powers. However, more elaborate training experiments with *Octopus* show that the visual system and the memory units connected with it have some powers of generalization and it is this which could depend upon the interactions and especially the feedbacks between the vertical lobes and the optic lobes within the brain. Given the differences of "scanning" and gravity-based orientation, the visual acuity of *Octopus* is comparable to that in man and, as M. J. Wells has shown, the capacity for perceptive generalization can involve the size-range relationships of the artist's perspective representation and appropriate distinctions between mirror images.

There are more important limitations both to sensory perception and to generalization in learning in the tactile system of *Octopus*. In this case, training experiments demonstrate convincingly and elegantly that the surface texture of objects held by the tentacles can be distinguished on the basis of the proportion

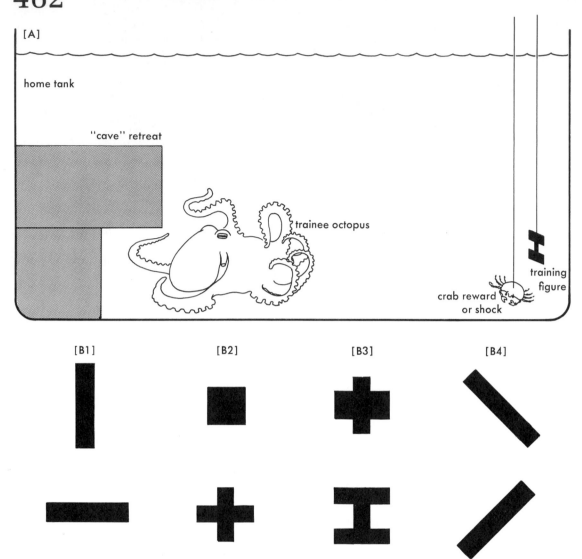

FIGURE 25-6. Octopus training experiments. A: Simple experimental tank. The octopus can retreat into a "cave" refuge or move to attack a crab presented along with a geometrical figure. A "correct" discrimination of the figure will be "rewarded" by having the crab to eat; an "error" will result in a mild electric shock (and no food). B1–B4: Pairs of geometrical figures (cut out of plastic) used in discrimination training. Training to distinguish between the upper and lower figure in each pair is readily accomplished, except for the pair of oblique rectangles (B4) with which the octopus has considerable difficulty. For further discussion, see text. [Derived from various publications of Young, Sutherland, and Wells.]

of the surface ridged or grooved, and the general fineness or roughness of texture. Although an octopus can be trained—and for once this may have some adaptational significance in wild Octopus—to sort bivalve shells of different shell sculpture as rapidly as a blind person could, it apparently remains incapable of distinguishing orientation of pattern, a feature clearly perceived by human touch. Octopuses are also unable to discriminate between different weights in objects picked up by the arms. This is not due to a lack of proprioceptors or other mechanoreceptors in the arms, but because the movements of the tentacles are regulated by local motor centers and in a crude oversimplification it may be said that the central brain of the octopus is "unaware of the relative positions of the sense organs in different parts of different arms." It may be a great disappointment to some potential writers of science fiction to learn that for this reason an octopus could never be able to learn to use its tentacles in totally *new* manipulative skills and thus is not a potential tool user, and future rival to primate evolution. This limited—and, to us, alien—perception of a tactile environment without any shapes determined by angles and edges, but with acute discrimination between the textures and tastes of different surfaces, is characteristically molluscan. As M. J. Wells has elucidated, the peculiarities of tactile perception in the octopus remain those almost inevitable in any soft-bodied animal, and result from a basic inability to relate proprioceptive to exteroceptive sensory information.

One feature shared by *Octopus* and those vertebrate brains which have been investigated is that there are characteristically two levels of memory storage in time: short-term memory and long-term. As noted earlier, careful surgical procedures of removal or lesion have allowed the sites of these faculties to be isolated in *Octopus* far more clearly than in any other animal. One particular surgical technique, the separation of the living cephalopod brain in the midline into right and left half-brains has been of particular significance in investigations of interactions between visual and tactile training and other aspects of the capacity of cephalopods to "generalize from experience." Since each half of the split-brain receives its input information from four arms and one eye, one-half can be trained in certain discriminations with the other half acting as a control in subsequent procedures and perturbations. Certain dubieties in complex learning experiments regarding the experiential background of different individual animals used as controls can thus be avoided.

In these ways, it can be clearly shown that visual learning is dependent upon interactions among the optic lobes, the superior frontal lobes, and the vertical lobes. Information from the retinal cells of the eye passes into the brain by way of the optic lobes and thence to the superior frontal lobes and on to the vertical lobes. Young's microanatomical studies show that fibers from the vertical lobes pass back to the optic lobes in essentially the fashion already set out in Figure 25-5. Even this input and recycling circuit involves enormous numbers of cells: 60 million in each optic lobe and some 25 million in the vertical lobes. If the vertical lobes are removed, the octopus almost entirely loses its previous training in visual discrimination (for example, between a horizontal and a vertical rectangle). However, if the training is repeated, it relearns considerably faster so that some aspects of the discrimination have remained in the damaged brain. There are considerable differences in retrainability depending on the frequency of conditioning; for example, relearning after damage is more effective if the conditioning is given at short intervals. Further, the damaged animal can still learn some new lessons but it does so more slowly and less accurately; and (as in humans with higher centers damaged) more complex kinds of training which are possible in the intact animal can no longer be carried out effectively.

From these and many more similar experiments it appears that long-term optic memory resides not in the vertical lobes but in the optic lobes themselves. The role of the vertical lobes is twofold. They seem to hold memories for a short time before, as it were, final storage in the optic lobes. Secondly, they play an important part in retrieving the memories from the optic store and correlating an appropriate response based upon them. Similar conclusions can be drawn regarding tactile learning. In this case, removal of the inferior frontal and subfrontal lobes completely abolishes the capacity to learn by touch. They are separate from the centers concerned with visual learning but in the normal entire animal both systems involve overlapping elements in the vertical lobes concerned with the interaction of short-term and long-term memories and with their retrieval and what Young has termed "maintenance of address." Once again, damage to the vertical lobes after some tactile learning will cause loss of training but a capacity for somewhat more rapid relearning will remain elsewhere in the system. The vertical lobe is also concerned in the transfer of learning from one side of an octopus to another.

Further experiments have thrown some light not only on the operation of separate short-term and long-term memory systems with parallel sets of inputs but also on the effects of different rates of training on capacity to learn under a variety of circumstances. In one set of experiments, octopuses were trained to discriminate between rough and smooth spheres. If each exposure to the training signal is termed a trial, then they learned faster with groups of thirty-two trials at five-minute intervals per session than with trials in groups of eight or sixteen. Animals with damage to the vertical lobes learned more slowly than normals at the higher rates of training. The final conclusion from this and similar work on the tactile learning system was that the vertical lobe was peculiarly concerned with ensuring that the signals of results (reward with a piece of sardine, or punishment with a mild shock, in this case) were correctly associated with the motor actions of acceptance and rejection which had been taken. In particular, damage to these higher centers in the touch-learning system increased the time required for the signals of a training trial to have each their cumulative effect on the total memory or "training experience of these parts of the brain." That these features are of general application to learning in cephalopods has subsequently been demonstrated by experiments on the suppression of the normal visual attack process in the cuttlefish Sepia. Cuttlefish shown prawns behind glass soon learn to stop striking at them, and this can be studied in relation to continuous and repeated presentations at different rates. Both the intensity of striking and its rate decrease with the level of negative reinforcement, and it appears that learning to suppress the earliest stages of the attack process may be contingent upon first having learned to suppress the subsequent stages. When prawns are provided without the glass protection, the natural striking response is allowed to recover, and this takes place in two stages which reflect again short-term and long-term memory systems.

The spatial separation of temporally different memory units in the brain of cephalopods is a demonstration of fundamental importance to a wide range of kinds of studies on memory and learning at all levels. Apart from the experimental octopus, another important source of data has been the work of neurosurgeons dealing with the results of "experiments" provided by unilateral or other selective damage suffered by careless or unlucky riders of motorcycles before the general use of crash helmets. One such Scottish neurosurgeon, W. Ritchie Russell, was enabled to give a lucid account of the sites of human processes of memory and of learning from such data. One highly relevant aspect is that retrograde amnesia after moderate accidental brain damage results from a destruction of neural tissue concerned with the short-term memory, and thus is most extreme for the most recent and hence least fully recorded events. Data from both octopuses and careless motorcyclists confirm that memory is a physical system spatially situated in particular parts of the brain; is not merely subjective; and does not require nonmechanistic explanation.

Future Research

Some kind of neural memory system is a necessary basis for all learned behavior in animals and for the "highest" neural processes, including human self-conscious thought. Memory systems have to be backward-looking in time. Even the simplest conditioned response involves a record of which events preceded a rewarding or unrewarding result for the animal. It is possible that all events detected by an animal's sense organs and passing through neural circuits, where there are classifier cells, will leave a memory trace in the short term. Long-term memory implies the retention of past information that can be integrated into a result-forecasting or decision-making neural process. Greater knowledge of the cellular mechanisms of mnemic units, and of the neural attributes of the whole animal's capacity to learn, will be sought in the near future using not only Octopus but also many other kinds of vertebrates and invertebrates.

It should be obvious that even if the cellular and biochemical arrangements of the units of memory were completely elucidated, there would still remain a whole range of problems of logic and mechanism concerned with the functioning of learning processes and other aspects of memory in higher animals and in humans. There are a whole range of problems connected with the encoding and classification of information; its representation and selective recall; and, most significantly, the continuity through time of these activities within a self-maintaining (in many senses, homeostatic) mass of metabolizing animal cells.

It is clear that some answers on memory mechanisms will certainly not come from man or mammals, and may not even be approachable in Octopus. The brain in cephalopods may be too big, with 300 million cells in the whole nervous system, to provide some answers. Investigators may have to ask certain questions about mnemic mechanisms using the simpler nervous systems of other invertebrates: arthropods, annelids, or lower molluscs. We noted in Chapter 17 work in progress on simple learning in insect ganglia, and it is appropriate to note here that ganglia with similarly small numbers of identifiable neurons are being profitably investigated in certain opisthobranchs and other gastropods. Experimental cellular systems in both Aplysia and Pleurobranchaea may soon allow analysis of processes of associative learning in terms of a reproducible switch of motor response in a group of only three or four neurons. During the next decade, there is some hope that using these opisthobranchs (and, perhaps even sooner, using roaches—see Chapter 17 and Figure 17-13) the development of reliable experimental systems with identifiable nerve cells will allow the framing of testable hypotheses on the unitary physicochemical processes involved in memory.

At the other end of the scale of investigations on brain function, including the problems associated with human self-consciousness, it is again probable that cephalopod studies can be of little *direct* help. What can emerge from the kind of studies we have discussed on *Octopus* and other cephalopods is a better estimate of how learning systems, perhaps even complex generalizing learning systems, first arose, rather than any direct assistance with all the logical problems which surround the development of a true linguistic ability, conceptual communication, and self-consciousness as properties of the human brain. It seems reasonably certain that future neural research will provide purely mechanistic explanations for the brain performance involved in intelligent actions of higher animals, indeed for all the processes which neurophysiologists can observe in human brains, that is, *other* human brains. There are obvious difficulties, now and in the future, for all scientific investigations carried on under those special circumstances on conscious human subjects who can contemporaneously report their conscious experiences. The knowledge gained in the last decade about human cortical function by such neurologists as W. Penfield and R. W. Sperry has clearly demonstrated the functional asymmetry in man's cerebral hemispheres which results from the dominance of the speech hemisphere (which is almost always the left one). Gross anatomical asymmetries of the posterior parts of the temporal surfaces of the human cortex have long been known, and recent detailed cytoanatomical studies on the auditory regions reveal that the temporoparietal cortex of the left side (concerned with language function) can have approximately *seven* times the volume of that of the right side. Thus observations made on human "split-brain" subjects would have many fundamental differences from those carried out on split-brain octopuses. Even if some of the functional divisions between the hemispheres in man which have been claimed (including conscious versus subconscious recall, verbalized logical thinking versus musical and pictorial pattern sense, and even reductionism versus holism) prove to be somewhat oversimplified, it is already clear, that while the minor hemisphere functionally resembles the brains of the highest anthropoids, the major hemisphere and its linguistic center marks the uniqueness of man. The difficulty of self-consciousness has been partly resolved by some philosophical neurologists (including John C. Eccles), who would claim that phenomenal experience is not identical with, but informationally coherent with, the neural events which can be observed by another. The unique dominance of the human speech center need not be surprising in retrospect. One obvious and long-recognized aspect of man as a species is the capacity for the transmission from generation to generation of large amounts of information by other than genetic means. Even before the discoveries on cerebral neurology, this vast nongenetic information transmission could almost have been used in a diagnosis for our species. The upright posture in man allowed the development of a long flexible pharynx, while the omnivore-carnivore habit did not necessitate the more complete separation of air passage from food channel required in grazing herbivores. These features of anatomy have allowed production of a variety of speech sounds unequalled by other animals, and this variety is the original basis for nongenetic communication in man. Here again, it may help to consider the limitations of the octopus system, which are dictated by other aspects of cephalopod physiology.

In Chapter 24, we noted that certain functional and anatomical features of cephalopods, including properties of their blood and excretory organs, along with their lack of adequate skeletal support, have prevented them from ever invading nonmarine habitats of fresh waters and of land. Similarly, the large numbers of relatively small eggs produced by cephalopods effectively has prevented the development of any parental care in these forms. This, in turn, has prevented the informational communication between generations which so characterizes man. As Jean Piaget has argued, a human parent turns an infant into a person by talking at it. No young octopus could ever experience the demonstrable intellectual benefits of being an only child—or even a first-born—of bourgeois human parents. This is perhaps the most fundamental reason why this book is not being read by an octopus to receive some nongenetic information about the diversity of other animals and, in passing, to consider recent experimental work on memory processes and learning in man. The use of such mechanistic comparisons need not demean the human condition. Even if one takes an extreme position and believes that human thinking (precisely like evolution by natural selection) proceeds by blind variation and recombination, followed by systematic elimination and selective retention, this need not diminish our awe of and reverence for the higher levels of living organization including the human intellect. It is not necessary to become a vitalist to admit to this awe. Is it not enough that symbolic language, capacity for environmental modification, complex social organization, systems of values and of ethics, and the probably unique capacity to define future goals and analyze motivations are all made possible by man's enhanced neural capacity, not merely to obtain information about the state of the human environment, but to organize it, and to *verbalize* regarding it. Consider the octopus: it cannot.

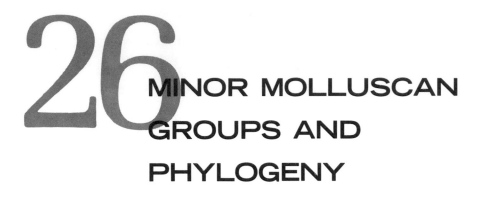

26 MINOR MOLLUSCAN GROUPS AND PHYLOGENY

APART from the three major molluscan groups already discussed, there are four minor groups, of which only the Polyplacophora or chitons have any numerical or ecological significance.

The class Scaphopoda or elephant's-tusk-shells are in some respects intermediate between bivalves and gastropods. There are less than three hundred species, all of which live at moderate depths in the sublittoral. The genus *Dentalium* is typical, with both the head, which bears prehensile tentacles, and the foot emerging from the broad end of the shell down into the sand. The narrower posterior end protrudes from the substrate, and both inhalant and exhalant water currents pass through it. There is a long mantle-cavity but no typical gills, the ctenidial structures being replaced by pallial folds. There is a strong radula used with the prehensile cephalic tentacles for feeding on Foraminifera. Scaphopods are clearly an unsuccessful pattern of molluscan construction, with no obvious adaptive radiation.

The class Monoplacophora was originally set up to accommodate certain early fossil molluscan forms, such as *Scenella* and *Pilina,* which had primitive limpet-like shells and multiple pedal-retractor-muscle scars and were assumed, on the basis of clever palaeontological deductions from circumstantial evidence, to be untorted "pregastropods." In 1952, the Danish oceanographic research vessel *Galathea* dredged some limpet-like ani-

468

mals from 3570 meters in the Pacific off Costa Rica which later proved to be living representatives of this group. A preliminary report on them appeared in 1957, and the detailed morphology of the species, named *Neopilina galathcac*, was beautifully and extensively described by H. Lemche and K. G. Wingstrand in 1959. For a few years, it seemed as though the discovery of *Neopilina* was going to revolutionize our ideas regarding the phylogeny of the molluscs and their relationships with other phyla. This has not proved to be the case, and we can now view *Neopilina* somewhat more objectively, as an interesting survivor of an aberrant molluscan stock, but not as a miraculously surviving "ancestor" of all other molluscs and "missing link" with the rest of the invertebrates.

Neopilina is almost perfectly bilaterally symmetrical, with a dome-shaped mantle and a flattened limpet-shaped shell bearing a spirally coiled protoconch (Figure 26-1). The foot is a muscular disc attached to the shell by eight pairs of pedal retractor muscles. There are five pairs of gills, which may or may not be homologous with ctenidia, in the pallial grooves on either side of the foot. The head bears elaborate branched postoral tentacular flaps, and contains a well-developed radula in a coiled radular sac. The gut has a simple stomach with a crystalline style-sac and a much-coiled intestine, and would in general suggest an unspecialized deposit-feeder. The heart has two pairs of auricles; there are six pairs of nephridia, and one or possibly two pairs of gonads. The nervous system closely resembles that of chitons, with a ladder-like organization without much ganglionation, involving ten sets of lateropedal nerve connections (Figure 26-1).

Within the molluscs, *Neopilina* shares certain characters with the chitons and the more primitive cephalopods. Both before and after the full description of the living monoplacophoran, all sorts of phylogenies were erected to include it, using it as a link between the arthropod-annelid stock and the rest of the molluscs. Lemche believed that several features in both *Neopilina* and the chitons are primitive metameric characters providing undeniable evidence of the segmental origin of the phylum Mollusca. This and other matters of phylogeny will be discussed below after a brief description of the chitons and one other minor group.

The remaining molluscs fall into two very distinct groups, which, however, until recently were associated in the class Amphineura. Further embryological studies have suggested that they are in fact sufficiently distinct to be regarded as separate minor classes. The first and smallest form the class Aplacophora, once called "Solenogastres," which are worm-like animals with a mantle but no shell. There are about twenty genera living in moderately deep water, all with fairly simple radular apparatus and a small posterior mantle-cavity, which in a few forms contains a pair of bipectinate ctenidia. Once again, the nervous system is a simple double ladder with four longitudinal cords, two pallial and two pedal, running back from a simple ring in the head and communicating by irregular lateropedal connectives. The mantle never secretes a shell but does contain numerous calcareous spicules which, in some species, protrude like a fine silvery fur. Aplacophorans are classified into two distinct orders: the Chaetodermatoidea, which live in burrows in bottom oozes and are probably detritus-feeders, and the more numerous Neomenioidea, which live among (and feed on) deep-sea hydroids and other colonial cnidarians.

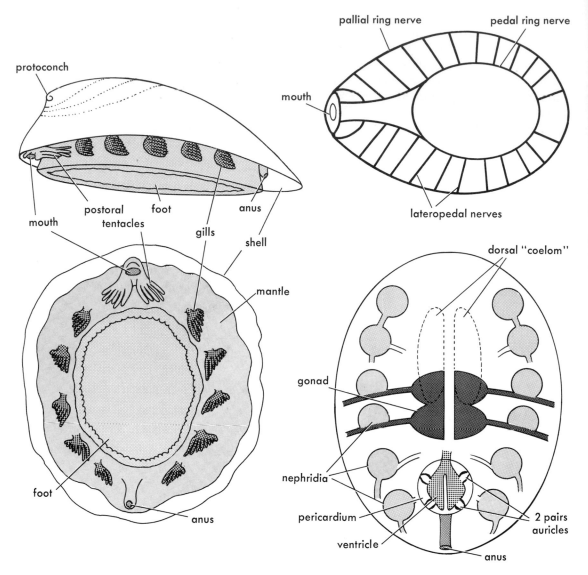

FIGURE 26-1. *Neopilina,* the living monoplacophoran. Lateral (**A**) and ventral (**B**) views of *Neopilina,* showing, surrounding the sucker-like foot, the pallial grooves containing the five pairs of gills, which may or may not be homologous with ctenidia. **C:** Plan of the nervous system in *Neopilina.* **D:** Dorsal diagram of the pericardial and urinogenital organs in *Neopilina* (compare with other molluscan patterns in Figure 20-8). The heart has two pairs of auricles, and there are six pairs of nephridia, two of which connect the gonads to the exterior. An additional pair of dorsal cavities may be coelomic like the peculiar pericardial extensions (Figure 20-8F) in neritacean gastropods, but renopericardial ducts have not been identified in *Neopilina.*

470

Chitons

The second group of the former Amphineura, the class Polyplacophora or chitons, are by contrast the most successful of the molluscan minor groups. The mantle bears eight articulated shell plates dorsally, and the group consists of animals clearly adapted for life on hard and uneven surfaces in the littoral zone. They have some features in common with the varied groups of gastropods which have become limpets, and these include the suctorial foot with its characteristic sense-organs and musculature.

There are nearly 600 species placed in about 43 genera; all but a few deep-sea forms live in the low intertidal. The species occurring on both sides of the Atlantic are relatively small and unobtrusive. A much greater variety, including much larger forms, is found around the Indo-Pacific. The Pacific coast of North America has about 130 species, many startlingly colored and over 5 centimeters long, including the massive species of *Cryptochiton* which may be up to 30 centimeters in length with the shell plates overgrown by a tough leathery mantle. The sucker foot and the articulations of the shell allow all chitons to adhere to the uneven rock surfaces of wave-swept shores and to resist the strongest surf action (Figure 26-2). In the unlikely natural event of a chiton being washed off its rock, or if removed by a human collector, the articulated shell allows it to curl up to protect the foot and gills of its underside, like a pill-bug (Chapter 15) or a fossil trilobite (Chapter 34).

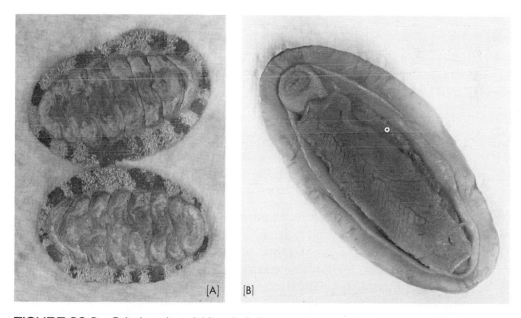

[A] [B]

FIGURE 26-2. Polyplacophora (chitons). **A:** Two specimens of *Chiton marmoratus* (about 4.5 centimeters long) from the dorsal side, showing the eight articulated shell valves surrounded by the mantle rim or girdle bearing small hard scales on its exposed surface. **B:** Ventral view of another large chiton, *Katharina tunicata* (8 centimeters long), showing the girdle surrounding the head and the elongate muscular foot, with a row of ctenidia visible in the pallial groove (between girdle and foot) of the chiton's left side. [Photos by the author.]

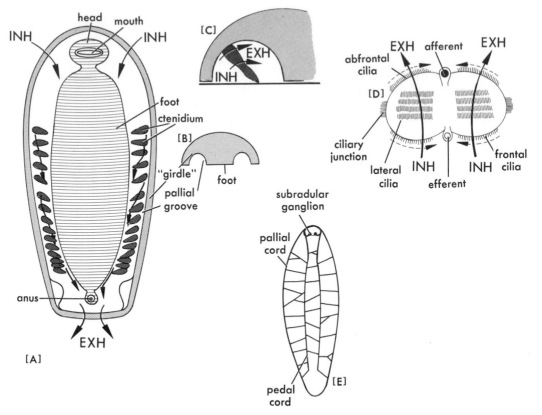

FIGURE 26-3. Polyplacophora (chitons). **A:** Ventral view of a chiton. **B** and **C:** Cross sections of chitons to show the arrangement of the pallial grooves and their functional division into inhalant and exhalant parts by the ctenidia. **D:** Ctenidial plates from a gill of a chiton. Compare with Figures 19-4, 23-1 and 24-3A. **E:** Plan of the nervous system in a chiton.

Compared with other molluscs, the body is elongate and the mantle-cavity has been drawn out into two narrow pallial grooves running between the foot and the broad mantle edge or "girdle" on each side. We still have the functional pattern of each of these pallial grooves being divided into an inhalant and an exhalant mantle-cavity, with the kidney and genital openings, along with the central anus, discharging in the exhalant stream. Here, the functional division is a ctenidial curtain forming a perforated screen by means of ciliary junctions between the tips of leaflets on adjacent ctenidia (Figure 26-3). Each ctenidium closely resembles the most primitive gastropod type (the aspidobranch gill of Figure 19-4 with alternating ctenidial leaflets on either side of a ctenidial axis in which run the afferent and efferent blood vessels). The respiratory water current is caused by the lateral cilia on the faces of the leaflets, and blood and water move in opposite directions with the usual molluscan physiological efficiency of a countercurrent system. However, the original ctenidia are multiplied to form lateral series numbering from four to eighty on each side, and functionally linked in the ctenidial curtain which divides the pallial grooves (Figure 26-3) into an anterior-outer inhalant mantle-cavity and a posterior-inner exhalant cavity. Temporary lifting of the mantle edge or girdle can result in a temporary inhalant current to the mantle grooves anywhere in the front of the animal. The exhalant current is always posterior, and the faeces, along with the excretory and reproductive products, are discharged into this posterior current. The alimentary canal and associated organs are clearly adapted for continuous grazing of attached algae and diatoms from rock surfaces. There is a long radula, and prominent salivary and sugar glands secrete into the posterior oesophagus. Amylase is prominent among their secretions, and digestion is largely extracellular, the digestive diverticula being also mainly secretory. There is a long posterior gut concerned with consolidation of faeces.

The nervous system consists of a double ladder of pallial and pedal cords, and the only ganglia are subradular. Connected with this very primitive nervous system are a surprising number of sense-organs. There are osphradia on either side of the anus, tactile organs in the mantle girdle and on the snout, taste receptors in the buccal region, and otocysts near the pedal cords. Also, to a varying extent in different species, there are peculiar sense-organs embedded in pits through the shell plates; these are termed megalaesthetes and microaesthetes. Little or nothing is known about their functioning, although the microanatomy of some suggests the structures of a simple eye.

The sexes are separate and external fertilization occurs after spawning, which usually shows some lunar periodicity. The zygote develops into a trochophore larva which becomes elongate and develops a mantle rudiment. About the time that it settles to the bottom, it secretes six shell plates. After an interval a larger plate is added anterior to the series, and still later a small last plate is added at the posterior end. At no stage in the development is there a "budding zone" for shell plates. In the young postlarva there are initially no ctenidia; after the first of these appear in the posterior parts of the pallial grooves, they are added to with growth of the chitons throughout life.

As noted above, the discovery and description of *Neopilina* has reopened discussion of possible metamerism in primitive molluscs. Thus the exact nature and developmental origins of the multiplied organ systems in chitons become of considerable significance. Lemche's phylogenetic proposals imply that chitons show metameric segmentation. Some years ago Stephen C. Brown and I published some data on these replicated structures, and we concluded that the evidences against metamerism in chitons are overwhelmingly strong. As growth continues in adult chitons, ctenidia are added anteriorly, irregularly, and independently on either side, but so that there is a broad correlation between gill number and adult size. Asymmetry in ctenidial numbers between the left and right sides of single specimens occurs in most chiton species which have been studied and, in some populations, involves more than half the individuals. Thus the ctenidia of chitons cannot be considered as paired structures. The mode of development of the shell plates was noted above, and like the other replicated structures such as heart auricles and renal lobes, these cannot be considered to be formed in a series as are the metamerically segmented structures of annelid worms already discussed. It is difficult to characterize as a metamerically segmented animal: a chiton with eight shell plates, with two, three, or four auriculoventricular openings, with 21 ctenidia on one side and 23 on the other (or 15 and 18), and with a ladder-like nervous system of irregular transverse connectives—all these arranged apparently independently of each other. If we consider the essential features of metamerism—particularly of its morphogenesis—which were discussed in Chapter 10, there is little of the serial succession of segments, each containing unit-subdivisions of the several organ systems, about the replicated organs of chitons. Thus, it seems most probable that the multiplied structures of chitons reflect functional replication rather than degenerated ancestral segmentation.

Molluscan Phylogeny

Unless some future physicist can provide a machine for time-travel, evolutionary biologists will never know with certainty the characteristics of the ancestor of *any* stocks of animals. As stressed in the introduction to this book and in Chapters 2 and 35, all statements regarding archetypes—and indeed all comments on the phylogeny and interrelationships of the existing animal phyla—are matters involving hypotheses. Many of the peculiar dangers which can invalidate the discussions of hypothetical ancestral types can be avoided if the attempt is made to create a working archetype—one in which the concert of organs and functions could operate as a whole in an integrated functional plan. That such a process is of value in attempting to comprehend the diversity of form and function in invertebrate animals is, of course, a major part of the *raison d'être* of this book. However, the difference between the real, but unknowable, ancestral progenitor and the hypothetical archetype, inferred from structural and functional homologies, should be remembered during the discussion of molluscan interrelationships which follows.

Extensive comparative anatomical studies on molluscs were carried on through the nineteenth century and in the first 25 years of this century, often by the same zoologists as were involved in comparative vertebrate morphology: ranging from Cuvier by way of T. H. Huxley, Ernst Haeckel, and Ray Lankester to E. S. Goodrich. By about 1930 the structural homologies, along with the diversity, were well worked out, and even before that date the next stage of functional morphology or comparative physiology had begun with the work of G. A. Drew and J. L. Kellogg at Woods Hole and J. H. Orton in Britain. The next twenty years saw the extensive and convincing work of the functional morphologists, such as C. M. Yonge, Alastair Graham, and Vera Fretter, whose studies of ciliary mechanisms, ctenidial blood vessels, and renopericardial and genital ducts in the more primitive gastropods and bivalves allowed a very convincing functioning archetype to be set up. As discussed in Chapter 19, this hypothetical animal was primitively bilaterally symmetrical, developed totally unsegmented, and possessed a posterior mantle-cavity enclosing a pallial complex of paired structures which included two ctenidia. All available evidence still indicates such an archetype as the most likely ancestor of the two major molluscan groups. From studies on *Neopilina*, Lemche has suggested that the ancestral mollusc must have shown relatively complete metamerism, that this is present to a somewhat reduced extent in *Neopilina*, and that this is still further reduced in chitons, and finally, that this metamerism degenerates so completely as to be undetectable in the rest of the molluscs. The broader phyletic significance of these alternatives should be noted. If no true metamerism occurs in primitive molluscs, the closest connections and possible origins of the phylum lie in the turbellarian-rhynchocoel phyla. If metamerism was a feature of the primitive mollusc, connections should be sought with the annelid-arthropod phyla. Earlier writers placed emphasis upon similarities of larval patterns and modes of cleavage in early development between the annelids (though not the arthropods) and the molluscs. As will be discussed in the next chapter (Chapter 27), attempts to link these and many other phyla in a supertaxon termed Protostomia are based partly on a long-outmoded view of the significance of embryonic recapitulation, and partly on inconsistent evidence. Once the *adaptational* aspects of embryonic and larval development are stressed, then the case for association of the phylum Mollusca with the phylum Annelida on these sorts of grounds becomes very weak. If true metamerism could have been demonstrated in *any* primitive mollusc, then the case for a basic relationship between annelids and molluscs would have to be reopened. In this regard, although few zoologists would accept the more extreme homologies proposed by Lemche, such as those between the primitive arthropod triramous appendage and the gill in *Neopilina,* and between this latter and the ctenidia in the rest of the Mollusca, a number of authorities have accepted in part the theory of metamerism in molluscs. As noted above, the evidence against the alleged metamerism of chitons is very strong. In these animals, at least, the replicated structures have a functional significance, and their morphogenesis certainly does *not* involve a regular production of metameric segments each with unit-subdivisions of the organ systems.

Even if we can thus dismiss true metamerism as regards chitons, the question of "segmentation," both in *Neopilina* and in a hypothetical molluscan ancestor, remains. Arguments similar to those presented about gills and shell plates in chitons can be used against the concept of metamerism as applied to the described structures of *Neopilina.* It has five pairs of gills, two pairs of auricles, six pairs of nephridia, one or possibly two pairs of gonads, eight pairs of pedal-retractor-muscles, ten sets of lateropedal nerve connections, and a single shell with a coiled protoconch. We do not yet have any functional studies on the organization of the gills in living *Neopilina,* which could establish whether or not these structures are homologous with the ctenidia of the other molluscs. Even more frustrating is the fact that we have no embryological studies on *Neopilina,* and only these could establish the significance of the allegedly metameric structures. At present, it seems unlikely that a "budding" morphogenesis of unit-subdivisions of the segmental organs, as in the Annelida, can occur in *Neopilina.* However, a reexamination of the question of the archetypic mollusc is in order. The evidence of functional morphology has led to the classical picture of an archetype with a posterior mantle-cavity containing one pair of ctenidia connected to one pair of auricles. However, even before the discovery of *Neopilina,* there were several difficulties in relating this model to conditions in the primitive cephalopod *Nautilus,* with two pairs of ctenidia and four auricles, and in the chitons, with many ctenidia and two elongate auricles usually with four auriculoventricular openings. There is striking resemblance in hearts (each with four auricles and a "single" ventricle) between *Neopilina* and *Nautilus*. There is a further, though less obvious, resemblance of both to conditions as they exist in the majority of chitons. Perhaps surprisingly, the arrangement of heart chambers and of their interconnections seems to form a relatively conservative feature in the evolution of other molluscan groups. In view of all this, it would seem as justifiable to set up a model ancestral mollusc, or archetype, with four ctenidia and therefore four auricles. Subjectively, a fourfold basic organization would seem a more reasonable starting point for two sorts of morphogenesis, either reduction to one pair or replication to many. Both a line of organisms with one gill on either side and a line with many could thus evolve from an ancestral stock with two gills on either side. It should be emphasized that, in this hypothesis, the former stock (that is, those with one pair of ctenidia, one pair of auricles, one pair of renal organs, and so on) would still be regarded as the stem-group of the two major groups of living molluscs: the gastropods and the bivalves.

A few points are worth recapitulating. All archetypes (including the last-mentioned) and all homologies involve hypotheses (see Chapters 2 and 35). However, the basic molluscan plan of structure and function is remarkably uniform throughout the group, and the homologies established in the major groups, particularly as regards the ctenidia and the organs of the pallial and pericardial complexes, are probably sound. Metameric segmentation—in the annelid sense—does not exist in chitons, and probably, despite *Neopilina*, has never occurred in an animal which could be called a mollusc. The weight of available evidence suggests that the stem-group of the molluscs was directly derived from animals like Turbellaria, and had no connections with the stock or stocks which gave rise to the annelid-arthropod phyla. But, again, archetypes are *not* ancestors.

27 DEVELOPMENT MODES AND SUSPECT SUPERTAXA

TOWARD the end of our synoptic treatment of the Protozoa (Chapter 6), we briefly considered competing schemes of classification above the level of phyla. We were there concerned with the significance and value of various classifications at the levels of kingdoms and subkingdoms to replace the simple separation into plants and animals of the early nineteenth century. A systematic matter of considerably greater practical and phyletic importance concerns the possible interrelationships between the various higher phyla of many-celled animals. Even for the empirical purposes of communication and understanding, it is not surprising that writers of zoological textbooks have often tried to set up assemblages of "related" phyla, in an attempt to reduce the thirty or so animal phyla (see Chapters 1 and 2) to three or so "major evolutionary stocks." Further, it is not surprising that characteristics of embryonic and larval development have been used in the logical diagnoses of "superphyla," "animal subkingdoms," or other higher taxa of doubtful significance. Like the artificial term "invertebrates" that defines the contents of this book, these assemblages may have some pragmatic value in easing human comprehension of animal diversity. Although consideration of a broad phylogeny of invertebrates will be postponed until Chapter 35, it seems appropriate to interpolate here a brief consideration of the use of data from morphogenesis in embryos, and from larval stages, in the development of hypotheses on phyletic relationships. Admission of

477

the utility of any scheme of classification need not lead us to suspend our scepticism regarding claims of "demonstrable common descent."

Pattern and Unity

Biology as a science has always depended on the perception of fundamental unities within the prodigious variations of nature. As regards the study of invertebrate animals, this statement is so obvious as to be trite. Classical studies on comparative anatomy reveal the unity of basic structural pattern in each phylum. Although varying in external characteristics, the members of a typical phylum form an assemblage, all constructed on the same ground plan with certain essential groupings of structural units. A major part of the *raison d'être* of this book, and of its predecessor paperbacks, is that the discussion of similar homologies of function and the use of working archetypes—each involving an integrated functional plan, in which the concert of organs and functions can operate as a whole—are of value in attempting to comprehend the diversity of form and function in invertebrate animals. However, as stressed elsewhere (see Chapters 2 and 35) *real* ancestry is ultimately unknowable, and statements on homologies, archetypes, and phylogenetic classification—including all comments on the interrelationships of animal phyla—are matters involving hypotheses, and subject to further testing and revision. The difference between the "hard" data on patterns of development, and the inferences therefrom regarding questions of degrees of interrelationship and common descent, should be remembered during the discussion which follows.

The phylogenetic value of embryological data was established early in the nineteenth century, but modern biology has taught us that the interpretation of developmental evidence is rarely as straightforward as some comparative embryologists then claimed. The resemblances between early embryos in a variety of metazoan phyla, and the even more spectacular similarities among various vertebrate embryos, had attracted attention by the time Cuvier was writing at the end of the eighteenth century. However it was Karl von Baer in 1828 who set forth certain fundamental principles of comparative embryology for the first time. Somewhat paraphrased, these state that the more general features of an animal, which are common to all members of a large group, appear earlier during embryonic development than the specialized features that characterize each particular subgroup and species. Thus, animals resemble *more* others more closely in the earlier stages of their development than they do in the later ones. An important qualification, stated clearly by von Baer but later ignored, was that the embryo of a "higher" form never resembles the adult of a "lower" form, but only the embryo of the latter. Although there were implications of degrees of interrelationship in von Baer's principles, for decades this had no connotation of evolution. Following publication of Darwin's theory of natural selection in 1859, however, it was almost immediately appreciated that this hierarchy of degrees of resemblance in embryos could be interpreted as measures of relative divergence from a common ancestor. This interpretation was amplified and codified by Ernst H. Haeckel into the theory of recapitulation (or Haeckel's biogenetic law) according to which each individual organism in its embryonic development retraces its ancestral history. The publicity and acclaim which this theory received in the middle of the nineteenth century, not only in scientific circles but also in those of popular education, were comparable to those for the discoveries in the middle of the twentieth century of the precise structure of nucleic acids and the functioning of the genetic code. Although actually misleading, Haeckel's law has a certain attractive simplicity: in development, the single-celled fertilized ovum would correspond to a protistan ancestor, the morula resulting after a number of cleavages and the hollow blastula to colonial forms like *Volvox* (see Chapter 6), and so on. Haeckel had to invent a hypothetical form, the Gastraea, to correspond more closely to the gastrula stage than any cnidarian would. According to Haeckel all questions of animal ancestry and evolution could be solved by appropriate study of the embryonic development of "higher" forms. Haeckel ignored the important *caveat* of von Baer by insisting that the animal recapitulates embryologically the *adult* stages of its ancestors. In the first two decades of this century, the standard textbook of invertebrate embryology by E. W. MacBride could still be based conceptually on this "law" that ontogeny recapitulates phylogeny. Modern knowledge of the genetic bases of evolution and of development makes Haeckel's biogenetic law, in its simplistic form, totally unacceptable. Perhaps the main reason, apart from historical interest, for discussing it here is its persistence as a cultural myth *outside* the science of developmental biology. It is worth noting, too, that even before knowledge of modern genetics forced its abandonment, recapitulation as an important factor in the evolution of larval stages had been seriously challenged by Walter Garstang (see Chapters 19 and 33). The major point which he pressed, and which has become a keystone in modern interpretation of data from embryos and larvae, is that adaptive modification is by no means confined to the adult.

It is most important to realize that the *special requirements of embryonic and larval life* are responsible for adaptive modifications involving both the structures and the time-pattern of development. Features exhibited by larval types can reflect *either* recapitulation of ancestral history *or* adaptational response to the immediate needs of larval life (see further discussion of these needs in Chapter 33). The responses to such needs of embryonic or of larval life have been termed caenogenetic. It is clear that caenogenesis may completely upset any simplistic interpretation of developmental evidence as mirroring phylogeny, and we owe this insight to Garstang.

The final rejection of Haeckel's law came with the incorporation of genetic theory into a synthetic evolutionary theory. By the 1940's, it was clearly recognized that evolutionary adaptation and change are based upon genetic mutation and the essential recombination in sexual reproduction. Thus new characters are not merely added on to the end of ancestral ontogenies but are products of the total ontogeny of organisms involving changed units as integral parts of their whole genome. The embryonic stages of a higher organism are never, in fact, similar in detail to the adult stages of its ancestors. However, von Baer was right and these stages often do resemble the *embryonic stages* of the ancestors.

Many examples could be set out but it will suffice to summarize the larval stages found in the higher crustaceans like crabs (Malacostraca—see details in Chapters 14 and 33). The earliest larval stage is characteristic of the class Crustacea as a whole (nauplius); there follow stages characteristic of eucarids (zoea) and then of the crab suborder Brachyura (megalopa). A lineage of *adult* ancestors is not represented by this sequence of larval stages. The basic reason for this is rather simple. Mutational changes that are likely to be preserved to form part of an evolutionary sequence are most likely to be relatively small ones occurring in the later stages of development. Mutational changes affecting early embryonic development are more likely to be harmful and thus unlikely to be preserved by processes of natural selection. Again, this relationship between mutational changes and embryonic development was anticipated before its genetic basis was understood. O. Hertwig and R. Hertwig, at the end of the nineteenth century, noted that some of the pattern of relationships between structures (unique to each phylum of animals) is "stamped in" early in development by the need for certain early sequences of differentiation to precede the more elaborate processes of morphogenesis of organs which must follow. In summary, provided it is used with caution and always integrated with evidence from functional morphology, and also from palaeontology if this is available, embryological data can provide important circumstantial evidence of evolutionary relationships.

Some modern biologists are convinced that accumulation of data from molecular biology will solve all phylogenetic problems, including those of relationships between phyla, within the next decade. At first sight, it would seem that all questions of doubtful homology might be solved by serial demonstrations of biochemical identity or nonidentity. Universal elucidation is, sadly, not that close. In particular, the optimistic claims that similarity in a given biochemical pathway—or in the mechanism of control of that pathway—would provide incontrovertible evidence of close phyletic relationship are somewhat suspect. Recent studies on the regulatory interactions of end-product metabolites upon the synthesis of various multiple forms of enzymes suggest that neither common enzymatic steps nor closely similar control patterns need reflect phyletic affinities. On the other hand, diversity in control patterns of biochemical pathways need not imply polyphylogeny. Other recent discoveries of extensive gene-enzyme polymorphisms add to the difficulties of drawing phyletic conclusions from biochemical data alone. Of course, data on actual amino acid sequences could provide evidence for phyletic relationships if obtained extensively, but the difficulties of collecting such crucial evidence of genetical and chemical identity make widespread utilization of this method seem unlikely. What we may obtain in the next decade or so in the way of protein sequence maps may resolve a number of crucial phylogenetic relationships (including some left open in this book). Much can be confidently expected to confirm the hypotheses set up from embryological, morphological, and palaeontological studies. Elucidation of much of the details and mechanisms of evolution will remain dependent upon data provided by these older disciplines. Meanwhile, it behoves us to remember (with F. R. Lillie and Walter Garstang) that there has been an evolution of developmental stages as well as an evolution of their adults.

Patterns in Embryos and Larvae

Many, perhaps the majority of, biology texts divide the more complex phyla of many-celled animals into two great assemblages: the Protostomia and the Deuterostomia. Platyhelminthes, Mollusca, Annelida, Arthropoda, and a number of minor phyla are classified as protostomes, while the Echinodermata, the Chordata, and at least two minor phyla are included in the deuterostomes. (A few texts even include the Cnidaria and Ctenophora among the protostomes, but this is a ridiculous extension based partly on misconceptions).

The features claimed as diagnostic of this division are largely developmental. The Protostomia are said to show: spiral cleavage of a mosaic egg, the mouth formed from the blastopore, the body-cavity schizocoelous in origin, the central nervous system ventral and deep-seated, and, if a free-living larva is formed, a trochosphere-like larval organization. The Deuterostomia are said to show: radial and indeterminate cleavage, the mouth formed as a secondary opening with the blastopore becoming the anus, the body-cavity enterocoelous in origin and usually showing tripartite subdivision, the central nervous system superficial and usually dorsal, and, if a free-living larva is formed, a "dipleurula-like" organization. Although based on facts of development, unfortunately the five diagnostic features are not nearly so consistently associated together as some authors suggest. Further, it is impossible to make a clear dichotomy of animal phyla on the basis of these five characteristics. Some of the protostome-deuterostome distinctions, where they occur, may reflect phyletic relationships, but they are neither so universal nor so consistent as to justify being used in major classification. However, it is worth examining in a little more detail some of the factual data of development involved. Perhaps the most significant developmental data concerns the occurrence of spiral cleavage. The eggs of annelids, molluscs, nemertines, and the polyclad platyhelminths show several patterns of this form of cleavage. Spiral cleavage is claimed to be mosaic in the classical sense, that is to say that each blastomere formed has a predetermined and fixed fate in the later embryo. With this determinate cleavage, experimental embryologists cannot produce twins from one fertilized egg. The contrasting indeterminate radial cleavage is seen best in some echinoderm eggs. In these, if the cells of the four-cell stage are artificially separated, each is capable of forming a complete gastrula and then a young larva.

The associations between spiral cleavage and mosaic development, and between radial cleavage and "regulative" development in which cells formed in the first few cleavages remain equipotential, are far from absolute. Modern developmental biologists do not categorize eggs as either mosaic or regulative. Experimental embryology has shown these to be at best relative terms. It is probable that in most early embryos (even including those of echinoderms) there is some relationship between cleavage pattern and subsequent differentiation. On the other hand, there are many cases where experimental alteration of cleavage pattern in a "spiralian" embryo does not interfere with subsequent normal development. The distinction between regulative and mosaic eggs does not reflect any fundamental difference in the mechanisms of development or in the genetic control of the processes of differentiation. It remains only of descriptive value with regard to relative rates of differentiation.

Among the arthropods only some barnacles show a form of spiral cleavage and a variety of patterns of early development is found in the minor phyla. Patterns of cleavage during early development are somewhat more varied than a simple dichotomy between deuterostomes and protostomes (or Spiralia) would suggest. In a recent review, my distinguished editorial predecessor, Donald P. Costello, has noted that the term "Spiralia" has no obvious significance in the interrelationships of animal phyla. As he had earlier summarized elsewhere, there are actually three main categories of cleavage: radial, bilateral, and spiral. Further, there are three basic types of spiral cleavage: by quartets, by duets, and by monets. Cleavage is a dynamic process in time, and these categories are not, in fact, fixed in the sense of persisting rigidly throughout the whole cleavage period of early development. In the majority of spirally cleaving forms there is usually a transition to bilateral cleavage, and in different groups of invertebrates this transition can happen relatively early in development, or relatively late.

After radial cleavage, if the developing egg is viewed from either pole, the blastomeres are arranged in a radially symmetrical form. This results from successive cleavage planes which have cut straight through the egg at right angles to one another and in a symmetrical disposition around the polar axis (Figure 27-1). Radial cleavage (and the accompanying postponement of differentiation of individual cells, which used to be called indeterminate cleavage) is seen best in some echinoderm eggs, but is also found in the sponges and in the phylum Cnidaria. In bilateral cleavage, the cleavages result from the centrioles and spindles being bilaterally arranged with reference to a dorsoventral plane of symmetry running down the midline of the resultant embryo. Bilateral cleavage is the earliest cleavage in many sea-squirts, in *Amphioxus*, and in most amphibians. Some other groups which appear to show bilateral cleavage relatively early in their embryology may actually start with a variant form of spiral cleav-

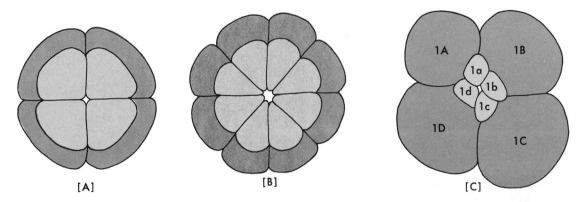

[A]   [B]   [C]

FIGURE 27-1.   Radial and spiral cleavage. **A:** Radial cleavage at the 8-cell stage, as shown in the development of many echinoderms. **B:** A similar egg after the fourth cleavage, now at the 16-cell stage. **C:** Classic spiral cleavage at the 8-cell stage, as shown in a mollusc such as *Crepidula*. (This is the four-quadrant spiral cleavage "by quartets" initially described for certain gastropods and for certain polychaetes. Other patterns of spiral cleavage occur: "by duets" in certain acoel flatworms and "by monets" in barnacles and possibly in a rotifer. These patterns are discussed in the text.)

age. As Costello has pointed out, spiral cleavage might well have been called oblique cleavage. It is always characterized by a rotational movement of cellular elements with relation to the egg axis so that both the division spindles and the resultant cleavage planes are inclined, or oblique, with relation to any axial planes of symmetry of the egg. In spiral cleavage of the quartet type, it is after the third cleavage that the so-called "spiral" orientation becomes obvious (see Figure 27-1). It would be inappropriate to discuss the nomenclature of spiral cleavage here, but it is important to realize that it was developed by the famous group of American embryologists (all with Woods Hole connections) including C. O. Whitman, E. B. Wilson, and E. G. Conklin, largely because they *were able* to follow cell lineages in developing annelids and molluscs. As is well known, a single cell originating around the 29-cell stage, and termed in the most usual nomenclature the 4d cell, was found to be the mesentoblast from which is proliferated all the coelomic mesoderm of the later embryo and of the adult animal (Figure 27-2). This pattern of spiral cleavage by quartets is found, as already noted, in polyclad flatworms, in annelids, in nemertines and in all the molluscan groups except the cephalopods. A simpler form of spiral cleavage involving duets of larger and smaller cells is characteristic of acoel flatworms, where the mechanics of alternating oblique spindles based on the pattern of replication of centrioles has been convincingly elucidated by Costello. Apparently uniquely among the arthropods, the barnacles show yet a third form of spiral cleavage. This form involves unit spiral cleavage instead of four-quadrant spiral cleavage, and Costello describes the process as involving monets of micromeres, instead of duets or quartets. Finally, it is possible that the earliest cell divisions in at least one rotifer and one member of the minor phylum Gastrotricha show spiral cleavage by monets in their early stages.

As regards patterns of cleavage, it would seem that grand-scale division of the invertebrate phyla into "Spiralia" and the deuterostome-rest could only be seriously proposed by a zoologist ignorant of the variety of cleavage patterns. In features like the position in the cell lineage of the 4d mesentoblast, there may be echoes of a universal ancestral feature, but these are overlain by numerous special features of the cleavage of each group of animals, and even of each species. At the end of last century, another great Woods Hole biologist, F. R. Lillie, stated his belief that special features of cleavage are adapted to the needs of the future larva in each form. Lillie's adaptational view of patterns in early development, like Garstang's for larval stages, has recently been championed by Costello, who has pointed out that only in species with direct development (that is, without larval stages) could ontogeny ever recapitulate (in Haeckel's sense) the phylogeny of adult form.

As regards the embryonic origins of the mouth, in some protostomes the blastopore becomes the mouth while in others the blastopore closes and a new mouth opens later close by. The origins of the anus in deuterostomes vary similarly. In many protostomes, the coelom arises as a schizocoel, by splitting of the mesoderm layer. The coelom in deuterostomes forms by a process called enterocoelic pouching in which the wall of the archenteron itself evaginates to form the mesoderm. The processes of coelom formation are not really consistent with the other characters. For example, the Brachiopoda have radial cleavage, protostomous mouth formation, and an enterocoelous body cavity. Indeed, taken as a group, the lophophore-bearing phyla (Bryozoa, Phoronida, and Brachiopoda) are in many respects intermediate between "true protostomes" and "true deuterostomes." There are other aberrations of pattern: the tripartite division of the coelom supposedly characteristic of the deuterostomes is also found in the Chaetognatha, Tardigrada, and other unrelated minor phyla.

Finally, we have the evidence provided by the free-swimming ciliated larvae where they occur. A trochosphere larva is shaped like a spinning top with a tuft

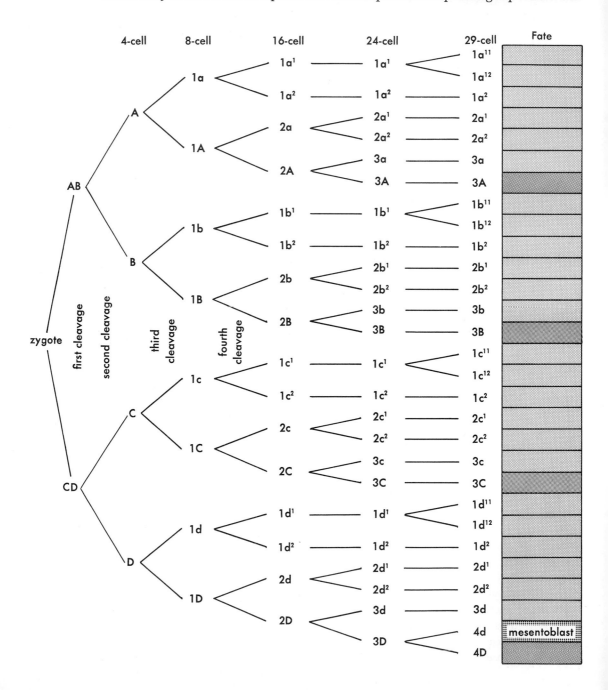

of cilia on the apex and an equatorial band of cilia through which the mouth opens (see Figure 10-2A; see also Chapters 10 and 33). Typical trochosphere larvae occur during the life-cycle of primitive polychaetes and archiannelids. These larvae seem well-adapted for feeding and locomotion in a planktonic environment and retain their characteristic form until the onset of metameric segmentation (see Chapter 10). A closely similar larva, the trochophore (see Chapters 19 and 33), is found in many molluscs, where it develops into the more characteristic veliger (see Chapter 19 and Figure 19-12). The flatworms and nemertines have markedly different ciliated larvae, as do most of the minor protostome phyla, although the Entoprocta have a "trochosphere." The characteristic larva of the deuterostomes has been claimed to have a "dipleurula" form, but this involves a synthesis from two distinct types of actual larval organization: the pluteus and the bipinnaria-auricularia (see Figures 30-6 and 30-7). Strictly, these larvae, which are clearly distinct from the trochosphere form, are characteristic of the phylum Echinodermata alone. No true chordate has such a larva, and the peculiar case of the hemichordate tornaria will be discussed later (see Chapter 31).

Like the trochospheres and veligers, the plutei and other free echinoderm larvae are all very well-adapted for continuous locomotion and for feeding on microflagellates in the plankton. Most cases of increasing complexity in these larvae—additional foldings of ciliated bands, exaggerated velar "wings" in some mollusc larvae, and the very elongate arms of some echinoderm plutei (see Figure 30-7A)—can all be attributed to the problems of increasing mass with growth and of maintenance of appropriate surface areas of the locomotory cilia. This increasing complexity will be further discussed in the consideration of invertebrate larvae in the marine zoöplankton in Chapter 33. All such larvae—no matter how bizarre their final appearance or how remarkable the parallels between their structures and those of larvae from other phyla—reflect the two competing selection pressures set out long ago by Garstang. First, there is the need of the larva to grow up rapidly into the adult stage and then reproduce its

FIGURE 27-2. (opposite) Cell lineage following spiral cleavage. Unlike radial cleavage, where cell differentiation is postponed through the first few cleavages, spiral cleavage normally involves mosaic development or a relatively early determination of the embryonic fate of specific blastomeres. In this system of nomenclature for cell lineage, the original four blastomeres give rise at the third cleavage (see Figure 27-1C) to a quartet of micromeres (**1a, 1b, 1c, 1d**) and four larger macromeres (**1A, 1B, 1C, 1D**). Successive quartets follow (for example, after the fifth cleavage **3a, 3b, 3c, 3d,** etc.), and the daughter cells of the original quartets are designated by exponents ($1a^1$, $1a^2$) and subsequent serial exponents ($1a^{11}$, $1a^{12}$, etc.). The letters continue to designate the original quadrant of origin; the number before the letter designates the quartet from which the cell was derived, and the exponent indicates not only the number of subsequent divisions but lower numbers should indicate proximity to the animal pole. This chart shows the start of the lineage in classic molluscan spiral cleavage, where the single micromere **4d** (or mesentoblast cell) will give rise to *all* of the coelomic mesoderm in the adult (and also a few enteroblasts). The complete nomenclature for the 16-cell stage is set out in the fifth column, as though the fourth cleavage had produced all possible daughter cells synchronously. In fact, later cleavages are *not* synchronous in most forms with spiral cleavage. In *Crepidula* (as first described by E. G. Conklin from Woods Hole in 1897), the successive cleavages after the 8-cell stage (Figure 27-1C) give 12-cell, 20-cell, 24-cell, and 29-cell stages. The last two columns list the 24-cell and 29-cell stages as they are manifest in *Crepidula*, with the **4d** cell separated in the 29-cell embryo.

own kind. Secondly, there is the larval need to remain as a floating organism as long as possible in order to disperse the species. As modern biologists, we might wish to particularize the second need and speak of wider gene flow and evenness of population density in addition to increased range. We might also wish to compare bioenergetics in nonfeeding and in normal planktotrophic larvae (see Chapter 33). None of these sophistications need detract from the realization that many of the adaptive modifications seen in embryos and in larvae result from the special requirements of embryonic and of larval life.

A Critical Look at Supertaxa

The inconsistencies in the characteristics claimed as diagnostic of the Protostomia and Deuterostomia may largely result from their basis in features of embryonic and of larval development. Haeckel's simplistic theories still have an appeal. In justice to most fellow textbook authors, it should be stated that the majority of systems employing these divisions would first segregate (within or without the subkingdom Metazoa) the Porifera and the Mesozoa. Better classification schemes would then separate the Radiata (for cnidarians and ctenophores) from the Bilateria before dividing the remaining metazoan phyla into protostomes and deuterostomes. The most self-critical schemes might admit that the four lophophore-bearing phyla (three of which will be considered in Chapter 28) show a mixture of deuterostome traits with protostome traits. Major classificatory units or supertaxa, such as subkingdoms, grades, and superphyla, are only valuable if they can be applied with some consistency, in which case they *can* aid the perception of fundamental unities even if phyletic relationships remain "not proven."

As noted in Chapter 6, broad systems of classification of living organisms into five or six kingdoms have had general acceptance. Two of these, Protista and Metazoa (sometimes called Animalia), are treated partly in this book. We have noted that the protozoans and the sponges, although referred to here as merely constituting two phyla, should be appropriately ranked as subkingdom Protozoa and subkingdom Parazoa, both of equivalent status to a kingdom or subkingdom Metazoa. Within the Metazoa we have set out thirty separate phyla. Much of the last few pages have been devoted to a sceptical consideration of one of the better-based attempts to reduce these thirty phyla to two or three "major evolutionary stocks." Despite this demurring, some superphyletic assemblages may have some value.

As noted in Chapter 10, the phylum Annelida and the phylum Arthropoda share many features of structural and developmental organization. With these two phyla, the minor metameric phyla of Chapter 18 form an assemblage of six metamerically segmented phyla, all clearly but distantly related to each other. As noted in Chapter 18, the four small groups of metamerically segmented animals are sufficiently distinct from one another and from the two major metameric phyla that it is best to consider each as a distinct minor phylum. However, this assemblage of six phyla constitutes perhaps the only good case for a composite supertaxon within the Metazoa. In Chapter 26 the case *against* associating the phylum Mollusca with these metameric phyla was emphasized. As discussed in Chapter 8, cases for the grouping of the minor pseudocoelomate phyla into composite supertaxa have to be based on much less convincing evidence than that linking annelids and arthropods. Neither the superphylum Aschelminthes (of certain modern textbooks), in which the six minor groups of pseudocoelomates are associated with the nematodes, nor the out-of-date assemblage Gephyrea, which also embraced some minor coelomate phyla, is likely to have any phyletic validity or even much pragmatic value. Lastly, the common possession by four phyla of a lophophore as a food-collecting organ may or may not be phyletically significant (see Chapter 28), but it is clear that these animals (including brachiopods, ectoproct Bryozoa, and phoronids) have little else which can be homologized in their structures, and clearly have less shared organization than have the annelids and arthropods. This class suit against supertaxa now rests. A consideration of the possible interrelationships between the thirty phyla of Metazoa is deferred to Chapter 35.

28 FURTHER MINOR PHYLA

In the second chapter of this book we set out some criteria which could be used to designate certain phyla as minor in terms of numbers of individuals, numbers of species, and basic ecological significance. In the plan of presentation adopted in this book's particular survey of invertebrate life, the minor phyla are not treated extensively. In Chapter 8 we considered synoptically six minor phyla which could be considered as triploblastic pseudocoelomates, and which are often associated with the successful pseudocoelomate phylum of nematode worms, without any clear evidence for interrelationships. In Chapter 18, we dealt briefly with four minor metameric phyla, where a somewhat better case could be made for association with the two major phyla Annelida and Arthropoda. We now have to deal, in the present chapter, with a group of six even more "independently assorted" minor phyla, united only in being triploblastic, coelomate, and not metameric. Three of them, bryozoans, phoronids, and brachiopods, have, in common with the pseudocoelomate phylum Entoprocta, a filter-feeding organ in the form of an encircling crown of ciliated tentacles around the mouth, called a lophophore. Some definitions of this extensile feeding organ, the lophophore (including that used by the late Libbie Hyman), would insist on its having a mesodermally lined, and thus truly coelomic, lumen. By this definition, the entoprocts do not bear a true lophophore. All four phyla are sedentary, and another shared anatomical feature is a U-shaped gut. From all this, it should be clear that if any "suspect supertaxon" were to be erected from any of the six minor coelomate phyla summarily dealt with in this chapter, it might combine bryozoans, phoronids, and brachiopods, perhaps even the sipunculids or the entoprocts, but probably never the priapulids nor the planktonic arrow-worms of the phylum Chaetognatha. As noted in Chapter 1, the arrow-worms form a minor phylum as distinct from all others, as the phylum Echinodermata is from all major animal phyla.

Although by our ecological definition (which is based on the proportion of annual energy-flow, or of organic carbon turnover, in the Earth's bioeconomy of the present day), all these six are minor phyla, two (Bryozoa and Brachiopoda) have some claims to other status. Nearly 20,000 species of ectoproct-bryozoans may have existed, and there could be 4000 of these alive today. Brachiopods, represented by about 300 living species, were a dominant faunal group in Palaeozoic and Mesozoic seas (see Chapter 34), and their fossil species number more than 12,000. However, in today's oceans, brachiopods and ectoproct-bryozoans make up only a minute fraction of the annual turnover of biomass.

The order in which these six phyla are summarily treated is arbitrary, and neither it nor the subhead "pairings" is intended to reflect any degrees of relationship.

Phylum Bryozoa

The ectoproct-bryozoans are the coelomate moss-animals (see Figure 1-16), with a superficial resemblance to the phylum Entoprocta (see Chapter 8) which have more than the nature of the lumen of the lophophore to distinguish them, and in fact are a totally distinct and truly pseudocoelomate minor phylum. There has long been some confusion, and a little controversy, in the name applied to the coelomate phylum of moss-animals. Specialists in the group have recently reverted to the name now used here, phylum Bryozoa, although the majority of general texts, including my own earlier paperbacks, used phylum Ectoprocta as originally suggested by Hyman. [Meanwhile a few biologists— mostly British—refer to the same animals as phylum Polyzoa. This name undoubtedly had priority, since in recognition of moss-animals as a separate group of animals from the marine hydroids, it was set up by J. Vaughan Thompson, the British naturalist of the early nineteenth century. We met Vaughan Thompson earlier (Chapter 14) as the first to elucidate the peculiar life-history of barnacles.] On balance, it seems best to follow the usage of the group specialists, and term the phylum Bryozoa. The microscopic, individual bryozoans, termed zooids, live in extensive, sessile colonies, each fastened within a secreted exoskeletal box, or zooecium, and feeding by means of an extensile lophophore bearing ciliated tentacles (Figures 28-1 and 28-2). Many genera, like *Electra*, form creeping mats, while others produce upright branching systems reminiscent of hydroid coelenterates. The majority of species are marine, although a few bryozoans inhabit fresh waters. The fact that they are all sessile organisms, microscopic as individuals, and mostly minute as colonies, means that they are easily overlooked. However, many species occur growing on sea-

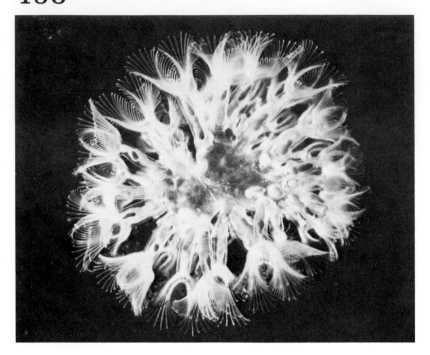

FIGURE 28-1. Colony of a freshwater ectoproct, *Lophopodella carteri,* showing the lophophores in their expanded feeding position. [Photo by Dr. Shuzitu Oda, 1960; courtesy of Dr. Thomas J. M. Schopf.]

weeds, rocks, and the shells of larger animals in the lower parts of the seashore. The great majority of species are in fact sublittoral and as many as thirty species have been recorded on a single large bivalve shell from below low water mark. Ninety species have been recovered from shelly debris of a single offshore dredge haul.

Although individuals are microscopic and colonies never massive, it may be incorrect to speak of the Bryozoa as a minor phylum. There are probably about 4000 living species and possibly about 20,000 different extinct forms. On the other hand, they are hardly a major group in the ecological sense discussed earlier in this book, and make up only a minute fraction of the biomass in the marine and freshwater habitats where they occur. Appreciation of the ectoproct-bryozoans is hampered by there being little in the way of good biological studies on living zooids, but an enormous literature on the details of the architecture of their box-like houses, the zooecia.

Growth of the mat-like colonies of forms like *Electra* (formerly *Membranipora*) may involve zones of actively budding zooids advancing to colonize new substrate, with areas of less active zooids and beyond them areas of persistent empty zooecia left behind as a dead crust over seaweed or rock (see Figure 22-9). On large brown laminarian seaweeds the bryozoan growth can be integrated with the algal growth, so that the colony expands predominately stipeward onto the zone of new frond and thus survives from season to season. Other observers must have felt that the steady spread of tiny box-like houses over suitable rocks and large seaweeds is analogous (in many ways) to the man-made suburban neoplasm creeping over our late twentieth-century countryside (as so memorably perceived in the 1960's by Pete Seeger).

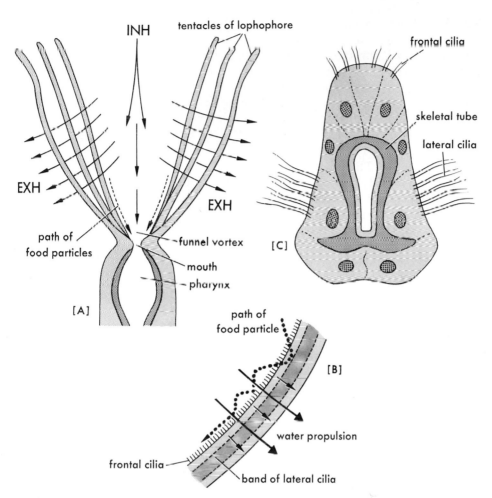

FIGURE 28-2. Feeding in ectoproct bryozoans. **A:** Part of the lophophore of a single zooid. Water enters at the top of the bell and passes centrifugally between the tentacles of the lophophore. **B:** Enlarged diagram of part of a single tentacle, showing the band of lateral cilia responsible for water propulsion. The bryozoans are impingement-feeders rather than true filterers. **C:** Cross section of a single filament. The skeletal tube consists of flexible collagen, and there are no cilia on the outer surface of the lophophore bell.

Feeding in ectoprocts is carried out by the lophophore, which may be horseshoe-shaped or circular. When fully expanded, the lophophore forms a funnel of diverging tentacles surrounding the mouth and leading into it. Each tentacle is ciliated with two lateral tracts of cilia (facing those of other tentacles) and a third tract of shorter cilia on the median inner surface (Figure 28-2). When the functioning lophophore is examined more closely, it can be seen to work in a totally different fashion from that of the entoprocts (Chapter 8). In the bryozoan lophophore, the feeding current enters at the top of the tentacular bell and flows centrifugally between the tentacles to the outside. This is the reverse of the entoproct pattern, where food trapped in mucus must be transported by ciliary tracts from the *outside* of the bell to the central mouth (see Figure 8-6). In the ectoproct-bryozoans the feeding current carries the particles in suspension directly toward, and in some cases into, the mouth. The main water current is caused by metachronal beating of the lateral cilia, with an effective stroke obliquely downwards and outwards in relation to the long axis of each tentacle. The median cilia on the inner face of the lophophore bell (Figure 28-2) are more strongly developed toward the base of each tentacle and they carry more particles (without much mucus) centripetally. The outer end of the gut, the pharynx, is ciliated and when the mouth is held open, particulate food is continually drawn in. The pharynx can also act as a muscular suction pump, becoming suddenly dilated at its inner end, and thus causing the ingestion of larger particles. As in the molluscs and other filter-feeding forms (see Chapters 19 and 22), ejection mechanisms for distasteful or undesirable particulate material are important to the life of sessile bryozoans. Sustained closure of the mouth results in some material being carried away by the outgoing current caused by the lateral cilia. In some forms there can be ciliary reversal in the pharynx used for ejection, and in others, including the common Atlantic sublittoral genus *Bugula,* there is a definite ciliary rejection tract leading from a sort of typhlosole groove within the pharynx across the lophophore to between the ventral pair of tentacles. It is not clear whether, in all bryozoans, trapped food particles are carried continuously into the mouth by the ciliary currents while the lophophore remains expanded, or whether in some forms, periodic retraction of the lophophore is necessary to carry some of the captured food into the mouth. In any case, among the various groups of suspension feeders in the sea, also including bivalve molluscs (Chapter 22) and primitive chordates (Chapters 31 and 32), ectoproct-bryozoans are perhaps better described as impingement-feeders rather than filter-feeders. This is also true of phoronids, but the distinction is less clear in the other lophophore-bearing phylum, the Brachiopoda, where the confines of the lampshell's mantle create a filter-feeding box unit. In all ectoproct-bryozoans the lophophore can be rapidly contracted into a bundle of tentacles and pulled down into the zooecium. The gut is U-shaped but the anus lies outside the lophophore. The trunk of the body has a peritoneum-lined coelom separating the gut from the body wall, and this coelomic cavity extends into the lophophore and tentacles which are extended by various hydraulic means. Extension and retraction of lophophores in ectoprocts having different forms of zooecia involve different elaborations of the musculature. There is always an important group of lophophore-retractors usually inserted on the coelomic septum at the lophophore base and originating on the aboral wall of the zooecium. Protrusion of the lophophore is carried out in a variety of ways, all essentially hydraulic. Some simpler forms of zooecia have the frontal wall (that is the roof of the box) made of a flexible membrane to which muscles are attached. Contraction of these muscles decreases the volume of the box, thus compressing the coelomic fluids of the zooid and extruding the lophophore. In some more complex forms, the frontal membrane still provides the compression to extrude the lophophore, but a perforated calcareous shelf, termed the cryptocyst, forms a second roof below the flexible one and serves for the better protection of the zooid. In yet another group of ectoprocts, the Ascophora, the zooecial box is completely calcified. In these forms a thin-walled sac, the compensation sac or ascus, has been formed inside the zooecium and has a separate opening through a pore to the exterior. In this case the contraction of muscles attached to the floor of the ascus brings about its dilation, thus sucking seawater in from the exterior. It also causes compression of the coelomic fluid in the zooid and thus extrusion of the lophophore (Figure 28-3).

Clearly the lateral cilia of the tentacles move water downward and centrifugally, but, apart from the rejection tract mentioned earlier, the median cilia and the other cilia around the mouth and lophophore base do not seem to be organized into directional pathways like those of bivalves or of ciliary-feeding sabellid polychaetes (see Figure 12-3). The similarities of ciliary pattern on the lophophore tentacles of Bryozoa, on molluscan ctenidial filaments, and on the pinnules of the tentacles of sabellid fanworms, with water currents caused by lateral cilia and "frontal," here median, shorter cilia moving trapped particles are similarities of common function and, as is obvious when their development and functional morphology are studied in detail, do not reflect any homology. Some ectoproct-bryozoans can respond to larger unwanted particles by muscular flicking of the tentacles, and all can respond by total retraction of the lophophore. There are cilia in the pharynx and throughout most of the gut, and in some ectoprocts the food is formed into a mucus-cord which is rotated as it passes through. It seems that in the commonest marine ectoprocts (class Gymnolaemata) digestion is partially extracellular and partly intracellular, as in the coelenterates and many molluscs. In common with many other filter-feeding organisms, the hindgut is concerned in consolidation of the faeces which in

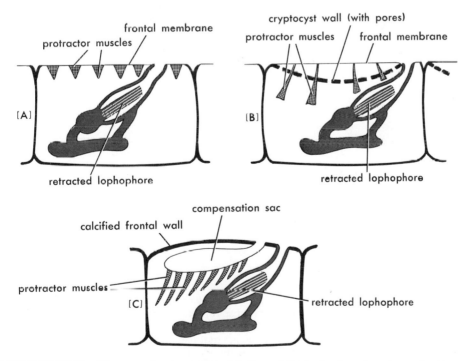

FIGURE 28-3. Three mechanisms of protraction of the lophophore in ectoprocts. For further explanation, see text.

ectoprocts are formed into mucus-covered balls and expelled away from the feeding mechanism of the lophophore.

The nervous system forms a plexus in the body wall with discrete nerves running into individual tentacles and around the lophophore, and a main ganglion between the mouth and the anus (which position is termed dorsal). Not unexpectedly, in view of the minute size of ectoproct zooids, there are no circulatory, respiratory, or excretory organs. Ectoprocts are mostly hermaphrodite, with the gonads developed on the peritoneum. In about 90% of the ectoprocts studied, fertilized eggs are retained in a brood-pouch. In the few forms where the ova develop externally a highly modified form of trochosphere larva—the cyphonautes larva is formed. This, or an even more modified larva, released from the brood-pouch, settles on a hard substratum, undergoes a degenerative metamorphosis, and forms the founder zooid of a new colony. Colonies are of course produced by asexual budding and in different ectoproct species there are different degrees of continuity between individual zooids in the colony. A few are polymorphic. In the freshwater forms there is also asexual production of balls of cells termed statoblasts which are like the gemmules of sponges and which can survive desiccation and freezing to develop in suitable spring condi-

FIGURE 28-4. Arborescent ectoproct colonies. Living colonies of *Bugula turrita,* some of which show the characteristic colony form in a spiral tuft of fan-like branches, attached to a drifting piece of the seaweed, *Sargassum.* Branching colonial forms, such as *Bugula,* are responsible for the name "moss-animals," although mat-like encrusting colonies, such as *Electra* (Figures 1-16 and 22-9), are probably more abundant. Green-dyed clumps of dried *Bugula* colonies are often sold as "everlasting plants." [Collected and photographed by Perry Russell-Hunter.]

tions. Once again, as in rotifers (see Chapter 8), the success of a relatively minor group of minute animals in temperate fresh waters depends on a modification of the life-cycle to suit the peculiar conditions of that environment.

As mentioned earlier, much of the detailed classification of ectoprocts depends on the characters of the zooecia which may be calcareous, gelatinous, or membranous—some of the latter being formed of chitin. The phylum is naturally divided into two classes. The Phylactolaemata are the ectoprocts of fresh water with the lophophore usually horseshoe-shaped. In these, the coeloms of the zooids are continuous through the colony, the zooids never show polymorphy, and the zooecia are never calcareous. Remarkably, and by unknown mechanical means, colonies of the freshwater genus *Cristatella* are motile. The other and much larger class, the Gymnolaemata, includes all marine ectoprocts plus a few in fresh and brackish waters. In this class the lophophores are circular, there are no coelomic connections between zooids, there can be polymorphy, and there is a wide variety of zooecial architecture. This larger class is generally divided into five orders, two of which are limited to the Palaeozoic (see Chapter 33), while the other three all have living representatives. Into the order Cheilostomata fall the great majority of living genera and species of marine bryozoans (see Figure 22-9 and 28-4) and significantly, this order is represented in the fossil record only from the Cretaceous onward. Some palaeontologists divide the phylum Bryozoa into three classes separating most of the Palaeozoic fossil groups from the Gymnolaemata in a third class Stenolaemata. Until recently, this has been less adopted by neontologists.

Phoronids and Priapulids

The minor phylum Phoronida—there are only about fifteen species—consists of worm-like animals with a lophophore (Figure 28-5). Although they share the possession of a lophophore with the Bryozoa and Brachiopoda, the phoronids are totally unlike these other groups in external anatomy and way of life. The small number of known species, which can probably all be assigned to the genera *Phoronis* and *Phoronopsis*, all live in the marine littoral as solitary individuals in secreted tubes, though many of these tubes may be aggregated together. Each tube is a permanent residence, formed of secreted chitinous material coated with pebbles and shells. Only the lophophore is thrust out for feeding and it can be rapidly retracted. The gut is U-shaped with an anus near the mouth, and there is a pair of nephridia with pores on either side of the anus. The coelom is subdivided with a separate lophophore cavity and there is a circulatory system with haemoglobin in corpuscles circulating in two longitudinal vessels with a ring vessel connecting them anteriorly. The nervous system consists of a diffuse net with a ring around the base of the lophophore and one giant nerve fiber running down the left side of the trunk, which functions to bring about the quick synchronous contraction which brings in the lophophore.

Some phoronids are hermaphroditic and some dioecious, but most brood their young for some time at the base of the lophophore. The larvae eventually released are the characteristic and peculiar actinotrochs, which some zoolo-

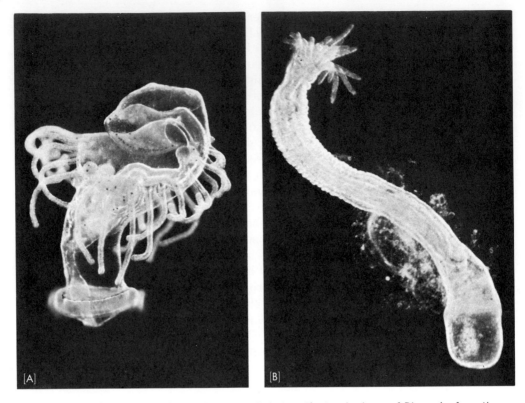

FIGURE 28-5.   Phoronid development. **A:** Late actinotrocha larva of *Phoronis,* from the marine plankton. The ciliated tentacles represent an extension of the metatroch (or equatorial ciliated band *below* the mouth), and the ciliated ring surrounding the anus (at the bottom of the photograph) is termed the telotroch. **B:** Juvenile settled stage of *Phoronis* from the sea bottom: metamorphosis is completed, the tentacles of the adult lophophore have developed, the gut has an elongated U-shape, and tube formation (of mucus and chitin) has begun. [Photos © Douglas P. Wilson.]

gists claim resemble the trochophore (Figure 28-5). Other phylogenies suggest that the phoronid worms are related to the stock which gave rise to the deuterostome phyla. The structure and functioning of the lophophore are the only features that link the phoronids to the ectoproct bryozoans. In relation to the use of modes of development in phylogeny discussed in the last chapter, it is worth quoting in full a sentence from a research report published in 1977 on the embryology of phoronids. It concerns cleavage pattern in the Phoronida, and runs: "The pattern is typically radial though biradial in some stages, but there are instances in which the blastomeres exhibit a spiral appearance." For a devoted follower of Haeckel, this statement could provide any desired premise.

The priapulids are another minor group of marine worm-like animals (see Figure 1-17). Only five species have been described. Their treatment in this chapter is somewhat arbitrary since, among other difficulties, it is not clear whether, they are coelomate or pseudocoelomate animals. In many anatomical features, they are intermediate between the coelomate phylum Sipunculoidea and the pseudocoelomate phyla, including the rotifers and acanthocephalans.

Like the sipunculids, the body is divided into a proboscis-like presoma and a trunk. This presoma is retractable and bears the mouth which is surrounded by a toothed cuticle whose spines are arranged in a series of concentric double pentagons. (Although no relationship is likely, echinoderms show similar circumoral pentaradiate symmetry.) Priapulids are thought to be carnivores, capturing slowly moving prey with the circumoral spines and swallowing it whole. The gut is a straight tube running to a terminal anus, that is, quite unlike conditions in the sipunculids. There are protonephridia made up of large numbers of nucleated solenocytes. The gonads form tangled masses of tubules. Early cleavage stages are radial, not spiral. The larva which results has the cuticle in the form of plates like the lorica of a rotifer, and the adult form is acquired through a series of cuticular molts.

Perhaps the most peculiar feature, and one whose significance is not yet clear, is that there is a spacious open body cavity. This is lined with a thin noncellular layer which also covers the viscera and forms double-sheet slings like the mesenteries of true coelomates. If it were not for its noncellular nature, this would be regarded as being homologous with the peritoneum in groups like the Annelida. In other ways the priapulids resemble pseudocoelomates and early in this century would have been incorporated in the now-discarded superphylum Gephyrea (see Chapters 8 and 27).

Phylum Brachiopoda

The lampshells, phylum Brachiopoda, again illustrate the phyletic unreality of assemblages involving combinations of animal phyla. In zoological textbooks where such groupings are used, the brachiopods are always placed among the protostomous phyla but their coelom is formed in the enterocoelous manner.

Lampshells show a superficial resemblance to bivalve molluscs, though the two valves of the brachiopod shell are always attached dorsal and ventral to the soft parts (shell valves in Bivalvia are lateral), and the feeding organ is a complex circumoral lophophore (Figure 28-6). They are all marine animals and the three hundred or so species presently living are merely a remnant of a once-dominant faunal group. The living brachiopods are placed in 68 genera, while the fossil genera described from Palaeozoic and Mesozoic seas number in excess of 1700. Various authorities place the number of fossil species between 12,000 and 30,000—they were undoubtedly abundant animals and are found as the dominant fossils in many limestone deposits. An extremely complex classification has been set up, but only two main divisions need be mentioned in relation to the presently living forms. The Ecardines are brachiopods whose valves are united by muscles only, without an elaborate hinge, and whose shell is largely calcium phosphate. They lack an internal skeleton for the lophophore and have a gut with an anus. As adults, they may be cemented down to the substratum like oysters (*Crania*), or they may have the shell valves borne on a very long pedicle as in *Lingula* (see Figures 1-18 and 28-7A). The latter is an extreme example of a persistent genus, being found in deposits from the Cambrian to the present—apparently with no change in the anatomical features preserved.

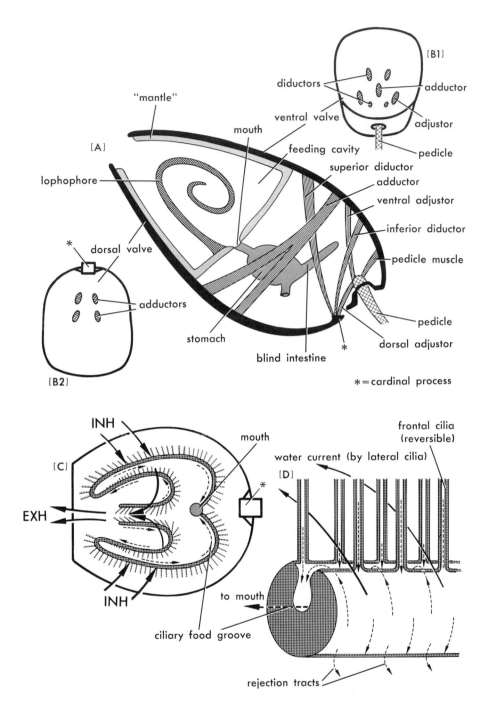

FIGURE 28-6. The functional organization of a brachiopod. **A:** Vertical section through a typical testicardine brachiopod, showing the elaborate systems of adjustor, adductor, and diductor muscles and the position of the lophophore within the feeding cavity. **B1** and **B2:** Ventral and dorsal valves, showing the muscle insertions. **C:** Major water currents employed in feeding, in relation to the coiled arms of the lophophore. **D:** Detail of water currents and ciliation of surfaces on part of a lophophore arm. For further explanation, see text.

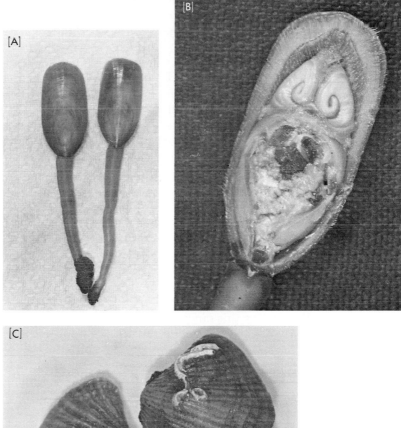

FIGURE 28-7. Modern Brachiopoda. **A:** Two specimens of *Lingula,* a living inarticulate brachiopod, removed from their burrows (compare with Figure 1-18B). **B:** A specimen of *Lingula* with the ventral valve removed to show the coils of the two-armed lophophore in the retracted condition and, closer to the pedicle (or stalk), the gonads and digestive diverticula. The valve length in these specimens of *Lingula* is about 36 millimeters. **C:** Three specimens of a living articulate lampshell, *Magellania* (or closely related genus of terebratulid). The largest has serpulid tubeworms as epizoöns, and is about 37 millimeters across. [Photos by the author.]

In contrast, the Testicardines (or Articulata) have hinged structures uniting the shell valves, an internal skeleton supporting the lophophore, always a short pedicle through an opening in the ventral valve, a gut without an anus, and a shell largely composed of calcium carbonate (Figures 28-6A and 28-7B). Typical genera such as *Terebratula* and *Waldheimia* live mostly in the deeper parts of the oceans.

As with the clams and oysters of the phylum Mollusca, there is a certain uniformity of anatomy and physiology which is imposed in sessile animals using a ciliary feeding mechanism. In lampshells, the feeding mechanism depends on water currents created by the cilia of the lophophore. (It is important to realize that although there are many parallels to the bivalve molluscs, the use of terms like mantle, lateral cilia, and so on does not imply any homologies with the molluscan systems—see Chapter 22.) Lobes of the body wall, termed the mantle, secrete the shells and enclose the feeding cavity with its lophophore. The mouth at the center of the lophophore opens, by way of an oesophagus, to a stomach with one or more pairs of digestive diverticula and a blind intestine (or one which opens to an anus in the part of the feeding cavity with an exhalant water current). There is an elaborate system of muscles which move the shells relative to each other and to the pedicle and also adjust the lophophore, their pattern varying from group to group. Brachiopods are coelomate, with an open circulatory system consisting of a middorsal vessel above the gut with open branches at each end. They may or may not have contractile vesicles in this system. There are one or two pairs of metanephridia which also act as gonoducts and may well be nephromixia. Brachiopods are mostly dioecious and either spawn, generally into the sea, or retain eggs in the feeding cavity to shorten the free-swimming larval life. The larva is trochosphere-like!

When the animal is feeding (see Figure 28-6C), the valves gape and there is an inhalant current from the sides and a median exhalant current. The main cilia creating this current are on the sides of the tentacles along the lophophore and are termed the lateral cilia. It is important to note that the double row of tentacles are placed so as to alternate on either side of the brachial groove (see Figure 28-6D). Food is strained off by the ciliated tentacles and carried by the cilia on their faces (or frontal cilia) into the brachial groove and along this groove on its circuitous route (see Figure 28-6C) to the mouth. These cilia on the faces, however, are reversible and can be adjusted to reject unsuitable food particles to the tips of the tentacles and off these onto the cavity wall and out in the exhalant current, thus providing an almost exact analog of the rejection of unwanted particles by bivalve molluscs as pseudofaeces.

There is another origin for rejected material. This lies in the tracts of cilia on the underside of the lophophore, that is, on the opposite face to the brachial groove. As shown in Figure 28-6D, particles impinging on the lophophore in this area of its surface are not carried to the mouth but again rejected onto the walls of the feeding cavity. It is perhaps one of the points where the efficiency of brachiopods is markedly less than that of bivalve molluscs, that lampshells do not seem to have evolved any elaboration of ciliary tracts for the later disposal of pseudofaeces, nor any neuromuscular patterns resembling the clapping of bivalves, to carry out the further disposal of pseudofaeces. Claims have been made that the lophophore in articulate brachiopods has surfaces which can serve for direct active uptake of dissolved organic nutrients in solution from seawater. It remains to be seen whether this involves a significant net uptake as may be the case in a few annelid worms and other forms, or whether it is negligible in relation to the major filter-feeding function (see discussion of this in Chapters 12 and 22, especially pages 213 and 412). It has also been claimed. perhaps on less doubtful evidence, that areas of the lophophore epithelium in these brachiopods can carry out intracellular digestion, rather like the cells lining the digestive diverticula in molluscs. This could account for the relatively small gut system, which can in some genera be blind. In broader terms, lamp-shells lack efficient mechanisms for the disposal of unsuitable particles, and the tentacles of their lophophores have never been fused (or even functionally coupled by ciliary junctions) into the perforated filter-curtain dividing the inhalant and exhalant parts of the mantle-cavity so characteristic of the eulamellibranch bivalves. Further, the sorting functions of the bivalve labial palps and the associated increase in efficiency in handling both rejected particles and those to be ingested have not been developed in any brachiopod. Despite all this, lampshells seem to be relatively efficient "box-machines" for ciliary filter-feeding, and their decline in importance in the fossil record parallels the increasing abundance of even more efficient filter-feeding machines—the bivalve molluscs such as mussels, clams, and oysters.

The details of structure and the finer aspects of functioning in lophophores are both more easily studied and already better described in brachiopods. More comparative studies could establish whether or not the lophophores in the other kinds of animals which bear them, including the Bryozoa and Phoronida, are structurally and functionally homologous with those in brachiopods. Obviously, lophophore homology would be of considerable phyletic significance.

Sipunculids and Arrow-worms

The minor phylum Sipunculoidea is rather distinct from the previous groups, and indeed from most other phyla. Some zoologists have tried to link these unsegmented, worm-like animals to the Priapulida, and thence to such acoelomate minor groups as the Rotifera and Gastrotricha, but this is indefensible. Though Sipunculids are externally similar to the Echiuroidea, which are probably related to annelid worms (see above, pages 345–346), there are no traces of metameric segmentation nor of setae. The sipunculids are unsegmented worms with the anterior third of the body as an introvert and with a fundamentally U-shaped gut running to an anus near the anterior (see Figure 1-19). This alone would suggest some relationship to the other more sessile minor coelomate groups such as the ectoprocts and phoronids. There are nearly three hundred species of sipunculid worms, all marine, but worldwide in distribution. A typical genus, *Golfingia*, is found burrowing in intertidal, muddy sandflats, in temperate seas. (These include the muddy shores off the "Royal and Ancient" golf-course at St. Andrews, Scotland, and the generic name results from an encounter of W. C. McIntosh and E. Ray Lankester.) As in other worms, the muscles of the body wall in *Golfingia* are organized around a hydrostatic skeleton of coelomic fluid. Extremely high coelomic pressures can be generated when grossly overstimulated, and moderately high coelomic pressures (of the order of 200 centimeters of water) are normally developed during protrusion of the introvert. In normal burrowing locomotion, however, the coelomic pressures generated are an order of magnitude less. The burrowing cycle consists of everting the introvert into the substrate, expanding its end to form a mushroom-anchor, and then pulling the body up to this by contraction of longitudinal body-wall muscles. It is likely that the highest coelomic pressures are normally only associated with faster escape reactions.

It is of some ecological and evolutionary interest that this representative of a minor phylum, *Golfingia,* commonly lives in the substrate of mudflats with a wide variety of other worm-like animals, and in probable competition with some of them. These include, besides a variety of annelid worms, burrowing sea-anemones, representatives of the Rhynchocoela and even of the Echinodermata (the worm-like holothurian *Leptosynapta*). *Golfingia* was long thought to be a detritus-feeder, and thus competing with the other species of this diverse faunal assemblage for finite supplies of organic detritus in the mudflats. However, Gunnar Thorson discovered that, in one Danish population of *Golfingia,* at least, these sipunculids were feeding as temporary ectoparasites on the large polychaete "sea-mouse" *Aphrodite* (see Figure 12-3). Thorson describes them as lying in wait partially buried until a sea-mouse passes, then attaching by insertion of the introvert into the coelom of the polychaete, and feeding while being towed along by the host.

The arrow-worms of the marine plankton form the minor coelomate phylum Chaetognatha. Although clearly deuterostomous in their early development and having a modified enterocoelous development of the body cavity, chaetognaths do not resemble the other deuterostomous phyla in any way in their adult anatomy (see Figure 1-19). Adult arrow-worms even lack a peritoneum lining the coelom. The phylum is clearly a minor one with sixty or so species, but local abundance of arrow-worms can be of considerable importance in the ecology of the marine plankton (Figure 14-11). The majority of the species in the phylum are placed in one genus, *Sagitta.* Coelomate and unsegmented, they have transparent, torpedo-shaped bodies showing perfect bilateral symmetry. The mouth is surrounded by a group of grasping spines, and arrow-worms are predaceous on other members of the zoöplankton. They have horizontal fins on sides and tail, and pursue their prey with surprisingly rapid darting movements. The body-wall musculature responsible for their swift swimming consists of contractile elements with undifferentiated cell bodies somewhat similar to those found in nematode worms. Indeed, this undifferentiated musculature and the lack of an appropriate coelomic lining make the body organization of chaetognaths surprisingly similar to that of the successful pseudocoelomates. However, chaetognath development makes it clear that this cannot be other than convergence of structural organization. The nervous system consists of a pair of cerebral ganglia, associated with a pair of eyes, and a large ventral trunk ganglion. The digestive system consists of a straight tube from the mouth to the anus near the tail and it is surrounded by a coelom with septal divisions separating it into head, trunk, and tail compartments. The trunk section is usually laterally paired and the tail section may be. Some comparative zoologists have claimed that this shows more than accidental resemblance to the three- or five-fold system of divisions of the developing coelom shown in the phylum Chordata and its assumed allies. There are apparently no organs or tissues specialized for excretion, circulation, or respiration. Arrow-worms are hermaphroditic with a pair of ovaries and a pair of testes. Self-fertilization is possible. The eggs are planktonic and, as already mentioned, development involves radial cleavage, gastrulation by invagination, and is deuterostomous and enterocoelous.

Although there are few species of arrow-worms, they can be among the commonest planktonic animals. They are often present in enormous numbers with significant effects on the marine economy, for example, by the destruction of whole broods of eggs or young stages of some of the commercially important fishes. Species of *Sagitta* are important as planktonic indicators of oceanic waters of different origins. Bodies of water which differ only very minutely in their physical and chemical characteristics are often typified by having, as biological indicators, specific arrow-worms. In many cases, these biological indicator species are more useful than microanalyses to oceanographers interested in the proportionate origins of bodies of oceanic water. A classical example of this is found in the temporal changes in the hydrography of the North Sea. To summarize a fairly complicated relationship, the presence of *Sagitta elegans* replacing *Sagitta setosa* is an indication of considerable influx of water from the northern Atlantic, and usually occurs in late summer and early fall of each year. Changes in micronutrients which accompany this influx of oceanic water are of great importance to the productivity of the area. Thus a series of census of *Sagitta* opens up a real possibility of detailed prediction of the returns from herring fisheries in the area. Arrow-worms appear in many of the figures of marine zoöplankton in Chapter 33.

A Partial Survey

In Chapter 34 of this book we will discuss the probable interrelationships of the major phyla of invertebrates. In it, we will conclude, as we did in Chapter 8, that the six minor pseudocoelomate phyla are best regarded as a series of separate stocks with some perhaps distantly allied to the nematode worms. There are a total of twelve minor coelomate phyla and of these one (phylum Hemichordata) has a clear connection with the chordates, and another (phylum Pogonophora) a much more tenuous one. Four other minor coelomate phyla have obvious annelid-arthropod relationships (Chapter 18). The remaining six surveyed briefly above are only obscurely, if at all, related to each other. Some authors have claimed that phoronids might represent survivors of a stem group both for the ectoproct bryozoans (whose peculiarities might be ascribed to miniaturization of lophophore-bearing along with the colonial "box" habit), and for some ancient ecardinate stock of brachiopods. Even if this were acceptable, this "stem-group" might spring from the Echinodermata-Chordata stock (the "deuterostomes"), or be related to the Annelida, however distantly. If the latter were true, then the sipunculids, perhaps along with the priapulids and even the phylum Echiuroidea (Chapter 18) might all share the relationship. A similar assemblage but leading back to minor pseudocoelomate forms like rotifers has been suggested, but it defies rational defense. Such features as a worm-like body with a large eversible presoma, or proboscis, or introvert, could well be independently evolved features of adaptive significance for burrowing and unspecialized microphagy in muddy-sand sea bottoms. Further work on fine structure and functioning in lophophores *might* establish homologies between these structures in ectoprocts, phoronids, and brachiopods, but it might establish instead that these lophophores (and those in phylum Entoprocta) are merely adaptational parallels like presomas and introverts. As already noted, the phylum Chaetognatha stands apart from these other minor coelomate groups. However, the tripartite coelom in their development has been used to link them to the echinoderms and chordates on the one hand and to the three lophophore-bearing phyla on the other. In my own partial view, the tripartite coelom is about as valuable a line of evidence on relationships as is possession of spiral cleavage, or of a trochophore planktonic larva—*and no more*.

29 ECHINODERMATA I: DISCERNING A UNIQUE ARCHETYPE

THE echinoderms form the most easily recognized phylum of animals. This has two implications. They show no clear relationship with any other phylum, except for the distant association with the chordates which will be discussed later. A second implication is that the features considered diagnostic of the phylum are universally possessed, and are both obvious and of functional importance.

The diagnostic feature most often stressed in textbooks for this group is the possession of radial symmetry with a pentaradiate basis. Two other unique features of the peculiar anatomy of echinoderms would seem to be of even greater physiological significance. Indeed the functional potentialities (and limitations) of the phylum Echinodermata would seem to hinge on the possession *both* of a water-vascular system (as a subdivision of the coelom) and of the skin ossicles which give the phylum its name. To anticipate later details, animals like sea-urchins and starfish are moved by numerous tube-feet (hydraulically operated by the water-vascular system), each of which is mechanically supported by passing between the numerous ossicles which closely approximate, or through a pore or pair of pores in one of them. It is difficult to visualize any way in which these many tiny locomotor units could create a gross movement of the starfish relative to the substratum, unless their muscles were inserted on and through such a skeletal system of dermal

505

ossicles. Another of the more unusual characteristics of echinoderms results from the fact that the interlocking system of ossicles (in forms like sea-urchins) consists of units which have to be enlarged during growth. This results in patterns rather more rectilinear than are found in most other living organisms. A glance at the test of a sea-urchin (or at Figure 30-3B) will show the nature of this rectilinearly integrated substructure.

Living Representatives and Classification

The starfish, sea-urchins, brittle-stars, and the like which make up the phylum (see Table 29-1) are entirely marine, but they are widely distributed throughout the seas and are a successful group by any measure. There are also huge numbers of echinoderm fossils and several groups of them have been established as successful marine animals over long periods of geological time.

They are triploblastic, coelomate animals (see Chapters 1 and 8), and show no trace of metameric segmentation. The five- or tenfold radial symmetry which they show as adults is diagnostic, although most echinoderms are bilaterally symmetrical during their larval development, and it would seem that the group shows secondary radial symmetry derived from originally bilaterally symmetrical ancestors. This will be discussed more extensively in the next chapter. Echinoderms have no heads, nor do they show an anterioposterior axis (Figure 29-1). The nervous system is without a brain and remains superficial throughout life, much of it in contact with the ectodermal epithelia from which it is derived. There is a calcareous endoskeleton of ossicles in the dermal layers (in some echinoderms this makes up the bulk of the animal), and these often are associated with external spines or bosses. The coelom in the adult supposedly derives from three divisions (or more correctly three pairs of divisions) in the embryo. In the adult echinoderm it is anatomically much subdivided, and of

TABLE 29-1
Outline Classification of the Echinodermata

Subphylum III ELEUTHEROZOA
  Class 6   SOMASTEROIDEA (fossil only)
  Class 7   ASTEROIDEA (at least five orders extant: starfish, sea-stars)
  Class 8   OPHIUROIDEA (at least two orders extant: brittle-stars)
  Class 9   ECHINOIDEA (sea-urchins)
     "REGULARIA" (at least two orders extant: sea-urchins, sea-hedgehogs)
     "IRREGULARIA" (at least two orders extant: heart-urchins, sand-dollars)
  Class 10 HOLOTHUROIDEA (seven orders extant: sea-cucumbers, trepangs)
  Class 11 HELICOPLACOIDEA (fossil only)
  Class 12 OPHIOCISTIOIDEA (fossil only)
Subphylum I HOMALOZOA
  Class 1   CARPOIDEA (fossil only)
Subphylum II PELMATOZOA
  Class 2   CYSTOIDEA (fossil only)
  Class 3   BLASTOIDEA (fossil only)
  Class 4   CRINOIDEA (at least two orders extant: sea-lilies, feather-stars)
  Class 5   EDRIOASTEROIDEA (fossil only)

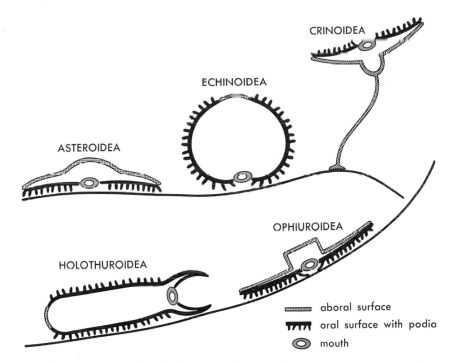

FIGURE 29-1. Characteristic body form and posture in the five major groups of living echinoderms: Asteroidea (starfish or sea-stars), Echinoidea (sea-urchins), Crinoidea (sea-lilies and feather-stars). Ophiuroidea (brittle-stars), and Holothuroidea (sea-cucumbers or trepangs).

varied functional significance. One part is the water-vascular system with its associated external podia (which can function as tube-feet in locomotion, or as tentacles in a food-collecting mechanism). Other subdivisions of the coelom surround the general viscera, and are associated with organ systems such as the so-called haemal system. The strands of lacunar tissue which make up this latter system are each enclosed in a tube of coelomic origin. Its anatomical layout (see Figure 29-2A) resembles that of the water-vascular system and of several other organ systems. The usual ground plan for any echinoderm organ system involves a circumoral element with five radial elements diverging from it, and thus the diagram of the layout of any system is similar to all the others. Since there is no head or bilateral axis of symmetry, the terms dorsal and ventral, anterior and posterior, are completely inappropriate. Anatomically there are two surfaces, which do have considerable physiological significance. One surface is the oral or ambulacral—bearing not only the mouth but the ambulacra or areas where tube-feet protrude from the surface. The other surface is termed aboral, and never has external projections of the water-vascular system as podia of any type. Of the five common living classes of echinoderms, four are more motile and keep their oral surfaces down to the substrate (see Figure 29-1). The fifth group, the crinoids (and perhaps many extinct groups of echinoderms), have the aboral surface directed downward and the mouth facing up. As shown in

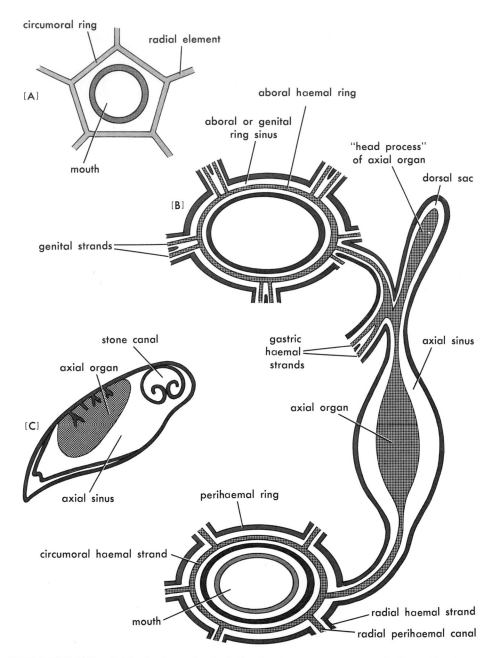

FIGURE 29-2. **A:** The basic anatomical plan for many organ systems in the echinoderms consists of a circumoral ring element and five radial elements. This diagram could serve for the central parts of the ectoneural nervous system, or of the oral haemal system, or of the water-vascular system. **B** and **C:** Stylized perihaemal coelomic and haemal systems in a starfish For further description, see text.

508

the outline classification (Table 29-1), which incidentally overstresses the comparative importance of the living groups, the orientation of the surfaces corresponds to the major subdivision of the phylum into the Eleutherozoa (including the four motile classes), and the Pelmatozoa (including several fossil groups and the Crinoidea). It should be noted that this outline, and our treatment of the living forms here, inverts most phylogeny, since the Pelmatozoa and Homalozoa are undoubtedly more ancient groups.

It should be realized that considerable controversy surrounds the classification of the echinoderms. The treatment adopted here, and in Table 29-1, represents a compromise, and may thus prove unacceptable to all authorities. To the most usual arrangement of living echinoderms into Pelmatozoa and Eleutherozoa is added a third subphylum Homalozoa. In the earlier paperbacks an additional class Somasteroidea (with one living representative, *Platasterias*) was added to the four extant classes of the Eleutherozoa. This group is now regarded as entirely fossil. Among H. Barraclough Fell's provocative and persuasive reassessments of echinoderm phylogeny was the claim that this group of "somasteroids" in some ways formed a link group between crinoids and asteroids, and thus between the major subphyla. More detailed consideration of ossicle organization in *Platasterias* by F. Jensenius Madsen and D. B. Blake among others has established that this form is related to *Luidia* (Figure 29-3) and is thus a true though primitive member of the Asteroidea. In a more sweeping—though more firmly based—rearrangement of classes proposed by Fell, the two living subphyla Eleutherozoa and Pelmatozoa were rejected and replaced by a division into four subphyla based largely on the organization of growth gradients in adult forms (i.e., distinguishing between those with meridional gradients and those with radially divergent gradients). Some of the phyletic aspects of this will be discussed in Chapter 30, but it is worth mentioning here that authors utilizing the customary classification (including Hyman) have long admitted that the Eleutherozoa probably do not represent a monophyletic grouping. In other words, it is freely admitted that the more motile echinoderms have originated from the more ancient crinoid-like ancestors along at least two differing lines. Inclusion of a separate subphylum Homalozoa to encompass the "carpoids" again departs from the "usual" classification. This is certainly justified (even in a neontological account) since the fossils of the class Carpoidea could represent the most ancient echinoderms. Found in Palaeozoic deposits, they are bilateral with dorsoventral flattening and show no evidence of radial symmetry. If it could be established that the carpoids were not antedated by radially organized echinoderms, and thus were not secondarily bilateral (like some forms discussed in the next chapter), then such "preradial" status would give them the utmost significance in echinoderm (and perhaps chordate) phylogeny.

Before we discuss the peculiar functioning systems of echinoderms, it is perhaps worthwhile to outline the characteristic features of these five living classes, from whose common peculiarities we can then try to deduce something of an archetypic echinoderm. The true starfish or Asteroidea have usually five arms which are not clearly marked off from the central disc and these fleshy arms contain lobes of the alimentary canal and gonads and other internal structures. The ossicles of starfish are embedded in a tough but flexible skin. The

FIGURE 29-3. Tube-feet on the oral side of a relatively primitive starfish, *Luidia ciliaris*. [Photo © Douglas P. Wilson.]

Ophiuroidea or brittle-stars have arms which are sharply demarcated from the disc and do not contain lobes of the alimentary canal (Figure 30-1A). The ossicles, particularly in the arms, form an almost continuous articulated armor. The sea-urchins or Echinoidea many of which tend toward a spherical shape, lack arms and have the aboral side reduced (see Figures 29-1 and 30-3B). They have a continuous armor of fused ossicles, and also long movable spines which in many cases assist the tube-feet in locomotion. The so-called irregular echinoids show yet another return to bilateral symmetry, in this case associated with a burrowing life in the sea-bottom. The sausage-shaped sea-cucumbers and trepangs of the class Holothuroidea lack arms and have a body drawn out oral-aborally, and a leathery skin with few small scattered ossicles (Figure 29-1). The only surviving Pelmatozoa, the sea-lilies and feather-stars of the class Crinoidea, have a greatly reduced disc and the arms showing great development of branching (Figure 30-1B). They are stalked, or at least attached, for the whole or part of their life. Only a few living forms survive of this once enormous class and subphylum.

General Morphology and Functioning of Systems

The eleutherozoan echinoderms represent one of the few groups of invertebrate animals where setting up a functioning archetype contributes less to an understanding of the physiology of the group than does a thorough appreciation of actual functioning in a form such as the starfish, *Asterias*. This is so for two reasons. First, the archetypic echinoderm must have had a physiology close to that of present day crinoids, and totally unlike that in the other groups. Secondly, most physiological investigations have been carried out on starfish and sea-urchins, and the former group represent in many respects the least specialized (or most "archetypic") of the Eleutherozoa. For once, looking at a "type," the starfish, is a better way toward an understanding of physiology in the group.

Beginning with the skin and body wall, the diagnostic characteristics of the phylum are shown and, particularly on the oral surface of the starfish, the epithelium is ciliated. This epithelium is supported by the ossicles formed in the dermis, linked together by dermal connective tissues. The ossicles occasionally penetrate the surface as plates, or spines, and some are movable, especially along the sides of the ambulacral grooves. This results in a range of appearance in different starfish from a warty or tuberculate surface to a prickly "exoskeleton." Besides the calcareous protrusions, pockets of thin skin are developed. These papulae are mostly aboral and have a respiratory function, because within them there is a prolongation of the general visceral cavity (part of the coelom) and they thus form regions of increased surface area and decreased thickness where respiration can go on. The similar protrusions within the ambulacral grooves (the podia) will be considered along with the water-vascular system. A fourth sort of skin appendage occurs on all surfaces except the ambulacral grooves. The pedicellariae are of various sorts (see Figure 30-2A) and their function appears to be protection of the papulae and the rest of the skin by capturing and holding small organisms. The pedicellariae in starfish are non-poisonous, though in sea-urchins they have associated poison glands. Once closed on an object, the pedicellariae can remain so for some hours. To some extent the pedicellariae resemble the nematocysts of coelenterates in that they are apparently independent units which respond in a local reflex to a local stimulus. Again, like nematocysts, they may in some cases require a double stimulation, both mechanical and chemical.

The nervous system is always superficial, and this is even more obvious in starfish than in the other groups. The most obvious system—the oral or ectoneural nervous system—lies immediately under the thin ectodermal epithelium, with a circumoral ring and five obvious radial nerves running closely applied to the ambulacral grooves (Figure 29-2A). These are, however, only the most obvious elements in a general subepidermal plexus which covers the entire body surface. Further, the ectoneural nervous system is only one of three nervous systems in the starfish. The entoneural nervous system, which is more important in crinoids, runs along the ossicles at the sides of the ambulacral grooves and provides motor innervation to the muscles between the ossicles. It is also connected to a nerve net in the coelomic lining of the body wall. The third system, the hyponeural nervous system, has radial elements on the sides of the perihemal canals and is connected to the local motor ganglia of the tube-feet. Perhaps the easiest hypothesis about the archetypic organization of nervous structures in echinoderms would be that there were originally networks connected to six centralized systems, that is an oral and aboral manifestation of ectoneural, entoneural, and hyponeural systems. Under this hypothesis, the different vestiges of these systems which are important in the different groups of living echinoderms would be those which had functional significance ap-

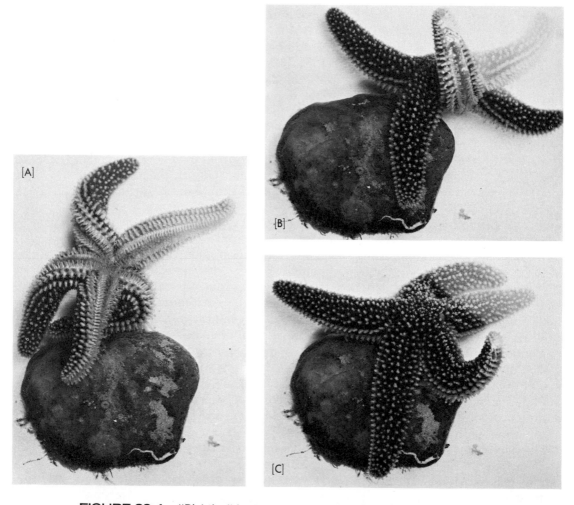

FIGURE 29-4. "Righting" in the sea-star *Asterias.* **A–C:** Three successive stages in the process of returning to the natural posture (with oral surface downward) in a young specimen of *Asterias,* taken at 4 second intervals. Such righting movements are much more rapid than the normal deliberate locomotion of sea-stars. Note the use of the tube-feet, and that in **A,** the sea-star has two alternative arms twisted to bring the tube-feet onto surfaces for attachment before the levering action of righting. The arm attached to the small stone was actually used in the "flip" (shown ending in **B**), and then the sea-star crawled back onto the stone (**C**). [Photos by the author.]

propriate to the group involved. If we consider only the neural aspects of loco-motion, the motor innervation to the muscles on the outside of the body wall immediately below the epidermis would be important in one group while the motor innervation to the muscles between the ossicles would be important in another. Similarly, *oral* centralization might seem to be adaptively more signif icant in starfish while aboral centralization might be appropriate to crinoids. Such reduction to one appropriate "system" does in fact occur.

As regards the functioning of starfish, much of their behavior involves local reflexes, for example the local responses of tube-feet, of pedicellariae, and of movable ossicles. These local reactions are not dissimilar to those found in coe-lenterates such as sea-anemones (see Chapters 3 and 4) and probably involve similar patterns of neural "spread," with decremental conduction and inter-neural facilitation both being involved. However, in starfish considerable "cen-tral" coordination is possible. The "righting" movements which are easily in-vestigated by any student with a healthy starfish must be centrally controlled (Figure 29-4). The normal processes involved in feeding on bivalves also re-quire overall coordination. The sense-organs are mostly diffuse and there are both tactile and chemoreceptors all over the surface. In most starfish the termi-nal tube-foot of each groove lacks a sucking disc and has an obvious pigment spot in its base which surrounds a group of ocelli.

Echinoderm Guts

As with other organ systems, in the alimentary canal of echinoderms, in spite of nomenclature, we are considering organs which are certainly not homolo-gous with those similarly named in other phyla. In typical starfish, the basic layout of the alimentary canal is that of a tube extending vertically from the mouth to the anus on the aboral, or upper, side. This last opening is almost nonfunctional. The minor elaborations of the system can be seen in Figure 29-5B. The stomach is completely eversible, both during feeding on small bi-valves and such and in extruding undigested waste. Most starfish are predatory carnivores, though a few species are known to be mucus-ciliary gatherers of particulate matter. Many textbooks suggest that toxins are secreted by the stom-ach wall during the feeding of forms like *Asterias* on small molluscs. It is most probable that only digestive enzymes are involved: originally secreted in the pyloric caeca, these lytic fluids can be passed into the folds of the everted stom-ach wall. Recent work has shown that in certain predaceous starfish, folds of the stomach wall are insinuated within the molluscan shell into its soft parts and release enzymes which begin the digestive process. This invasion of the prey by folds of the stomach wall can take place within the starfish in the case of small molluscs, or totally externally, so that the prey is digested outside the body of the starfish. Digestion is, of course, entirely extracellular (except possi-bly for some fatty substances). The ciliation of the central parts of the gut is all directed orally and is concerned not only with expulsion of undigested particu-late matter, but also with bringing to the folds of the stomach wall the powerful digestive enzymes which have been produced in the glandular pockets of the pyloric caeca. Little, if any, enzymes are produced by the gastric-wall epithelia

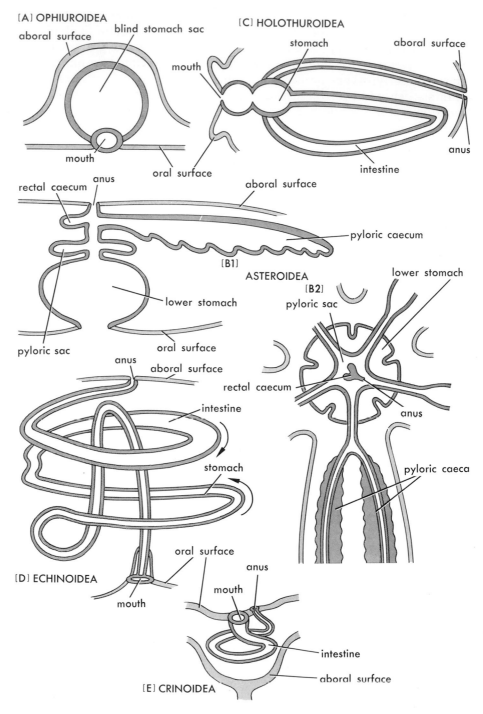

FIGURE 29-5.   Gut patterns in echinoderms. **A:** The simple blind sac of brittle-stars. **B1** and **B2:** Lateral section and aboral view of the gut of starfish with its elaboration of diverticula. **C:** The elongate tubular gut of sea-cucumbers. **D:** The similarly elongate gut of regular sea-urchins with its peculiar apposed loops. **E:** The relatively minute tubular gut of crinoids, which terminates in an anus alongside the mouth.

themselves, and the functional significance of the intimate contact with prey tissues which seem necessary to successful digestion remains obscure. Digestion and absorption go on in both stomach and pyloric caeca. The latter are also concerned with enzyme secretion, with storage of nutritional reserves (masses of glycogen but also surprisingly high quantities of lipids), and, in some forms, with ciliary "sorting" of partly digested food material.

Something of the variation found in the guts of echinoderms is shown in Figure 29-5, the ophiuroid pattern being simplest as a blind sac without an anus and without any diverticula. The other three groups, crinoids, echinoids, and holothurians, have rather more elongate guts lacking diverticula. Recent work with radioactive tracers has shown that the main nutrient transport within star fish is through the coelomic fluid in the general visceral coelom. Such transport is probably also true of echinoids and holothurians; little is known about transport in ophiuroids or crinoids. It seems almost impossible in the latter group where the visceral coelom is open through a pore system to the sea. It is worth mentioning here that the coelomic fluids of the visceral cavity also serve as the main respiratory transport system in the more massively bodied echinoderms: starfish, sea-urchins, and holothuroideans. There is no true circulation, but food components and oxygen are carried along diffusion gradients, aided to some extent by movements of the coelomic fluid. It is likely that this method of nutrient and respiratory tranport represents one of the physiologically limiting features of the echinoderm functional plan. Faster moving animals require a true circulatory system for these purposes.

Water-Vascular System

As stressed earlier, many of the unique features of the functioning of echinoderms depend on the existence of the part of the coelom termed the water-vascular system (WVS). The system (Figure 29-6) consists of a circumoral canal and five radial canals with paired side branches, each leading to an internal ampulla and its externally protruded tube-foot. Each ampulla connects between ossicles (starfish), or through pores in an ossicle of the ambulacral groove (sea-urchins), to a single tube-foot externally. There is also a single asymmetric canal leading from the oral ring toward the aboral surface. This "stone" canal opens into a small dorsal ampulla which connects through many pore canals to the external surface. The plate bearing these pore canals is termed the madreporite. As shown in Figure 29-6, there are also nine small sacs inside the ring canal, termed Tiedemann's bodies, and some starfish, though not *Asterias*, have larger sacs (presumably for water storage), termed Polian vesicles, in approximately the same anatomical position. The numbers for radial canals and for sacs given here refer to the large number of forms of sea-stars which have—like *Asterias*—five arms. However, forms like the sun-stars, *Crossaster*, with ten or eleven arms, and like some specimens of *Luidia* (Figure 29-3) with seven, will have appropriate numbers of radial canals and Tiedemann's bodies. As discussed below, each ray has a unit set of organ structures.

If we examine the functional organization of the system in *Asterias*, a slight

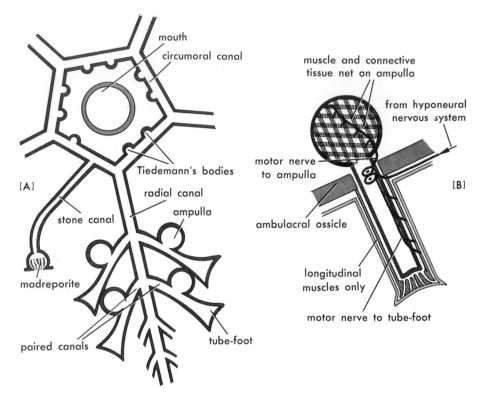

FIGURE 29-6. **A:** Stylized water-vascular system of a starfish. All the elements shown lie near the oral side except for the stone canal running to the aboral madreporite. **B:** Stylized organization of a single tube-foot with its ampulla. For account of general functioning and mechanics, see text.

water current can be observed to pass continuously in through the madreporite and this is thought to make good any water losses to the exterior through damaged tube-feet. There is, however, no clear experimental proof of this. In fact, it is claimed that some starfish with the madreporite surgically removed, or experimentally blocked, can live and move using the tube-feet as usual. Actually, each ampulla—tube-foot system seems to be functionally separated by valves from the rest of the WVS and experimental preparations can be made of a single ampulla with a single tube-foot which can be caused to continue to function. There is some variation in the microanatomy of tube-feet. In *Asterias* and other forms, there is a muscle-operated vacuum-cup sucker at the outer end. Other forms have long gland cells that produce a sticky secretion, and still others have neither and possess tentacular or pointed tube-feet. All podia, however, have a similar basic microanatomy (which we know largely from the work of J. E. Smith), the walls of the podium, or tube-foot itself, having only longitudinal muscle (Figures 29-3 and 29-6B). Thus stimulation of the tube-foot can bring about only its contraction or withdrawal. The ampulla is well-muscularized with a roughly rectangular meshwork of muscles and connective tissue surrounding the bag. Contraction of the ampullar muscles, with the valves to the rest of the WVS closed, causes extension and elongation of the tube-foot. We thus have a local effector unit working around a hydraulic skeleton. Local nervous arcs were demonstrated by J. E. Smith, and also connections of these to the hyponeural system through a ganglion cell near the neck of each ampulla, which acted as an individual motor center.

As well as starfish, echinoids and holothurians use their tube-feet in "stepping" locomotion. The forces applied to the substratum can involve both levering and pulling (or traction). In starfish, it seems that in normal horizontal movement the tube-feet are used mainly as levers, while in near-vertical movement, the traction method is used. Levering and postural bending is accomplished by contraction of the longitudinal muscles on one side of the tube-foot which work against the hydraulic pressure of the system, the pressure being maintained by partial contraction of the muscles of the ampulla. As already emphasized, it is difficult to imagine the myriad tiny locomotor units of a starfish creating any gross movement relative to the substratum unless they were supported by the unique skeleton of dermal ossicles, appropriately placed.

Another important function of the ampulla–tube-foot systems is to serve as respiratory organs. It has been demonstrated by A. Farmanfarmaian that the main sites of ingress of oxygen into the inside of sea-urchins is by way of exchange from the fluid within the ampulla to the perivisceral coelomic fluid. It is significant that in echinoids the ampulla is connected to the tube-foot by two canals passing through two pores in each ambulacral plate. In an expanded system a current of water passes continually out through one canal to the tube-foot and back to the ampulla via the other. It has recently been shown that even in starfish, similar ciliary currents cause a circulation of fluid to be maintained within each system, with water entering each tube-foot on the outside of the starfish arm and leaving it on the side toward the ambulacral axis. A number of workers believe that respiration must have been the archetypic function of the WVS and its podia. This attractive hypothesis suggests that these structures, unique to echinoderms, were first used by pelmatozoans for respiration and that they lay on either side of the food groove to derive advantage from the ciliary water currents along it. Later they became muscular tentacles which assisted in feeding, and still later became the organs of locomotion in the eleutherozoan echinoderms. Because of the mechanical importance of the recognizable ambulacral ossicles, stressed above, this is one of the few phylogenetic problems which may eventually be solved by palaeontological work on the most ancient echinoderm fossils. It is worth noting here that the unit set of organ structures which makes up an arm in a sea-star has itself serial features analogous to the sequence of metameric units in an annelid. As John M. Anderson has pointed out to me, in each asteroid ray, the skeletal, muscular, nervous, and WVS structures are serially repeated in a bilateral fashion in relation to the ray axis—a pattern functionally integrated into the "units of locomotion" discussed above. Even in its mode of growth the ray mimics the annelid; with the oldest part, the advancing tip, generating new units in a budding zone just behind it (as in the posterior budding of annelid-arthropod metamerism, see Chapter 10). The capacity in starfish for regeneration of rays broken off by accident or autotomy is also based on this kind of repetitive morphogenesis as in polychaetes (see Chapter 12).

Haemal System and Gonads

As seen in starfish, the haemal system consists of strands of spongy tissue, which never form distinct vessels with walls, each one enclosed within a closed coelomic channel or perihaemal space. The main layout consists (as usual) of a circumoral ring and five radial haemal strands on the oral side (Figure 29-2B). There is a single axial organ enclosed within the axial canal which runs in close association with the stone canal of the WVS to the aboral surface where there is an aboral, or genital, ring sinus containing the aboral haemal ring and its genital branches. There is some slight contractility of the axial sinus, particularly where it surrounds the head process of the axial organ near the aboral side. A few years ago this was claimed to constitute, in echinoids, a heart within a circulatory system. There is no evidence whatsoever of true circulation or even of a connected system of lumina within the axial complex, or anywhere else in the system of haemal, spongy tissues. It may be that the contractility is a functional vestige of the contractile structures, long known in the larval stages of echinoderms. More recently, it has been demonstrated that surgical removal of the axial complex does not affect the respiratory rates in some echinoids, thus disposing of one aspect of circulatory function. Further, recent work by Norman Millott has demonstrated that the functional significance of the axial organ itself lies in its capacity to respond to invasive particles or organisms within the visceral coelom of the echinoderm. By producing large numbers of amoebocytes in a sort of "immune response," it plays an important part in the defense of the echinoderm against both mechanical injury and disease organisms.

Another great importance of the haemal system lies in its developmental connection with the gonads of echinoderms. In starfish, the genital system consists typically of ten gonads, two in each arm. In young stages, each gonad is enclosed in a genital sac which is an outgrowth of the genital, or aboral, sinus of the perihaemal-coelomic system. The sexes are normally separate and usually each gonad has a separate, small gonopore to the exterior near the base of each ray. Of course, there will be sun-stars with 22 gonopores and 22 gonads, and other variants on the basically tenfold replication. In certain primitive genera we have, as again pointed out by John M. Anderson, a serial arrangement of small gonad units down both sides of each arm. In no echinoderm is there any great elaboration of secondary sexual structures or auxiliary duct systems, and most spawn eggs or sperm through simple gonopores directly into the sea.

In many cases, a stimulus to spawning appears to be provided by the presence in the water of gametes of the other sex. It is clear that a hormone-like substance produced in the radial elements of the ectoneural system is responsible for the actual expulsion of eggs in spawning, and probably also concerned in the earlier processes of egg maturation. It is possible, but much less clearly established, that tissues connected to the madreporite (perhaps in the axial organ) serve as the chemoreceptors for the detection of "pheromonal" substances accompanying the gametes of the opposite sex, and that such detection triggers the release of radial-nerve substance which induces contractions in the gonad walls and hence spawning. In the majority of echinoderms, fertilization is external, and some aspects of the pelagic larval development with its metamorphoses will be discussed in the next chapter.

30 ECHINODERMATA II: VARIANT FUNCTIONAL PATTERNS

JUST as the diagnostic anatomical features of the echinoderms as a whole have a basis in a unique pattern of functioning—involving the dermal ossicles as skeleton and the hydraulically operated podia as effectors in a pentaradiate organization—so each of the five major living classes has a rather clearly defined structural pattern of functional significance. Within each class, more extensive homologous features, both structural and functional, can be discerned, and there is also a "way of life" or general pattern of ecology which is common to the members of each major subdivision.

Brittle-stars

The significantly distinctive features of form and function in the Ophiuroidea (brittle-stars) seem to spring from the use of the whole arms as the locomotor organs with a corresponding reduction in the importance of tube-feet. The long, slender, muscular arms imply not only peculiarities of the ossicles as a skeleton for these muscles, but also that there are no lobes of the gut or gonads outside the central disc. Further, the podia are without suckers, are often reduced, and they are protracted by a system of vesicles and head-bulbs which is probably less efficient than the ampulla–tube-foot system already described (which is possessed by all the remaining motile echinoderms).

519

[A]

[B]

FIGURE 30-1.   Brittle-stars and feather-stars. **A:** *Ophiocomina nigra,* the black brittle-star, and *Ophiothrix fragilis,* the European "common" brittle-star. **B:** *Antedon bifida,* speci-mens of the rosy feather-star, with at the bottom a small regular sea-urchin covering itself with shells. [Photos © Douglas P. Wilson.]

Brittle-stars are surprisingly abundant animals. Their powers of homeo-stasis being slight, they do not occur to any extent between tidemarks and are thus rarely seen by shore-collecting students. However, throughout the oceans of the world, on shallower and deeper bottoms, brittle-stars are among the most abundant, successful animals (Figure 30-1). For a time after the Second World War, when techniques of submarine photography and underwater television were being developed, it seemed that almost every photograph or television probe of the bottom showed heaps of brittle-stars or regularly spaced patterns of them like carpets. Most ordinary brittle-stars can be described as carnivore-scavengers. They are predatory upon all suitable small organisms, but, on the soft substrata where they abound, spend much of the time sorting through the detritus and selectively picking out organic material for ingestion. The aberrant group of Euryalids (Figure 30-3A), and perhaps some other brittle-stars, feed by waving their arms in the water and catching planktonic organisms—mainly small crustaceans.

In ophiuroids, the axis and much of the arm is occupied by a series of mas-sive ossicles which have been termed vertebrae (Figure 30-2B), to which are attached four series of longitudinal muscles. These can carry out different movements depending on the articular surfaces of the ossicles, but in most ophiuroids simultaneous contraction of all four muscles between two adjacent vertebrae will bring about autotomy or breaking off of the arm. Most brittle-stars use this as an escape reaction. All have great capacity for regenerating not only arms but parts of the disc and internal organs.

Living species of the class fall into two very unequal groups. The Ophiurae or typical brittle-stars have vertical articular surfaces in contact between the vertebrae, and this allows mainly movements in the horizontal plane. There are about 1900 species in this group, which makes it the most numerous order of the phylum Echinodermata. The other order—the Euryalae or basket-stars—consists of forms where the arms can move vertically as well as horizontally and can twine around objects, as a result of the articular faces having "hour-glass" surfaces (Figure 30-2D). Typical basket-stars like *Gorgonocephalus* (Fig-ure 30-3A) have branched arms although some allied genera, including *Astro-toma*, have simple arms. There are only a few genera and species of euryalids and they live mostly on hard substrata at considerable depths.

The other anatomical modifications of the arm in ophiuroids (see Figure 30-2B) are best understood if it is realized that the vertebrae are the ambulacral ossicles from the floor of that groove moved into the anterior of the arm. The ambulacral groove has become closed, that is the radial nerve is no longer at the surface but lies on the inner surface of a new canal formed by the closure of the groove—the epineural canal—and the general visceral coelom is squashed up aboral to the vertebrae. This was probably the way the ophiuroid type of arm evolved, and it is certainly the way in which the structures develop embryoni-cally. During development the open ambulacral groove is replaced, and some ophiuroid vertebrae show throughout life the "suture" line of the fusion of the two ambulacral ossicles which gave rise to it. Above the epineural canal and the radial nerve lies the radial perihaemal canal with a haemal strand and hy-poneural nerves. In brittle-stars the relatively thin ribbon of the hyponeural system lies just above, and closely applied to, the thicker strand of the ecton-

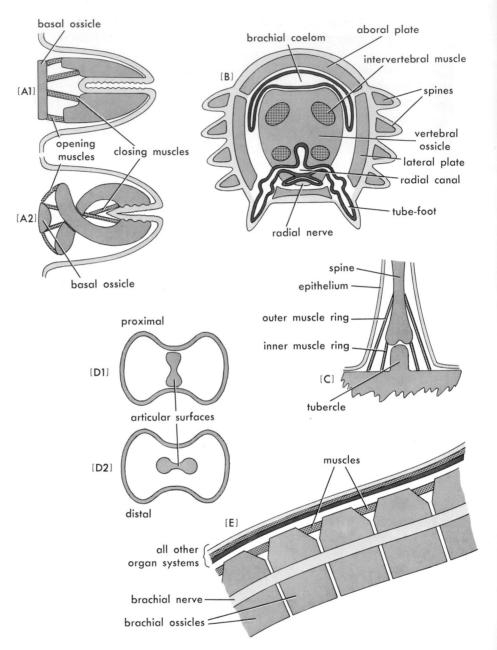

FIGURE 30-2. Functions of echinoderm ossicles. **A1** and **A2:** Vertical sections through two kinds of pedicellariae, showing the antagonistic muscles which open and close the "jaws." **B:** Cross section through the arm of a brittle-star, the bulk of which consists of the massive vertebral ossicles. **C:** Vertical section through a movable spine from a sea-urchin. Note that the epithelium is continuous over the spine. **D1** and **D2:** Proximal and distal faces of a vertebral ossicle taken from the arm of a euryalid or basket-star. **E:** Longitudinal section through a crinoid arm, showing the massive brachial ossicles with muscle masses near the oral side and the contained brachial nerve (entoneural system).

[A] [B]

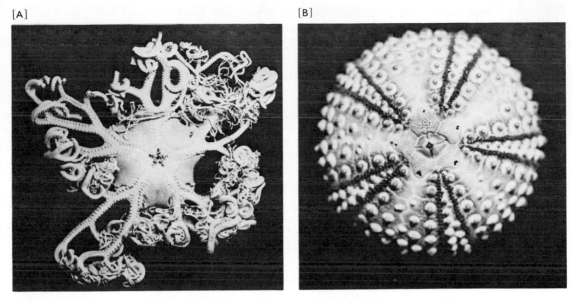

FIGURE 30-3. **A:** Oral view of a basket-star, the euryalid ophiuroid *Gorgonocephalus*. **B:** Aboral view of the test of a regular sea-urchin, *Arbacia,* showing the alternation of the ambulacral (darker) and interambulacral double rows of plates arranged meridionally. The periproct (anal) area is surrounded by five large "basal" plates at the ends of the interambulacral rows, four of which show clearly the genital pores; the fifth (at the top) is the madreporite, with numerous minute pores in addition. [Photos by the author.]

eural radial nerve. It can be demonstrated that the hyponeural is pure motor nerve while the ectoneural involves both sensory and motor elements. Corresponding to each vertebral ossicle is a ganglion on the compound radial nerve, but there is a much greater degree of central nervous control than in the asteroids. For example, if the circumoral nerve ring is cut in two interradii, so as to isolate an arm, that arm cannot show any coordinated movement and is simply towed behind limply as the brittle-star moves along. In view of the relatively simple arrangement of the arm muscles (which in normal locomotion involve the muscles of one side as antagonists of the muscles of the other), it is significant that, in addition to autotomy, gross overstimulation of brittle-stars causes them to freeze or stiffen. All these escape reactions—autotomy, freezing, etc.— involve simultaneous contraction of all the intervertebral muscles.

Another major modification of brittle-star anatomy is that the aboral surface grows down interradially and around on to the undersurface. Thus the madreporite is moved to the oral surface at one side of the mouth. On either side of each arm as it nears the mouth are slits in this underturned aboral surface and these ten slits lead into the genitorespiratory bursae which are lined with respiratory tissue and into which the ten gonads discharge. As already stated (see Figure 29-5A) the gut is extremely simple. A short wide buccal region with teeth leads through an even shorter oesophagus to the blind sac of the stomach. There are no lobes, no intestine—not even an anus. Five independently moving jaws with teeth surround the mouth, and the first pair of podia on each arm form sensory buccal tentacles. Obviously these tentacles and delicately moving jaws are important in the selection of valuable food material from the detritus of the sea bottom. The tiny gut with its blind stomach, simply occupying all the visceral cavity not occupied by other organs, is obviously unsuitable for wholesale ingestion of the entire substratum regardless of nutritive value, as is done by holothurians and certain echinoids.

Sea-urchins

The functional pattern of the class Echinoidea is based on the fact that the ossicles of the endoskeleton form an external armor, or test, of closely fitted plates arranged around a globose body, and this allows movable spines operated by dermal muscles to be used along with the podia in locomotion. Characteristically they have a reduced aboral surface and an extended oral surface (see Figure 29-1), and the test consists of plates arranged in twenty meridional rows, ten being ambulacral with pores for the tube-feet, and ten being interambulacral and usually larger (Figure 30-3B). The more rigid skeleton thus formed provides a basis for the action of the movable spines. Most regular sea-urchins are macrophagous herbivores, feeding on attached algae or other large plants for preference, but they seem to be potentially omnivorous. Their preference for grazing on living larger algae, means that they are typically a group of the immediate and illuminated sublittoral. In life, a ciliated epithelium, which is often very thin, covers all the outer surface of the test including the spines and pedicellariae. Recently Grover C. Stephens' reinvestigations of direct uptake of organic molecules from seawater (discussed more fully for polychaetes in Chapter 12) have shown that, in the sand dollar *Dendraster,* there can be a nutritionally significant uptake of amino acids. It is even probable that, in some other echinoids, the epithelial and other tissues *outside* the test are nourished by direct uptake, in contrast to the more massive internal tissues which rely upon "normal" nutritional uptake through the gut wall. In discussing the evolution of the ampulla-tube-foot system in Chapter 30, we noted how the lack of a true circulatory system along with the dermal armor of ossicles modified respiratory physiology in echinoderms. Here, in echinoids, the nearly continuous dermal test imposes similar constraints upon nutritional physiology.

The ossicles of the echinoid test are less fused around the two poles. This results in two circular leathery areas with separate ossicles termed the peristome and periproct. Besides the pores of the tube-feet, the ossicles bear patterns of bosses to which are attached both spines, which are often of two sizes, and pedicellariae, which can be of four kinds. These pedicellariae often have a three-jawed arrangement of movable ossicles, and some have a poison gland in each jaw. Each spine has a double muscular system with the outer muscles causing the movements and the inner serving as catch muscles (Figure 30-2C). Thus the inner ring of tonic muscles holds the spine erect on the tubercle of the test ossicle, and the outer ring of phasic muscles is responsible for pointing the spine in any particular direction. Although most knowledge of the nerve supply to these muscles is inferential, it is obvious that the spines are capable of considerable coordination in their more delicate movements. Apart from locomotion, the work of Norman Millott and others has revealed elaborate patterns of photic responses, which are used by some sea-urchins (notably the black *Diadema* of coral reefs) for aggressive defense, and by others for constructing concealing camouflage.

The water-vascular system is similar to that in starfish with a circumoral ring with five Polian vesicles and five radial canals running meridionally, supplying the paired vessels to the ampullae. As already mentioned, each ampulla has two canals running through two pores in the ambulacral ossicle to each tube-foot and this is thought to have considerable respiratory significance since the water current circulates in each unit under the influence of the cilia lining the lumen of the tube-foot and ampulla. The haemal system and axial gland are again similar to starfish, and in spite of recent claims, there is no evidence of true circulation in the haemal system. The five gonads are suspended by mesenterial strands in the interambulacral regions. In the adult, there is no trace of the genital stolon ring, though it is present as a coelomic derivative during early adult development. The main nervous system is ectoneural with circumoral and radial elements, but this is connected to a subepidermal plexus running all over the surface outside the test. As in ophiuroids, the ambulacral groove in sea-urchins is closed over to form an epineural canal. Much experimental work on the movements of spines has involved experimental cutting of this subepidermal plexus and of the radial nerves. In crude summary of this work, the spines point toward the source of stimulus in detached pieces of sea-urchin body wall, and it can be demonstrated that the stimulus is transmitted through tissues on the outside of the test.

As already noted, the alimentary canal is more complex and elongated (see Figure 29-5D) and the mouth is surrounded by a complicated skeletal framework which has been termed Aristotle's lantern. This supports five chisel teeth, each attached to a complicated system of muscles and coelomic spaces. This complicated structure (the lantern itself consists of forty skeletal pieces) allows regular sea-urchins to protrude and turn the whole lantern, to open and close teeth, and to move individual teeth in all appropriate directions. The teeth themselves are secreted continually from sacs at their internal ends and their outer ends are hardest.

The urchins which are termed irregular, and which belong to at least two distinct orders, show a secondary bilateral symmetry. The anus and the periproct area is no longer apical but lies in the posterior interradius and the aboral parts of the ambulacra are expanded like petals and bear flattened respiratory tube-feet. These forms are all sand-burrowing and this may partially explain their development of a real anterioposterior axis. Forms like the heart-urchin *Echinocardium*, have the mouth placed anteriorly and lack the jaw apparatus of regular urchins. The sand-dollars (including *Clypeaster* and *Mellita*) are much more flattened but have the mouth placed centrally below and retain the lantern structures. Zoologists have long used an oversimplified classification of the urchins: into two orders of regular urchins and two orders of irregular urchins. However, the Echinoidea constitute a group where neontologists can no longer afford to ignore the fossil record, here so abundant and so well studied. H. Barraclough Fell (one of the rare individuals who has worked on both living and fossil forms) has proposed a classification of the Echinoidea which involves nineteen orders, fifteen of which have recent representatives.

Sea-cucumbers

The class Holothuroidea are characterized by two features which have allowed these animals to become the only really successful burrowing forms among the echinoderms. First, they have a muscular body wall with very small, widely separated ossicles (without spines or pedicellariae externally). Secondly, they are sausage-shaped with an elongation of the disc part of a starfish body in an oral/aboral direction but without any arms (Figure 30-4). The new worm-like patterns of locomotory musculature have allowed degeneration of a major archetypic feature: the intimate association of the podial system with dermal ossicles is no longer necessary. In some—including many deep-water forms—there is a secondary bilateral symmetry which arises from one side of the body becoming a sole on the ground. When this happens, usually three radii of the ambulacra have more functional tube-feet and become the ventral sole. The tentacles around the mouth—which are modified tube-feet—may be peltate (so-called shield-shaped), or pinnate with a limited number of lateral branches, or very much branched or dendritic. Their retraction may involve a hydraulic system with large ampullae, as in the peltate ones, or retractor muscles as in the dendritic ones, or either or both, as in the forms with pinnate tentacles. Certain sea-cucumbers, including *Cucumaria* and its allies, are plankton-feeders, catching minute organisms on the sticky mucus secreted from the expanded tentacles and rhythmically wiping the loaded tentacles through the mouth in turn. Almost all other holothurians ingest masses of the substratum, like certain worms, and extract organic food material from it (Figure 29-5C).

FIGURE 30-4. A living sea-cucumber. A small (5 centimeters long) specimen of *Cucumaria saxicola*, with extended tube-feet in five double rows along the worm-like body and a circlet of five much-branched tentacles around the mouth (right) which are used with sticky mucous threads in the capture of zoöplankton as food. [Photo © Douglas P. Wilson.]

In most, there are two branched long ducts, the so-called respiratory trees, running in from the cloaca. Muscular pumping action of the cloaca results in exchange of seawater into the tree system and there is an oxygen exchange with the fluid of the visceral cavity. In *Holothuria* and its allies, the more posterior branches of the system are termed Cuvierian organs and are extrusible for defense as sticky threads. They always wriggle actively after extrusion, and in some cases may be toxic. Under gross stimulation, many sea-cucumbers show an escape reaction or sacrificial trauma involving total evisceration. In some, this is accomplished by expulsion of most of the viscera through the cloaca like the Cuvierian organs, but in others, notably *Thyone*, the evisceration can be anteriorly through the mouth and involve all the feeding tentacles, buccal structures, and alimentary canal as far back as the cloaca. Expulsion of viscera may be more than a simple reflex response to threatened predation, since it is reported to occur in response to foul reducing water and even with seasonal regularity in some forms. In most forms which eviscerate, regeneration of the alimentary canal and other structures can be accomplished in a few weeks.

Most of the water-vascular system is anatomically similar to those already discussed. In some sea-cucumbers there is a secondary multiplication of the stone canals, and in most the madreporites are internal, opening within the visceral cavity in adults. It is clear in development that each tentacle is an enormously enlarged podium, or tube-foot, and in some sea-cucumbers it is extruded by an ampullar hydraulic system. The haemal system is more greatly developed than in other echinoderms though the axial organ and its sinus are reduced. There are two large lacunae with elaborate branches lying alongside the S-shaped alimentary canal, and the subdivided branches of these become involved with the respiratory trees in some sea-cucumbers. Once again, there is no real evidence of a circulatory system, although there are characteristically enormous numbers of amoebocytes associated with this haemal system.

Normal and Aberrant Starfish

As already discussed, starfish or sea-stars are normally regarded as predaceous carnivores, feeding especially on bivalve and other molluscs. Such prey if small is ingested, but if large is consumed extraorally. This involves extrusion of the stomach lobes as already discussed, but never involves toxins, although these are invoked by most textbook writers. The forces applied to the bivalve shell by the numerous tube-feet have been the subject of recent investigations, and forces of the order of five thousand grams can be applied by starfish such as *Asterias* and *Pisaster*. These forces are applied by a contraction of the longitudinal muscles in the tube-feet as is used in "pulling" locomotion.

A number of "aberrant" starfish are apparently ciliary mucus-feeders. Over sixty years ago, J. F. Gemmill in Scotland demonstrated ciliary feeding in *Porania*. This was largely forgotten, until in recent years John M. Anderson, and subsequently B. N. Rasmussen, have clearly demonstrated that *Porania*, species of *Henricia*, and other forms are ciliary particle-feeders. The ciliation in the pyloric stomach and caecum of *Henricia* (as described by Anderson) resembles nothing so much as the gut-sorting mechanisms in particle-feeding molluscs. The "giant" reticulate sea-star of clean sand bottoms in the West

FIGURE 30-5. A giant micropha-gous sea-star, *Oreaster*. A dried speci-men of *Oreaster reticulatus* from Flor-ida. This specimen is 23 centimeters across, but individuals of this species can reach 48 centimeters. *Oreaster* could be termed a facultative particle-feeder, and probably still functions as a macrophagous carnivore on occa-sion. [Photo by the author; courtesy of Dr. John M. Anderson.]

Indies and Florida, *Oreaster reticulatus* (Figure 30-5), dried specimens of which are often sold as curios, has now (1977) been shown by Anderson to be another microphagous particle-gatherer, using a large, elaborate and eversible cardiac stomach for ciliary-mucus collection. Both *Oreaster* and *Porania* seem to have retained the capacity to feed as macrophagous carnivores, and thus might be termed facultative particle-feeders. In contrast, *Henricia* seems to be exclusively microphagous.

With regard to the evolution of feeding patterns within the echinoderms as a whole, it would be satisfactory to propose that the "normal" carnivorous and macrophagous starfish were derived from ciliary-feeding, more primitive, forms. Structurally, the most archaic living starfish are species of *Luidia* (see Figure 29-3) and related genera, most of which have been observed to be vora-cious carnivores and relatively active. However, *Platasterias*, which is now re-garded as a luidiid asteroid rather than a somasteroid (see page 509), is a parti-cle-feeder, and the family is characterized by the possession of a blind gut with no anus and powerful ciliary water currents around the edges of the rays. Thus, on available evidence, we cannot state confidently that the earliest starfish *were* microphagous forms with a sedentary life-style. Living crinoids *are* sessile sus-pension-feeders, and similar feeding habits probably sustained the totally ex-tinct blastoids and cystoids.

Feather-stars and Sea-lilies

Obviously, the living species of the class Crinoidea represent the survivors of the most archaic stocks of echinoderms. As their position in this discussion, and the inversion of Table 29-1, are meant to indicate, unfortunately less is known about their physiology than that of all the other groups. This is because most of them survive in relatively deep water, and they are very difficult to maintain alive in aquaria.

Of the group, more than 5000 species have been described as fossils, but only about 620 living forms, of which 80 are sea-lilies (i.e., are stalked and permanently sessile). Their near limitation to the deeper parts of the oceans suggests a group going to extinction, but they were once abundant as the thick beds of certain Carboniferous limestones, formed from crinoid ossicles, indicate. *Antedon* is typical of the more numerous comatulids, which can swim and attach only temporarily (Figure 30-1B).

The body wall is mainly ossicles and the epidermis is not continuous over these. The ten arms are made up of the brachial ossicles (see Figure 30-2E) with muscle masses near the oral side and a big brachial nerve running through them. The podia, which are directly connected to the radial elements of the water-vascular system without ampullae, form a double row of tentacles on the sides of the ciliated grooves which are the feeding mechanism. These split radial canals lead to a circumoral canal, but there is no external madreporite: openings from numerous stone canals are within the perivisceral cavity. The haemal system is reduced in the adult, but there is a ring (or genital rachis) from which genital strands run down each arm and are enlarged into gonads in the pinnules. There is an ectoneural nervous system (as in starfish) with radial nerves lying immediately below the ciliated grooves of the food-collecting mechanism. Much more important for movements is the aboral entoneural nervous system which forms a cup-shaped mass below the aboral calyx and has branches from the ring nerve around it into the ten arms. These ten brachial nerves supply the motor fibers to the muscles between the ossicles, as can be demonstrated by destructive experiments. The alimentary canal runs from a central mouth through a short oesophagus into the stomach and coils once round the disc, where, in *Antedon*, arise peculiar small diverticula of unknown function. The gut runs then through the short rectum to an anus placed interradially on the oral side. Crinoids apparently live on a mixed diet of detritus, microorganisms, and plankton. The larger elements appear to be trapped by an interlocking net formed of the pinnular tentacles, while a mucous secretion, produced by the small pointed podia, traps smaller food particles, which are then transported by the ciliated ambulacral grooves which run together into the mouth.

Echinoderm Larvae

No other group of animals has such complicated metamorphoses in the course of development. A few echinoderms brood their young but, in many, fertilization is external and the zygote develops in the sea. The first cleavages are relatively regular and indeterminate, thus providing the famous material used by the earliest experimental embryologists. The embryo remains holoblastic to the blastula stage. Gastrulation is followed by various patterns of larval development, which always involve bilaterally symmetrical stages for varying periods. This is then followed by a metamorphosis during which the left anterior part of the larva gives rise to most of the radially symmetrical adult structures. The

[A]

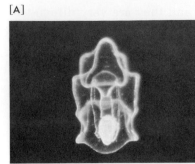

[B]

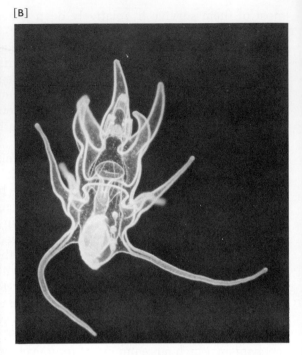

[C]

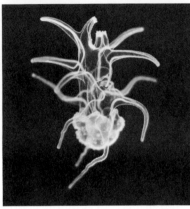

[D]

FIGURE 30-6. Starfish larval development: stages of *Asterias*. **A:** Bipinnaria larva, a ventral view showing the complete bilateral symmetry. **B:** Early brachiolaria larva in ventral view. **C:** Later brachiolaria larva in lateral view, showing the radially symmetrical rudiment of the starfish body beginning to form. **D:** Young starfish after metamorphosis: [Photos **A–B–C** © Douglas P. Wilson. Photo **D** by Marvin E. Snow of a specimen prepared by Dr. Albert J. Burky.]

major larval types are illustrated in Figures 30-6 and 30-7: the bipinnaria of starfish (Figure 30-6A) resembling the auricularia of sea-cucumbers (Figure 30-7B), and the ophiopluteus of brittle-stars (Figure 30-7A) resembling the echinopluteus of sea-urchins.

In the echinoderm gastrula, a wide blastocoel, or segmentation cavity, separates the enteron wall from the ectoderm. The blastopore becomes the anus, and a mouth is formed by a breakthrough of the stomodeum in a "typical" deu-

[A] [B]

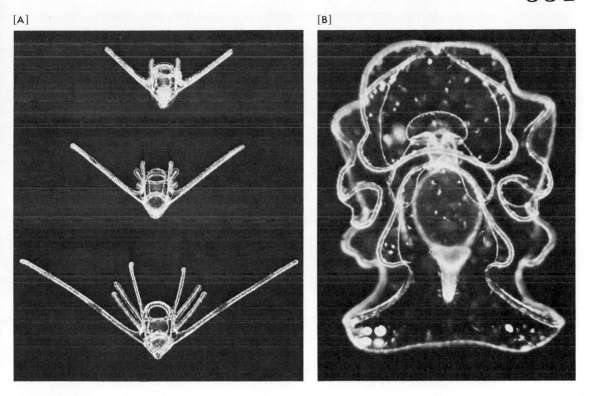

FIGURE 30-7. More echinoderm larvae. **A:** Three successive ophiopluteus stages of *Ophiothrix fragilis,* in ventral view. The elongate ciliated *larval* arms are bilaterally symmetrical and completely lost at metamorphosis to the adult brittle-star. **B:** The auricularia larva of the sea-cucumber *Labidoplax digitata,* in ventral view. Compare with the starfish bipinnaria larva of Figure 30-6A and the hemichordate tornaria larva of Figure 31-3; for further discussion, see text. [Photos © Douglas P. Wilson.]

terostomous fashion. Many texts state that, at this stage, all echinoderm embryos develop three pairs of segmentally arranged coelomic sacs from enterocoelic vesicles. This idealized development has been elevated to give a hypothetical ancestral form for the echinoderms, the dipleurula. In its simplest —and perhaps its only useful—manifestation, this theoretical organism is regarded merely as an idealized larva combining many of the features common to all eleutherozoan larvae. Thus it shows three pairs of coelomic spaces, but omits both the specializations of the pluteus larval types and those of the auricularia-bipinnaria group. In more extreme—and largely indefensible—hypotheses, the dipleurula is enlarged, given gonads, complex sense-organs, and a benthic crawling habit, and is set up as a hypothetical preradial echinoderm ancestor. When larval types are surveyed throughout the Echinodermata, there are several examples of close larval resemblances between unrelated species (reflecting evolutionary convergence), while other closely related forms exhibit major differences of larval types (divergence). Detailed phylogenetic relationships cannot be extrapolated from larval similarities in echinoderms. In broader questions of lineage, larval features can perhaps provide corroborative evidence but cannot be accepted as the sole proof of relationships. Actually, a twofold and threefold subdivision of the larval coelom does occur in several different groups, but not with the regularity that certain books suggest. There is, however, considerable constancy (except in the case of the axial organ) in the derivation of the adult coelomic structures from the larval segmental sacs, particularly those of the left side. In a few sea-stars, there *does* occur the development of the coelom which is claimed as archetypic, and of the three pairs of coelomic sacs in the larva: the posterior spaces give rise to the visceral coelom and perihaemal coelom, the left anterior gives rise to the axial sinus and stone canal, and the left intermediate coelom of the larva gives rise to the rest of the water-vascular system of the adult.

At the blastula stage of development, cilia cover the surface of the embryo evenly, but as the bilaterally symmetrical larva develops further, they become restricted to characteristic bands. In crinoids, the barrel-shaped larva has five ring bands of strong cilia surrounding the body. In the other four groups, the ventral surface of the larva becomes concave and this depression is outlined by the major band of cilia which is the main organ both of locomotion and of feeding. This grows much faster than the rest of the surface ectoderm, and becomes thrown into folds on special projections. These are the larval lobes, or arms, which are always arranged in bilateral symmetry (Figures 30-6A and 30-7A). (These lobes bear no relation to the adult arms.) In the sea-cucumbers, the elongate larva has short blunt lobes. In this auricularia larva, the ciliary band forms a figure-8 shape before dividing into a smaller preoral ciliary band and a larger band posteriorly (Figure 30-7B). In Asteroidea, there is a similar larva, with somewhat greater lobe development, termed a bipinnaria (Figure 30-6A), which shows complete separation of the preoral band from the rest of the ciliation. In brittle-stars and sea-urchins, there is an enormous development of long temporary arms with internal calcareous skeletal supports. Both show relatively slight preoral development and the ophiopluteus (Figure 30-7A) and echinopluteus show remarkable resemblances to each other. In the later development of some sea-cucumbers, the auricularia becomes a pupa larva with five barrel rings like a crinoid larva. Both crinoids and asteroids become attached preorally before metamorphosis. In crinoids a new mouth is developed posteriorly and there is always a miniature stalked sea-lily stage, even if the adults (like *Antedon*) are free feather-stars. In the other four major classes, a new mouth develops on the left side of the larva and the radially symmetrical organs form around it. For example, in the brachiolaria, the starfish-rudiment forms inside, on the left of the gut (Figure 30-6C).

Phyletic Thoughts

In spite of our much more sophisticated knowledge of the fossil groups of echinoderms, and the realization that the Eleutherozoa form an unnatural assemblage, it still seems justifiable to separate the crinoids and the several fossil groups allied to them in a division which may still be called the Pelmatozoa (or else the Crinozoa; and see Chapter 34). All these forms seem to be clearly distinct, probably more ancient, and are undoubtedly not only basically sessile in habit but almost certainly microphagous with ciliary or mucous feeding on the outstretched arms. Among the mobile echinoderms, or Eleutherozoa, the Somasteroidea are clearly closest to the crinoids, but the Asteroidea also share certain possibly primitive characters. These include attachment at metamorphosis, and the exposed ectoneural nervous system. If we accept the customary classification as reflecting natural relationships, then we emphasize the features which link the echinoids and ophiuroids, including the larval similarities, and the common features of anatomy involving the canals associated with the ambulacral groove and its ossicles. If we accept the relationship suggested by Fell, which stresses the affinities between holothuroideans and echinoids on the one hand, and ophiuroids and asteroids on the other, then we give greatest phyletic significance to his distinction between the meridional gradients of the former and the radially divergent growth gradients of the latter. Whichever interpretation of the interrelationships of the living groups of echinoderms is more correct (and it may be very difficult to settle this), considerable evolutionary convergence is implied. In the one case, the resemblances between the ophiopluteus and echinopluteus must be due to convergent larval evolution, and the anatomical similarities of the ambulacra result from similar processes. Alternatively, the two dominant gradient patterns established by the rudiments of the water-vascular system, which subsequently control the development of the dermal ossicles and all other structures, must each have arisen twice (again in a startling example of convergent evolution).

However difficult it is to sort out the interrelationships within the echinoderms, there are some important phyletic conclusions regarding the group as a whole which—although hypotheses like all phylogeny— are accepted by the majority of zoologists. The following seem the most significant points. To a great extent, radial symmetry with the mouth up is functionally useful to sessile animals which have the same relationship to their environment on all sides. On the other hand, bilateral symmetry with a front and rear end, upper and lower surfaces, left and right pairs of effector organs, is functionally suited to the needs of a traveling animal. Apart from the echinoderms, the only other group of metazoan animals in which radial symmetry can be the rule is the coelenterates, many of which are fixed. Now, in more archaic living echinoderms there is attachment to a hard substratum at metamorphosis. Further, in addition to the six groups with living representatives, there are at least five other major groups, at least as distinctive, found only as fossils and now completely extinct. It is probable that many of these were sessile organisms, and it is also most probable that the earliest echinoderms were all fixed. Thus it seems likely that all echinoderms were at one time fixed and those now mobile retain the radial symmetry of these ancestors. If we go a little beyond this hypothesis, the ontogeny of an irregular echinoid, for example, involves a radially symmetrical gastrula giving rise to a bilaterally symmetrical larva, giving rise to a radially symmetrical preadult, giving rise to a secondarily bilaterally symmetrical adult. Apart from the secondary return to bilateral symmetry, which is special to the case of irregular echinoids, this ontogeny suggests a phylogeny involving a bilaterally symmetrical ancestor before the radially symmetrical one which is also common to all groups.

It is worth stating first that the concept of the dipleurula larva is grossly overemphasized in most discussions, and secondly that the metamorphosis of typical living echinoderms is so complete that it is very difficult to speculate on presettled ancestors. Once again, it is important to note that certain characters common to several larval types need not reflect ancestral characters, but can simply be characters required by all larvae with that particular mode of larval life. Further, as several authors have noted, it is unlikely that a fossil, common preradial ancestor will ever be discovered since it is likely that only after settlement and the acquisition of radial symmetry that endoskeletal ossicles developed. However, the enigmatic carpoids could represent this unlikely fossil stock (see Chapters 31 and 34). As already stressed here, only when the group became echinoderms, by the development of such ossicles, was the further development of the characteristic hydrocoel, or water-vascular system, with its multipurpose podia, functionally likely. As regards the phylogeny of this diagnostic feature of the group, in all living forms this unique division of the coelom is found, and it is concerned with extending the podia hydraulically out into the environment. Whether these podia were first most important as respiratory, excretory, locomotory, or sensory structures is an unresolved question. It seems clear that very early in the history of the echinoderms, they were involved in the food-collecting apparatus of the sessile forms, both as mobile food-collecting tentacles and as the sources of a mucous net. They function thus to this day in the crinoids and perhaps elsewhere. Almost certainly, the evolution of internal ampullae isolated from the rest of the system by valves, the development of strong postural musculature, and the development of suctorial discs were later developments in the evolution of these organs. Such true tube-feet allowed the efficient locomotion of starfish and the forceful opening of bivalves for food. The fantastic elaborations of the feeding tentacles of sea-cucumbers, and the funnel-building tube-feet of irregular echinoids, were obviously still later elaborations.

31
INVERTEBRATE CHORDATES I: DIAGNOSTIC PHYSIOLOGY

DISCUSSION of vertebrates—animals with a jointed axial skeleton of vertebrae—lies outwith the scope of this book. Vertebrate animals—fishes, amphibians, reptiles, birds, and mammals—make up over 95% of the species of the phylum Chordata. The remaining forms which fall within the diagnostic definition of the chordates are usually described, along with their allies in the phylum Hemichordata, as "the primitive chordates" or "invertebrate chordates." Some of the great biological interest in these forms lies in the hypothesis that they represent preverte brate protochordates, examination of whose physiology should give us some ideas of functioning in the stocks of animals ancestral to the vertebrates. In fact, understanding of vertebrate physiology and evolution *has* profited greatly from recent detailed work on the physiology and ecology of primitive chordates.

The Protochordata do not form a natural assemblage, and in the classification used here fall into one phylum, Hemichordata, and two distinct subphyla of the phylum Chordata (Table 31-1). These are the subphylum Cephalochordata or Acrania, for *Amphioxus* and its allies, and the subphylum Urochordata or Tunicata for the more numerous sea-squirts. (There is, of course, a third subphylum in the phylum Chordata: the Craniata or Vertebrata for the approximately 43,000 species of backboned animals.)

535

TABLE 31-1
Outline Classification of the "Protochordates"[a]
and Their Allies

Phylum POGONOPHORA
Phylum GRAPTOLITHINA (fossil only)
Phylum HEMICHORDATA
      Class ENTEROPNEUSTA
      Class PTEROBRANCHIA
Phylum CHORDATA
  Subphylum CEPHALOCHORDATA (= ACRANIA)
  Subphylum UROCHORDATA (= TUNICATA)
      Class ASCIDIACEA
      Class LARVACEA
      Class THALIACEA
        Order Pyrosomatida
        Order Salpida
        Order Doliolida
  Subphylum CRANIATA (= VERTEBRATA)
    Superclass AGNATHA
    Superclass GNATHOSTOMATA
      Class CHONDRICHTHYES
      Class PLACODERMI
      Class CHOANICHTHYES
      Class ACTINOPTERYGII
      Class AMPHIBIA
      Class REPTILIA
      Class AVES
      Class MAMMALIA

[a] The term "PROTOCHORDATA," encompassing phylum HEMICHORDATA and subphyla CEPHALOCHORDATA and UROCHORDATA, does not form a natural assemblage, but covers the "invertebrate chordates" of Chapters 31 and 32.

Functional Significance Behind the Diagnostic Features

The diagnosis of the Chordata is relatively clearcut, and can be stated—in anatomical and embryological terms—in about five main items. First, the notochord, a dorsal, axial skeleton, is present at some stage in the life-cycle. Secondly, pharyngeal gill clefts open to the exterior from the gut and, having certain consistent features of anatomy and development, occur at some stage. Thirdly, there is a dorsal central nervous system, derived from the surface epithelium of the embryo, usually in the form of a dorsal hollow nerve cord. Fourthly, the coelom is said to be enterocoelous in origin, and primitively divided into three divisions. Fifthly, there can be—as in no other invertebrates— a postanal propulsive tail, that is, blocks of muscle around an axial skeleton posterior to the anus. The other features which could be used, including those of muscle physiology and ciliary control, do represent what are probably significant differences in the chordate stocks, but are not sufficiently consistent to be involved in a precise diagnosis. Of the five features of true chordates mentioned above, three (the second, third, and fourth) are part of the diagnosis of the phylum Hemichordata.

The "unknowable" ancestor of the chordates must have been a working, efficient machine involving most of these structural features. Thus in setting up a working archetype for the chordates—one in which the concert of organs and functions can operate as a whole—it is important to survey the functional significance of these diagnostic anatomical features in the more primitive living forms. While we survey the nature and extent of each feature in the main primitive chordate groups, it is worthwhile also to examine the physiological significance of each feature, especially in regard to such mechanical aspects as feeding and locomotion.

The notochord is an axial rod of peculiar vacuolated cells, which runs from end to end of the body in the Cephalochordata where it is present throughout life, and is the longitudinal skeleton of the larval tail in the Urochordata. It is no longer claimed as a feature of the Hemichordata, although some cells involved in the proboscis neck may be developmentally homologous with it. In the subphylum Craniata it is, of course, an embryonic feature replaced in the adult by the jointed, bony, or cartilaginous vertebral column. Functionally, the notochord is a skeletal structure, with the mechanical properties of being laterally flexible while being incompressible from end to end. These mechanical properties arise from the turgidity of the large vacuoles of the cells. In the functional integration of the archetypic chordate, this type of skeleton must always have been associated with laterally placed blocks of muscles which could act as antagonists to each other, thus producing the characteristic lateral swimming flexures of the group. Propulsion by a laterally flexing tail with an internal skeleton is unique to the chordates and structurally and functionally homologous from sea-squirt larvae to newts.

The gill clefts, which are openings from the pharyngeal part of the gut to the exterior, are present in all groups of chordates, though only in the embryos of the higher vertebrates. There is a tendency for their numbers to be reduced in the higher craniates, while they tend to be multiplied in the cephalochordates and urochordates. In all the protochordate groups, they multiply by splitting by tongue bars (Figure 31-1). Anatomical features such as their skeletal supports, their ciliation and the blood vessels between, seem to be homologous throughout the chordates. Functionally, recent work has made it clear that they act as food-filtering organs, as in the bivalve molluscs, and this will be discussed in more detail below. Meanwhile, it is worth noting that the feeding function of the pharyngeal gill slits in the urochordates, hemichordates, and cephalochordates has been replaced by a respiratory function in most fishes and larval amphibians. However, reversion of function to filter-feeding is found in several distinct groups of fishes, including the largest living forms—basking sharks and whale sharks. The pharyngeal gill clefts are of course only embryonic, transitory structures in the higher vertebrates, but the reader should remember that when one equilibrates the pressure in ones middle ears on change of altitude in an aircraft, one is utilizing the Eustachian tubes, the last functional vestiges of the food-filtering slits of the archetypic chordate. The dorsal hollow nerve cord is diagnostic, though only present to a limited extent in the hemichordates. The only other group of many-celled animals with nerves developed from the sur-

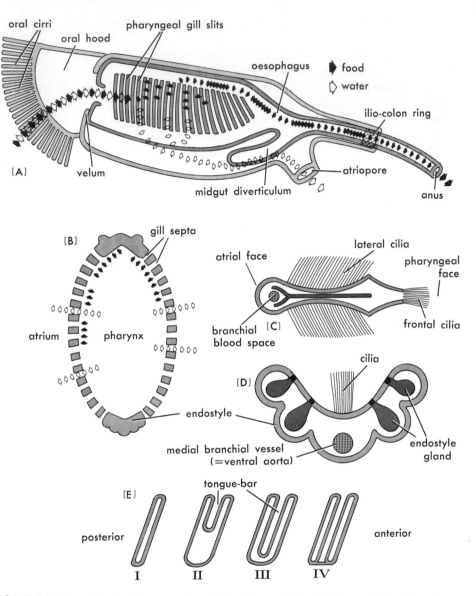

FIGURE 31-1. The feeding apparatus of *Amphioxus*. **A:** Sagittal view of the alimentary canal, and the outer tube of the atrium, with the major food and water currents. **B:** Cross section of the pharynx. **C:** Enlarged view of one gill septum in section. **D:** Enlarged view of endostyle in section. **E:** Four successive stages in the multiplication of gill slits.

face epithelium is the echinoderms. One functional consequence of this arrangement is that, as is well-known, several vertebrate neurosensory structures, including the vertebrate eye, develop centrifugally as outgrowths of the central nervous system. The inverted retina is a consequence. This curiosity of development, with the sensory ends of the receptor cells turned away from the source of sensation and toward the central nervous system, also occurs in some primitive chordates. Some textbooks suggest that this primitive dorsal hollow central nervous cord shows metameric segmentation. As we shall see in Amphioxus, this is not so in its primitive condition.

The coelom is of somewhat doubtful structural homology throughout the three protochordate groups, and it is certain that no *functional* homology throughout the chordates can be set up. In fact, it is only in the hemichordates that the coelom is a large body-cavity in the adult with any mechanical significance. The tripartite division of the coelom can be seen in the pouches of the archenteron during the development of cephalochordates. It is an adult feature of the hemichordates where the proboscis collar, and trunk sections each contains a coelomic cavity.

The possession of a postanal tail is another clearly chordate characteristic. Among the primitive chordates, only the Hemichordata have an anus which is terminal. In urochordate larvae, in cephalochordates, and in vertebrates, a muscular tail can be developed posterior to the opening of the alimentary canal. This tail typically consists of blocks of muscles on either side of the central axial skeleton, which, as we noted above, is laterally flexible while being incompressible in length. The muscle segments of tailed chordates arise directly from the myotomes or mesodermal segments of the embryo. It is of some relevance in relation to theories of segmentation in chordates that in *Amphioxus* the myotomes are arranged alternately on either side of the axial notochord, *not* in the left and right pairs which would be expected with segmentation.

Other chordate characteristics concern the physiology of muscle action and ciliary patterns. The process of contraction in fast striated muscles involves the translation of chemical energy into mechanical energy by means still not finally established. However, ATP (adenosine triphosphate) is clearly involved, and ADP (diphosphate) reacts with the substances called phosphagens to produce ATP. Phosphagens, such as arginine phosphate and creatine phosphate, are thus reserves of energy-rich phosphate bonds. The main phosphagen found in invertebrates is arginine phosphate, and that characteristic of vertebrates is creatine phosphate. This was once thought to be a biochemical diagnostic feature of the chordates, with the echinoid echinoderms (which have creatine) and the hemichordates (which have both) being intermediate between the typical chordate animals and the invertebrates. However, the consistency of this biochemical characteristic proved unreal and based on ignorance of conditions in many other groups. Creatine phosphate—and other phosphagens—have been identified in certain annelid worms, in coelenterates, and in sponges. Although the absolute distinction no longer can be claimed, the principal phosphagen of most invertebrates remains that of arginine, and of vertebrates that of creatine.

Another physiological characteristic of the group was noted relatively recently. The cilia around the gill slits in the pharynx show metachronal waves, that is to say the cilia of one cell are slightly out of phase with those of the next and a wave appears to pass over the surface in the opposite direction to the effective stroke of the cilia. When viewed from outside the animal, the metachronal waves in the cilia of the pharyngeal gill slits seem always to run counterclockwise; that is, there is a functional asymmetry, (think of this in relation to the wheels of a car or train viewed from outside). This functional asymmetry of the gill-slit cilia is found in *Amphioxus*, in several sea-squirts, and in some hemichordates. The developmental implications of this functional asymmetry are not yet clear: though they could imply derivation of the paired gill slits from a single series of openings, either ventrally or on one side. As was first pointed out by E. W. Knight-Jones and R. H. Millar, such functional asymmetry is unique among bilaterally symmetrical animals and seems to be a functionally homologous feature characteristic of the chordates.

Nearly Archetypic: Lancelets

Although the anatomy of *Amphioxus* is almost universally studied in courses on vertebrate morphology, aspects of its physiology, including its archetypic feeding mechanism, are less often documented. It is treated here as an invertebrate chordate. The anatomical features so important to comparative studies of vertebrates are neglected here; the features thought to be functionally homologous throughout the primitive chordates are, in contrast, emphasized.

Lancelets of the genus *Amphioxus* (sometimes known by the sub-generic name *Branchiostoma*) are not uncommon marine animals (Figure 31-2). They are found in shallow water over cleanish sand substrata, in restricted localities, but in all the world's oceans. They have been called "headless fish"; they are laterally flattened, spindle-shaped, about 5 centimeters long, and nearly translucent in life. Most of their life is spent half-buried, with the anterior end protruding above the surface of the sandy substratum. They can swim quite efficiently by fish-like, lateral flexures of the whole body. Both while semiburied and while swimming, *Amphioxus* continues to feed by drawing a current of water in through the mouth and out via the pharyngeal gill clefts. Actually, the burrowing of *Amphioxus* is carried out by the same muscular movements used in swimming and really amounts to a vigorous swimming into the substratum. Ecologically, the sedentary feeding position taken up varies with the nature of the substratum: lancelets being more completely buried in coarse-grained sands but having most of the body (from oral hood to atriopore) protruding when in fine sand.

The great bulk of the tissues of the body are the muscles arranged on either side of the laterally flexible notochord. They form a series of units—which are

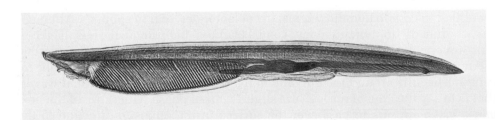

FIGURE 31-2. *Amphioxus.* A stained and cleared whole mount of *Amphioxus* shows the notochord and dorsal nerve cord running the full length of the body, the oral hood with cirri, the large pharynx with diagonal gill slits, and the midgut with its anterior diverticulum. Note the positions of the atriopore and the anus, and compare with Figures 1-1 and 31-1. [Photo courtesy Carolina Biological Supply Company.]

not like metameric segments—termed myotomes, each V-shaped block being separated from the next by a myocomma of connective tissue and myotomes of each side alternate as can be clearly seen in cross sections of Amphioxus. The left-hand myotomes are the antagonists of those on the right, and swimming involves waves of contraction. The muscles of the anterior end contract first and are followed in regular order by muscles further and further back toward the tail along one side. The waves of contraction of the left-hand myotomes are out of phase with those of the right and these waves passing back along the animal create a forward motion through the water or sand. The central nervous system which controls this is of course a dorsal, hollow cord lying above the notochord. It has some enlargement anteriorly, with two so-called cranial nerves, and throughout most of its length, it gives rise to dorsal sensory roots and ventral motor roots running to the myotomes. Unlike the spinal cord and its paired roots in vertebrates, the sensory and motor nerves here alternate on either side of the nerve cord.

The mouth is protected by an oral hood with cirri and a velum. More than half the length of the gut behind it is the pharynx perforated with about one hundred gill slits on each side (Figure 31-1A). There is a short midgut with midgut diverticulum and a hindgut with a ring of prominent cilia termed the ilio-colon ring. The anus is asymmetric on the left side of the ventral fin in front of the tail. In adult Amphioxus most of the gut is enclosed in an outer tube, the atrium, which opens through the atriopore (Figure 31-1A). The filter-feeding system is basically simple: water and food particles pass in through the mouth, the water passing on through the gill slits into the atrium and out the atriopore while the food particles, entangled in mucus, pass posteriorly into the midgut section of the alimentary canal. The water current is created by the lateral cilia on the sides of the gill slits (Figure 31-1B,C), all other cilia in the system being concerned with moving particles, or mucus. The details of the process show many analogies to filter-feeding in bivalve molluscs, and indeed were first worked out by J. H. Orton (one of the earliest workers on the functional morphology of molluscan feeding structures—see Chapters 20 and 22), and further details of both feeding and digestion have been elucidated more recently by E. J. W. Barrington and Q. Bone. The feeding mechanism in Amphioxus derives its efficiency partly from the fact that, as in the bivalves, there is considerable capacity for sorting particles, and rejecting, largely on a basis of size, those unsuitable for food. There is some preliminary sorting on the oral hood: larger particles being arrested on cilia of the hood tentacles and rejected by muscular flicking. The main water current created by the lateral cilia pulls water in through the mouth opening in the center of the velum, which water passes on through the atrium to the atriopore. Particulate material is collected on the frontal cilia and on mucus secreted largely by the glands of the endostyle. This last structure is formed by the imperforate central, ventral gutter of the pharynx (Figure 31-1D), which is ciliated, contains four lines of endostyle glands, and is almost certainly homologous (both developmentally and functionally) with the thyroid of the vertebrates. The frontal cilia all beat in a dorsal direction and carry the captured food material and the mucus to the dorsal groove where it collects in a mucous cord. This is passed posteriorly by ciliary action to enter the digestive part of the alimentary canal. Digestion in Amphioxus is partly extracellular and partly intracellular. Enzymes are secreted by the midgut but principally by the diverticulum, and include protein-breaking and fat-breaking enzymes as well as amylase. The food cord passes slowly posteriorly, being relatively rapidly rotated by the cilia of the ilio-colon ring. Enzymatic globules secreted in the midgut diverticulum are carried by ciliary tracts to the main lumen of the gut where they impinge upon, and are incorporated into, the mucous rope. Small particles of partially digested food are broken off from the rotating rope and carried by other ciliary tracts into the dorsal side of the midgut diverticulum and onto its lateral walls. The cells here carry on intracellular digestion after taking up the fine particles. There is some absorption and consolidation of the mucous food cord in the hindgut and the faeces are discharged through the anus. The continuous rotation of the mucous rope thus involves trituration of the particulate food, the intimate mixing of food particles with digestive enzymes, and a sorting of partially digested material from more massive food. Thus it is almost exactly analogous in function to the rotation of the crystalline style in the bivalve gut (see Chapter 23). In anatomically entirely different systems, cephalochordates and bivalves have found means of transmitting the effects of ciliary movement to the food material in the gut and achieving thereby trituration, intimate mixing, and sorting. Since most vertebrate textbooks refer to the midgut diverticulum of *Amphioxus* as the equivalent of the liver in vertebrates, it is worth pointing out here that functionally it has a closer equivalence to the pancreas of vertebrates, and an even closer functional similarity to the multiple digestive diverticula of filter-feeding molluscs.

In *Amphioxus*, the outer tube around the pharynx, the atrium, appears relatively late in development. It is structurally an invagination of the body wall which comes to surround the pharynx and to become the only other large internal space other than the gut: largely occluding the coelom in adult *Amphioxus*. Larval specimens of *Amphioxus* with about twelve pairs of gill slits live in the plankton and have a feeding mechanism exactly similar to that of adults, except that the water current which passes in the mouth to the pharynx passes out through the individual gill slits to the exterior directly. Taking an ecological view of the functioning of feeding mechanisms, it is significant that the atrium, or outer tube, protecting the outer openings of the gill slits is developed at the time of so-called metamorphosis when the young *Amphioxus* settles to the bottom and exchanges its planktonic life for one involving burrowing in the substratum. This distinction seen in the "two ages" of *Amphioxus*, turns up again and again in many groups of primitive chordates. Thus, many primitive chordates living a planktonic existence have gill slits which open directly to the exterior, while those which burrow in the substratum have the outer openings of the slits protected in some way by structures which arise relatively late in development. If this ontogeny accurately reflects phylogeny, then atrial and other structures protecting the gill slits externally are in an evolutionary sense neomorphic.

Earlier circumstances in the development of the gill slits are also of considerable interest, and probably of evolutionary significance. When the embryonic gut first becomes functional, after about one hundred hours of development, the embryo is remarkably asymmetric. There is a very small mouth lying on the left side and a single gill slit behind it, considerably larger than the oral opening. Ciliary action moves water and food particles in through the mouth and out through the gill slit, and it seems that many particles pass straight through the system without being filtered. A few however are caught on the peripharyngeal band of cilia on the left side of the pharynx and these move past the slit and eventually back into the midgut. At this stage, the anus is nearly terminal but again asymmetric. The endostyle appears at an early stage but remains for some time on the right side of the pharynx as a gutter of secretory and ciliated epithelium. Further gill slits appear successively, but in two series. The primary series are ventral, initially, but then move up on the right side of the larva. Then the secondary series appears dorsal to the primaries again on the right side. After further growth, there is rearrangement with the primaries becoming the slits of the left side, the secondaries, those of the right side, and the endostyle gutter coming to lie between them. In the most studied species of Amphioxus, the larva at this stage has eight pairs of slits. The branchial slits then begin to multiply by the downgrowth of tongue bars (Figure 31-1E) giving the opening of each slit as it divides a characteristic horseshoe appearance. This method of subdivision of gill slits is found in other primitive chordates.

A few other anatomical characteristics of cephalochordates should be mentioned. The coelom is reduced in adult Amphioxus, but has been claimed as showing tripartite division embryonically. Actually, at its greatest extent in the embryo, the coelom consists of five spaces: three anteriorly, two lateral and one dorsal; and two lateral trunk spaces which run back on either side of the gut (compare with Figure 31-4C). As already mentioned, these are squashed dorsally by the ingrowth of the atrium. The circulatory system is of immense interest to vertebrate zoologists, being almost part-for-part, structurally and functionally, homologous with the arterial and venous system of the most primitive fishes. There is no true heart, but a contractile ventral aorta from which afferent branchial vessels lead between the gill slits on the sides of the pharynx. The blood from these is carried by efferent vessels to paired dorsal aortae which unite behind the pharynx, the single posterior dorsal aorta giving rise to most of the arteries supplying the posterior gut and muscles. The blood returns to the posterior end of the ventral aorta (the "heart" position) through systems of cardinal veins and also through a "hepatic portal system," in a fish-like fashion. Functionally the direction of circulation, and the position of the respiratory organs, is clearly archetypic for the vertebrate pattern, blood passing anteriorly in the ventral blood vessels and posteriorly in the dorsal blood vessels. From a phyletic viewpoint, the excretory organs of Amphioxus are somewhat enigmatic. They consist of a series of protonephridia, closely similar to those found in some marine polychaetes, with bunches of solenocytes turning into short curved tubes opening into the atrium. There are about one hundred pairs of protonephridia associated with the occluded lateral coelomic cavities. The sexes are separate, and the gonads are developed on the atrial body wall. The gametes pass to the exterior through the atriopore. The small but yolky eggs, with regular holoblastic cleavage, are classic embryological material.

The subphylum Cephalochordata consists of about twenty species, all very similar in form. In their simplest classification they all fall in one genus Amphioxus, the typical bottom-dwelling forms which have already been described falling in the subgenus Branchiostoma. The subgenus Asymmetron differs in having a single row of gonads and some of the characteristics of the so-called Amphioxides late larval form of *Amphioxus*. It seems probable that the species which show greater asymmetry, and the gill slits opening directly to the exterior, are forms which remain planktonic as adults.

Digression on Doubtful Allies

The only major group of primitive chordates that show any real ecological success is the subphylum Urochordata, the tunicates or sea-squirts. Before we discuss them (Chapter 32), we must review a number of distinctly minor groups, of greater or less chordate affinities.

The minor phylum Hemichordata is divided into two unequal classes: the Enteropneusta for the numerous worm-like animals such as *Balanoglossus*, and the Pterobranchia for some rare but related forms which are sessile and have some resemblances to the lophophore-bearing minor phyla. The more typical enteropneusts, or acorn-worms, of such characteristic genera as *Balanoglossus* and *Saccoglossus* are worm-like animals which live a sedentary existence in semipermanent burrows on sandy shores between tide marks in temperate waters, but which have a ciliated planktonic larva called a tornaria (Figure 31-3). Their burrows and sand-castings are remarkably similar to those of the lugworm, *Arenicola*. Their body is divided into three regions (see Figures 1-20 and 31-4A): a proboscis, a collar and a trunk region. The mouth opens behind the proboscis at the anterior end of the collar, branchial gill slits are represented by a series of pores in the anterior section of the trunk, and the anus is terminal. The general surface of the body is covered with a ciliated epidermis, in places forming ciliary tracts, of considerable importance in the way of life of the animals. There are also numerous mucous glands. The body wall is some-

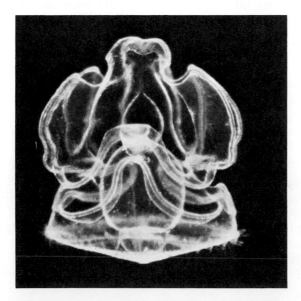

FIGURE 31-3.   Tornaria larva of the hemichordate *Glossobalanus*, in ventral view. Compare with Figure 30-7B; for further discussion, see text. [Photo © Douglas P. Wilson.]

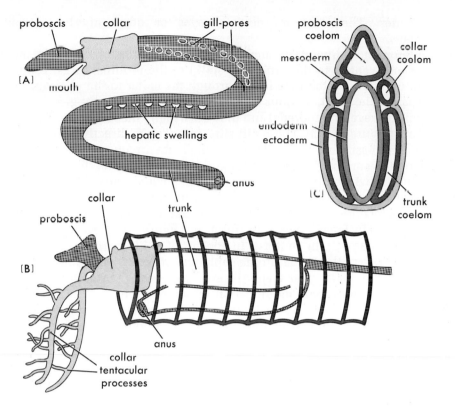

FIGURE 31-4. Hemichordate organization. **A:** Stylized body form of an enteropneust. **B:** Stylized body form of the minute sessile colonial pterobranch *Rhabdopleura*. **C:** Late embryo of an enteropneust, showing the threefold (actually fivefold) division of the embryonic coelom.

what weakly muscularized, with no trace of myotomes, but with a tendency to an annelid-like arrangement of circular and longitudinal muscles. Perhaps related to their relatively simply organized body wall, most enteropneusts have great powers of regeneration.

The nervous system consists of a nerve plexus with certain definite tracts becoming consolidated into nerve trunks. It is peculiar in lying immediately under the ciliated epidermis and *above* its basement membrane. The primitive level of its organization is comparable only with those of the minor phylum Pogonophora, and the coelenterates; even the echinoderm system being more highly organized. In the trunk region of the body there are two consolidated tracts, one dorsal and one ventral, which are united anteriorly, that is in the boundary zone between trunk and collar, by a circular nerve. The circular nerve in the posterior end of the collar is also connected to the tubular dorsal nerve cord (the chordate diagnostic character) which runs only in the collar region. There is also another circular plexus on the surface of the proboscis stalk. Recent physiological and histological work suggests that the dorsal hollow nerve cord of the collar region seems to be essentially a through-conduction tract. There is no functional or microstructural evidence for major integrative activities. There are no obvious eyes or other highly organized sense-organs.

The enteropneust alimentary canal is a relatively simple tube from the mouth (posterior and ventral to the proboscis stalk) to the terminal anus. From the mouth an undifferentiated tube leads through the collar to the pharynx in the anterior part of the trunk. Various species vary in the number of gill slits (from twelve pairs to several hundreds) and they are usually incompletely divided by tongue bars (Figure 31-5D). These gill slits do not open directly to the exterior but through gill-sacs and their gill-pores which are arranged in two laterodorsal lines on the surface (Figures 31-4A and 31-5D). The slits themselves have lateral cilia, as in *Amphioxus*, and are the agency which causes a water current to be drawn in continuously through the mouth and pass out through the gill-pores. In *Glossobalanus* and *Ptychodera*, the pharynx is modified into an upper and a lower duct by lateral ingrowths (forming a figure-8-shaped lumen in cross section—Figure 31-5E). In contrast, in species of *Schizocardium* the pharyngeal condition is similar to that in the cephalochordates; there is no such division and only the dorsal groove and the endostyle are unperforated by the gill slits. There is an additional modification in some species of *Ptychodera* and *Schizocardium* where lateroventral fins of the body wall become greatly developed from the genital bulges or fins of most hemichordates. In these species genital wings are formed and enclose a false atrium. In a further development, in some species the gill-sacs may become reduced so that the U-shaped gill slits open from the pharynx directly into the space inside the genital wings. Behind the pharynx in the trunk there is an oesophagus which leads into a midgut extending for about one-third of the length of the trunk. This midgut has blind sacs, the so-called "hepatic" sacs, which show in some species as swellings on the outside of the body wall. There is then a short undifferentiated intestine to the terminal anus.

The feeding of enteropneusts involves three distinct mechanisms, of which the primitive ciliary filtration feeding of the other primitive chordates is probably the least important. The principal feeding mechanism is undoubtedly the collection of food particles by mucus on ciliary tracts of the proboscis and anterior edge of the collar. Our knowledge of this mucus-feeding mechanism results largely from the recent researches of E. J. W. Barrington, C. Burdon-Jones, and E. W. Knight-Jones. Since some misunderstandings have arisen concerning this feeding, it is worth noting that the ciliated surfaces of the trunk and most of the collar are *never* involved in the feeding mechanism. The third method of food collection involves the worm-like habit of engulfing sand and passing this through the alimentary canal for the extraction of suitable organic material.

The principal feeding mechanism involves the proboscis in secretion of mucus. The cilia upon it then carry the food particles in mucous threads backward and ventrally. These strands collect together (Figure 31-5C) in the region of the preoral ciliary organ on the posterior face of the proboscis and becoming bound together into a mucous rope, pass from the underside of the proboscis stalk into the mouth. One of the few fast muscle reflexes found in these sessile worms is here involved in a rejection mechanism which allows the animal to interrupt the continuously acting food-gathering process. When larger particles, for example, big sand grains, or distasteful substances come in contact

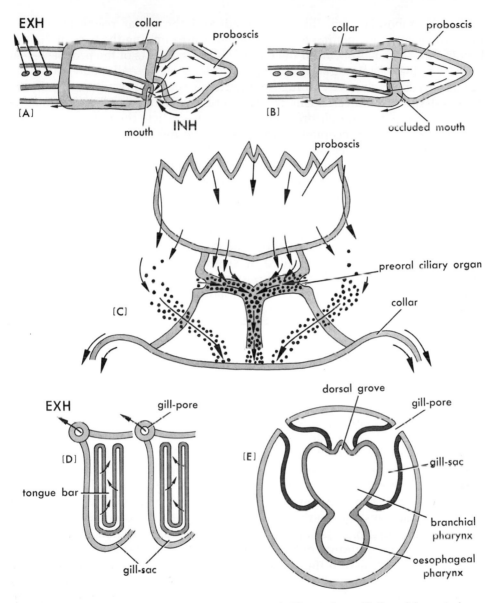

FIGURE 31-5. Ciliary feeding in enteropneusts. **A:** The surface ciliation of the anterior parts of an enteropneust while suspension-feeding is going on. The water current into the mouth (INH) and out of the gill-pores (EXH) is created by the cilia on the side of the concealed pharyngeal gill slits. **B:** The rejection mechanism, in which withdrawal of the proboscis occludes the mouth, stops the pharyngeal water circulation, and aligns the ciliary tracts of the proboscis for rejection of particles posteriorly over the collar. **C:** The ciliary collection tracks of the proboscis and preoral ciliary organ of the proboscis neck, viewed ventrally. **D:** The gill-sacs which enclose the outer side of the U-shaped pharyngeal gill slits. **E:** Cross section of the pharynx in *Glossobalanus,* to show the arrangement of the gill-sacs and the almost divided lumen of the pharynx. [**C** adapted from C. Burdon-Jones, *Bol. Fac. F.C.L. Sao Paulo,* **261,** *Zoologia,* **24:**225–280, 1962.]

with the proboscis base, the anterior edge of the collar reacts by moving forward closing the opening to the mouth and surrounding the base of the proboscis. After this has occurred the ciliary collecting tracts of the proboscis lead directly to the rejection tracts running posteriorly on the collar and trunk (Figure 31-5B). The selectivity in ingested particles seems to result mostly from a bivalve-like sorting mechanism. No distinction is made except on mechanical grounds of size, but this sorting mechanism gives rise to the rejection of larger sand grains, and the acceptance of the smaller animal and vegetable particles from the gathered particulate material. It has been suggested, however, that the ciliary preoral organ is capable of some tasting and that this results in the selection of particles on a qualitative basis quite apart from size.

The food material reaching the anterior part of the oesophagus is thus of three origins. In the most studied species, the most important source is the mucous rope deriving from the gathered strings of the proboscis, but there are also contributions from the filtering action of the gill slits in the usual chordate fashion and from material engulfed by swallowing substratum. It seems that the last component passes along the unperforated ventral part of the gut in those forms where the pharynx is subdivided (see page 546 and Figure 31-5E). From what knowledge we have of the variety of forms of hemichordates, it seems likely that the different proportions of these three types of food vary with ecological circumstances and perhaps correspond to the varieties of pharynx found. Behind the pharynx, the wall of the midgut, and more especially the cells of the hepatic sacs, secrete the digestive enzymes in globules which are whirled by cilia into the continuous strand with the ingested food. The food itself never enters the hepatic sacs. The enzymes include amylase, protein-breaking and fat-breaking enzymes, and it is of interest that amylase is also secreted by the cells amid the ciliary mucus tracts of the proboscis so that, to some extent, digestion begins externally. Absorption takes place in the intestine and the faeces are consolidated before extrusion through the anus.

As a whole, the digestive system does not show any high level of specialization in structure or in function. There seems to be relatively little of the ciliary sorting mechanisms which are well-developed in other microphagous animals, for example, Amphioxus (see page 541) and the bivalve molluscs (see Chapter 22). However, longitudinal ciliated grooves have been described in the oesophagus and intestine of different species, and in different forms these may be single, asymmetric, paired dorsolateral, or dorsal and ventral. Further investigation may show that the specific patterns of these grooves, and their potential for particle sorting, and the variant patterns of gill-sacs and genital atria, do correspond to particular feeding habits and/or particular ecological niches.

It should be noted that the cilia and mucous glands of the proboscis, collar, and trunk also play an important role in burrow formation.

Some other aspects of structure and function in hemichordates are worth mentioning. In an embryonic Balanoglossus (Figure 31-4C), the coelom has five divisions, one in the proboscis, two collar cavities, and two trunk cavities. These correspond to the three head and two body-cavities in Amphioxus, but in adult hemichordates they tend to become obscured by muscles and connective tissue. Excretion seems to be largely by the proboscis gland in the neck region, which gland surrounds the tissues once thought to be homologous with the notochord. This region has a glomerulus of blood vessels connected to both the main dorsal blood vessel and the main ventral one. In adults, the two main vessels are both contractile and are responsible for all the circulation. The other blood vessels are mere crevices between tissues and organs. There is a so-called larval heart lying immediately posterior to the glomerulus on the dorsal blood vessel. There is some functional interest in that the blood circulation is annelid-like; that is, the propulsion is anterior in the dorsal blood vessel and posterior in the ventral blood vessel, with the afferent branchials leading dorsally from the ventral blood vessel and the efferents from the tongue bars of the gills into the dorsal blood vessel. In other words, the longitudinal circulation is in the opposite direction to that in *Amphioxus* and in fishes.

The sexes are separate with two rows of simple gonads in the body wall of the anterior part of the trunk. As already mentioned, these may develop into genital wings. There are no gonoducts or other accessory structures and each gonad opens separately to the exterior. Fertilization is usually external, but there is considerable variability in the type of development. In several species, there is complete segmentation leading to a pelagic larva termed a tornaria (Figure 31-3), which has close resemblances to the auricularia larva of holothurians, though it has an additional perianal band of cilia. Unlike the echinoderms, however, there is no drastic metamorphosis, the tornaria growing into a little tripartite larva with proboscis, trunk, and collar. Thus bilateral symmetry is retained from the larva into the adult. The pulsating vesicle, corresponding to the madreporite of echinoderms, becomes the rudimentary heart of the young adult. In other species, there is a more direct development with the hatching embryo being already worm-like and never passing through a planktonic phase as a ciliated larva. This is the case in *Saccoglossus kowalevskii*, which is the only common hemichordate of the Atlantic coast around Cape Cod. Some evolutionary zoologists attach great importance to the tornaria larva as evidence of chordate-echinoderm affinities. This has perhaps been overstressed, as in other cases where larval similarities have been used as a basis for phylogenetics (see Chapters 27 and 33).

The remaining hemichordate class, the Pterobranchia, involves a few species placed in two genera, *Cephalodiscus* and *Rhabdopleura*, both minute sessile colonial forms, dredged from deep water of the oceans, and relatively rare, though worldwide in distribution. Although the individual zooids in colonies of both genera show the three divisions of the hemichordate body—proboscis, collar, and trunk—the trunk is reflexed on itself so that the gut is U-shaped. The collar in *Rhabdopleura* has two arms (Figure 31-4B) and in *Cephalodiscus* many pinnate ones. Feeding is presumably by the ciliation on these and on other parts of the anterior surface, but the only account of a living pterobranch dates from 1915, and contains several ambiguities and unexplained complications. Further study of the patterns of ciliation and of feeding is long overdue, and with increased numbers of deep-sea research vessels, the coincidence of living pterobranchs and suitable investigators may occur soon. *Cephalodiscus* has one pair of pharyngeal gill slits, and *Rhabdopleura* has a pair of probably homologous dorsolateral grooves in the pharyngeal wall. Of course, if one disregards possible feeding significance, there can be no need for respiratory circulation in animals so small (*Cephalodiscus*: 1–5 millimeters; *Rhabdopleura*:

0.1–1.0 millimeter). Both genera have a single pair of gonads in the anterior part of the trunk, and thus there is no replication of gill slits or of gonads or of muscles in these primitive chordates. Asexual budding is prevalent and probably is responsible for colony formation. Sexual development has not been well worked out, but late larval forms are known with a pronounced resemblance to young enteropneusts, that is having a worm-like tripartite body with a terminal anus.

There are some other, even more doubtful, "chordate allies." The minor phylum Pogonophora consists of worm-like tube-dwellers, again living in the deepest waters of the oceans. The group has claims to be the most recently discovered phylum of animals—the twentieth-century phylum—since they were first dredged early this century and thoroughly investigated only about thirty years ago.

Over one hundred species of pogonophorans are now known, mainly described by Russian zoologists. They are truly "thread-like" worms, mostly less than 1 millimeter in diameter but from 6 to 36 centimeters in length (see Figure 1-20). They apparently live upright in thin secreted tubes like bristles placed in the soft-bottom muds at depths of around 1 kilometer and more in all the oceans (Figure 31-6B–E). It now appears that they are among the commonest of deep-water organisms and were probably ignored as nonbiological material by many earlier deep-sea dredging expeditions. Alister Hardy has speculated that on earlier Antarctic oceanographic expeditions, tons of them "must have been shovelled overboard by some of the leading marine biologists of the day." The extremely long, thin body is basically tripartite, with the anterior part bearing a single tentacle or many tentacles, and they have some other deuterostome characters including an enterocoelic body cavity with tripartite division and a pair

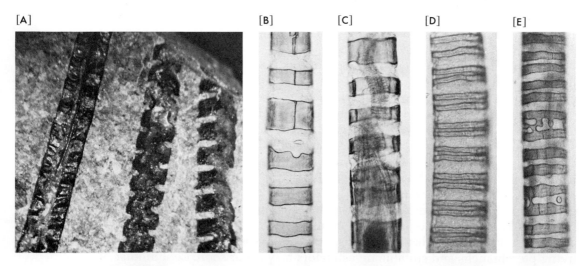

[A] [B] [C] [D] [E]

Figure 31-6. Two phyla of doubtful affinities (Graptolithina and Pogonophora). **A:** Three "colonial strands" of fossil graptolites (? *Monograptus* sp.) from limestone in Ayrshire, Scotland. The regularly spaced thecae are individually about 1 millimeter deep. **B–E:** Four sections of tubes of two species of pogonophorans: *Siboglinum ecuadoricum* (**B** and **C**) and *S. albatrossianum* (**D** and **E**). The tubes are from 4 to 9 centimeters long, but only about 0.12 millimeter wide. In **C**, the anterior part of the animal can be seen, including the attachment of the single tentacle characteristic of this genus. [Photo **A** by Frank A. Romano III of material collected by the author; photos **B–E** courtesy Dr. Edward B. Cutler.]

of coelomoducts. They have a closed circulatory system with a heart, but remarkably for animals about 20 centimeters long, no trace whatsoever of an alimentary canal. Details of feeding and digestion remain obscure, but it seems likely that the "beard" of tentacle filaments could function like the body wall of the similarly "gutless" cestode tapeworms (see Chapter 7) in a direct uptake of amino acids and carbohydrates from the external medium (in the case of pogonophorans from the interstitial water of organically rich muds). The body shape of all pogonophorans ensures a very high surface:mass ratio. No secretion of enzymes into the zone between the secreted tubes and the body or into the substrate surrounding the tubes has been demonstrated, and it would seem that pogonophorans rely upon the activities of microorganisms in the organic muds to provide "digested" glucose, amino acids, and fatty acids. Further, there is no evidence whatsoever of any phagocytosis of particulate material into epithelial cells. Nutrition seems dependent upon direct uptake, and pogonophorans thus are the only group of animals for which Pütter's theory holds true (but see discussion in Chapter 12).

The chitinous tubes which they secrete are annulated and have some resemblance to those of *Rhabdopleura* (Figure 31-6B,E), and it is to the fact that the eggs of some species are brooded within the tube that we owe our knowledge of the holoblastic cleavage and enterocoelous body-cavity formation. To confuse this issue, one investigator has claimed that the paired posterior coelomic spaces show schizocoelic development. Recent investigation of new samples of pogonophorans has shown that, in some species at least, there is a slightly dilated anchor formed as a fourth body region posteriorly, and that this has regular rows of setae projecting from its surface. This would also suggest annelid affinities, but it is claimed that the setae have a totally different mode of formation from those in segmented worms. Obviously, consideration of this minor phylum among doubtful chordate allies is a matter of convenience, and if there is any "chordate alliance" it is a remote one. In our discussion of interrelationships of phyla in Chapter 35 (see Figure 35-1), we make the timid decision of placing the phylum Pogonophora not with the hemichordates close to the main chordate line but with the assemblage of six other minor coelomate phyla which were discussed in Chapter 28.

Early Chordate Fossils

The wholly extinct phylum Graptolithina (see Chapter 34) has skeletal organization which resembles the tubes secreted by *Rhabdopleura*, and although the great bulk of fossil graptolites show no traces of internal structures (Figure 31-6A), there are a few well-preserved petrifactions for which some other pterobranch characteristics are claimed. (Some palaeontologists, however, relate this relatively extensive fossil group to the phylum Cnidaria.)

Certain other groups of Palaeozoic fossils are more clearly chordate and, in addition to "experimental" stocks, include the forms considered to be the earliest known vertebrates. Perhaps the most significant of these from our viewpoint, though not the earliest in time, is the genus *Jamoytius*, discovered and described from Scottish Silurian deposits by Erroll I. White about thirty years ago, who named it to honor J. A. Moy-Thomas, a brilliant young British palaeon-tologist then recently dead. *Jamoytius* was jawless, and had simple myotomes like *Amphioxus*, lateral fin folds, a persistent notochord, and no bony internal skeleton. The chordate stocks from which vertebrates are derived must have included similar forms. Armored, jawless fish (sometimes collectively referred to as ostracoderms) are found in Silurian and Devonian deposits, with fragmen-tary remains in the Ordovician. Among these are the cephalaspids which have been meticulously reconstructed (initially by E. H. O. Stensiö and his Scandin-avian colleagues), and show a truly vertebrate organization of skeleton, brain, and sense-organs. Their pharyngeal apparatus conforms to the primitive chor-date pattern we have described as archetypic and it is almost certain that they fed by stirring up bottom deposits, sucking in organic detritus through the jaw-less mouth and filtering it on the ten pairs of gill slits. Among one stock of such agnathous fishes, biting jaws were later evolved, and thence came the bulk of vertebrates.

It is worth noting here that at least two workers consider the fossil carpoids as primitive chordates, basing this on supposed branchial slits, and even more doubtful homologies with the cranial skeleton and neural cavities of primitive vertebrates. (In this book the carpoids were treated as primitive echinoderms—see Table 29-1 and Chapters 29 and 34.) That controversial allocation to these two distinct phyla is possible could be taken as highly suggestive. An al-most equally bizarre, but more certainly "chordate" fossil is the "unassigned" species *Ainiktozoon loganense* from Silurian rocks in Scotland. Over thirty

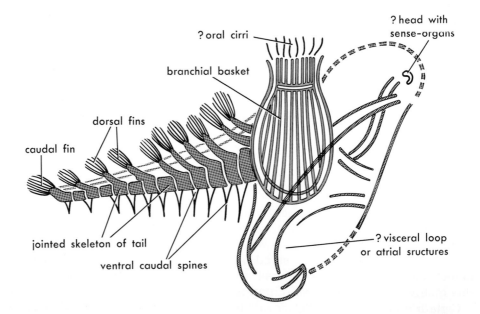

FIGURE 31-7. An enigmatic "unassigned" fossil chordate, usually neglected in phylo-genies, a diagrammatic reconstruction of *Ainiktozoon loganense* from Silurian rocks in Scot-land. This curious assemblage of sea-squirt and vertebrate characteristics is discussed in the text. [Largely based on D. J. Scourfield, *Proc. Roy. Soc. London,* **121**:533–547, 1937, and original specimens in the Hunterian Museum of the University of Glasgow.]

specimens have been found and they are all similar, consisting of a "branchial basket" like that of a sea-squirt with an attached, jointed, and finned tail like those of craniate chordates. A reconstruction is shown in Figure 31-7, and it is immediately obvious that although probably chordate, this organism could not be assigned to any subphylum as they are presently diagnosed. *Ainiktozoon* was first described in detail by the late D. J. Scourfield, a distinguished worker on crustaceans, and this eponymously enigmatic fossil organism seems to have been largely ignored by zoologists constructing phylogenies of the chordates. Of course, there are many other enigmatic fossils. Although *Ainiktozoon* is included in this book partly because I have handled most of the extant specimens (in the Hunterian Museum of the University of Glasgow) and been disproportionately impressed by them, the species has real evolutionary significance. The possible interrelationships of the chordates will be considered later (Chapter 35), but the assemblage of characteristics in this fossil form suggests that there were several "experimental" chordate stocks in the Palaeozoic, from which there survive the minor but important group of cephalochordates, and the two widely divergent "successful" lines—the sea-squirts (Urochordata) and the vertebrates.

32 INVERTEBRATE CHORDATES II: THE SUCCESSFUL SEA-SQUIRTS

IF the vertebrates are excluded, the subphylum Urochordata includes the most successful animals built on the chordate ground plan. They are successful by all our usual measures; there are over two thousand species, which make up a considerable part of the animal biomass in some marine environments. On suitable rocky substrata in the lower littoral and sublittoral, sessile tunicates (solitary or colonial) are among the commonest marine invertebrates. Less commonly, and for limited periods of time, pelagic urochordates (both solitary and colonial) can be the dominant animal organisms of the marine plankton. The great majority of species are allocated to the class Ascidiacea, the true sea-squirts or ascidians. There are two other classes of urochordates: the Larvacea and the Thaliacea, less numerous groups, both highly specialized for permanent life in the marine plankton. Solitary forms of typical sea-squirts (class Ascidiacea) will be discussed first.

Functional Organization

Solitary ascidians are cylindrical or spherical animals of moderate size (rarely over 10 centimeters long), attached at one end to rocks or manmade hard substrata like pilings, dock walls, and ships' bottoms, and bearing at the other two openings—oral

554

and atrial. The genus *Ciona* (Figures 32-1A and 32-2A) can be regarded as show-ing an almost archetypic simplicity of structure and function. The protective integument, termed the tunic or test, is a cuticular secretion of the ectoderm but has in it wandering mesodermal cells and blood vessels. It was formerly thought unique among animals because the ground substance, tunicin, includes cellu-lose as a constituent, though this supporting material, typical of plants, has now been found to occur in the connective tissues of the dermis of mammals. The muscles lie in the body wall beneath the test in the form of longitudinal and circular strands. There are also sphincter muscles around the openings. There is no real skeleton, the muscles partly being antagonized by the elasticity of the test and partly working around an enclosed hydraulic skeleton of the seawater in the pharyngeal and atrial spaces. There is a plexus of nerves on the body wall and this is attached to a single central ganglion, lying between the oral and atrial openings on the side of the pharynx (Figure 32-2C). There are no com-plex sense-organs, but mechanoreceptors are well-developed around the oral and atrial openings. Some experimental work has suggested chemoreception, and there can be pigment spots which have been termed ocelli although there is no physiological evidence of photoreception.

Internally, the pharynx is proportionately enormous, with the gill slits subdi-vided, forming an elaborate basketwork. The atrium forms an outer bag later-ally and dorsally around the pharynx, and the anus and genital openings lie within it. In life, when a sea-squirt is expanded, water is continuously passing in the oral opening to the pharynx, through the multiplied gill slits to the atrial cavity, and out through the atrial opening. The oral opening leads into a short, wide buccal cavity and thence into the prebranchial zone of the pharynx. This is bounded posteriorly by the peripharyngeal band, two ciliated ridges which lead to a dorsal tubercle on the atrial side of the pharynx. Here there is a ciliated funnel leading through a duct to a gland below the ganglion—the subneural gland. All these structures are relatively well-innervated and a sensory func-tion has been proposed. Apart from this sampling of the branchial current, in a fashion analogous to the osphradium of molluscs, other functions have been proposed for the subneural gland. In some forms, it is supposed to secrete the hormone which induces spawning. In others, including *Ciona,* it is known to be a site of phagocytic action, dead blood cells being removed from it and passed out the duct into the pharynx and the alimentary system. The ascidian subneural gland has been homologized with the preoral ciliary organ of en-teropneusts, and with the preoral pit of the developing *Amphioxus,* and thence —though the homology may be latent at best—with the pituitary gland of ver-tebrates. (This last of course has a twofold origin: the adenohypophysis is a preoral invagination and the neurohypophysis is a neural downgrowth from the floor of the brain.) Behind the peripharyngeal band and subneural gland lies the enormous branchial basket. All the bars are hollow, with blood lacunae running through them, and are covered with a ciliated epithelium which does not show much differentiation. However, the cilia of the sides of the openings (the stigmata) are somewhat longer, could be termed lateral cilia, and cause the main water current through the sea-squirt. There is a ventral imperforate tract forming an endostyle, homologous with that in *Amphioxus,* and there is also the dorsal hyperpharyngeal band from which, in *Ciona,* hang a row of proc-

[A]

[C]

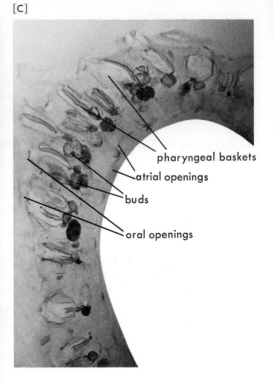

pharyngeal baskets

atrial openings

buds

oral openings

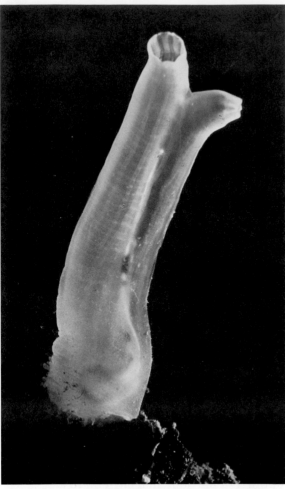

[B]

FIGURE 32-1. Sea-squirts. **A:** Living specimen of *Ciona intestinalis,* a solitary sea-squirt. **B:** Living specimens of two color varieties of the colonial sea-squirt *Botryllus schlosseri,* growing on a mooring line. Each "flower" system consists of a group of seven or eight zooids with their separate oral openings surrounding a single common atrial opening. **C:** Part of a cross section through the pelagic colonial form *Pyrosoma.* In the wall of the tubular colony the oral openings of the individual zooids face outward, the atrial openings toward the internal "lumen" of the colony. [Photo **A** © Douglas P. Wilson; Photos **B** and **C** by the author.]

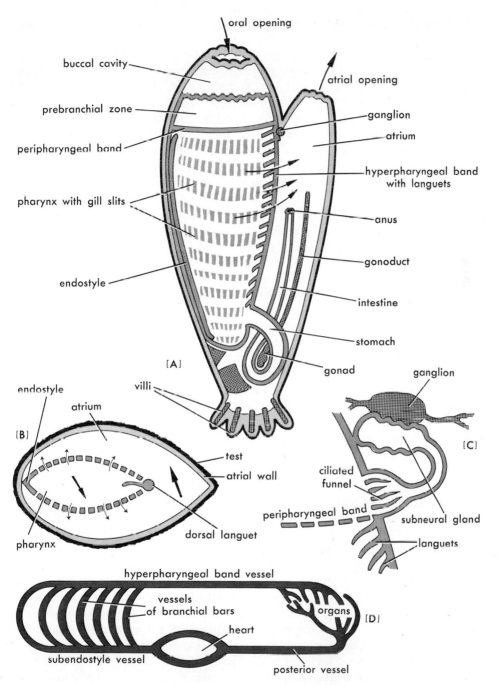

FIGURE 32-2. Sea-squirt organization. **A:** Stylized vertical half of a solitary sea-squirt showing the direction of the feeding water current through the enlarged pharyngeal basket. **B:** Cross section through the pharynx and atrium. **C:** The ganglion and subneural organ. **D:** Diagrammatic circulatory system of sea-squirt. Direction of circulation can be clockwise or counterclockwise. For further discussion, see text.

esses termed the languets. (In some other genera there is a continuous dorsal lamina.) The pharynx leads through a small oesophagus to the stomach, thence to the intestine, and through the rectum to the anus in the atrium (Figure 32-2A).

Appropriately enough, the processes of feeding and digestion in sea-squirts have been best worked out by investigators also concerned with filter-feeding molluscs (including J. H. Orton, C. M. Yonge, and R. H. Millar). Unlike the filtration in bivalves (see Chapter 22), feeding in sea-squirts is dependent on the secretion of a mucous net. This is produced by the mucous gland cells of the endostyle which secrete on to the inner face of the basketwork, the cilia of which beat dorsally, thus pulling a sheet or net of mucus across to the dorsal row of languets. Meanwhile, a continuous water current is drawn in through the oral opening, down through the pharynx, and moved through the stigmata into the atrium by the lateral cilia. Food particles picked up on the mucous net are carried dorsally. On the row of languets, the mucus is worked posteriorly as a cord or rope into the oesophagus. The oesophagus itself is ciliated and the stomach is the main region of enzyme secretion. Some ascidians, though not *Ciona*, have secretory diverticula in this region. The cilia of the grooves of the stomach are arranged in spiral rifling so that the food is churned with the enzymes in a mucous-bound mass. Absorption takes place in the midgut and the remaining material is carried out to the anus by ciliary action, since there is no peristalsis. The filtration in urochordates can be fantastically efficient: the mucous net in some forms being capable of retaining particles of between 1 and 2 microns in diameter. In contrast to *Amphioxus*, to the enteropneusts, and indeed to most ciliary mucus-feeders, the sea-squirts have little in the way of rejection mechanisms for disposing of large particles. Ecologically, this is connected with their life on harder substrata, filtering relatively clean seawater.

Although, as chordates, the sea-squirts are classified as coelomate animals, there is no definite evidence of any true coelom existing in adult urochordates. The small epicardial sacs which lie beside the heart in some forms, and which are "enterocoelic" in that they communicate with the posterior end of the pharynx, are thought to be vestiges of a once more extensive coelom. The evidence, however, is not conclusive, and functionally the epicardium is concerned in excretion in some sea-squirts, and is part of the stolonic asexual budding system in others. The circulatory system, and the blood within it, are equally peculiar. The blood consists of a colorless plasma with various corpuscles, some nucleated. It may contain fantastically high concentrations of unusual elements, such as vanadium and niobium. (The concentration of vanadium in the sea is of the order of 0.0002 milligram per liter, while the concentration in the plasma of *Ciona* is 400 milligrams per liter, and in some other ascidian species where it is contained in corpuscles reaches concentrations of 1.8 grams per liter.) The smaller ducts of the circulatory system really lack walls and are essentially the spaces between organs. The heart at the base of the pharynx is a muscularized fold in the pericardial wall giving rise to a vessel running anteriorly under the endostyle, and another posteriorly to the digestive organs and gonads directly. The vessels in the branchial basket arise from the vessel under the endostyle and unite on the other side of the pharynx in a hyperpharyngeal band vessel which sends branches to the digestive organs, gonads, etc. Structurally, this is diagrammed in Figure 32-2D, but functionally the contractions of the ascidian heart are peristaltic for about ten contractions driving blood in one direction, then, after a pause, contractions start again, but in the opposite direction. Thus alternately, the blood system resembles that of *Amphioxus* and fishes (i.e., when it leaves the heart by the subendostyle vessel and returns by the posterior vessel), and alternately, that of *Balanoglossus* and annelid worms when it leaves the heart by the posterior vessel and returns by the subendostyle vessel. It was earlier claimed that the regular reversals of direction of heart beat were due to the jamming of corpuscles in the branchial bar vessels, causing a back-pressure, which in turn caused reversal of the peristaltic action of the cardiac musculature. This process certainly occurs, but the isolated heart has since been found to reverse its beat. In itself this reversal would be evidence against back-pressure as a control, but M. E. Kriebel has now demonstrated that the two pacemakers controlling the contractions are themselves affected by changes in intracardiac pressure and can show adaptation.

The Ascidian Tadpole

Many tunicates, including *Ciona*, are hermaphroditic: all mature animals have both a compact ovary and a ramifying testis within the loop of the alimentary canal in the base. The two genital ducts have their external openings within the atrium, but fertilization—normally cross-fertilization—takes place in the sea. In the most typical cases, the small egg undergoes a regular segmentation, developing to a larva rather like that of *Amphioxus*. Figure 32-3A illustrates the peculiar appendicularian larva or ascidian tadpole. Both the embryonic development and the free-swimming tadpole larval stage in ascidians are relatively brief. There are also many forms where the free-swimming larval stage is lacking, being retained during development within the maternal atrium, and in some cases, showing little or no development of the tail. However, in those forms, presumably archetypic, where the free-swimming tadpole is fully developed, it exhibits the full integrated concert of structures and functions which we regard as diagnostic of chordates. Dorsal to the gut with its pharynx and gill slits lies a hollow, central nervous system, and there is a propulsive tail with, as an axial skeleton, a typical notochord. Unlike *Amphioxus*, the tail muscles are not arranged in myotomes.

After a varying period of larval life, the ascidian tadpole attaches by the adhesive papillae lying below the mouth to a suitable substratum for adult life, and almost immediately the processes of metamorphosis begin. There are two processes of change which occur simultaneously: the tail is reduced, and the internal organs of the body undergo a rotation of 180 degrees (Figures 32-3A–C). Most accounts suggest that the tail is absorbed by phagocytosis. This is incorrect; the method, recently investigated by Richard A. Cloney and by James W. Lash, is a controlled shrinking of the outer epidermis which pulls the notochord and muscles into the body where reorganization takes place. The rotation brings the oral opening (Figure 32-3C) to a position opposite the attachment papillae. N. J. Berrill has presented evidence that suggests that the tail epidermis reacts to buildup of metabolites after the muscular activity of disper-

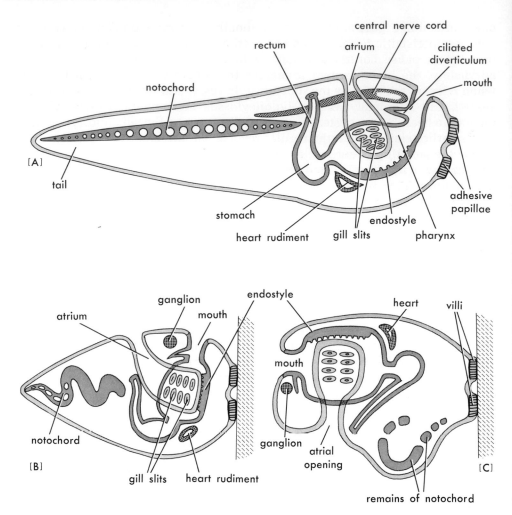

FIGURE 32-3. Larval form and metamorphosis in an ascidian. **A:** The "tadpole" or appendicularian larva, whose locomotory tail is lost at metamorphosis. **B** and **C:** Successive stages in the process of metamorphosis which follows attachment to a suitable substratum; a controlled shrinking of the outer epidermis pulls in the tail, and the internal organs undergo a rotation of 180°.

sal. However, the manner in which rotation of the organs, due to increased growth between the mouth and the adhesive papillae, and the reduction of the tail, occur simultaneously immediately after successful settlement, suggests some kind of overall regulatory control.

Sea-squirt Diversity

The class Ascidiacea is a large group of animals, encompassing considerable diversity of form and of habitat. They are surprisingly difficult to subclassify since they exhibit very complex interrelationships, in many cases involving considerable convergence of structural patterns. They are all sessile animals but show various degrees of colonial habit, associated with various types of budding. The so-called simple ascidians can be divided into two main groups using diagnostically what seem to be anatomical characters of some phyletic significance. There is a *Ciona* group with dorsal languets, a perivisceral cavity developed from the epicardial pouches, and having associated compound forms, mostly arising from stolonial budding. In contrast, the *Ascidia* group show a continuous dorsal lamina in the pharynx, no epicardial pouches, and have their associated compound forms mostly resulting from pallial budding, that is, budding of the so-called mantle, or pallium—the outer body wall.

Three principal types of budding are found: two stolonial and one pallial. In the most complex stolons a double tube arrangement is found, with a median ventral outgrowth of the visceral region forming the outer tube with an inner tube derived from the epicardium and with some mesenchyme cells lying between the two tubes. A second simpler form of stolonial budding involves an outer tube with no inner one, but merely a strip of mesoderm cells. Pallial budding usually involves an outgrowth of the atrial wall, each bud having a lining corresponding to the atrial lining and an outer coating of the outer body wall and its integument. It is worth noting that the concept of three determinate cell layers in triploblastic animals breaks down in the processes of this budding. For example, in the stolonial budding of forms like *Aplidium*, the parental endoderm of the epicardial tube forms not only the alimentary canal in the bud, but also the ectodermal central nervous system and atrium, and the mesodermal gonad. In the pallial budding of Botryllidae, the parental ectoderm of the atrial wall forms not only the central nervous system in the bud, but also the endodermal alimentary canal, and the mesodermal gonad. In the budding and regeneration processes of most other groups of many-celled animals, the germ layers are usually deterministic: mesodermal structures being derived from mesoderm, and so on. However, other exceptions to the concept include ectoproct budding and holothurian regeneration.

In ascidians, all degrees of colonialism—from loose associations to highly organized interdependence—are found. Apart from the solitary forms which separate completely immediately after budding, there are the so-called social ascidians, like *Clavelina* and *Perophora*, where the only link is a common stalk, and *Polyclinum*, where the individual zooids are completely separated but are embedded in a common test which is well-developed and massive. Perhaps the commonest colonial forms are those like *Botryllus* and *Didemnum* which form spreading patch colonies. As shown in Figure 32-1B, the individual zooids produced by budding—the blastozooids—are arranged in flowerlike systems, each group of seven or so sharing a common cloaca or united atrial opening. Forms like botryllids grow all over suitable substrata as rather indefinitely organized patches. In contrast, forms like *Distaplia* show a more highly organized colonial habit, with budding in the colony limited to one region so that the colony as a whole has a definite shape and shows some division of function between groups of zooids.

It seems clear that the various levels of colonial habit have evolved several times in the ascidians, so that the old-fashioned classification into simple ascidians and compound ascidians is unreal. However if one tries to base a classification of the ascidians on the basis of the three types of budding which occur, then it quickly becomes apparent that this will not match with the basic structural features which provided a dichotomy between the *Ciona* group and the *Ascidia* group. Many phylogenies and "natural" classifications of the sessile tunicates have been proposed. None is entirely satisfactory. As a whole, the diagnosis of the class Ascidiacea would run: "tunicates where the tailless adult is sessile, with the mouth and atrial opening subterminal at the same end." This sufficiently distinguishes the largest group of the urochordates from the remaining planktonic forms.

Four Patterns of Planktonic Life

Perhaps the most fantastic way of life found in primitive chordates is that of the urochordate class Larvacea, diagnosed as tunicates in which the sexually mature animal retains the organization of the larva including a locomotory tail with a notochord. The "head" of the tadpole, propelled by the tail, is like a much simplified sea-squirt, with a simple pharynx with two gill slits and no atrial cavity. Both gill slits and anus open directly to the exterior. Species of such genera as *Oikopleura* (see Figure 33-5), *Megalocercus*, and *Fritillaria* occur in the marine plankton in many parts of the world. The most bizarre aspect of their biology is that they build an external house of hardened mucous secretions. As shown in Figure 32-4A, this house is used as a filter-feeding apparatus. Continuously repeated flexures of the tail cause a water current to be drawn in through the coarse filter and thence through the internal net. Particles trapped on the internal net are ingested. Internal pressure opens the flap valve at the "posterior" end of the house, and the jet drives the animal forward. When coarse particles accumulate on the outer filter, the animal discards its house, escaping through the spring-loaded trap door at the anterior end and constructing an entirely new house in about an hour. In a few species, including those of *Fritillaria*, the adult tadpole lives outside its secreted house during the feeding process. Larvacean tunicates live their entire life in the plankton, and they have a simple life-cycle involving only sexual reproduction. No form of budding occurs in the group.

It is generally agreed that the Larvacea is a group of urochordates which have become specialized by the retention of an essentially larval organization after the onset of maturity and the development of gonads. Thus, they are one of the classic examples of the evolutionary process which is termed neoteny. The hypothesis implicit in the concept of neoteny is that during the evolution of a stock of animals, the developmental processes and life-cycle have been so modified that sexual maturity becomes attained in association with larval organization. Like so many other brilliant—but essentially intuitive—hypotheses regarding larvae, the concept that neoteny has occurred several times in the evolution of chordates was first developed by Walter Garstang (see, for example, the discussion of larval torsion in gastropods in Chapter 20).

The third and last class of the urochordates is the Thaliacea, diagnosed as tunicates where the tailless adult is pelagic with the mouth and atrial openings

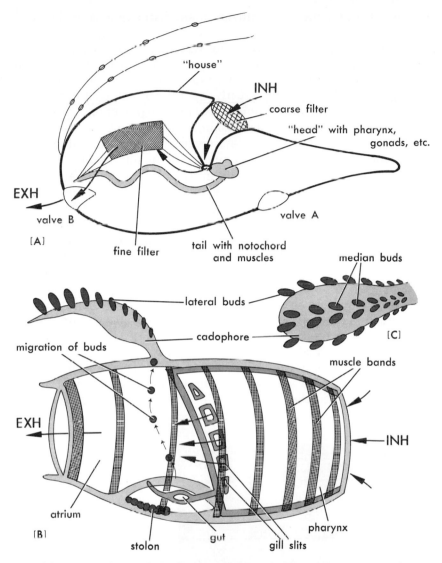

FIGURE 32-4. Two forms of planktonic chordates. **A:** The feeding "house" of *Oikopleura*, a larvacean urochordate. For full description, see text. **B:** The adult asexual oözooid of the thaliacean urochordate *Doliolum*, with its characteristic barrel-shaped body and muscle bands. Note the blastozooid buds migrating to the cadophore. Details of the complex life-cycle, with its alternation of sexual and asexual reproduction, are given in the text. **C:** The cadophore of *Doliolum*, viewed from above.

at opposite ends. In these oceanic forms, the water current through the body is not only used for feeding and respiration but also as a means of locomotion. The class splits into three rather distinct orders: Pyrosomatida, Salpida, and Doliolida. The first order, with its single genus *Pyrosoma*, consists of tropical species in pelagic colonies, with structure which is intermediate between that of typical sea-squirts (Ascidiacea) and those of the salps and doliolids.

Certain points of functional organization are distinctive of the class Thalia-cea, apart from the planktonic habit and positions of the openings, and these must first be discussed. There is budding which is basically stolonial but the complex stolon may contain elements of different cords of cells: from the atrium, from the gonad, from nervous tissues. Individuals which develop from fertilized eggs—oözooids—differ greatly from the asexually produced blasto-zooids. In all three groups the oözooids have lost the power of sexual reproduc-tion. In salps and doliolids the blastozooids have lost the power of budding. Thus, in these two groups there is a regular alternation of generations. Throughout the Thaliacea, there is a reduction in the importance of cilia, water currents being assisted in *Pyrosoma* by muscular movements and caused by them in salps and doliolids. These groups could be considered as analogous to the Cephalopoda and Septibranchia among the molluscs (see Chapters 22 and 24) where muscles have replaced cilia in creating the water currents through the mantle-cavity. Particularly in doliolids, the muscle strands form strong complete rings encircling the barrel-shaped body and by their contraction, cre-ate the water current which serves, not only for feeding but for jet-propulsion (Figure 32-5).

In *Pyrosoma* there is no larva and the oözooid is retained within the parental atrium where it produces the first four buds of the blastozooid generation. These are freed and the oözooid regenerates while the four primordial buds form the typical cylindrical colony by further budding (Figure 32-1C). The col-onies are not uncommon in tropical waters and are brilliantly luminescent, re-sponding readily to tactile stimulation. The list of oceanographers who have written their names "in letters of fire" on the sides of newly captured *Pyrosoma*

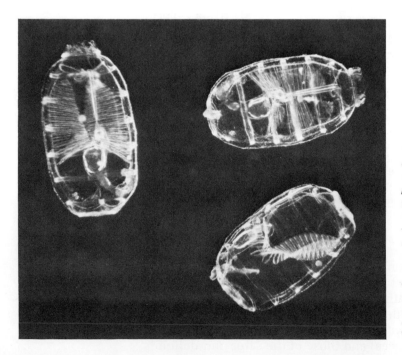

FIGURE 32-5. A planktonic chordate. Specimens of the bar-rel-shaped planktonic tunicate *Doliolum,* showing the propulsive muscle bands and the diagnostic pharyngeal gill slits. ×18. A fan-tastically complicated life-cycle (involving the alternation of asex-ual and sexual forms, a tadpole-like larva, and a complex process of stolonic budding of polymor-phic zooids) is carried out entirely in the plankton of the open ocean. [Photo © Douglas P. Wilson.]

colonies begins with Wyville Thomson, Murray, Moseley, and company on the *Challenger* Expedition, and has not yet ended.

In the order Salpida, there is again no larva, but the oözooid leads a separate existence as an asexual form, with a coiled stolon from which blastozooids are regularly budded. These blastozooids, which are sexual individuals unable to bud, are shed in relatively short chains, the form of which differs in different species. There is thus a regular alternation of generations. Salps are among the few primitive chordates important in the economy of the sea. In the plankton, they both compete with and prey on young fish and other elements of the zöoplankton. They appear irregularly, but in enormous numbers, in certain areas of the North Atlantic, and the years when salps occur represent very poor years for young fish. Some investigators have attempted to correlate variations in the occurrence of salps with the sunspot cycle.

Doliolum shows the most complex life-cycle of thaliaceans, indeed of all chordates. The solitary sexually mature adult, which is a blastozooid, produces three eggs, each of which develops to a tadpole larva. Each larva metamorphoses to the adult asexual oözooid in which there is a stolon which continuously buds off blastozooids. These blastozooid buds (Figure 32-4B, C) migrate through the atrial wall to a structure called a cadophore where further budding results in the formation of groups of polymorphic blastozooids forming a colony. There are usually two lateral lines of gastrozooids or feeding individuals, and a median row of phorozooids or nurses, and gonozooids or sexual individuals. Eventually, solitary blastozooids which are sexually mature adults break away and give rise to the eggs again. Unlike the case in the salps, the solitary sexual blastozooid does not differ much in gross anatomy from the adult asexual oözooid.

It is almost certain that these four groups of planktonic urochordates have evolved from more typical sessile ascidians, with a tadpole larva. The Larvacea have probably evolved by neoteny. On the other hand, in *Pyrosoma* and in the salps, the tadpole larval stage has been suppressed. As N. J. Berrill has pointed out, the ascidian larval tail could hardly propel the large, barrel-shaped body of a thaliacean, and as a consequence, the powerful muscle bands have been evolved from the circular muscle strands which are present in all adult ascidians. However, the limited capabilities of the ascidian tail may have been surpassed in at least one extinct organism (*Ainiktozoon*—see page 552). Certain other aspects of chordate phylogeny are discussed in Chapters 31 and 35.

33

INVERTEBRATE
DOMINANCE
AND DIVERSITY:
THE MARINE PLANKTON

ALL students of invertebrate biology should have an opportunity to *explore* the world of marine zoöplankton. This is not only because the zoöplankton of the temperate and tropical oceans presents an assemblage of invertebrates more diverse and more comprehensive than is found in any other environment. Larval stages or adults of every major phylum and class in the animal kingdom (except insects, arachnids, and higher vertebrates) are represented. There is also the felicity, like that of earlier generations of naturalists, which comes from what can correctly be called "exploration" of a living world in which the unknown (at least to textbooks) is still not uncommon. (Ornithologists and similar kinds of late-twentieth-century naturalists have no offshore islands or other unexplored territories left on this planet to yield comparable bird discoveries). Even *preserved* samples of marine plankton are worth working through, but today, thanks to refrigeration, thermos bottles, and air freight, it is possible for an interested biologist in a midcontinental laboratory (in Denver, in Prague, or in New Delhi) to receive *living* marine zoöplankton for study.

Knowledge of systematic zoology is not all that can be provided by plankton studies. Feeding, swimming, and escape mechanisms await investigation by the mechanistic physiologist, and a microcosmos of food-webs awaits bioenergetic analysis by the ecologist. In functional and ecological relationships, too, the unexpected is always turning up in the plankton.

566

The term plankton, meaning "that which is made to wander or drift," emphasizes the passive nature of the horizontal movements in the sea of these organisms. (Many are capable of considerable and effective locomotion, but generally this swimming is used in vertical movements rather than in directed horizontal ones.) The plankton can be contrasted with the nekton (actively swimming organisms, usually of greater size, like fish and squid) and with the benthos (now taken to cover all bottom-dwelling organisms in all depths of the sea, originally restricted to deep-sea-bottom forms). A general term for both plankton and nekton of the open sea is the *pelagic* fauna and flora.

The photosynthesizing microorganisms of the plankton are referred to as the phytoplankton, the animals as the zoöplankton. The latter, except for a few larval fish and fish eggs, are all invertebrates, ranging in size from about 50 microns to a few millimeters in length, and encompassing—as already noted—the majority of phyla and classes discussed in this book. In a more functional division, the zoöplankton can usually be divided into the more numerous "permanent" animals, which live and grow to maturity and breed as drifting organisms, and the more varied "temporary" zoöplankton, larval stages of the many benthic and nektonic animals which spend a short or long larval period in the plankton. Another functional division is into the larger biomass of herbivore grazers, feeding on the world's largest crop of green plants—the marine phytoplankton—and the considerably smaller (in individual numbers *and* in biomass) group of carnivorous zoöplankton forms which prey upon the herbivores and upon each other. These categories will be more fully discussed below; meanwhile, any understanding of the zoöplankton must be based on an appreciation of the basic aspects of phytoplankton production in the sea.

Basic Pasturage

The phytoplankton of the open oceans makes up the largest annual crop of green plants in the world. These primary producers in the seas form a mixed population of photosynthesizing microorganisms of which diatoms (single-celled algae with tests of silica) are usually the most important. In addition, there are always extremely small naked green flagellates (termed microflagellates) and, especially in temperate seas, numbers of the armored dinoflagellates (see Chapter 6). In some tropical oceans, the last are replaced by coccolithophores, which have a calcareous test. In some parts of the oceans at specific seasons, more than 99% of the green pasturage will belong to a single species, which is usually a species of diatom. Living phytoplankton is shown in Figures 33-1 and 33-2. The minute size of the individual units is worth emphasizing. Most individual diatoms lie in the size range from 10 to 200 microns, dinoflagellates run up to 100 microns, while the microflagellates are much smaller, actually about 5 microns. Therefore the invertebrates of the zoöplankton which feed on this pasturage are themselves mostly small and are usually filter-feeders.

Diatoms, and all other producer forms in the phytoplankton, have physiological requirements like all green plants. To grow and reproduce, they require inputs of four main categories: water, carbon dioxide, sunlight, and certain in-

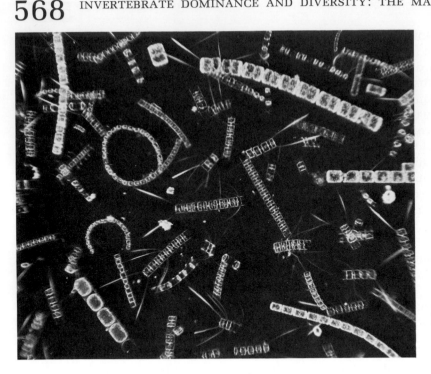

FIGURE 33-1. Living diatoms from the marine phytoplankton, mostly species of chain-forming diatoms of the genera *Chaetoceros* (with lateral spines), *Thalassiosira*, and *Lauderia*. ×80. [Photo © Douglas P. Wilson.]

organic nutrient salts. In the sea, water and carbon dioxide are readily available, and solar radiation alternates with certain inorganic salts (notably the so-called nutrients, phosphates and nitrates) as the limiting factors for plant growth.

Much social behavior—and indeed much religious practice, including the pre-Christian festival taken over as Easter—has been founded on the succession of seasons as it is displayed in the land vegetation of the temperate regions of the world. In the seas of the same latitudes, a seasonal succession of diatoms takes place, which is less obvious, but of even greater ecological significance. In all but certain tropical areas of the oceans, the same seasonal pattern of change occurs year after year. Typically, a sudden outburst of diatom growth occurs every year within a few days of its average occurrence for that locality, and this is termed the spring diatom increase (SDI). The fantastic reproductive capacity of the diatoms concerned (in sequential asexual divisions of these single cells) is responsible for the SDI. Dividing at rates greater than once every eighteen hours, the diatoms show exponential increase in numbers, and areas of the sea may become green or brown with diatoms within a week. Repeated divisions build up the number of diatoms per unit volume of water to levels usually from 500 to 2000 times greater than their population densities through the winter in the same regions. In a few local areas, maxima of the SDI (at 50,000 times winter densities) persist for only a few days, and in some cases the decline in numbers which follows can be as dramatic as the rapid increase itself. The occurrence of this startling SDI is not really difficult to understand if two groups of features in the physical environment and one additional biotic factor are clearly recognized. First, the light-energy required for photosynthesis

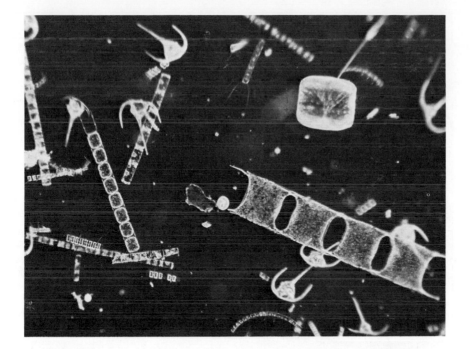

FIGURE 33-2. Living phytoplankton from the English Channel, including *Ceratium tripos* and related dinoflagellate species (anchor-like cells), as well as several diatom species. The single large barrel-like diatom cell is of *Coscinodiscus concinnus;* the chain of four large cells below it of *Biddulphia sinensis;* and the chains of smaller diatom cells include species of *Rhizosolenia, Stephanopyxis,* and *Chaetoceros.* ×60. [Photo © Douglas P. Wilson.]

is available only in the surface layers of the oceans. Secondly, when a body of water like an ocean (or a large lake) is heated by solar radiation by way of the surface, a temperature (and density) stratification of the waters results. The third factor is biological and concerns the vertical distribution of bacteria and other decomposers in the marine environment. With all green plant growth limited to a small superficial fraction of the ocean's volume, and with the great bulk of decomposition of organic materials by bacteria and subsequent recycling of nutrients occuring in waters below this euphotic zone, the occurrence of a discontinuity layer in which temperature decreases rapidly with depth (a thermocline) may permit little or no exchange of water or dissolved nutrients. In temperate seas in summer, the thermocline forms a physical barrier to water circulation at depths of about fifteen meters.

The constant annual cycle of phytoplankton productivity in temperate oceans is illustrated in Figure 33-3. As a result of the breakdown of the thermocline in winter, there is an enriching of the surface waters with nutrient salts, so that in early spring the concentrations of phosphates and of nitrates in the superficial layers are higher than at any other time of the year. Thereafter, only when the available solar energy rises to an appropriate level does the outburst of the SDI occur. The limiting factor involved in ending the SDI is nutrient depletion. However, increased grazing by a new generation of copepods and

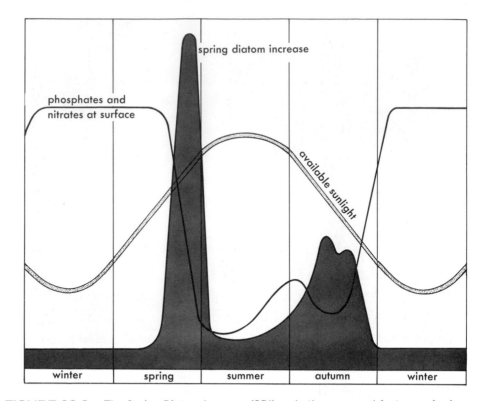

FIGURE 33-3.   The Spring Diatom Increase (SDI) and other seasonal features of primary productivity in temperate seas. In the surface waters, nutrient salts occur in greatest concentrations in winter, being depleted in late spring and remaining at low levels through most of the summer and autumn. In contrast, the availability of solar energy for diatoms and other phytoplankton is lowest in winter. In the stylized sequence shown here, the exponential plant growth of the **SDI** begins when a threshold level is reached in available energy, and the growth phase is ended by nutrient depletion. The less extensive (and less universal) diatom increase in autumn begins after a late-summer rise in available nutrients and ends as the available sunlight declines.

other herbivorous zoöplankton brings about an ever-increasing rate of consumption of the phytoplankton. This brings the standing crop of plant production down to the lower level which lasts throughout summer. The occurrence of an autumn diatom increase (as shown in Figure 33-3) is not so universal as that of the SDI. If it occurs, it begins as a result of increased availability of nutrients and ends with the autumnal fall in levels of solar radiation. Finally, during winter, reduced illumination again results in lower plant growth, while the continued mixing of deeper waters with those of the surface levels restores the nutrient concentrations in the surface waters to their high winter levels.

It is easy to see that, at any single geographic locality in middle latitudes of the Atlantic or Pacific, the zone of green plant productivity will move slowly to greater depths in the water column as the days lengthen in spring. Thus if a euphotic zone 5 meters deep is required in a particular locality to produce an SDI of a particular diatom species, then we can see how, along a north–south

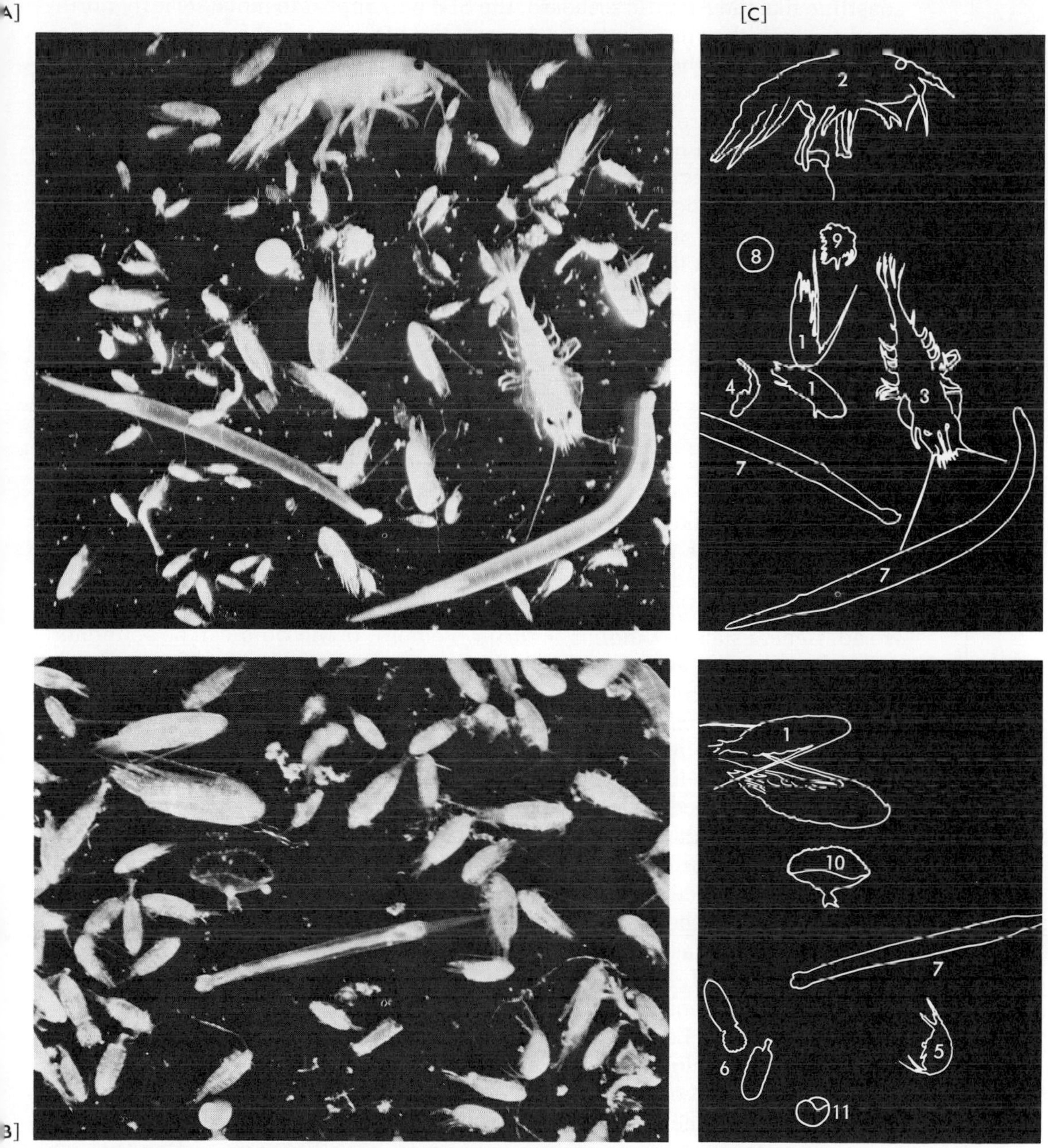

FIGURE 33-4. Late spring zoöplankton from inshore waters near Woods Hole. **A** and **B:** Samples from tow-nettings preserved in formalin (**A** ×6; **B** ×8). **C:** Key diagrams outline specimens of the following forms: adult copepods, *Calanus* (1); an adult amphipod, *Calliopius* (2); a subadult shrimp, probably *Crangon* (3); younger "mysis-stage" larvae of a carid shrimp (4); smaller adult copepods (5); mostly unidentified but possibly including *Centropages* (6); adult arrow-worms *Sagitta* (7); a fish egg (8); a free polyp head of the hydroid *Tubularia* (9); a leptomedusan jellyfish (10); and a late veliger larva of a gastropod withdrawn within its shell (11). [Photos by the author; carids and *Calliopius* kindly identified by Dr. Eric L. Mills.]

571

coastline like the Atlantic seaboard, the SDI will appear to move steadily north-ward through the spring months. In fact, the SDI reaches the highest latitudes of open waters in the Arctic in midsummer (and the corresponding waters of the Antarctic around December 21).

This pasturage provided by the planktonic diatoms, and to a lesser extent by dinoflagellates and microflagellates, is primarily grazed by the animals of the zoöplankton (see Figures 14-11, 33-4, 33-5, and 33-6). They, too, show seasonal changes in occurrence, in growth patterns, and in reproduction, which fit them to best exploit the seasonal changes in primary productivity. Such seasonal cycles involve both the permanent plankton, such as copepod crustaceans like *Calanus*, and the wide variety of temporary planktonic larval stages of benthic invertebrates and of fishes.

Permanent Zoöplankton

Herbivores of the zoöplankton must themselves be relatively small animals, and microphagous filter-feeders. Thus, with few but important exceptions, the permanent zoöplankton consists of animals of between 0.5 and 5.0 millimeters in length, and there may well be reasons of energetic efficiencies for this size limitation. As noted in Chapter 14, calanoid copepods are the dominant zoöplankton herbivores and are probably the most numerous animals in the world. Genera such as *Calanus*, *Acartia*, *Temora*, or *Metridia* can be abundant, and the single most abundant species—in terms both of numbers of individuals and of fraction of total animal biomass—is almost certainly *Calanus finmarchicus* or *Calanus helgolandicus* or a closely similar form. The feeding mechanisms used by *Calanus*, and the sequence of stages (nauplius, metanauplius, copepodites, and adult copepod) in each generation of its life-cycle have already been described (Chapter 14). These trophic and reproductive functions are essentially similar in all the other calanoid copepods. A rather stereotyped morphology and uniformity of physiology characterizes this animal group where a highly successful pattern of functional machine has been successfully exploited. One of the earliest detailed studies of the seasonal aspects of life-cycle in *Calanus* was carried out by A. P. Orr and Sheina M. Marshall in Loch Striven in Scotland. Here, and over much of the seas of northwestern Europe, the annual populations of *Calanus* are made up of three successive generations (see Figure 14-10). Each generation cycle takes about two months in summer, one month for development from egg to adult, and up to one month for maturation of the eggs. In Loch Striven and elsewhere, the typical overwintering stock consists of fifth-stage copepodites, and the slow development of this preadult stage means that the overwintering generation has a life-cycle lasting through seven or eight months. Egg production and the early development of individual broods are fitted to the seasonal changes in plant productivity, including the SDI. In these temperate waters at those latitudes, a three-generation annual cycle is probably best fitted to exploit a plant-growing season extending from late March to September. There are certain advantages in having a subadult form as the overwintering stage.

At higher latitudes, the growing season is obviously shorter, and, as we noted earlier, the phytoplankton outbursts are moved nearer to midsummer. Under these conditions—for example, in the waters of the east Greenland fjords and in the Barents Sea—there is only one generation of *Calanus* in each year. In the coastal waters of Norway and in the open ocean of the Gulf of Maine, conditions are intermediate, and there are two generations per year. This may prove to be the commonest pattern of life-cycle in the central parts of the north Atlantic. In contrast, as we move into warmer waters, we find four generations of *Calanus* per year in parts of the southern North Sea and the English Channel, and possibly off the Georgia–South Carolina coast. The same species of *Calanus* are not found in the subtropical waters of the Caribbean and central south Atlantic; but if they did occur, in theory, given the two-month minimum period required for life-cycle and egg maturation, *Calanus* could have an annual cycle in such waters of five or even six generations per year. This variable life-cycle pattern in *Calanus* (and in several other copepods of the permanent zoöplankton) allows exploitation of the very different seasonal patterns of phytoplankton productivity in Mediterranean or Floridian waters from those in the high Arctic off Greenland and in the Barents Sea. It allows for the maximum possible buildup of animal biomass by the timing of these generations. At their peaks, concentrations corresponding to two hundred adult specimens of *Calanus* per cubic meter (that is, a hundred times that in the same locus in the winter months) can be met with. To put this another way, if fished at the right time in most temperate seas, a 1-meter-diameter net towed for fifteen minutes would yield about 2.5 million specimens of adult *Calanus*, or about 1.5 gallons (U.S.) or 5.7 liters packed solid.

In Arctic, and more conspicuously in Antarctic, waters, rather larger crustaceans, the shrimp-like euphausiids (see Chapter 15), may also be important as herbivores. They are of moderate size (2–5 centimeters long) and, except in the colder regions of the oceans, are strictly deep-sea planktonic animals, feeding on detritus rather than on the crop of living green plant cells. When they occur in the surface plankton of boreal seas, they may be found in relatively dense swarms many miles in horizontal extent; known as the "krill," these swarms are the principal food of whalebone whales during the brief Antarctic summer season.

Other crustaceans which are members of the permanent zoöplankton include cyclopoid copepods, bivalved ostracods, and a few peculiar Cladocera. Representatives of this last group dominate the zoöplankton of most smaller lakes, but are represented in the sea only by small raptorial carnivorous species in the genera *Evadne* and *Podon* (see Figure 14-5). Among higher crustaceans, there are a few smaller shrimps and mysids, and rather rarer isopods and amphipods (Figure 33-4), represented in the permanent zoöplankton.

Apart from crustaceans, at least six other phyla have representatives which grow to maturity and breed as permanent members of the marine zoöplankton. The phylum Cnidaria is represented by many species of small jellyfish which, by suppression of certain attached stages or by other means, spend their entire life-cycle in the plankton (see Figures 4-3 and 33-5), as well as the smaller members of the "oceanic Hydrozoa" (see Chapter 4) whose polymorphic colonies are in the smaller forms truly planktonic (see Figure 4-8). The allied phylum, Ctenophora (see Chapter 5), consists almost entirely of marine planktonic

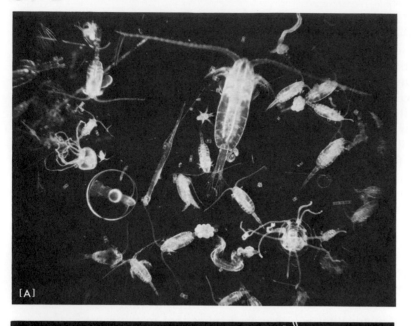

FIGURE 33-5. **A:** Living zoöplankton from the English Channel. × 12. **B:** Specimens of the following forms are represented in the key diagram: adult copepods, *Calanus* (1) and *Pseudocalanus* (2), including some females of the latter genus with eggs (3); a predaceous cladoceran, *Podon* (4); planktonic tunicate (primitive chordate), *Oikopleura* (5); a fish egg (6); a young arrow-worm, *Sagitta* (7); two different anthomedusan jellyfish (8); and copepod nauplius larvae (9). A few diatom chains are also visible. [Photo © Douglas P. Wilson.]

forms. The comb-jellies—which, when alive, are among the most beautiful of marine organisms (Figure 5-1)—are all carnivores, feeding avidly on zoöplankton like *Calanus* and often outcompeting stocks of young fish for this resource.

Apart from the dinoflagellates and other photosynthetic single-celled "plant-animals" of the last section, several groups of Protozoa (see Table 6-1) are represented in the permanent zoöplankton of the oceans. From the subphylum Sarcomastigophora, both sarcodine classes, Rhizopodea and Actinopodea, are prominent; and there are several different groups from the subphylum Ciliophora represented. Some dinoflagellates, including the luminescent genus *Noctiluca*, are heterotrophs—feeding largely on diatoms, but occasionally on other zooplankton. The calcareous-shelled Foraminifera (class Rhizopodea) can occur in vast numbers in the plankton of warmer oceans today. (Their past abundance is attested by deep-sea ooze-deposits and certain fossil limestones.) Most forams ensnare prey by their webs of pseudopodia (see Chapter 6) and form food vacuoles *outside* their shells. A few species of forams, and rather more radiolarians, have symbiotic algal cells living within their cytoplasm (see Figure 6-4). Some radiolarians (class Actinopodea) can be almost as abundant in tropical seas, and most produce delicate and beautiful skeletons of silica. Many can adjust their density by balloon-like vesicles, and thus carry out vertical migrations in the plankton. A variety of ciliates live in inshore marine plankton, where they feed mainly on bacteria, but one group, the Tintinnidae, can be truly pelagic in the open ocean. They construct bell-, flask-, and vase-shaped tests, which may appear glassy or may be completely coated with fine sand grains (or graded detrital particles like miniature caddisfly cases). The ring of cilia which emerges from the "mouth" of the test in life functions not merely as a food-gathering peristome as in many other ciliates, but also (like the pitched rotor of a helicopter in its "lift" phase) to thrust the tintinnid upward through the water. Tintinnids feed actively on green microflagellates and often appear more numerous after local algal "blooms" (see Figure 6-15).

The phylum Annelida, whose adults are typically benthic animals, is, however, represented by a few forms like *Tomopteris*, with parapodia developed as swimming paddles, which pass their entire lifespan in the plankton. Similarly, the phylum Mollusca has only a few kinds of snails which live as adults in the plankton. (Both molluscs and annelid worms are very well represented by their ciliated larval stages in the temporary zooplankton.) Among the planktonic snails are the opisthobranch pteropods, or sea-butterflies, where the middle parts of the foot are drawn out into ciliated wing-like extensions used both for locomotion and for the collection of suspended food particles. Some are shell-less, but others have beautiful little conical glassy cones which can form distinctive deposits (the pteropod oozes) in certain parts of the ocean floor. As noted in Chapter 20, the planktonic snails are polyphyletic, pelagic life having evolved in the major group Opisthobranchia in at least three different stocks.

The minor coelomate phylum Chaetognatha or arrow-worms are characteristic and important members of the permanent zooplankton. The majority of the species in this phylum are placed in one genus, *Sagitta*, and as noted in Chapter 28, species have been used as biological indicators in the identification of oceanic waters of different origins. Sampling of arrow-worms is important to productivity predictions for certain fisheries. Ecologically, they are predaceous carnivores, feeding on forms like *Calanus* and occasionally destroying whole broods of larval stages both of invertebrate species and of commercially important fishes. Their transparency and moderately large size (see Figures 14-11 and 33-4) make them one of the more interesting objects for study in living marine plankton.

Finally, there are representatives of the phylum Chordata in the permanent zooplankton (described in Chapter 32). Four groups of planktonic urochordates have evolved from the stocks of more typical sessile sea-squirts (the attached

[A]

[C]

[B]

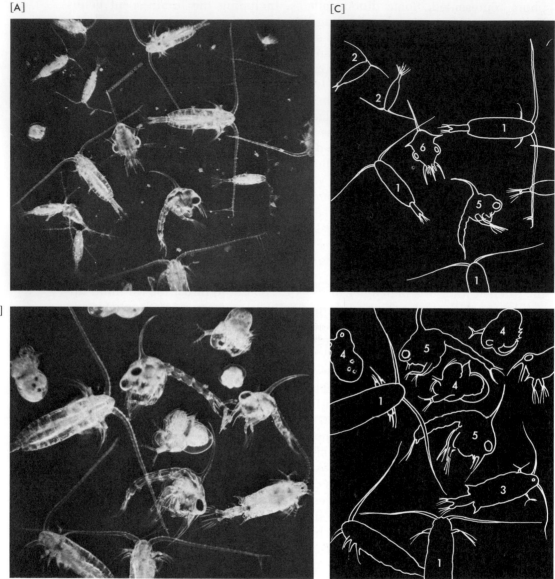

FIGURE 33-6. **A** and **B:** More living zoöplankton, almost entirely crustacean (**A** × 10; **B** × 12). **C:** The key diagrams outline specimens of the following forms: adult copepods, *Calanus* (1), *Acartia clausi* (2), and *Centropages* (3); predaceous cladoceran, *Podon* (4); and zoea larvae of crab in side view (5) and from behind (6). [Photos © Douglas P. Wilson.]

ascidians which have planktonic "tadpole" larvae). The class Larvacea has become specialized by the retention of an essentially larval organization after the onset of maturity and the development of gonads. Specimens of such genera as *Oikopleura*, *Megalocercus*, and *Fritillaria* occur in the marine plankton in many parts of the world (see Figures 32-4 and 33-5). As already noted, the bizarre biology of these forms involves the construction of houses which are actually elaborate filter-feeding apparatus. The life-cycle is simple, with only sexual reproduction, and larvaceans live their entire lives in the plankton. In the other three groups of planktonic chordates—*Pyrosoma*, salps, and doliolids—we have barrel-shaped forms like adult sea-squirts propelled by water currents created by muscle bands (see Figure 32-5). The water current passing through a simplified group of gill slits within the "barrel" serves for feeding and respiration as well as for locomotion by jet-propulsion. In the case of these three orders, the entire life-cycle is spent in the plankton, but it is very complicated (involving both sexual and asexual phases, and a regular alternation of sexually produced forms or oözooids and budded blastozooids). As already noted, salps are important in plankton ecology since they compete with, and prey on, young fish and the larger elements of the zoöplankton. At intervals of five or six years, "outbursts" of salps occur in some areas of the North Atlantic to the great detriment of most other zoöplankton forms and of the stocks of young fish.

Temporary Zoöplankton

When we turn to the temporary members of the zoöplankton, we find a much greater diversity and note representatives of every major phylum of the animal kingdom and also of the majority of minor phyla and of major classes. The temporary plankton is thus made up of the dispersal stages from the life-cycles of a great variety of benthic invertebrates, which are all otherwise sedentary or sessile. The majority of these dispersal forms—though not all of them—can be called larvae, defined as *immature stages* basically dissimilar to the adults which they will become. It is the sheer diversity of larval representation that almost mandates for all invertebrate biologists some exposure to living marine plankton. The survey of larval types now attempted here must necessarily be synoptic.

The phylum Cnidaria is extensively represented by larvae in the strict sense, including the planulae of a wide variety of hydroids, sea-anemones, and corals, and by even more specialized larval forms such as the actinula, of *Tubularia* and its allies, which looks like a free-swimming juvenile hydroid individual (see Figures 4-5 and 33-4), and the characteristic ephyra stage of the larger jellyfish (see Figure 4-11C). Perhaps even more numerous in the temporary zoöplankton are the medusoid stages of the enormous number of benthic hydroids, little jellyfish which function as minute dispersal stages, but are actually the gonad-bearing sexual adults (see Chapters 3 and 4). In these forms, the relatively massive and extensive hydroid colonies, living benthically attached to stones and seaweeds, are by most definitions the larval stages.

Sponges (phylum Porifera) are represented by a few flagellated larvae, actually little more than motile blastula stages being dispersed in a brief planktonic phase. Flatworms (phylum Platyhelminthes) are represented in the plankton by the ciliated larvae (Müller's and Götte's) of polyclads, which larvae gradually become more complex (with eight ciliated lobes in Müller's larva, and four in Götte's, serving to increase the surface area) as they grow more massive. Eventually they become more flatworm-like in shape, their ciliated lobes shrink, and they sink to the bottom and assume the adult form. The nemertine worms (phylum Rhynchocoela) also have a characteristic ciliated larva, termed a pilidium, somewhat like a helmet-shaped version of the earliest annelid and molluscan larvae (see Chapters 10 and 19). In this case, an ingrowth of four pockets essentially separates off a central mass, which then differentiates into the juvenile nemertine worm, which is subsequently dropped like a mini-bomb to the sea floor. The ciliated larval husk now floats off in the plankton and, if not eaten, soon dies. At this point it is worth re-emphasizing a matter discussed in Chapter 27, that the special requirements of larval life (rather than recapitulation of ancestral history) have determined the evolution of many adaptive features of structure and function—and even time-pattern of development—in pelagic larvae. As already emphasized, common features in these relatively simple dispersal units need not imply close phyletic relationships. The resemblance of the planktonic trochosphere larva of a primitive member of the Annelida to an early pilidium larva of the Rhynchocoela, and to the earliest trochophore larva of the Mollusca, may well be just that, and reflect merely similar solutions to the problems of temporary life in the plankton as a minute dispersal and developmental stage. The common possession of an apical tuft as a sense-organ derived from modified cilia; of nearly equatorial double bands of metachronally beating cilia, which combine the functions of locomotion (top-spinning) and of food capture; and of the folding of these ciliated bands into lobes of greater elaboration as the larval body becomes more massive reflects common adaptive mechanisms, but not necessarily common descent.

At least nine other minor phyla of invertebrates have benthic adults and planktonic larvae. In the phylum Entoprocta, some species have a "trochosphere-like" larva, which differs in cellular detail from the annelid form, while other species have a ciliated planktonic larva of completely different form with a peculiar creeping sole used after settlement. Of the other minor phyla: the Echiuroidea and the Sipunculoidea have trochosphere larvae more like those of typical annelid worms; the phylum Brachiopoda has larvae like miniature ciliated brachiopods with mantle lobes and lophophores already formed as they drift in the plankton; some ectoproct bryozoans have a peculiar triangular, bivalved, ciliated larva termed a cyphonautes; the phylum Priapulida has a "rotifer-like" larva; the Phoronida has a long-lived tentaculate actinotrocha larva (see Figure 28-5); and, finally, in the phylum Hemichordata, some species have a tornaria larva (see Figure 31-3) superficially resembling certain echinoderm forms.

The major phylum Annelida is represented in the temporary zoöplankton not only by larvae but by other dispersal stages (see Chapters 10 and 12). We have already described how the archetypic trochosphere larva can remain planktonic during later stages when metameric segments are being budded off posteriorly (see Figures 10-1 and 10-2). Many of the more advanced polychaetes have replaced the trochosphere by larger eggs and more elaborate dispersal larvae. For example, in some species of Nereis a swimming larva of head plus three segments bearing paddle-like parapodia with long setae, termed the nectochaeta larva, swims for a long time in the plankton before settlement. Walter Garstang pointed out the resemblance of the three-paired appendages to the limbs of a crustacean nauplius larva. Another modification is shown in the larva of the polychaete genus *Owenia* where the trochosphere continues to grow, with great elaboration of its anterior ciliated band from a simple equatorial girdle into a much folded and sinuous set of lobes. In addition to the larval forms discussed in Chapter 10, several kinds of polychaete worms also produce epitokes which are budded off, contain mature reproductive organs, and possess modified parapodia which allow for more efficient swimming during a temporary prefertilization dispersal stage (Chapter 12). In a number of other cases, whole adult worms metamorphose to sexual swarming forms which also temporarily enter the plankton.

The various larval stages of bottom-dwelling crustaceans are also well-represented. As already noted (Chapters 10, 14, and 15), a typical nauplius larva occurs in the development of some species in almost all groups of crustaceans. Some of these resemble the larvae of planktonic copepods. Others, including those of barnacles (see Figure 14-14), have a characteristic body shape and can be more readily identified. At times, the coastal plankton of temperate seas is dominated by the later larvae of barnacles, including the characteristic triangular metanauplius with horns, and the penultimate stage before settlement, the bivalved cyprid larva. A few of the higher crustaceans have a nauplius larva, but many more show the typical zoea (Figures 33-6 and 33-7), which, in crabs and similar forms, is succeeded by a megalopa larva. It is in the later presettlement stages, like the megalopa, that mysids, euphausiids, mantis-shrimps, prawns and true shrimps, marine crayfish, lobsters, and crabs all have their own distinctive late larval stages. Both the Euphausiacea (krill) and the penaeid prawns (Macrura-Natantia) go through a long series of larval stages after hatching: nauplius → metanauplius → protozoea → zoea → postlarva → adult. The euphausiid postlarva was named the cyrtopia, and the prawn one, the mastigopus stage. Both postlarvae more closely approach the adult form than any earlier stages, and both have been termed "mysid-like." In contrast, most true crabs (Brachyura) and hermit-crabs (Anomura) omit the earlier larval stages and hatch as a protozoea or a zoea (Figure 33-7A), their later postlarvae being termed the megalopa (Figure 33-7B) and the glaucothöe, respectively. Just as in the major divisions of insects (see Chapter 17), the less specialized malacostracans (of the "caridoid facies") have more gradual changes in their instars, while the "higher" crabs have more drastically metamorphic patterns of development. Although not reflecting "recapitulation of adult ancestry," as Ernst H. Haeckel's superseded "biogenetic law" would have stated the concept, it is clear that, in each developmental succession, the more general larval types of the big group (nauplius and metanauplius) precede those of the more specialized subdivisions (zoea and megalopa). Some more specifically larval recapitulation does occur: for example, the megalopa is much more "macruran" (with a large, movable abdomen bearing full appendages —see Chapter 15) than is any adult crab. Finally, it should be noted that when such names as Megalopa, Cyrtopia, and many others were first applied to these planktonic forms, all were thought to belong to distinct genera. Careful rearing to establish each major larval series came considerably later, and much still remains to be discovered about the life-histories of higher crustaceans. There are still many bizarre larvae ("of unknown malacostracan") awaiting study in the marine zoöplankton.

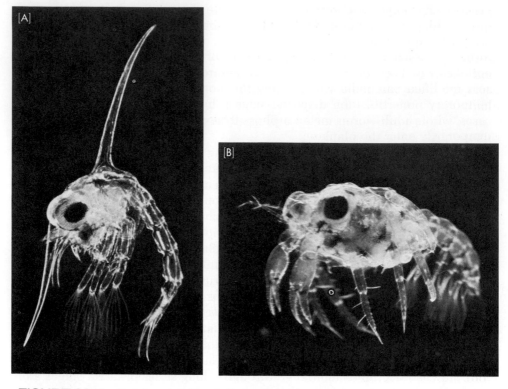

FIGURE 33-7.   Crab larval stages. **A:** Second *zoea* larva of the velvet swimming crab, *Portunus puber*. **B:** Megalopa larva (a few molts later) of *Portunus puber*. As discussed in the text, the succession of larval stages in higher crustaceans like crabs is of evolutionary significance, though it does *not* recapitulate any phylogeny of *adult* forms. [Photos © Douglas P. Wilson.]

The major phylum Mollusca is represented in the temporary zoöplankton by numerous ciliated larvae of a few rather stereotyped patterns. In a large number of more primitive molluscs, external fertilization in the sea gives rise to a trochophore larva with a characteristic double ring of locomotory cilia which also serve to collect particulate food. This is unlikely to represent shared ancestral recapitulation with the trochosphere of annelid worms; rather, it seems to reflect a parallel adaptational response to the immediate needs of larval life. Particularly in the marine prosobranch gastropods, there is a later characteristic larva, the veliger, in which the trochophore ring has become developed into a typical pair of wheel-lobes on either side of the head, termed the velum, and bearing long cilia (see Figure 19-12). Still later, a mantle rudiment appears and secretes a characteristic shell, and the minute shelled snail larva may drift for a variable period in the plankton before settling to the bottom to grow up as an adult snail (see Figure 33-4). As we noted in Chapter 21, many gastropods living between tidemarks have taken a step which can constitute a preadaptation to terrestrial life by producing larger eggs from which hatch either very late veligers or miniature adults. Once this or viviparity is achieved, they no longer have a planktonic phase in their life-cycle. Among the opisthobranch snails there are species with planktotrophic larvae, which feed, grow, and metamorphose as members of the plankton; there are others with larvae which swim briefly, but, having emerged from larger eggs, still contain sufficient stored food materials that they do not require to feed during their brief planktonic existence; and, finally, there are species with direct development which are not planktonic.

In the other major group of molluscs, the Bivalvia—including clams, mussels and oysters—the majority of species have small eggs and ciliated larvae which live for relatively long periods in the plankton and feed on microorganisms. Bivalves have a trochophore-like larva which soon develops a bivalve shell. This miniature and well-ciliated bivalve will then drift for a variable period in the plankton before it undergoes a relatively slow process of metamorphosis, which involves enlargement of the foot and reduction of the velum. Thus the trochophore, by way of a bivalved veliger stage, becomes a nonswimming form which settles to the bottom, termed a pediveliger. At this stage all young bivalves have a large mobile foot with a thread-producing byssus gland (even if these are not retained in the particular bivalve form of the adult). When the young bivalve develops into a miniature of the adult form, it is generally called a spat. The process of settlement and metamorphosis for certain bivalve species (termed spatfall) can be relatively synchronous through an extensive area of the sea, and, for common littoral forms such as mussels, can involve a spectacular, simultaneous descent of myriads of tiny mussels onto the seashore, during a high tide.

Certain other molluscan groups can have young stages in the plankton, but they are never abundant. Scaphopods have a bivalve-like symmetrical veliger which gradually assumes the mantle-shell shape of the adult tuskshell, while chitons (see Chapter 26) have an elongate trochophore, whose mantle rudiment secretes six shell plates, about the time that the juvenile chiton settles to the bottom. Hatching cephalopods all closely resemble the adults in miniature, and a few drift for a time as members of the plankton. They are never common and are always instantly recognizable.

Another major phylum extensively represented in the marine zoöplankton is the Echinodermata. As noted in Chapter 30, no other group of animals undergoes such complicated metamorphoses in the course of development; starfish, sea-urchins, sea-cucumbers, and their allies all have diverse and characteristic planktonic larvae. There are two basic larval types (see Figures 30-6 and 30-7): the plutei of brittle-stars and sea-urchins and the bipinnaria or auricularia of starfish and sea-cucumbers. In all cases the body of the larva and particularly the larval lobes or arms (which are arranged in bilateral symmetry) bear bands of strong cilia which serve both for microphagous feeding and for locomotion. Growth and increasing larval mass are accompanied by increasing length and/or complexity of these ciliated lobes or arms (see Figure 30-7), in a fashion similar to that of planktonic larvae in other groups already discussed, including flat worms, certain polychaete larvae, and those snail veligers which have a longer planktonic life and which develop ciliated "wings" like butterflies (see Figure 19-12). Within the Echinodermata, if larval types are surveyed, there are numbers of examples of close larval resemblances between unrelated species (reflecting evolutionary convergence), while other closely related forms exhibit major differences of larval type (divergence). Neither among the echinoderms, nor in the broader questions of interrelationships between phyla, can detailed phylogenetic relationships be extrapolated from larval similarities. Crinoids have a peculiar barrel-shaped larva with five annular bands of strong cilia surrounding the body; this changes into a miniature adult form called the penta-crinule before settlement. In all echinoderms, there is always a radical process of metamorphosis involving asymmetric rudiments developed in the larva, which rudiments give rise to the radially symmetrical adults, normally before settlement. Sea-cucumbers have a very late larval stage, the pentactula, already completely metamorphosed and with adult features before settlement; while the older larva of true starfish, the late brachiolaria, shows the crinoid feature of attachment at metamorphosis, and for this purpose has temporary specialized suckers (see top of larva in Figure 30-6C).

As already noted, primitive members of the chordates are represented in the plankton by the "tadpole" stages of many benthic sea-squirts. Before they settle and undergo their metamorphosis (described in Chapter 32), they may live in the plankton for only a few hours or for many days; the longer-lived tadpoles filter-feed on microorganisms. Sea-squirt tadpoles, like many other planktonic larvae, tend to swim upward (negative geotropism) and toward higher light intensities for their initial period in the plankton, and then show a reversal of these responses (for example, migrating to shaded areas on the shore or bottom) just before attachment.

The pattern of appearance of this great diversity of larval forms in the zoöplankton is even more seasonal in the temperate oceans than is that of the copepods. Many bottom-dwelling invertebrates exploit seasonal or temporary fluctuations in the basic productivity of the phytoplankton by spawning or by liberating young larvae. Control of this is discussed on pages 587–588.

A Live View

Every invertebrate biologist should have an opportunity to work through some samples of marine zoöplankton. It is regrettable that the education of most modern biologists neglects the opportunities which this can present at all levels. As already noted, marine zoöplankton *can* be transported alive to inland teaching laboratories, but something can be gained even from working through (identifying, classifying, and counting) a preserved sample of marine plankton. Such work should always proceed from the largest plankton organisms to the smallest, and the process is best begun at low magnification with a diluted portion of the sample spread in fluid in shallow containers like petri dishes. Initial examination should be with the lowest power of the dissecting microscope, or, better still, with a large, low-power simple lens like a reading magnifier clamped or mounted in some way to leave the hands free. Instruments for sorting plankton (alive or dead) should include various sizes of eye-droppers and Pasteur pipettes, and mounted needles and fine forceps whose tips have been extended by attachment of barbs (from a large quill feather) with some kind of cement or hard wax. Such tips of soft but microscopically rough keratin are excellent for lifting tiny plankters without damage.

Any such work will immediately provide an introduction to the methods and units of practical systematics. Identification of even the dominant forms like copepods will involve the use of dichotomous systematic keys. Any counting of relative numbers of forms will lead to verification of simple ecological principles (including the Eltonian pyramid, in which animals at the base of the food-chain, second-trophic-level herbivores, are the most numerous, and there is a progressive decrease in numbers with each successive trophic level of carnivores). Even in making such counts and assessments in a zoöplankton sample, the intelligent student is driven to consider the complications introduced into such simple concepts by the fact that most zoöplankton communities form food-webs rather than parallel series of discrete food-chains. This in turn can lead to considerations of "feedback relationships" and other density-dependent controls upon the interactions of predators and prey. Even the simple enumeration of numbers of species and of numbers of individuals can lead to consideration of the significance of different levels of species diversity and of community stability. In spite of continued efforts by several distinguished theoretical ecologists, we do not yet have a rigorous predictive model whose terms can explain the general empirical observation that higher degrees of complexity of food-webs are associated with greater community stability. In a large number of natural animal communities, including most zoöplankton assemblages, in relation to *any* predictive model so far deployed, the common species are excessively abundant, and the uncommon species excessively rare. In these, as in certain other evolutionary problems, we need more hard data on levels of species diversity (for example, from the zoöplankton) as well as further efforts by the theoretical ecologists. Indeed, given the past history of biological advances, it is not improbable that the next refinements of theoretical models of community structure will come from someone currently engaged in the "dirty-handed" work of actually counting plankton samples.

Much more is added to all this if the zoöplankton can be studied alive. Several biological supply houses can arrange for marine zoöplankton to be delivered anywhere in thermos bottles. Appropriately diluted, samples can be kept alive for a number of days, even without aeration, in a domestic refrigerator at temperatures of around 4°C. For classes or individual students within driving distance of the sea, the solution is even simpler: personal collection.

All early—and many continuing—investigations of the plankton used a relatively simple collecting device called a tow-net: essentially a small bag of fine silk mesh shaped like an elongated cone (Figure 33-8). Tow-nets were invented, and first used in plankton investigations in 1816, by J. Vaughan Thompson, the British army surgeon and great amateur naturalist, whom we already encountered as the first to elucidate barnacle life-histories (Chapter 14) and as one of the earliest to recognize bryozoans (his "Polyzoa") as complex animals (Chapter 28). A towing rope is attached to cord bridles which diverge onto a hoop supporting the sieve, a fine fabric cone. In most cases the netting cone does not taper completely to a point, but to a narrow tube to which a small tube or jar can be attached. In earlier years, both professional and amateur investigators of plankton had to use the silk "bolting cloth" made for grading flour mill products to make plankton nets. Professional ones are now mostly made of nylon mesh, and the interested amateur can prepare an excellent substitute from part of a leg of a pair of discarded pantyhose. The finest mesh

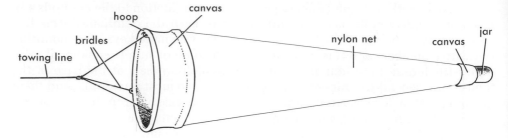

FIGURE 33-8. A simple plankton net. Fine silk or nylon mesh (at about 80 cross threads per centimeter) is used to sample phytoplankton, while coarser mesh (about 25 threads per centimeter) filters better for the capture of the more active zoöplankton. The optimum towing speeds for nets of the latter mesh size lie between 1.5 and 2.5 miles per hour.

plankton nets, with about 80 cross threads per centimeter, are required for the sampling of the finest phytoplankton. Coarser mesh, of about 25 threads to the centimeter, filters rather better for the capture of the more active zoöplankton. Towing a small plankton net requires no great effort; indeed, the most efficient speed for mesh sizes appropriate to zoöplankton is actually a little less than two nautical miles per hour, well within the capacity of an amateur oarsman in a small rowboat. In many coastal areas, sufficient tidal currents along the shore, or through channels, provide for the even lazier or less nautical biologist who can let his or her plankton net stream at an appropriate time of the tidal cycle for ten minutes or so from a pier or dock. The most practical advice to the plankton collector who wishes to study a live sample is: "Don't be greedy." If the plankton is at all abundant, the sample collected even in ten minutes will be relatively thick. It is important to pour most of this away, and only take back to your laboratory the smaller part of it, diluted with more seawater (if possible, precooled). It is worth remembering that a cubic meter of seawater in summer may contain between 2000 and 5000 individual animals of the zoöplankton and from five to two hundred times those numbers of phytoplankton organisms.

A live view of zoöplankton allows for the possibility of simple experimental work on, for example, responses to light of various forms, and the mechanics of locomotion and feeding. In particular, the reader with access to living material can confirm or correct any statements describing the mechanics of feeding by copepods or by larval stages as set out in earlier chapters of this book. The action of "flap-valves" in the recovery strokes of locomotion in some Crustacea, or the action of larval ciliary bands in collecting bacteria and minute flagellates as food, becomes more memorable when seen in living plankters rather than merely read about. There is also an aesthetic aspect. Try to have a view of living marine zoöplankton.

The Basic Marine Economy

The open ocean as an ecosystem can be assessed in terms of standing-crop biomass, or in the rate terms of calorific energy flow (productivity rate). In either

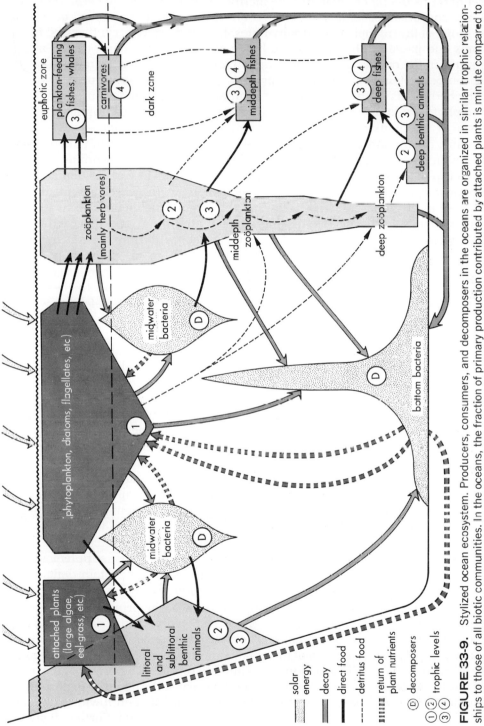

FIGURE 33-9. Stylized ocean ecosystem. Producers, consumers, and decomposers in the oceans are organized in similar trophic relationships to those of all biotic communities. In the oceans, the fraction of primary production contributed by attached plants is minute compared to the immense productivity of the phytoplankton (the largest annual crop of green plants in the world). Thus the dominant fraction at the second trophic level consists of zooplankton organisms feeding as microherbivores—in the open sea, often mainly calanoid copepods. These copepods are thus probably the most numerous animals in the world. Note the importance of detrital food-chains and the distribution of the principal concentrations of bacteria below the levels of greatest plant production. The proportions of the stylized volumes do represent the approximate depth distributions of the organisms concerned in a portion of the ocean of moderate depth but are otherwise not strictly proportional.

set of terms the invertebrates of the marine zoöplankton at the herbivorous primary consumer (second trophic) level are the most important such unit in the ecology of this planet. The carbohydrates, fats, and proteins synthesized in diatoms and passed to higher trophic levels are mostly first converted to animal tissues in the crustaceans of the zoöplankton. Figure 33-9 illustrates a stylized ocean ecosystem. The fraction of primary productivity contributed by all the attached plants in the sea is minute compared to the immense productivity of the phytoplankton. Thus the dominant fraction at the second trophic level consists of zoöplankton organisms feeding as microherbivores—in the open oceans, often mainly calanoid copepods. Apart from the fraction of zoöplankton which passes to bacterial decay and then recycling, the enormous secondary crop of zoöplankton passing through food-chains at various depths in the sea provides the direct food of such fishes as herring, mackerel, and menhaden, as well as of the largest true fishes and the largest mammals. The two largest fishes in the world are the basking shark (*Cetorhinus*) and the whale-shark (*Rhineodon*) both of which swim near the surface with their mouths open and use fine, comb-like bristles arranged in their gills to filter the zoöplankton. Basking sharks, unlike their smaller, more notorious predatory relatives, swim slowly near the surface at about 1.5–2 nautical miles per hour, the best speed for zoöplankton filtration. The world's largest animals are whalebone whales, such as the blue-whale and the fin-whale or common rorqual; they employ a cyclic pumping of many tons of seawater at a time. These enormous gulps with their contained plankton are forced sideways through the baleen or whalebone (the fringed, horny plates hanging from the roof of the whale's mouth). Besides providing for such direct feeding, the zoöplankton constitutes an important indirect food source for many other commercial fishes, including: tuna, swordfish, haddock, cod, halibut, sole, plaice, and many others. As noted in Chapter 14, even relatively inefficient human exploitation of the vast productivity of zoöplankton—by making protein-supplement feeding-stuffs for intensive agriculture, from fishmeal processed from menhaden caught by traditional fishing methods—is a multimillion dollar enterprise. The even larger amounts of fishmeal originating in the Peruvian anchoveta fisheries, despite their effects on trading in soybean futures, represent a remarkably inefficient and energetically indirect utilization of the proteins assembled by anchoveta from the marine zoöplankton. It is only to be hoped that the last decades of this century will see the more humane use of such fishmeals as direct dietary supplements for human nutrition in those many tropical countries where protein deficiencies already occur (see my earlier book, *Aquatic Productivity*).

Figure 33-9 shows, for a model section of the sea, the principal trophic categories along with the energy exchanges of direct feeding and of decay (with consequent recycling of plant nutrients). The stylized volumes do represent approximately the depth distribution of the organisms concerned, although they are not strictly proportional. The section is intended to represent a portion of the ocean of moderate depth; in the deepest parts of the oceans, the quantity of decomposing bacteria would be greatly attenuated in the middle and lower depths and would only increase in bulk very close to the abyssal bottom. Further, it should be realized that while the relative volumes could represent the biomasses of the various constituents at a single point in time, seasonal changes in the relative phytoplankton:zoöplankton biomass could make hundredfold differences in this ratio alone. If we attempted to represent, not standing crops, but actual rates of turnover (annual productivities) for each group as stylized volumes, then the disproportion between those for the primary producers and those for the animal groups would be such as to hinder visualization on the page. Similarly, for the oceans as a whole, the proportion of attached plants to the phytoplankton would be very much smaller than that shown. All of this should serve to reemphasize the supreme importance of the invertebrates of the marine zoöplankton in passing along the productivity of the largest annual crop of green plants in the world, the oceanic phytoplankton, directly and indirectly (think of the food-chain positions of man, and of birds like gulls, in this respect) not merely to the rest of the marine economy but to all life on this planet.

Ecology and Evolution in Plankton

Although investigations of the overall bioenergetic relationships (or of community ecology) tend to dominate ecological studies of the marine plankton, much physiological ecology is also of interest, and some of it of great significance to evolutionary concepts. The physiological mechanisms which bring about appropriate ecological synchrony in the reproduction of benthic animals to yield the swarms of the temporary zoöplankton, and the factors similarly involved in appropriate departure from the plankton by successful settlement, are intriguing. As noted above, the larval and similar stages making up the temporary zoöplankton (in teleological terms) "appear in time to exploit temporary fluctuations in the basic productivity of the phytoplankton." In some cases, such as the spawning of starfish and of clams and the liberation of young larvae by barnacles and by various crabs, clearcut seasonal patterns can be seen. In many cases this is clearly arranged to allow the best utilization of temporary peaks of phytoplankton regularly occurring in the oceanic seasons, or to correspond to the best possible time in these seasons for successful settlement to the bottom. For the great majority of known larval life-cycles, we can readily perceive the adaptational advantages of spawning at a very specific time, but relatively little is known in detail about how such timed spawning is controlled in nature. In a small number of cases which have been investigated, including certain molluscs like *Melampus* (see Chapter 21), the trigger for the onset of the breeding period is changing day-length. Spawning in other forms is clearly temperature-controlled. In still other cases, chemoreception of food organisms or of other appropriate environmental conditions has been implicated. Perhaps the simplest form of control is found where there is an apparent synchronization of spawning, or of the release of larval stages from the parent, with the spring diatom increase (SDI). There is evidence for a number of benthic invertebrates that spawning or larval release occurs early in the spring in lower latitudes, and progressively later to the north, to take place in midsummer in the highest latitudes. Such synchrony of larval production and the SDI, in the Arctic particularly, has been elucidated by Gunnar Thorson. In at least one barnacle species, direct "chemical perception" of the SDI is almost certainly involved. Harold Barnes has demonstrated that substances present in the water of a concentrated suspension of *Skeletonema*, which probably result from the metabolic activities of these diatoms, have a marked effect in promoting the hatching of nauplii of the common barnacle *Balanus balanoides*. It is theorized that perception of the substances by the barnacles' sense-organs is signalled to the reproductive organs by way of a neurosecretory change. Barnes was also able to demonstrate that over several years at Millport in Scotland, local variations in the initiation and progress of the SDI could be correlated with hundredfold changes (within a few days) in the numbers of barnacle nauplii present in the plankton.

Thus we can have a broad latitudinal synchrony of larval release determined by geophysical signals like changing day-length to ensure a "fit" with the peaks of primary productivity in those latitudes. However, there can also be local timing control, related more directly to fluctuations of the phytoplankton crop, which is involved in assuring optimal conditions for the relatively short larval lifes in such forms.

Temporal control of the processes of metamorphosis followed by settlement (or, in other invertebrate groups, settlement followed by metamorphosis) is similarly complex. Again, a number of geophysical time signals are used. Detection of, or "internal clocks" set to, the phases of the semilunar tidal cycle of about fourteen days, and of the circadian tidal periodicity, can be of immense importance in allowing settlement and return of the larvae of intertidal benthic animals to the appropriate vertical zones occupied by their parental stocks. Apart from such time-cues, most planktonic larvae of benthic invertebrates develop a new (or modify their existing) repertoire of responses to light, to gravity and pressure, and to tactile and chemical stimuli in their environment. One such signal, utilized independently by a variety of phyla, is larval detection of a "taste" of dead, empty shell material of adults of their own species. Such specific chemoreception occurs in such diverse forms as the pediveliger larva of oysters, the cyprid stage of certain barnacles, and the budding late trochosphere of some tubicolous polychaete worms. In all these cases, a simple experiment can show that species-specific organic substances are involved: heat treatment of the old shells in question effectively destroys their attractiveness to the appropriate settling larvae.

Not only larval bioenergetics, but also behavior in relation to settlement, is markedly different when planktotrophic larvae which actively feed are compared to those which subsist upon the energy stores provided in larger eggs by the parent. Particularly in some polychaete larvae, as beautifully documented by Douglas P. Wilson, and in the veliger larvae of certain prosobranch snails, capacity to feed for continued maintenance after arriving at full presettlement size, allows remarkable postponement of metamorphosis if suitable bottom substrata are not available. This capacity is of special interest in relation to the patchy distribution of most benthic animals on the seabed. Facultative delay also allows suitable planktotrophic larvae to survive passage across the Atlantic ocean from one continental shore to another, as has been convincingly worked out by Rudolph S. Scheltema for certain gastropods. In contrast, non-feeding larvae often show dramatic changes in their susceptibility to settlement stimuli as their energy stores are used up. As D. J. Crisp has shown, the "urgency to settle" in barnacle cyprids increases steadily with age, and E. W. Knight-Jones has shown that larvae of the tubicolous annelid *Spirorbis* will metamorphose indiscriminately in quite unsuitable places if aged without access to the normal chemical "releaser" for settlement. The length of planktonic life obviously affects the range of dispersal of the young produced by a benthic parent. Even in closely related forms, such as the species of *Littorina* discussed in Chapter 21, longer planktonic life is associated with wider gene flow. As Edward M. Berger has demonstrated from electrophoretic studies on polymorphic enzyme systems, there is significantly less local geographic differentiation in *Littorina littorea* (with free-swimming veliger larva) than in *Littorina saxatilis* (viviparous) or in *L. obtusata* (with miniature adults hatching from attached egg-masses—see Figure 21-1).

It is worth briefly considering the advantages of long-lived planktonic larval stages in an evolutionary framework. The adaptive advantages of larger eggs which provide sufficient energy stores to each larva to enable it to be nonfeeding and independent of its surroundings obviously reduce any risks from predation or excessive dispersal away from suitable settlement areas or into areas of the sea with little or no suitable food supply. The balancing "disadvantages" are that the larvae have only a limited time to reach their settlement zone (and may be forced to abort their specific behavior patterns, if delayed), and, since the eggs are larger and more yolky, each female can produce fewer of them. On the other hand, forms producing long-lived and feeding larvae can gear their reproductive output to much larger numbers of eggs (individually smaller) since the young can accumulate food during their planktonic life for both maintenance energy and the "building blocks" of growth to carry them through metamorphosis and settlement. Such planktotrophic forms have adaptive advantages, including increased geographic range with its possibilities of wider and more homogeneous gene flow and evenness of population density, but "disadvantages" are the increased risks of predation or inappropriate dispersal resulting from the longer period of planktonic life. These two reproductive strategies are, of course, mutually exclusive in any species—but congeneric species may differ in the "evolutionary choice" they have made. Among closely related forms, of similar adult size, there is a precise inverse relationship between egg size and numerical fecundity which reflects the maximum possible production of egg-biomass by the parent. My research associates and I have reported many instances of this relationship for snails, and it clearly holds for a wide variety of other invertebrates. Some other generalizations can be made about different kinds of "large" eggs. If, when analyzed, large eggs have relatively low carbon:nitrogen ratios (reflecting a high protein fraction), then they usually show truly direct processes of development, giving rise to miniature adults which are not usually planktonic. Similarly large eggs with relatively high carbon:nitrogen ratios can give rise to fully formed and active ciliated larvae carrying food stores which enable them to avoid feeding for a period of planktonic life after hatching and before settlement. Such differences can be taken as reflecting the differing significance of "building blocks" for development, and of provision for energy requirements in different species. Further study of the various evolutionary trade-offs is required. In the meantime, it is clear that the evolution of larger eggs in nonmarine environments (see Chapter 21) has been influenced not only by a need to suppress free larval stages but also by pressures to reduce the temporal extent of immature growth. The "start in life" of being born large is very important in all nonmarine environments (with their marked seasonal changes) and also in some marine habitats, and this determines the most efficient patterns of size and composition of eggs for species in these places. However, certain long-term genetic advantages must be sacrificed along with higher numerical fecundity.

As already discussed (see Chapters 19 and 27, and pages 479–486) Walter Garstang emphasized that form and function in planktonic larvae are determined less by recapitulation of the evolution of adults (as Ernst Haeckel and others had claimed) and more by the need for special larval adaptations (often developed in parallel fashion by totally unrelated stocks). From his studies of planktonic larval forms, Garstang developed another evolutionary hypothesis, which he termed paedomorphosis, which claims that, however specialized the *adults* of a stock may have become, it is still possible for their *larval* stages to be modified in new ways and then by the assumption of an earlier sexual maturity (neoteny—see Chapter 32) produce a stock of totally new evolutionary potential. It is generally agreed by most biologists that one class of the phylum Chordata, the Larvacea, are sea-squirts in which neoteny has resulted in the retention of an essentially larval organization by mature adults with gonads. Less generally accepted examples of neoteny have been put forward for siphonophores, ctenophores, certain cladocerans, ostracods, insects, and even the true vertebrates. In the last case, supporters of neoteny would point to the common use by *Amphioxus,* by larval sea-squirts, by young lampreys, and almost certainly by the ostracoderms (almost the first vertebrates in the fossil record) of a filter-feeding mechanism utilizing pharyngeal gill slits (see Chapters 31 and 35). Within the vertebrates, the occurrence of neoteny, in the evolution of certain newt-like forms from true salamanders within the class Amphibia, can be confirmed *experimentally* by administration of appropriate hormones controlling the relative rates of developmental processes. If Garstang's theory of paedomorphosis holds more generally, then the phyletic diversity of the marine zoöplankton may already contain in various genomes the bases of new animal stocks (equivalent to existing orders, classes, or perhaps even phyla) in the more distant *future* of evolution.

34

INVERTEBRATES OF
TIME PAST:
THE FOSSIL RECORD

ALL worthwhile questions in the biological sciences involve the
dimension of time. This rather trite statement may serve to
focus attention on the fact that even in a short book centering
on the mechanistic-physiological and adaptive-functional levels
of explanation for the invertebrates, we have had again and
again to shift our conceptual level to the evolutionary-
historical. Not just in our inferential constructs in phylogeny,
but in considerations of cephalopod learning, larval development, and even the mechanics of locomotion, we have had to
ask: "What is the history of this organ or process in time?" Biologists vary greatly in their views of the significance of the fossil record in modifying concepts and interpretations from their
own studies. One extreme but understandable view arises subjectively from handling fossils. To recognize by sight and by
touch the still-organized remains of what was a living organism
some hundreds of millions of years ago cannot but arouse
wonder. To the ecologist there is the additional awe that this
organization has continued to exist as a chance exception to
that part of the essential recycling within the living biosphere
which Louis Pasteur characterized as the necessary inevitability
of dissolution. The countless millions of fossils can only represent an infinitesimally small fraction of all the organisms which
have lived. Thus we can have the extremist view that fossils are
facts and that there is no value in evolutionary hypotheses

which are based on any other kind of evidence of biological changes in time. In contrast, there are some modern experimental biologists who would say that since one cannot set up a null hypothesis and test it on a fossil population, palaeontology has little relevance to biological science. Even in the case of comparative studies on the invertebrates, they would point out that, since all the major phyla of invertebrates had become distinct *before* the beginning of a usable fossil record, little of consequence to understanding of the invertebrates can emerge from fossil studies. Although a neontologist by training and vocation, I trust the readers of this book can perceive that both extreme views are not merely wrong but stultifying.

One aspect of modern palaeontology requires emphasis. It is that over the last 25 years methods of physical assessment (mostly by study of radioactive transformations in certain minerals) of the age of rocks have provided time-scales which are both more precise and less disputable than were previously available.

It is important to realize that, until relatively recently, the fossils themselves provided geologists with a clock. Moreover, determination of the mutual relations of the majority of rock formations or strata is still based upon the occurrence of certain characteristic fossils. The relative ages of the units of the standard geological column were first identified on the basis of superimposition and characterized by each typical fossil assemblage. Thereafter, correlation by fossils was possible, allowing the age (in relative rather than absolute terms) of any isolated sequence of strata to be determined by the fossils found in it. Over a century and a half, this *relative* time-scale became more and more finely elaborated and, by correction and further correlation, ever more reproducible and applicable to further newly investigated sequences of sedimentary rocks. It was possible for large classic texts of palaeontology to be republished in the 1950's using only such *relative* ages, and almost completely eschewing any reference to time in absolute units.

Time-scales

Evidence for that part of the geological time-scale which concerns animal fossils—the so-called Phanerozoic time-scale—was formerly almost entirely derived from stratigraphy and fossil correlations. However, over the last four decades, it has increasingly been given scale-points from detailed radiometric studies of rocks. Such nonbiological methods of establishing the ages of specific rock formations were pioneered by Professor Arthur Holmes of Edinburgh University. Even by the 1930's, Holmes had already replaced earlier age estimates based on salinities or rates of sedimentation (which, with different authorities, could vary tenfold) by the first assessments based on the radioactive decay of uranium to lead and to helium. At the present day, the age of many minerals and rocks can be determined within fairly narrow limits, not only by finer and less laborious studies on the uranium $\rightarrow$ thorium $\rightarrow$ lead series but also independently, and in some cases in even shorter time units, by the decay of the isotope potassium-40 to argon, and of rubidium-87 to strontium. By 1964

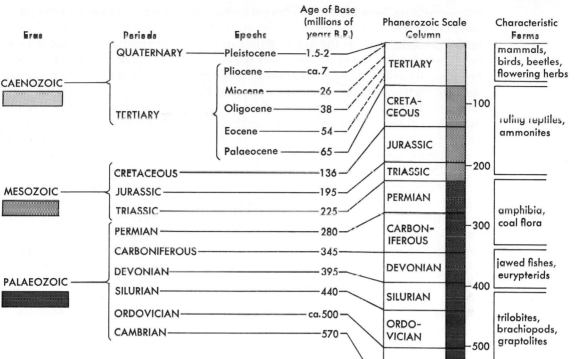

FIGURE 34-1. The geological (Phanerozoic) time-scale. Radiometric methods (see discussion in text) have now established, with considerable certainty, standard age values for what were previously merely proportionate scales deduced from fossil correlations in biostratigraphy. The *base* of each stratigraphical unit is given in millions of years B.P. (before the present).

it was possible to publish standard age tables for the potassium → argon method in intervals of a few thousand years from 1 million to over 4000 million years before the present (B.P.). A few minor difficulties remain in linking the radiometric determinations to details of the stratigraphy of fossil-bearing rocks in a few localities, and more recent datings (less than a half-million years B.P.) have to depend on other means, such as use of the isotope carbon-14, which has somewhat less certainty. Controversy on detail will continue among geologists, but it is unlikely that the broad summary of the Phanerozoic time-scale as presented in Figure 34-1 will require much future modification.

In Figure 34-1 the age of the *base* of the stratigraphical unit is given in millions of years B.P., and it is important to note that in many cases the recent radiometric data have merely confirmed proportionate scales already deduced by use of fossil correlations in biostratigraphy. Note also that the labels attached to entire periods, such as the age of fishes or the age of reptiles, are useful mnemonically but, as we shall see when we consider the major groups of invertebrate fossils in the next two sections of this chapter, distort a great deal to fit the human vertebracentric viewpoint. As Stephen J. Gould has pointedly commented, there have been vastly more beetles than mammals in the age of mammals, and there were vastly more brachiopods in the age of fishes. That the "age" labels also evoke a simplistic view of evolutionary progress again reflects certain temporary biases of pedagogy.

Another problem with the geological time-scale arises from human difficulties in comprehending it in relation to either the time-scale of human history or that of individual life experience. Starting with brave efforts by H. G. Wells early this century, teachers have tried to present in textbooks and elsewhere "scale-models," as it were, of the geological time-scale. Our attempt at a scale presentation appears in Figure 34-2. If we take as a standard age for the modal reader of this book 25 years, and somewhat arbitrarily set the origins of life at 2 billion B.P., then we have the main fossil record beginning seven years ago when you were age 18. On this modal scale (closer to Proust's "temps perdu"), plants first grew on land five years ago, and insects began at least four years ago; the giant reptiles were dominant around two years ago, and mammals originated about one year ago; the first true man appeared about five days ago, several bronze age civilizations were under way by a half-hour ago, and the Greek city states and their colonies flourished seventeen minutes ago. These scale values would be halved at least if we used the probable age of the earth's crust as our starting point, and they would be doubled if the life of the author, instead of the modal reader, were used for scale. In several of his attempts at representation, Wells used linear models. If we use his style of model, but with somewhat changed values derived from radiometry, we could represent the approximately two thousand years of the conventional Christian dating system as extending 1 centimeter back from the point designating the present. On this scale, man arose about 5.5 meters (about eighteen feet) back, mammals extend from 400 meters (about a quarter-mile) back, the fossil record starts about 2.9 kilometers (or 1.8 miles) back, and the origins of life would lie 10 kilometers (or 6.2 miles) back.

The names given to the major divisions of the geological time-scale are long-established. The three eras of the fossil record (Caenozoic, Mesozoic, and Palaeozoic) are each divided into periods, and these in turn into epochs. In the two older and longer eras we will be concerned only with the periods, such as Jurassic or Silurian, but in the shorter time-span of the Caenozoic, we will be concerned with the epoch subdivisions, such as Miocene or Eocene. It should be noted that in the United States many geologists subdivide the Carboniferous period into two, named Mississippian and Pennsylvanian.

Before the three eras of fossils, together termed the Phanerozoic time-scale, we have the Precambrian or Cryptozoic era going back more than another billion years and encompassing the time-span during which we believe some life occurred on the earth's surface, although the fossil traces are much less obvious than those of the Cambrian period and later. It is likely that many forms

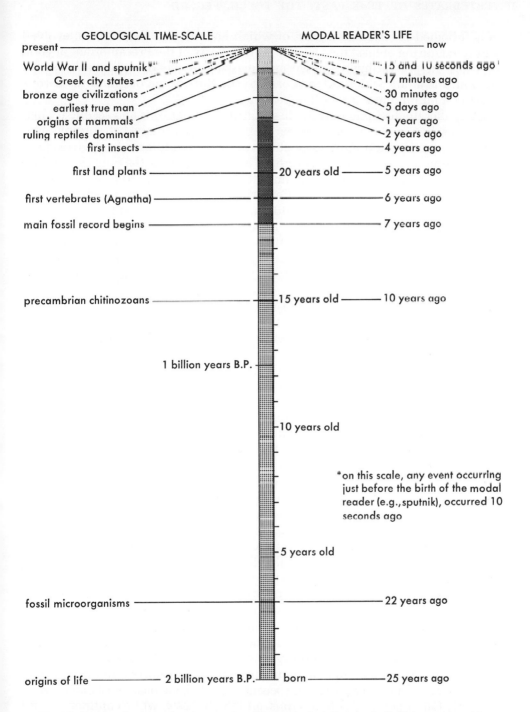

FIGURE 34-2. Comparative time-scales. The geological time-scale (left) is compared to a scale of past time (right) for the life of a modal reader of this book (aged twenty-five years). The origins of life are somewhat arbitrarily set at two billion years B.P., so that the scale values would be halved at least if we had used the probable age of the earth's crust as our starting point, as they would be doubled if the author's, rather than the modal reader's, life were used for scale.

595

of soft-bodied invertebrates, most of which have left no fossil traces, evolved and lived during at least the last 500 million years of the Precambrian. Beyond this lies the era variously known as the Azoic or the Archaeozoic, possibly without life at all, which extends back to the time of the origin of the earth's crust at a minimum of 3.6 billion years B.P.

Long after the first photosynthetic autotrophic organisms and secondary heterotrophs had evolved from the earliest acellular primary heterotrophs and chemosynthetic autotrophs, some eucaryotic forms similar to protozoans (and thus appropriate to this book) must have arisen. However, the only certain Precambrian fossil protozoans are foraminiferans (see Chapter 6) from only a few tens of millions of years before the Cambrian period. There are also more or less undisputed records of Porifera, Cnidaria, Annelida, Arthropoda, and Echinodermata from the late Precambrian in several parts of the world. Among the more important of these are formations in the Grand Canyon of Arizona, in Southwest Africa, in Siberia, and in the Ediacara sandstone of south Australia. The marine assemblage in the last locality is represented by some hundreds of specimens, assigned to some 25 species, apparently all soft-bodied invertebrates (cnidarians, annelids, and, less certainly, echinoderms and arthropods). Much earlier microfossils, found in the Arizona Precambrian, include chitinozoans (from around 800 million years B.P.), planktonic animals which may have been case-building protistans (see Figure 6-15), or even primitive metazoans, but which were almost certainly heterotrophic. Much more recognizable invertebrate fossils appear at the beginning of, and at points during, the Cambrian period, largely as a consequence of the development of calcareous shells or other protective skeletal structures in previously soft-bodied forms.

Before we pass to a consideration of the fossil record for the major invertebrate groups, it is worth reviewing briefly a few nonbiological events of geochronology in relation to the Phanerozoic time-scale. Cycles of crustal activity appear to have occurred with a periodicity of about 30 million years; thus there have been about twenty of these since the Precambrian. Periods of more intense continental mountain-building, or orogenesis, have occurred at longer intervals, and correspond to major crustal movements. On the denser material of the earth's mantle "float" the less dense rocks of the continental plates like gigantic rafts [the crust being less than 8 km (5 miles) thick under parts of the ocean bed, but the continental rafts displacing the mantle to depths of 40–72 km (25–45 miles)]. Major movements of these continental plates produced periods of intensive mountain-building, four of which have occurred within the Phanerozoic record. The first of these was ending as the Cambrian began; another occurred about 160 million years later in the Silurian and lower Devonian; a third occurred after another 140 million years, near the end of the Carboniferous; and the last, or Alpine, period of mountain-building occurred in the Tertiary period, 220 million years later. The oceans seem to have been most extensive in the early Cambrian, the Carboniferous, and the Jurassic, while continental land caused restriction of the seas in the Devonian, Permian, and late Cretaceous. Each restriction of the seas seems to have corresponded to some faunal extinctions and there were alternations of time-spans of evolutionary diversification with waves of slower change and even extinction through the entire Palaeozoic era. A time of high productivity and of biological diversification came in the

Carboniferous period, and left us the coal measures as fossilized energy-containing biomass. The Mesozoic was a time in which great diversification both of marine faunas and of land plants and animals occurred. It also encompassed the crustal movements which produced (embryonically, so to speak) the features of the present continental land masses. At the close of the Mesozoic, the marine retreat and the beginning of the last period of mountain-building were accompanied by extinction of a number of major and previously dominant fossil lines, including not only the ruling reptiles of our labelling system but also an incredibly successful invertebrate marine group, the ammonoid cephalopods.

Major Stocks: Extant and Extinct

It is obvious that certain major phyla with hard skeletal parts, including the molluscs, arthropods, echinoderms, and chordates, are more likely to leave fossil remains in suitable conditions of deposition than are other phyla including flatworms, nematodes, and annelids. By its nature, the fossil record of any stock of animals is erratically incomplete, but for those groups whose representatives are rarely preserved—including most soft-bodied invertebrates without shells or skeletons—the record is almost useless for phylogeny.

Although there are doubtful remains of medusae from the Precambrian and scattered hydrozoans throughout the fossil record, the main fossils of the phylum Cnidaria are those of anthozoan corals. There have apparently been three main time-spans when coral reefs were extensively formed and preserved; in Silurian/Devonian seas; through much of the Mesozoic era, most extensively in the Jurassic; and in the Tertiary, from the Miocene, continuing to some extent in our present tropical oceans. Little can be deduced from the fossil record regarding the separation of the major groups of the phylum Cnidaria. Fossil sponges (phylum Porifera) occur from the Cambrian to the present and show a marked peak in the number of named genera in Cretaceous deposits. Again, little can be said about evolution of major groups. Some protozoans show their greatest diversification relatively late in the fossil record, particularly in groups like the Foraminifera, in which diversity of species and genera builds up from the Jurassic through the present. Fossil members of the phyla Platyhelminthes and Nemathelminthes, and of several minor acoelomate and pseudocoelomate phyla, have not been identified with certainty in Phanerozoic deposits.

The fossil record of the major phylum Annelida is also relatively poor. Jaws, setae, and worm traces have been reported from the Precambrian up through the Palaeozoic, but the record is again of little phylogenetic value. There are exceptionally well-preserved annelid worms in one middle Cambrian deposit, the Burgess shale of British Columbia, which represent several already distinctly separated groups. As noted in Chapter 12, the evolution of oligochaete earthworms must have taken place during and after the Cretaceous, in close association with the rise of a deciduous vegetation of flowering plants.

In the phylum Arthropoda, we have one of the six phyla with an abundant and extensive fossil record. Unfortunately, even the arthropodan fossil record can tell us more about sequences, and implied relationships between different subdivisions *within* the major classes, than about the more problematic relationships *between* the four successful classes: Trilobita, Crustacea, Arachnida, and Insecta. In broad general terms, we can only say that the trilobites were more diverse and abundant earlier than the other three classes and that the insects were last to appear in the fossil record (in the Carboniferous). The three "modern" classes all show increasing diversification of forms up to the present, whereas the trilobites come into the record already considerably diversified and, after a peak in the Ordovician, show declining diversity until their final extinction in the Permian.

Of the seven well-established orders of trilobites, five occurred in the Cambrian and two arose in the Ordovician; no new orders appeared between Ordovician and Permian (see Figures 16-10 and 34-3). Trilobites are one of only three groups abundant in the earliest faunal assemblages of the lower Cambrian, making up on average 32% of the species lists. All early Cambrian fossils are small, except for trilobites and representatives of the extinct phylum Archaeocyatha (a group of sponge-like organisms, abundant and diverse in the lower Cambrian but apparently limited to that period, hence not discussed in this book). Precambrian and Cambrian many-celled organisms were limited to the sea, and, as already suggested, most of them were probably small, soft-bodied, and planktonic. It is not impossible that the beginnings of a benthic habit for metazoan animals, the onset of skeletal mineralization, and the development of predaceous animals of more than a few grams in mass, occurring together in the late Precambrian, brought about the inauguration of a real fossil record. At the level of families and genera it would seem that nearly half the trilobite extinctions occurred in the late Cambrian and Ordovician. The decline continued through the Silurian, and by the end of the Devonian about 90% of their diversification was over. Only one new family group appeared in the Devonian and one in the lower Carboniferous. Thus, it is very misleading to regard the extinction of the trilobites in the middle Permian as a sudden event (as some textbooks do). From an evolutionary point of view this arthropod group was a class of decreasing abundance (and probably dominance) and certainly

FIGURE 34-3. Trilobite protective behavior. Two specimens of the Silurian trilobite, *Calymene celebra,* from the Niagara Beds, at Grafton, Illinois. The one on the right completed the "roll-up" protective reaction (also seen in modern pill-bugs) before its death and subsequent fossilization. [Photo by Perry Russell-Hunter; see caption for Figure 16-10.]

decreasing diversity from at least the Silurian to its final extinction in the Permian. It is important for us neontologists to realize that this group's was no transitory dominance or extent. In terms of millions of years, six of the seven orders of trilobites have fossil records spanning more than 100, one group reaching 205 and another 340 million years. (For comparison, mammals and birds have existed for about 160 million years, and almost all modern orders of mammals originated less than 65 million years ago.)

One last matter about fossil trilobites is of significance to arthropodan phylogeny. As noted earlier (Chapter 10), the trilobite body consisted of three tagmata (cephalon, trunk, and pygidium) made up of the many tripartitely divided segments, each with one axial and two lateral lobes which gave the group its name. This body form was not modified during the long evolution of the trilobites. The trunk segments were so articulated that many genera of trilobites could roll up into armored balls (as do isopod pill-bugs today) for protection of the softer underparts and limbs (see Figure 34-3). Details of trilobite limbs were first worked out from sections made through rolled-up fossils. Subsequently, very perfect trilobite specimens with the appendages extended in the normal position were found in the Utica shales of upstate New York. The basically biramous form of the trilobite limb (see Figure 14-1E) and its relation to the archetypic limbs of primitive crustaceans were discussed in Chapter 14. The remarkable point is that the basic plan of trilobite appendages and their occurrence in a uniform and nondifferentiated series on all trunk segments did not change from the genus *Olenoides* in the lower Cambrian through about 200 million years to such forms as *Phacops*, dominant in the Devonian. The contrast with the fantastic evolution of crustacean limbs, starting with a similarly uniform and multipurpose phyllopodous form, is remarkable.

In the Burgess shales of British Columbia, of middle Cambrian age, are found what are probably the earliest fossils of the class Crustacea. There are possible conchostracans and ostracods (see Chapter 14), as well as true trilobites, and a few peculiar genera usually placed in a separate class Trilobitoidea. (The majority of modern palaeontologists set up the enormous, but stereotyped, successful class Trilobita and this small group of more diverse, generalized, and probably primitive arthropods, class Trilobitoidea, as the two unequal classes encompassed by the arthropod subphylum Trilobitomorpha. This has not yet been adopted by many neontological texts.) These trilobitoids of the Cambrian are regarded by some workers as the remnants of a Precambrian protoarthropod stock which gave rise to three more successful classes, Crustacea, Arachnida, and the Trilobita proper. This concept of a stem-group will be further discussed in the next chapter. In the Cambrian, the trilobitoids were more diverse than trilobites in their body shapes and in their segmental and limb anatomy, some being apparently planktonic and some benthic with biramous limbs; and they may have included forms intermediate between the class Trilobita and some protocrustacean arthropods. Apart from this unusual Burgess shale deposit, ostracods are the most abundant of early fossil crustaceans, with many families being represented in the Ordovician and Silurian, and other families including immensely useful indicator species for biostratigraphy in the Carboniferous and Permian. Their diversity may have declined somewhat in the Mesozoic, but ostracods remain relatively abundant in every geological period up to the present. Since it is their bivalve shells rather than their limbs which are preserved, little can be deduced from the fossil record to relate them to the ecological life-styles of living Ostracoda, or to relate their earliest representatives to the rest of the crustaceans. One of the earliest crustacean fossils for which we have detailed knowledge of limbs has already been mentioned. It is *Lepidocaris* from the Rhynie chert of the Scottish Devonian. This rock contains incredibly well-preserved plant and animal fossils resulting from a sort of embalming by silica-saturated water. As D. J. Scourfield first showed, even the limb bristles of *Lepidocaris* can be distinguished. As we noted in Chapter 14, this form is placed in a separate group of the branchiopod crustaceans and in some features appears to be more advanced than the living cephalocarids. The early fossil history of such crustacean groups as the barnacles remains uncertain, and that of the copepods completely unknown. As regards the higher crustaceans, phyllocarids, closely similar to the living leptostracan *Nebalia*, are known from the Ordovician onward, and the true shrimp-like "caridoid facies" begins with such genera as *Eocaris* and *Devonocaris* from the Devonian. Similar "stem-malacostracans" become abundant and diverse in the later Devonian and Carboniferous. The earliest true peracarids (the stock of isopods and amphipods) and the earliest eucarids probably arose early in the Carboniferous. By the beginning of the Triassic, separation of the higher groups of the eucarids into swimming forms (Natantia) and crawling benthic forms (Reptantia) had occurred. On the other higher stem, the earliest mysids are from the Triassic, and the earliest cumaceans and isopods from the Permian, but the earliest true amphipods are from Eocene deposits. Lastly, the earliest true crabs and anomurans are from the Jurassic, while the aberrant hoplocarids (like *Squilla* and related genera, the so-called mantis-shrimps) have remained a separate line since their earliest fossil representation in the Carboniferous, as has the division Syncarida from the same period. In summary, the fossil record for the higher crustaceans is more complete than for most of the lower groups, and our classification of living forms, based on other features including limb patterns, reproductive methods, and feeding mechanisms, still closely reflects the separations in time of the various lines of the Malacostraca.

The second major group of living arthropods, the Arachnida, is also well, but sporadically, represented in the fossil record. The major fossil subclass, the Merostomata, represented today only by the three surviving genera of horseshoe crabs, *Limulus* and its allies, were the most formidable creatures of the lower Palaeozoic. Some small forms appear as early in the fossil record as the lower Cambrian; the larger eurypterids first appeared in the Ordovician and were dominant marine organisms well into the Carboniferous. Some, reaching lengths of 3 meters, may have been the largest animals ever built on the arthropod plan, and a few forms became adapted to life in estuaries and fresh waters in the later Palaeozoic. The Xiphosura, the group including *Limulus*, arose in the Devonian and has continued for 350 million years. Among other arachnids, the earliest scorpions are found in the Silurian, but were probably marine. The earliest spider is Devonian, again from the Rhynie chert, as is the earliest acarine mite. Later records of mites and spiders come from Baltic amber of Eocene and Oligocene age. These chance preservations in cherts and ambers of such enormously important constituents of present-day arthropod faunas can only emphasize the general incompleteness of the fossil record for some forms. Like nematode worms, mite species in vast numbers must have escaped preservation, certainly from the Mesozoic onward.

The myriapodous arthropods are first represented in the fossil record by millipedes from the late Silurian or early Devonian and by true centipedes from the Carboniferous, although again some kind of myriapod may have been present in the Burgess shale. A group of enormous myriapods called arthropleurids appear briefly in the coal measures of the Carboniferous and are not readily relatable to any of the other myriapodous groups. They reached lengths of nearly 2 meters, and if they lived on land rather than in the waters of the coal swamps, they may have been the largest terrestrial arthropods ever in our planetary history.

It has been appropriate to leave the class Insecta to be considered last of the true arthropods, not only because of their great present success but also because they were clearly the latest of the major arthropod groups to arise in the fossil record and to diversify. The earliest fossil insects are apterygotes from the early Devonian, and related forms are preserved in the Rhynie chert. The earliest representatives of winged insects all appear in the upper Carboniferous and include giant dragon-flies, cockroaches very similar to those of today, and forms which may have been ancestral to may-flies. Nearly half the insect fossils of the Carboniferous and Permian periods are cockroaches. Insects with more complex metamorphoses (see Chapter 17) probably arose before the Permian and by the end of that period beetles (Coleoptera) made up 40% of the insect fossils. This dominance, at about this percentage level, has been retained up to the present. The first Lepidoptera and the first Diptera, or true flies, probably arose in the Triassic and continued to diversify through the Tertiary. These and the Hymenoptera, arthropods exclusively found in the late Mesozoic and Tertiary, evolved in association with land plants particularly in relation to flowering plants and modern deciduous trees. A large number of the presently living superfamilies of both Diptera and Coleoptera appear in the fossil record in the Eocene. This could represent real evolutionary radiation in the Tertiary, but it may be a case where the groups are not likely to be recognized under the usual conditions of fossil formation in the Mesozoic. Palaeontology of all arthropod groups has to consider the Devonian Rhynie chert and the Tertiary Baltic amber not only as standards of the excellence which some processes of fossil preservation can reach but also as salutary reminders of the rarity of such processes through the whole of Phanerozoic time.

The related metameric phylum Onychophora (see Chapter 18) is somewhat enigmatically represented in the fossil record. A middle Cambrian genus, *Aysheaia*, has been described, again from the Burgess shale; and a more doubtful form, *Xenusion*, may be an onychophoran of the late Precambrian. Unfortunately, there are no other representatives of this "missing-link" metameric group until forms closely related to present genera and species appear in Quaternary deposits.

Rivalled only by arthropods and echinoderms among invertebrate phyla still extant, the phylum Mollusca has an extensive, and at times abundant, representation in the fossil record. Not unexpectedly, molluscan palaeontology is unlikely to elucidate any questions of the origin of the phylum, or even of the interrelations between its three major classes: gastropods, bivalves, and cephalopods. It does, however, provide some useful sets of time relationships for our deduced phylogenies of many orders and superfamilies *within* the major groups. In a few cases, it provides data which are more valuable than, though similar to, those we have discussed for the higher malacostracan crustaceans in their fossil record.

The four minor molluscan classes are not well represented when compared to the fossil series for the three major ones. As already discussed (Chapters 19 and 26), the extant genus *Neopilina* is the sole living representative of the class Monoplacophora, untorted limpet-shelled molluscs with paired ctenidia. There are no monoplacophorans found in Mesozoic or Tertiary rocks, and fossils are limited to a few genera, including *Pilina* and *Scenella,* which were not uncommon in Cambrian and Ordovician periods, but which do not appear in the fossil record more recently than the Silurian. The class Polyplacophora (or chitons) are moderately well represented from rocks of practically all ages from upper Cambrian to the Tertiary. It must be deduced that the earliest chitons (like so many other stem stocks) were of Precambrian age, but no fossils of them have been discovered. Some Palaeozoic chitons were characterized by less complex shell plates without articulating structures. In contrast, most Mesozoic and Tertiary chitons are closely similar to modern forms. However, one living genus, *Lepidopleurus,* is known from the Carboniferous. The fossil record gives no clue as to any relationship between chitons and either the Monoplacophora or the major class Gastropoda. The third minor class of living molluscs, the Aplacophora or solenogastres, has never been reported as fossils. The last minor group, class Scaphopoda, or elephant's-tusk shells, appears in the fossil record from the Devonian (perhaps from the Ordovician) as moderately rare members of offshore marine benthic faunas. There are relatively few fossil genera and species in the Palaeozoic and Mesozoic, and the number of living species is probably in excess of the total number recorded as fossils from the Tertiary.

In their fossils as in their living representatives, the class Gastropoda constitutes the largest and most varied group of genera and species built to the molluscan plan. Even more clearly than with other groups of fossils, our definition of Gastropoda as a group name will determine where we begin to record these snails in the fossil record. If we follow the diagnosis employed in Chapters 19 and 20, then we limit the class to those molluscs with univalve shells whose palliovisceral organs have undergone torsion, and consequently to snails with their mantle-cavity placed anteriorly above the head. By this definition, the earliest gastropods are bilaterally symmetrical forms with planospiral shells, classified as Bellerophontids, which first appear in the lower Cambrian. Careful work on exhalant shell-slits and muscle scars shows that many forms (including some with turbinate, and some with limpet-shaped, shells) similar to the two-gilled archeogastropods of today, such as the Pleurotomariids (see Chapter 20), lived throughout the earlier Palaeozoic. Gastropods with a single gill (Monotocardia) first appeared in the Ordovician, and, by the Carboniferous, several divergent superfamilies of mesogastropods, opisthobranchs, and pulmonates are being recorded. Some of the fossil pulmonates from the late Palaeozoic occur in nonmarine formations, although the allocation of particular species among them to land or to fresh-water habitats remains uncertain. By the lower Jurassic, families (and in some cases genera) assignable to typical land pulmonates of today, and to the higher freshwater pulmonates, have appeared. The majority of presently living families of snails are represented in Tertiary deposits, and the first true pteropods (planktonic opisthobranchs) appear in the lower Eocene.

For the second largest living class of molluscs, class Bivalvia, there is an abundant fossil record, but one which is not particularly helpful in dealing with those questions of parallel and convergent evolution in this group presented in Chapter 22. The earliest true bivalves, from the middle Cambrian, are symmetrical and dimyarian and have heterodont hinge dentition. In Chapter 22, we noted the occurrence of the minor fossil class Rostroconchia from the lower Cambrian to the Permian. With an adult bivalved shell and a single larval protoconch, rostroconchs are regarded by some palaeontologists as intermediate between Precambrian untorted pregastropods and the earliest Bivalvia. Among Ordovician forms of true bivalves, the genus *Babinka* has a series of pedal-retractor-muscle scars, as well as others which apparently are comparable to the ctenidial muscle scars of Monoplacophora. Unfortunately, we can tell nothing from fossil bivalves regarding their gill condition, and only rarely is it possible to deduce that a particular fossil group was protobranchiate or septibranchiate, much less to distinguish between the filibranch and eulamellibranch conditions (see Chapter 22). Until a few years ago, all works on bivalve palaeontology attached great importance to the taxodont condition of hinge teeth. This occurrence of many small teeth with alternating sockets is found in the most primitive of living bivalves, such as *Nucula,* and also in certain true lamellibranch families, such as the arcids. It was formerly thought that the taxodont condition was the most primitive and that various forms of heterodont dentition had evolved from it. It now appears that recent arcids with taxodont dentition were probably evolved from Mesozoic and earlier forms with a distinctly different dentition from that of the protobranchiate forms (represented today by *Nucula*). It seems best to assume that comparable dentition occurs in different lineages of bivalves, and that it merely corresponds to similar conciliation of similar mechanical stresses as each particular set of muscle and shell growth features becomes adapted to closely similar habitats and life-styles. Despite these reservations, it is possible in the fossil record to trace the evolution both of asymmetric heteromyarian bivalve stocks (such as mussels), and of fixed and free-swimming forms (respectively, the lines of oysters and of scallops) with the monomyarian condition of a single adductor muscle. Nonmarine bivalves from brackish and fresh waters are known from the Devonian onward, and some stocks of them are important as indicators of stratigraphy in the coal measures of the Carboniferous. In general, bivalves are abundant from the Silurian onward, with considerable radiation in the Permian and Triassic; they remained abundant and important marine benthic animals throughout the Mesozoic and up to the present. It might be noted that the long slow decline of the other filter-feeding, two-shelled group, phylum Brachiopoda, corresponds to this increasing dominance through time of the molluscan bivalves in the seas. It also might be noted that bivalve fossils often consist of completely separated shell valves, or of paired valves spread widely open. In contrast, brachiopod shells are most usually found in closed-pair sets. This difference results from the different modes of shell opening (see Chapters 22 and 28), and from the fact that the elastic ligament in the bivalves will commonly cause postmortem gape. One peculiar group of bivalves, the rudistids, is known only from fossils of the late Mesozoic which, although they were derived from more typical fixed bivalves like oysters, lived in colonial masses and formed reefs by the growth of the fixed valves into pillar-like structures.

The last major molluscan class, the cephalopods, provides one of the most intriguing groups of fossils. There are at least 9000 known fossil species of cephalopods, although only about 350 species remain alive in the modern oceans. The earliest cephalopods represented in the fossil record were nautiloids from the upper Cambrian. Many orders of nautiloids differentiated during the Ordovician but the group had begun to decline by the Devonian. Nevertheless, several genera and families are found to the end of the Triassic, and the single living genus, *Nautilus* (which had appeared in the early Triassic), persisted alone from the Jurassic to the present. The most successful group of the cephalopods in terms of numbers of species and of abundance as fossils was the Ammonoidea (see Figures 24-8 and 34-4). Although some related forms made their appearance in the Ordovician, this group began to diversify in Devonian seas and was clearly one of the most important groups of animals being fossilized by the close of the Palaeozoic. True ammonites are the dominant marine fossils throughout the Mesozoic with great diversification in the Jurassic and Cretaceous. They do not disappear gradually from the fossil record; rather, though still an enormously diversified and apparently successful group, they suddenly vanish at the close of the Cretaceous. The third major subdivision of cephalopods is the Coleoidea, to which belong all presently living forms (including squids, cuttlefish, and octopods) except *Nautilus*; this group is represented as fossils largely by the belemnites, which can be thought of as ancient squids. These occur in the fossil record as chambered shells which were internal and which are generally shaped like straight cigars, rather than being coiled. Convincing series of forms link such existing internal cephalopod

FIGURE 34-4.   A small ammonite. A specimen of *Douvilleiceras mammillatum* from the Lower Cretaceous of California. Erosion of the fossil allows us to see the internal structure of the "juvenile" chambers of the shell. The whole fossil is about 5.1 centimeters across, so that the initial shell chamber was about 0.6 millimeter, comparable to the first shell of many modern spat snails. [Photo by the author; see caption for Figure 16-10.]

shells as those of cuttlefish, squid, and of *Spirula* to a belemnite ancestral form. The first belemnite appears in the Carboniferous, the first shell of a sepiid cuttlefish in the Jurassic, and the first "pen" like that of present-day squid also in that period. The beak-like jaws of the "modern" cephalopods have been found as fossils in late Mesozoic and in Tertiary rocks. In a very general sense, the fossil record shows that the three major groups of cephalopods serially replace each other: the nautiloids dominating the middle Palaeozoic, the ammonoids dominating Mesozoic seas, and the coleoids, which also flourished in the Mesozoic as belemnites, surviving in the three main stocks of presently living cephalopods.

The phylum Bryozoa (or Ectoprocta) is a minor order only in our ecological sense. Behind the 4000 or so living species there lie possibly 15,000 different extinct forms. The small class Phylactolaemata, including the present bryozoans of fresh waters, is poorly represented in the fossil record, having membranous rather than mineralized skeletons. They appear in the Cretaceous. The other major group of marine ectoprocts (class Gymnolaemata) has an abundant fossil record which extends from the upper Cambrian to the present. Two major groups which are now extinct, Cryptostomata and Trepostomata, became well-established in the Ordovician but died out by the end of the Permian. Two other orders, the Cyclostomata and the lesser group Ctenostomata, extend from the Ordovician to the present, while the fifth order, Cheilostomata, arose in the Cretaceous and its families and genera have been the commonest bryozoans from then until the present. Thus, as in so many other fossil groups, fossil bryozoan faunas change markedly at the beginning of the Mesozoic, but many of the forms established in the Cretaceous persist to the present. Both living bryozoans and their fossils are commonly found as epizoites on the shells of molluscs and of other invertebrates. Bryozoans were involved in reef-building by cementing together other calcareous organic remains, such as crinoid stems and molluscan shells, in the Silurian and again in the Permian.

As was true of cephalopod molluscs, presently living forms of lampshells (the phylum Brachiopoda) constitute a mere remnant of that once-dominant faunal group. Both of the major divisions of the phylum, the Ecardines (or Inarticulata) and the Testicardines (or Articulata), have representatives in the Cambrian and a few genera living in today's seas. The inarticulates started in the lower Cambrian, flourished in the Ordovician, and had been markedly reduced in numbers by the Devonian. However, *Lingula* (Figure 28-7A) survives as a representative of a family probably established in the Silurian and *Crania* as one of a probably Ordovician stock of oyster-like cemented inarticulates. The inarticulates seem always to have had shells made of calcium phosphate and rather high amounts of organic materials, probably proteins, in contrast to the articulate lampshells made of calcium carbonate with relatively smaller amounts of organic components. Articulate brachiopods appear in the fossil record in the lower Cambrian, but their greatest diversification occurs a little later, in the Ordovician, and again in the Silurian and Devonian. Numbers of brachiopod genera and species reach striking peaks in the middle Ordovician, in the Devonian, and lastly in the Permian. The later two peaks are almost entirely made up of articulate genera. The Permian dominance particularly involved the superfamilies Productacea and the Spiriferidina, and there are limestones in all the periods of the upper Palaeozoic which are known by the names of genera of brachiopods abundant within them (Figure 34-5). During these time-spans of dominance, brachiopods were the most common benthic marine animals, and estimates of the total numbers of species involved range from 12,000 to 30,000: clearly, they *were* a major phylum. Most of the living articulate brachiopods belong to a group of families (the Terebratulida) which first arose in the Triassic and apparently diversified mainly in the Cretaceous and Tertiary. Although the details are difficult to investigate, the decline of the Brachiopoda to the status of a minor phylum among living invertebrates is undoubtedly connected to the rise of the more efficient filter-feeders, the eulamellibranch bivalve molluscs. This would suggest circumstantial support for the deduction that many of the bivalve molluscs of Palaeozoic seas, when brachiopods were dominant, were in fact protobranchiate and not specialized filter-feeders (see Chapter 22). The difficulties of relating the phylum Brachiopoda to any other invertebrate group (see Chapter 28) are not resolved in any way by the abundant fossil record, although most palaeontologists hypothesize that the various Cambrian stocks of lampshells originated in a Precambrian inarticulate form with uncalcified valves of tanned protein.

The third of the major phyla of living invertebrates to have an abundant fossil record is the Echinodermata. As already discussed in Chapter 29, it is clear that the more typical mobile forms of today (placed in the subphylum Eleutherozoa) were long antedated by representatives of the other two subphyla, Homalozoa and Pelmatozoa. One of the classes encompassed in the last group is the Crinoidea; the sea-lilies were abundant in late Palaeozoic seas. The Carpoidea, the wholly extinct group which we placed in the Homalozoa, appears in the middle Cambrian and dies out in the Devonian. As already discussed, they are of interest for their bilateral symmetry and because of the claims by some workers of features held in common with chordates. However, members of the Crinoidea are certainly more ancient in the fossil record (lower Cambrian), and from that stock of sessile sea-lilies came a first flowering in the Silurian. Another group of crinoid families were prominent in the late Palaeozoic, and then were replaced by still a third group in the early Mesozoic, representatives of which have persisted till today. Comatulid forms like *Antedon,* which are mobile, occur in the fossil record only from the Jurassic onward. Three other to-

FIGURE 34-5. A characteristic fossil brachiopod. A fine specimen (4.8 centimeters across) of an articulate brachiopod of the family Spiriferidae, extensive beds of which occur in limestones and shales of Carboniferous age. [Photo by Perry Russell-Hunter; see caption for Figure 16-10.]

tally extinct classes are placed in the subphylum Pelmatozoa: Cystoidea, Blastoidea, and Edrioasteroidea. Cystoids and edrioasteroids are limited to the lower Palaeozoic and have their greatest flowering of genera in the Ordovician and Silurian. The blastoids seem to have been a more successful group, with a few early representatives from the Ordovician. They are among the most numerous echinoderms of the Carboniferous (genera occasionally giving their names to particular limestone strata), and the last of them are found in Permian rocks. Turning to the more mobile subphylum of echinoderms, the Eleutherozoa, we find, according to the classification used in this book (see Table 29-1), four well-known classes of living forms and three important fossil groups. The smallest class, Somasteroidea, which arose in the Ordovician, was never an extensively diversified group, being relegated by many neontologists to the status of a subclass of the Asteroidea. (The living genus, *Platasterias*, is no longer regarded as a somasteroid—see page 509) The class Helicoplacoidea had spirally pleated tests which were probably flexible and are known from the lower Cambrian only. The class Ophiocistioidea is structurally similar to the central discs of brittle-stars but with very large scale-covered tube-feet emerging in pairs from five short ambulacral regions; they occur from lower Ordovician to middle Silurian. Of the presently living groups, true Asteroidea are known from the Ordovician with a considerable increase in genera and families occurring in the Mesozoic; Ophiuroidea, or brittle-stars, have an essentially similar fossil record; while Holothuroidea, or sea-cucumbers, are known from the Devonian, but many modern groups date only from the early Cretaceous. Finally, the class Echinoidea, or sea-urchins, are known by a few regular forms (that is, with near-spherical symmetry) from the late Ordovician, remain relatively rare as fossils through the later Palaeozoic, and become abundant and diversified in Mesozoic and Tertiary times. As noted in our earlier discussion of the secondary acquisition of bilateral symmetry in sea-urchins, the irregular forms (those which are bilaterally symmetrical) mostly arose in the Cretaceous and later. Many Cretaceous limestones have extensive beds of both regular and irregular sea-urchin fossils. In relation to echinoderm classification, it might be noted that some authorities would place the Edrioasteroidea in the Eleutherozoa, and the Helicoplacoidea in the Pelmatozoa, reversing the assignments made earlier in this book. This disagreement merely serves to illustrate the difficulty of applying an arrangement based on life-styles and feeding patterns even to well-preserved sets of calcareous ossicles which belong to clearly separate groups at lower levels of classification.

Numbers of fossils are placed in phyla not represented by *any* living forms. We have dismissed the sponge-like Archaeocyatha, and we will not consider the Conodonta, tiny tooth-like fossils from the Ordovician onward which may be vertebrate or invertebrate, but we have to give brief consideration to the phylum Graptolithina. As fossils, graptolites superficially resemble colonies of hydroid coelenterates (Chapter 4); they flourished in the Ordovician and Silurian and had all died out by the lower Carboniferous (see Figure 31-6A). Graptolites are important index fossils in correlation of Ordovician and Silurian beds, but they are mostly represented by flattened, two-dimensional remains which give little evidence of internal structures. A few excellently preserved, three-dimensional specimens have been sectioned and the results seem to relate the graptolites to the aberrant hemichordate genus *Rhabdopleura* (see Chapter 31). Within the last few years, analysis of the exoskeleton of graptolites has shown that it is proteinaceous—not chitinous—and has some equivalence in amino acid composition to the colonial skeleton of *Cephalodiscus,* the other genus of atypical hemichordates.

Although the phylum Chordata is represented in the fossil record by magnificent series of fishes, amphibians, and higher vertebrates, sea-squirts and the other "invertebrate" chordates of today are very poorly represented. If we exclude the graptolites and some other even more doubtful *Cephalodiscus*-like fossils from the Ordovician, there remain a few fossil ascidians from the Permian onward, a possible cephalochordate (like *Amphioxus*) from the Scottish Silurian, and the enigmatic fossil *Ainiktozoon* discussed at the end of Chapter 31. In contrast, the magnificent fossil record of the vertebrate classes, beginning with the jawless fish collectively referred to as ostracoderms which appear in the Ordovician, gives us sound palaeontological evidence of the origins and of the diversification of such groups as the reptiles, the birds, and the mammals. As noted earlier, the details which can be recovered of the earliest cephalaspids confirm that after the acquisition of a vertebrate skeleton, brain, and sense-organs, these forms retained the pharyngeal apparatus with filtering gill slits which clearly relates them to the more primitive chordate pattern of life-style we have described for *Amphioxus* and for the larval stages of sea-squirts. From these ostracoderms, as the fossil record clearly tells us, came forms with biting jaws, and thence the rest of our phylum (see Table 31-1).

Some Implications

As already noted, the fossil record of any stock of animals, even those with readily preserved shells or skeletons, is erratic and incomplete. When, as evolutionary biologists, we turn to the fossil record for evidence, we must continually remind ourselves of this chance element in preservation, and of the appropriate time-scales in terms of generations of the animals we are concerned with. Apart from questions arising out of these stochastic processes and extended time-scales, there are biases which arise from our training as modern biologists. Since the last four decades of work on population genetics have been grafted onto, and have strengthened synthetically, the Darwinian theory of natural selection, we have been compelled to regard evolution as involving processes of gradual change in time (or phyletic gradualism). Many book pages have been devoted to expounding why apparent examples of macroevolutionary change must be explained by the processes of microevolution which can be observed in natural and experimental populations of living organisms, processes whereby new species can arise through slow and steady transformation of the genomes of entire populations over some 10^5 or so generations. In this light, the fossil record is very imperfect: there are almost no gradually and continuously changing fossil series available to us. Again it is important to emphasize time-scales: local areas of sedimentary rocks have frequent temporal interruptions in deposition which may last around 10^6 years (or a similar number of generations for the majority of invertebrate stocks).

It is particularly important not to confuse time-scales when discussing extinction and replacement of major fossil groups in relation to such "crises in earth history" as major crustal changes in conditions of deposition and of mountain-building. The end of Palaeozoic conditions marked by the passage from the Permian to the Triassic apparently involved the extinction of many strikingly different groups of fossil invertebrates: certain corals, trilobites, eurypterids, two major groups of Bryozoa, two major groups of Brachiopoda, certain crinoid superfamilies, and all blastoid echinoderms, among others. In this case, the general absence of Permo-Triassic transitional marine beds has certainly exaggerated our view of the "catastrophic" nature of this faunal change. It is even more important to note that there was no "instantaneous extinction of large numbers of organisms"; the pattern of extinction in different groups shows little uniformity in time or in geographical distribution. The critical period of upper Permian time probably extended over 16 million years, and, in most groups, the evidence is rather of piecemeal extinction of stocks at different times within that subdivision, rather than a sudden mass extinction at its end.

Another feature of biological evolution which is difficult to assess in the fossil record is the occurrence of coevolutionary processes involving two groups of organisms within a food-chain or similar relationship. Changes in what we may regard as the biotic environment of a stock of organisms may have profound effects on its evolution, and they have occurred relatively rapidly in terms of the time-scale of the fossil record. The coevolution of flowering plants with suitably "designed" attractive colors and nectars and of successive stocks of insect pollinators has already been mentioned. We also hinted earlier that the apparently "sudden" beginning of a fossil record may have been the result of the evolution of certain predatory animals creating a need for shells, tests, or other external armor in the Cambrian, which had not existed for the invertebrates of the Precambrian. About 95 million years ago in the upper Cretaceous there was an ecological "revolution," a major shift in the dominant vegetation, which affected all animals living on land. At this relatively late stage in geological history, the previously dominant tree ferns, gymnosperms, and horsetail-relatives were replaced by the broad-leaved angiosperms dominant today. None of the "superior" features claimed for angiosperms, including the broad leaves, better vascular systems for fluid conduction, and the presence of defensive compounds such as toxic alkaloids, is exclusive to the group. Although many highly specialized insect stocks, including Lepidoptera and Hymenoptera, became diversified after, and as a result of, the angiosperm replacement, these and the Diptera and beetles *did* antedate the vegetation revolution and coexisted with the more ancient plant stocks. Modern interpretation of this would again stress the exaggerations introduced by confusions of time-scale (the "revolution" may have taken about 10 million years), and also emphasize the importance of coevolution with the angiosperms, not only of certain insect groups but also of the "new" birds and mammals as important agents of seed dispersal. The fruits and edible seed cases of angiosperms were not evolved to attract insects to maintain cross-fertilization, but rather to employ birds and mammals in achieving the genetic consequences of efficient distribution in a patchy environment. Highly specific instances of insect-plant coevolution are clearly demonstrated in certain relationships today. For example, the larvae of a beetle of the genus *Caryedes* feed exclusively on seeds of a leguminous species which contain 13% canavanine, a potent insecticide which is a nonprotein amino acid analog. This beetle has developed unique biochemical pathways which allow it to detoxify the canavanine and use it to provide dietary nitrogen for growth. In most instances (though not with *Caryedes*), there are clear differences between those plants which "defend themselves" against herbivores by small quantities of highly toxic substances, such as alkaloids, and those plants which decrease their edibility by having tough leaves or large quantities of tannins or resins which make them of low nutrient value. The former category are more likely to occur as early successional plants or in species-rich climax communities, while the latter are characteristic of species-poor stands in low-fertility areas. Corresponding differences have developed by coevolution in the animals—both invertebrate and vertebrate—which live in these plant communities.

Other difficulties in the interpretation of fossil series can be viewed in the light of modern studies on gene mechanisms and on population genetics. Many genes are known with more than one phenotypic effect, and others may demonstrate the same phenotypic effect at the cellular level with markedly different results in the differentiation of organ structures. Selection may act on only one phenotypic manifestation, and the other effects may be "carried along" in functionally nonadaptive changes. An even more important example comes from the contribution of modern population genetics to our understanding of how new species may arise very rapidly in small isolated local populations placed geographically on the periphery of the species' main geographic distribution—by allopatric speciation.

One further aspect of the fossil record requires emphasis. It is that extinction and replacement of stocks are not the exception but the nearly universal rule. When figures are calculated for the percentage extinction of new and holdover genera in well-fossilized and reasonably well-studied groups like the crinoids, we find that the percentage extinction of the late Permian was not much greater in its severity for each group than similar "catastrophic" reductions in diversity near the boundaries of most rock systems (corresponding to the periods of the time-scale). The startling fossil change of the Permo-Triassic is exaggerated by a lack of immediate Triassic replacement and diversification of stocks. In general, it is unnecessary and inappropriate to invoke catastrophic hypotheses of instantaneous mass extinction of stocks on a global scale (even though the extraterrestrial overtones are attractive to some kinds of the human imagination). At the other end of the scale, the search for fossil sequences showing a continuous process of phyletic gradualism from species A to species B to species C, and so on, resembles the pursuit of a mirage. To be sure, honest confession by scientists of the rarity of continuous transitional series in the fossil record provides arguments for antievolutionists of all kinds who claim that complete chains of transitional forms must be demonstrated to "prove" natural selection.

In recent years, two thoughtful palaeontologists, Stephen J. Gould and Niles Eldredge, have introduced the concepts of allopatric speciation into their consideration of fossil lineages and concluded that the occurrence of speciation in small peripheral populations will *automatically* result in gaps in the fossil record. As a consequence of speciation being largely allopatric, they claim that new fossil species do not usually originate in the place where the principal successful populations of an ancestral species lived. Thus it is extremely improbable that we would be able to trace the gradual separation of a stock into two species merely by following that stock's representative fossils up through a local rock column. More fundamentally, Gould and Eldredge believe that the fossil record is a relatively good representation of what they call "punctuated equilibria" as a characteristic of evolutionary history rather than the supposedly smooth continuous process of phyletic gradualism. This conveniently accommodates the notoriously "gappy" nature of most fossil lineages to another concept of modern genetics, that of genetic homeostasis as a property allowing the self-regulation of populations and the consequent "conservatism" of larger interbreeding populations.

As we noted at the beginning of this chapter, biologists vary in the emphasis they place on palaeontological evidence. Palaeontology and neontology can complement each other in evolutionary studies, but few individual scientists are sufficiently competent in both fields. The value of the Gould and Eldredge concept of punctuated equilibria derives from their unusually promiscuous standing as palaeontologists also involved in modern biology. In the past, there has been a tendency for the separation of disciplines to lead to some very uncritical syntheses. A palaeontologist would accept one possible theory derived from comparative physiology uncritically if it "fitted" his concept of deduced palaeo-ecology for the animal stocks involved. Similarly, an animal physiologist might uncritically accept one of the alternative interpretations of the fossil record. Ideas on the environment of the earliest true vertebrates were much colored by scientists of both kinds relinquishing the tool of scepticism when they happily accepted theoretical conclusions from the other discipline. The strengths and weaknesses of evolutionary conclusions from neontology are different from those from palaeontology. The best phylogeny still depends on input from palaeontology, and this will come increasingly from palaeo-ecology (including reconstruction of habitats and trophic relationships with predators and competitors) and probably less from attempts to reconstruct complete lineages of successive speciations. Further, the only *dated* data for evolutionary studies come from the fossil record. However, it is unlikely that any future palaeontological studies will yield much additional evidence on questions of the origins of, and interrelationships among, the major invertebrate phyla.

35
A PARTIAL PHYLOGENY
OF INVERTEBRATES

AMONG the definitions given by Webster for the adjective "partial" are "relating to the part rather than the whole" and "biased." Both are applicable to this short chapter.

As stressed in Chapter 2 and elsewhere, all statements regarding homologies, archetypes, and phylogenies involve hypotheses which cannot readily be tested, each by a single crucial experiment based on a "null" statement. The hypotheses *can* be tested (in an *objective* fashion), against further work on the fossil record, the embryology, the functional morphology, or even —in a few cases—the biochemistry and molecular genetics of the organisms concerned. But all good phylogeny involves carefully controlled and objectively cross-checked processes of inductive reasoning, essentially similar to those required in the detection and prosecution of a murderer when there are no direct eyewitnesses of the crime available. Only bad phylogeny, like bad detection, will depend upon subjective impressions (or, worse, upon an appeal to the opinion of a higher authority). Close *coincidence of independent evidence* from palaeontology, from biochemistry, from comparative anatomy, and from developmental biology can build a better case than even extensive data from any one of these alone can ever do. Occasionally, certain experimental biologists, on the basis of a dogmatic (they would claim "revolutionary") opposition to descriptive science, and a parallel adherence to a very narrow view of the proper design of "null" experiments and their controls, state polemically that they regard all phylogeny as nonsense. Like Henry Ford's reputed statement, "history is bunk," this rejection is immediately obscurantist, and potentially totalitarian. There are other pressures on the study of phylogeny from "fashionable" areas of biology. The fact that Jacques Monod found it necessary to state "I don't know of any *good* molecular biologist who says that organismic biology is useless" suggests that there are others in the field who do just that. Some of the biases in this chapter are already apparent. Discussion in Chapter 27 made it clear that any simplistic form of Haeckel's theory of recapitulation was unacceptable. Used with caution and integrated with evidence of other kinds, however, embryological data can provide important circumstantial evidence of evolutionary relationships. A middle-of-the-road position on the value of palaeontology was set out in Chapter 34. The best phylogeny will always depend on input from palaeontology which provides the only kind of evolutionary data incorporating measurable time. However, this remains a book based on, and biased toward, a curiosity about modes of life in whole organisms.

Although this chapter is concerned with phylogenetic theories, there are many other biological hypotheses throughout the book which are more directly testable. Most of them concern whole animals discussed at the mechanistic-physiological and adaptive-functional levels of explanation. In other words, the reader can turn (with Louis Agassiz and Wordsworth) "to the solid ground of Nature" and confirm or correct any statements describing the mechanics of feeding in *Artemia* by careful observations on living brine shrimps. Thus also can be verified or modified any descriptions of restorative regeneration in sponges or in polychaetes; of ciliary sorting in bivalves, in sabellid worms, or in *Amphioxus*; of suspension-filtering by sponges, by gastropods, by barnacles, by lampshells, or by sea-squirts; of locomotory forces and their application in ribbon-worms, in leeches, in brine shrimps, in millipedes, in starfish, or by the tails of chordates. But these studies of living mechanisms are not all of biology, even after restricting consideration to the organismic grade of complexity (and thus ruling out almost all concern with ecosystems on one hand, and with "molecular" biology on the other). Even such a miniature attempt at a pandect of invertebrate biology as this volume has involved the use of systematics and phylogeny as linking material. In the study of biology at any level, both the arbitrary distinctions of higher taxonomic categories and the inductive concepts of homology in organ systems or in functional processes are of pragmatic value.

Such theoretical matters as homology and phylogeny *do* aid human comprehension of animal diversity. Their implicit biological significance, however, cannot readily be tested. There can be neither verification by observation of the living animals nor any appropriate application of "the scientific method" in a crucial experiment. Essentially this is because the hypotheses involved in archetypes, homologies, and phylogenetic classifications all concern past time more distant than Proust's "*temps perdu*." No matter how extensive and how convincing the evidence for any statement on the phylogeny of an animal stock, that statement remains merely a "reasonable inference." It can never be turned (lacking a time-machine) into a notarized genealogy. Similarly, such animal phyla as the Mollusca and Echinodermata must each have had *real* ancestral progenitors. Although the synthesis of studies in comparative anatomy, functional morphology, palaeontology, and embryology allows us to set up more or less convincing functioning archetypes for these two phyla, such hypothetical types are not ancestors. Current optimism that molecular data on actual amino acid sequences will solve all phylogenetic problems within a year or so was critically examined in Chapter 27. The difficulties of collecting such evidence of chemical identity and relationships are real. A number of crucial phylogenetic relationships, including some left open in this chapter and book, may well be resolved by appropriate protein sequencing within the next decade or so. However, future work on the mechanisms of evolution, as well as on many of the details of phylogeny, will remain dependent upon data provided by embryological, morphological, and palaeontological studies. The evolutionary-historical level of explanation (so necessary to our science) will increasingly be set within a framework of stochastic prediction arising from modern work in population genetics and in ecology rather than from that in molecular biology.

The above statements may seem to involve some overemphasis on a simple distinction between the immediately verifiable facts of structure and mechanism and the less directly testable hypotheses of common descent. Phylogeny can be objective, but only if based on the continuous process of questioning that marks all good science. However, any acquaintance with the literature of any branch of biology can demonstrate that biologists—from ecologists to biochemists—have need to be periodically reminded of this distinction, and hence of the more speculative and possibly biased nature of all phylogenies.

Homologies and Phylogenies

The specific bias of the phylogenies presented here depends on a belief that natural selection has promoted functional interdependence so that a stereotyped pattern of whole animal physiology has emerged in each major group of animals (see Chapter 2). It follows that perception of functional homologies within a group—along with the more widely accepted homologies of structure—can lead to the concept of an archetype as a working, efficient machine with such a pattern of functional integration. Such a perception is not merely concerned with counting similarities (as in some numerical systematics) but rather with establishing the functional relationships between the parts. Thus the model—the temporary hypothesis for testing—we call an archetype can only be of value if it involves a concert of organs and functions that could operate as a whole, in an integrated functional plan, as in all *living* organisms. As noted in Chapter 2 and elsewhere, perceiving functional homologies within a group and setting up archetypes are much more difficult in the phyla of *less complex* animals.

After these disclosures of its subjective and biased nature, the purpose of this chapter is the construction and defense of Figure 35-1 as a rough phyletic "tree," which involves the possible interrelationships among the 31 phyla of many-celled animals set out in Table 2-1. The perceptive reader will have noted that a preliminary question has thus been bypassed. The grouping of the million or so "species" of many-celled animals into these 31 phyla has itself involved acceptance of ideal arbitrary categories. Defense of each of these phyla as reflecting some objective reality of common descent and interrelationship has been set out in the preceding chapters; in some cases, we have mashalled extensive and convincing evidence of "common descent," and in other cases, otherwise. Once the 31 phyla are accepted, then the despairing evolutionary biologist might be tempted to draw, instead of a "tree," 31 separate diverging lines originating in a blank "unknown." Moved by pragmatic needs, or by the seductiveness of intellectual exercises in matters ultimately unprovable, many biologists venture further. Some would regard Figure 35-1 as a timid and retrograde step in the continued evolution of phylogenics.

Most biologists accept that many-celled animals evolved from protistans—whether, as in the majority view, from flagellate stocks or, as in a vocal minority opinion, from multinucleate protozoans—need not involve us in extensive discussion here. The alternative theories were mentioned at the end of Chapter 6. The most frequently encountered hypothesis, that of flagellate origins, has usually been termed the colonial theory because in its most classic statement by E. Metschnikoff it involved as an intermediate stage a colonial association of flagellate cells like *Volvox*. The development of multicellular interdependence would have followed, and an advantage to this theory is that the usual sexual reproduction characteristic of Metazoa (involving true sperm and eggs) had probably already evolved in such flagellates. The alternative theory, termed the syncytial theory, is that metazoans were derived from some kind of multinucleate protozoan, almost certainly a ciliate or protociliate, in which there were already many nuclei in a continuous mass of protoplasm. According to this theory, the development of multicellular construction and interdependence involved the isolation and specialization of parts of that protoplasm by the interpolation of cell membranes. Advocates of the syncytial theory, including J. Hadži, would maintain that the earliest true metazoans were already bilaterally symmetrical and could possibly have been classified as flatworms belonging to the order Acoela. The facts that some acoels have partially syncytial tissues and that they are generally ciliated are regarded as supporting this theory, but the syncytial tissues in acoels actually arise *secondarily* by breakdown of cell membranes after normal cell cleavages in their *earlier* embryonic development. Further, no ciliate ever produces a flagellated sperm, such as occur generally throughout the metazoans. A few recent authors have attempted to compromise between these theories and suggest that the cnidarians were evolved independently from colonial flagellates, while the flatworms (and presumably all the higher bilateral phyla as well) were derived by the syncytial route from a ciliate stock. We here adopt the majority view of a flagellate origin. It seems clear that the Parazoa (phylum Porifera, the sponges—see Chapter 5) evolved independently of the rest of the Metazoa, but again most likely from a flagellate stock of protistans. The relationships of the phylum Mesozoa remain obscure. The present phylogeny regards the rest of many-celled animals as having a common origin. In this stock, the phylum Cnidaria is clearly the most simply constructed of many-celled animals and is here regarded as encompassing the most primitive. This does not necessarily imply that the triploblastic phyla were derived from forms like presently living adult coelenterates. The probable

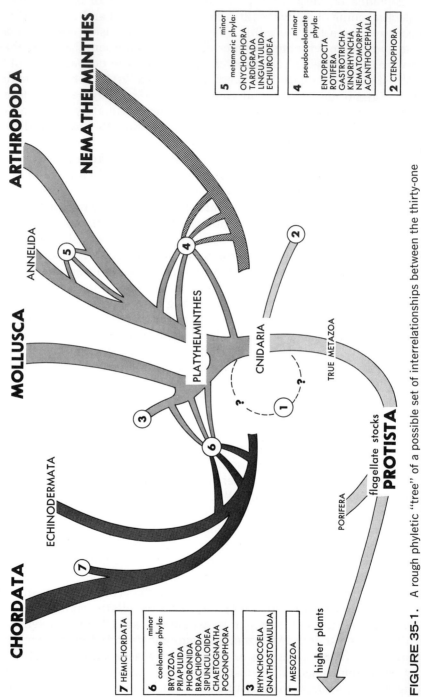

FIGURE 35-1. A rough phyletic "tree" of a possible set of interrelationships between the thirty-one phyla of many-celled animals. Pragmatically motivated, this figure involves several timid compromises, and its defense is outlined through Chapter 35. It should be noted that there is no attempt to represent *time* but only to depict possible degrees of interrelationship.

phylogeny within the cnidarians is discussed in Chapter 4, and it can be argued that the first coelenterate could have been a ciliated, nematocyst-bearing organism resembling a simplified version of the actinula larva found in some hydrozoans. The ctenophores are obviously allied with the cnidarian coelenterates (see Chapter 5), though it is impossible to derive the ctenophores from any of the present stocks.

The simplest triploblastic animals are free-living flatworms belonging to the phylum Platyhelminthes. Within this phylum, many authorities believe the most primitive living group to be the order Rhabdocoela rather than the order Acoela. The latter are almost gutless, but some of their supposedly primitive features could well result from a highly specialized mode of nutrition and others represent simplifications resulting from small size. Some rhabdocoels certainly have equally "primitive" features, including reproductive systems without discrete ducts. Whichever flatworms are the more ancient stock of free-living forms, it is generally supposed that they were derived from planuloid ancestors, that is, motile, gutless, larval coelenterates with bilateral symmetry. A contrary hypothesis, not accepted here, is associated with the name of J. Hadži, and suggests that the coelenterates are not primitive metazoans but are secondarily derived from the flatworms. Apart from other difficulties, this implies that the most primitive cnidarians were forms like sea-anemones, a phylogeny of coelenterates not readily acceptable (see Chapter 4).

The ribbon-worms of the phylum Rhynchocoela are obviously closely allied to the flatworms, and can be regarded as comprising several experimental patterns of worm organization. If we ignore for the moment the various groups of minor phyla, three or possibly four stocks of triploblastic animals with body cavities originate from the flatworm stock. The phylum Nemathelminthes stands apart from all other stocks as the only successful group of pseudocoelomate worms. Nematodes, though enormously successful, show no obvious relationship to any of the other phyla (see Chapter 9). There is a group of six minor phyla, the rotifers being the most successful, which are built on the pseudocoelomate plan. They have often been grouped together in various ways (see Chapter 8), but the case for a superphylum like Aschelminthes is not a good one. The minor pseudocoelomate phyla are best regarded as all separate stocks, some perhaps distantly allied to the Nemathelminthes.

As with the origins of the Metazoa, there are a number of alternative theories regarding the evolution of the true coelom and of metameric segmentation. There is, however, little or no dispute that the triploblastic coelomate phyla were evolved from acoelomate stocks like flatworms and that the pseudocoelomates like nematodes were evolved completely separately. The true coelomates are those whose body cavities are lined with mesoderm (see Chapters 1 and 8 for definitions and for discussion of the nature of body cavities). The hypothesis on the origin of the coelom which is probably most favored today has been termed the gonocoel theory, and it maintains that the coelom originated as a persistent sac-like cavity with a short duct to the exterior, primarily surrounding the gonad. Most supporters of this theory would place some protoannelid ancestor as an intermediate form between flatworms and all the coelomate phyla. As a result of his detailed studies of the evolution of excretory systems, E. S. Goodrich (see Chapter 12) favored this theory, and also its subsequent modification to suggest that multiple or pseudometameric gonads might have given rise to a metameric series of coelomic compartments. The principal alternative hypothesis on coelomic origin is the enterocoel theory, which maintains that the coelom evolved as a series of pouches separated as outpocketings from the primitive gut-tube. The majority, though not all, of the supporters of this theory regard cnidarians as the stock from which the coelomate phyla were directly derived, and the flatworms and similar forms are thus said to be secondarily acoelomate (by the loss of coelomic spaces present at one time in their ancestors). Among the most distinguished early supporters of the enterocoel theory was E. Ray Lankester, and modern supporters include Marcus and Lemche. Obviously, this theory "fits" best with the mode of development exhibited by the echinoderms and the other phyla termed deuterostomes (see Chapter 27). A powerful objection to phylogenies based on the enterocoel theory is that most of them require anthozoans (sea-anemones) to be considered as the most primitive cnidarians. Two related theories, termed the cyclomerism and the corm theories, have been elaborated to account for the origins of metamerism. They involve essentially enterocoelic origins for the coelom in anthozoans and in flatworms, respectively. What is essentially a third hypothesis on coelomic origins has been termed the schizocoel theory; it holds that coelomic cavities arose as slit-like spaces within the mesoderm for a variety of functional reasons. Supporters of this theory would suggest that the most primitive coelom is found in phyla like the Annelida and Mollusca, which have a schizocoelous ontogeny, and that *they* represent an intermediate stage between acoelomate flatworms and the enterocoelous phyla which are secondarily derived. A recent variant on the schizocoel theory comes from the persuasive arguments of R. B. Clark that both coelomic spaces and metamerism have been evolved as adaptations for locomotion. Emphasis has been placed in Chapter 11 and elsewhere in this book on the functional significance of segmentally arranged coelomic spaces as hydrostatic and hydraulic skeletons in annelid worms. In our present concern with the possible interrelationships between phyla (Figure 35-1) we have adopted an arrangement which would conform either to the schizocoel theory or to most elements of the gonocoel theory, but follow R. B. Clark's emphasis on the importance of locomotory and other hydraulic mechanisms in postulating that coelomic body cavities may have evolved independently, probably in three stocks, from the acoelomate flatworm stem. [The claim that the more complex phyla can be divided into two assemblages, Protostomia and Deuterostomia, on the bases of modes of embryonic development and of planktonic larval types, has already been discussed (Chapter 27), and largely rejected. However, a relatively modern statement of this and other claims from comparative embryology can be found in *Principles of Comparative Anatomy* by the great Russian morphologist W. N. Beklemishev, which belatedly became available in English translation in 1969.]

The case for regarding the major phylum Mollusca as being derived from the turbellarian-rhynchocoel phyla, directly, is set out in Chapter 26. It hinges on the conclusion that metameric segmentation (as it occurs in annelids and arthropods) does not occur in primitive molluscs. The discovery of *Neopilina* provoked a reconsideration of ideas on the phylogeny of the molluscs and their relationships with other phyla. A number of zoologists have used it as a link between the arthropod-annelid stock and the rest of the molluscs, a major premise being that several features in both *Neopilina* and the chitons are primitive metameric characters. However, the evidence against the alleged metamerism of chitons is very strong, and probably, despite *Neopilina*, annelid-like segmentation has never occurred in an animal which could be called a mollusc. It should be noted that the basic molluscan plan of structure and function is remarkably uniform throughout the group, and the homologies established, particularly as regards the mantle-cavity and the gills, are extensive and convincing.

Apart from metamerism, there are undoubtedly some features in which the annelid-arthropod stock shows a pattern of specialization from the platyhelminth-rhynchocoel stem-group which paralleled that of the molluscs. The difficulties of this are bypassed in Figure 35-1. The case for a close relationship, though distant in time, between the annelids and the arthropods has already been set out (see Chapters 10 and 16). Annelid metamerism must have arisen in rhynchocoel-like worms which became coelomate. By the development of a metamerically divided body cavity, which could be used as a hydrostatic skeleton, annelids became the most successful group of worm-like animals. The question of the origin of the arthropods in some primitive stock of annelid-like animals has already been discussed (see Chapter 16). The possibility that the arthropods were of multiple origin in protoannelid stocks was there considered and dismissed. The diverse stocks of arthropods, including the enormously successful crustaceans, arachnids, and insects, do show a uniquely integrated pattern of organs and functions and can be best considered as a single monophyletic group. There are four minor phyla (see Chapter 18) more or less closely connected to the annelid-arthropod stock. Given our ignorance of their true relationships, it is best to treat the Onychophora, the tardigrades, the linguatulids, and the echiuroid worms as separate minor phyla, though clearly metameric.

There remain eight minor coelomate phyla, of which two are probably connected with the stock of the echinoderms and chordates. The other six coelomate stocks are discussed in Chapter 28; and as with the pseudocoelomate minor phyla, interrelationships are obscure. Already discussed, the common possession of a lophophore as a food-collecting organ may or may not be phyletically significant. The phylum Bryozoa is a moderately successful group, as was the phylum Brachiopoda at an earlier period of the geological time-scale.

In some ways, the phylum Echinodermata stands disparate in a similar fashion to the nematode worms from the other groups of many-celled animals. However, as has been discussed (see Chapters 30 and 31), echinoderm origins and those of the chordates can be postulated as close. The fact that some fossils, described by some investigators among primitive echinoderms, can be regarded by others as chordates, is itself significant (see Chapter 31). It seems likely that there were several "experimental" chordate stocks in the Palaeozoic, from which three stocks survive. Apart from the minor group of *Amphioxus* and its allies, the successful chordate groups are the sea-squirts and the vertebrates.

As has been often discussed (see Chapters 32 and 33), the probably neotenous origin of the Larvacea among the urochordates is significant in relation to the origin of true vertebrates. Whatever the exact historic sequence, it is obvious that the first vertebrates possessed many of the features which we have attributed to the chordate archetype. Both the excellent fossil record (see Chapter 34) and several physiological characteristics confirm that a group of agnathous fishes—the ostracoderms—comprised the most primitive truly vertebrate animals. The beautiful reconstructions of ostracoderms by Stensiö and his associates have convinced most zoologists that these armored jawless fishes used a filter-feeding mechanism involving the pharyngeal gill slits, as do *Amphioxus* and larval sea-squirts on the one hand and the young stages of the presently living parasitic cyclostomes on the other. In addition, there are several cases of structural and functional homologies which link the vertebrates to the primitive chordate stocks. There is anatomical, developmental, and biochemical evidence for homologous origins of the endostyle of sea-squirts and the thyroid gland of vertebrates. Several other homologies are mentioned in Chapters 31 and 32. In fact, to conclude this discussion of Figure 35-1 and invertebrate phylogeny in general, it can be said that the evidence linking the vertebrates with the invertebrate chordates is both more extensive and more convincing than that which can be used to link any two invertebrate phyla.

Envoi

Some recapitulation may be allowed to a Scots pedagogue. Discussion of structural and functional homologies is of value in trying to comprehend the prodigious diversity of form and function in invertebrates. This book is intended for use, not as a comprehensive reference volume of invertebrate systematics and comparative anatomy, but as an introductory or intermediate text on invertebrate life-styles. Its use should normally be accompanied by some personal observations of, and simple experiments upon, a variety of living invertebrates (and by gaining some comprehension of invertebrate life from these observations and experiments). These constitute, as J. Eric Smith more elegantly puts it, "the science of looking at animals and learning from the looking."

Most of the book takes the viewpoint of the bioengineer and employs reductionist methodology; thus it is concerned with whole animals considered mainly at the mechanistic-physiological and at the adaptive-functional levels of explanation. In answering questions at these levels ("how does it work?" and "what is the biological value to the individual or population?"), we develop hypotheses which are directly testable in the most obvious fashion. Only rarely will there be any danger of confusing the "model" hypothesized and the real biological system under study at these conceptual levels. It is when we move (as biologists invariably *must*) to the evolutionary-historical level of concept (dealing with "what is the history of this process or organ-system in time—how has it evolved?") that we have difficulties and must cherish our objectivity. All scientific understanding (including that of evolutionary biology) depends upon the systematic organization of knowledge by developing testable hypotheses from reductionist explanations. When we turn to these peculiar models which are phylogenies or archetypes, we must stay within the same methodology of science. There is no need to sustain any stultifying polarization between experimental biology, on the one hand, and phylogeny with descriptive biology on the other. In considering the origins of the phylum Mollusca, we saw how a "modern" experimental approach using electronic force transducers to elucidate the mechanics of snail locomotion (Chapters 20 and 21) could increase our understanding of what might be regarded as a "classical" problem in phylogenetics: molluscan relationship directly with a turbellarian-rhynchocoel stock rather than by way of a protoannelid one (Chapter 26). Further, when we move to considerations of phylogeny (just as when we think about the workings of brains—Chapter 25), we need not—and should not—move away from our usual mechanistic reductionism to any vitalistic interpretation or final cause.

In discussing in this book the various patterns of functional machinery found in invertebrate animals, we have stressed what we term the "functional homologies" revealed when comparative physiology is studied at the level of the integration of functions in the whole animal. Thus consideration of the functional unities perceived at the organismic level ("of architectural plans") rather than of those at the cellular and molecular levels ("of building blocks") has led us to treat invertebrate biology largely phylum by phylum. From the bias of a faith in the primacy of education (characteristic of Scots and Ashkenazim), I have tended to emphasize the obvious pragmatic value of such homologies in aiding comprehension of invertebrate diversity, and have probably understated here their genetic and evolutionary implications. Discussion of structural and functional homologies leads to the utilization of the working archetype, in which the deduced concert of structures and functions forms an integrated functional plan, as a basis for phylogeny. Despite some "experimentalists," phylogeny is *not* "bunk." However, it is important to remember that each archetype, as well as each phylogeny, constitutes a temporary model, set up from a reductionist explanation, to be questioned by the collection of new data and perhaps replaced. Here at the evolutionary-historical level of biological explanation, there is a greater danger of misunderstanding of the model's status. *Real*, but unknowable, ancestral progenitors are mirrored, with variable distortion, in our hypothetical archetypes. *Actual*, but untraceable, genealogies are reflected—less or more efficiently—in our circumstantially determined classifications. Once more, archetypes are *not* ancestors.

FURTHER READING

ALL textbooks are out of date when they are published. All text-books copy each other's errors. Both of these trite assertions are true only in part. Insofar as any new textbook succeeds in com-municating its author's conception of what is significant in the subject-matter at that particular time, they are false, and any such textbook will have its place in the history of ideas (whether merely as one of many temporally modish examples, or as a truly seminal work, will emerge considerably later). It remains true that the matter to be conveyed in textbooks must largely be selected and summarized from secondary reviews and tertiary sources (including other textbooks), with somewhat less material being derived directly from primary research pub-lications. Thus, a complete set of references is both inappro-priate and impossible at this intermediate level of text. The value of a partial list of primary literature consulted by the au-thor would be even more transitory than that of his textbook.

Really useful "further reading" from the vast and steadily in-creasing literature on invertebrate biology can only be selected for a student who has already chosen a particular group or lim-ited theme to explore in depth. For more general purposes, only brief annotated citations of four classes of works follow. These are other general textbooks (mostly modern, one-volume works), comprehensive surveys (mostly older, and largely anatomical), accounts of specific groups, and modern surveys of invertebrate physiology. Almost all of the last three classes contain extensive bibliographies, but any student reading up on a topic in depth should learn to use *Biological Abstracts* and *Zoological Record* in the preparation of his own reference list.

Two good, relatively up-to-date, general texts on the invertebrates are R. D. Barnes, *Invertebrate Zoology* (Philadelphia: Saunders, 3rd ed., 1974) and P. A. Meglitsch, *Invertebrate Zoology* (New York: Oxford University Press, 2nd ed., 1972). Also valuable are C. P. Hickman, *Biology of the Invertebrates* (Saint Louis: C. V. Mosby, 2nd ed., 1973); and, employing an arrangement by general concept rather than by systematics, M. S. Gardiner, *The Biology of Invertebrates* (New York: McGraw-Hill, 1972). The natural history of invertebrates is treated in a series of modern essays in J. E. Smith *et al.*, *The Invertebrate Panorama* (London: Weidenfeld and Nicolson, 1971), and a functional approach to animal machines (similar to that adopted in this book) is used in an excellent volume for more advanced students by V. Fretter and A. Graham, *A Functional Anatomy of Invertebrates* (London: Academic Press, 1976). Useful paperbacks include M. S. Laverack and J. Dando, *Essential Invertebrate Zoology* (New York: Wiley, 1974); W. D. Russell-Hunter, *A Biology of Lower Invertebrates* and *A Biology of Higher Invertebrates* (New York: Macmillan, 1968 and 1969, respectively); and M. J. Wells, *Lower Animals* (London: Weidenfeld and Nicolson, 1968). Among the older textbooks which remain valuable (mostly quoted in revised editions) are R. Buchsbaum, *Animals Without Backbones* (Chicago: University of Chicago Press, 1948); F. A. Brown, Jr., *Selected Invertebrate Types* (New York: Wiley, 1950); R. W. Hegner, *Invertebrate Zoology* (New York: Macmillan, 1933); L. A. Borrodaile, F. A. Potts, L. E. S. Eastham, J. T. Saunders (and G. A. Kerkut), *The Invertebrata* (Cambridge: Cambridge University Press, 1961); and G. S. Carter, *General Zoology of the Invertebrates* (London: Sidgwick and Jackson, 1951).

The finest modern comprehensive survey in English is, unfortunately, incomplete. Libbie H. Hyman's *The Invertebrates*, Volumes I–VI (New York: McGraw-Hill, 1940–1967), provide a superb survey in five volumes of protozoans, coelenterates, flatworms, nematodes, echinoderms, and several minor phyla, but do not cover annelids or arthropods. The sixth volume (dealing with only part of the Mollusca) falls below the high standard set earlier, and the death of Dr. Hyman means that this magnificent series will never be completed in its original form. (It may be continued as a multiauthor series.) In French, the excellent and up-to-date series *Traité de Zoologie* (Paris: Masson et Cie., 1948 and subsequent dates) is edited by P.-P. Grassé, and though still incomplete, covers many of our groups. In German, a series more useful in some sections than others, *Klassen und Ordnungen des Tierreichs* (Leipzig: Friedlander und Sohn, 1873 and subsequent dates to present), was originally edited by H. G. Bronn. Another useful work, comprehensive and complete, although dated, is the *Handbuch der Zoologie* (Berlin: Walter de Gruyter, 1923), which was edited by W. Kukenthal and T. Krumbach. Two older series in English are still valuable: E. Ray Lankester's *Treatise on Zoology* (London: Adam and Charles Black, 1900–1909) and the *Cambridge Natural History* (London and New York: Macmillan, 1895–1909). Finally, three modern compilations prepared for other purposes include fine up-to-date surveys of some invertebrate groups: *Treatise* on *Invertebrate Paleontology*, edited by R. C. Moore (Lawrence, Kansas: Geological Society of America, 1952 and subsequent dates to present); *Treatise on Marine Ecology and Paleoecology*, edited by J. W. Hedgepeth (New York: Geological Society of America, 1957); and *Chemical Zoology*, edited by M. Florkin and B. T. Scheer (New York: Academic Press, 1967 and subsequent dates to present).

In our third category—accounts of specific invertebrate groups—many useful books have been published in the last fifteen years, and most of them have extensive bibliographies. Excellent examples of this kind are *Molluscs* by J. E. Morton (London: Hutchinson, 4th ed., 1967); *Living Marine Molluscs* by C. M. Yonge and T. E. Thompson (London: Collins, 1976); and *The Biology of Hemichordata and Protochordata* by E. J. W. Barrington (Edinburgh: Oliver and Boyd, 1965). For the major phyla we also have R. D. Manwell's *Introduction to Protozoology* (New York: Dover, 2nd ed., 1968); F. M. Bayer and H. B. Owre's *The Free-Living Lower Invertebrates* (New York: Macmillan, 1967); T. Goodey's *Soil and Freshwater Nematodes* (New York: Wiley, 1951); H. D. Crofton's *Nematodes* (London: Hutchinson, 1966); R. Phillip Dale's *Annelids* (London: Hutchinson, 1963); J. Green's *A Biology of Crustacea* (London: H. F. and G. Witherby Ltd., 1961); W. L. Schmitt's *Crustaceans* (Ann Arbor: University of Michigan Press, 1965); A. D. Imms's *A General Textbook of Entomology* (London: Methuen, 10th ed., 1975); K. R. Snow's *The Arachnids: An Introduction* (New York: Columbia University Press, 1970); J. L. Cloudsley-Thompson's *Spiders, Scorpions, Centipedes, and Mites* (Oxford: Pergamon, 1958); R. D. Purchon's *The Biology of the Mollusca* (Oxford: Pergamon, 1968); and D. Nichols's *Echinoderms* (London: Hutchinson, 1962).

Similar recent accounts of minor phyla include R. Gibson's *Nemerteans* (London: Hutchinson, 1972); J. S. Ryland's *Bryozoans* (London: Hutchinson, 1970); M. J. S. Rudwick's *Living and Fossil Brachiopods* (London: Hutchinson, 1970); and A. V. Ivanov's *Pogonophora* (London: Academic Press, 1963). In addition, *The Cnidaria and Their Evolution*, edited by W. J. Rees (New York and London, Academic Press, 1966), and *Echinoderm Biology*, edited by N. Millott (New York and London: Academic Press, 1967), contain several valuable reviews and bibliographies. Certain older books remain valuable, and these include the *Handbook of the Echinoderms of the British Isles* by Th. Mortensen (London: Humphrey Milford, 1927); *The Arachnida* by T. H. Savory (London: Edward Arnold, 1935); and *A Textbook of Arthropod Anatomy* by R. E. Snodgrass (Ithaca: Cornell University Press, 1952). Dealing with more circumscribed groups are certain recent monographs, including M. W. Jepps's *The Protozoa, Sarcodina* (Edinburgh: Oliver and Boyd, 1956); J. O. Corliss's *The Ciliated Protozoa* (Oxford: Pergamon, 1961); F. S. Russell's *Medusae of the British Isles* (Cambridge: Cambridge University Press, 1955); K. H. Mann's *Leeches (Hirudinea)* (New York: Pergamon, 1962); P. E. King's *Pycnogonids* (London: Hutchinson, 1973); and T. E. Thompson's *Biology of Opisthobranch Molluscs* (London: Ray Society, 1976). Two outstanding examples of such monographs conclude this third list: *The Biology of a Marine Copepod* by S. M. Marshall and A. P. Orr (Berlin: Springer-Verlag, 2nd ed., 1972), and *British Prosobranch Molluscs* by V. Fretter and A. Graham (London: Ray Society, 1962).

Lastly, there are a number of relevant modern surveys of invertebrate physiology, of which *Comparative Animal Physiology,* by C. L. Prosser and F. A. Brown, Jr. (Philadelphia: Saunders, 1961), was justly regarded as a seminal classic. Its most recent edition is in two multiauthored volumes, *Comparative Animal Physiology:* Volume I, *Environmental Physiology,* and Volume II, *Sensory, Effector and Neuroendocrine Physiology,* edited by C. L. Prosser (Philadelphia: Saunders, 1973). E. J. W. Barrington's *Invertebrate Structure and Function* (Boston: Houghton Mifflin, 1967) covers some aspects of comparative physiology excellently and concisely. In German, W. von Buddenbrock's six-volume *Vergleichende Physiologie* (Basel: Birkhauser, 1953) is most useful. Two valuable short books for students are J. A. Ramsay's *Physiological Approach to the Lower Animals* (Cambridge: Cambridge University Press, 1952) and E. Baldwin's *An Introduction to Comparative Biochemistry* (Cambridge: Cambridge University Press, 1948). In its more circumscribed scope, T. Bullock and A. Horridge's *The Structure and Function of the Nervous System in Invertebrates* (San Francisco and London: Freeman, 1965) is extensive, detailed, and excellent. Similarly, C. M. Yonge's *The Sea Shore* (London: Collins, 1949), A. C. Hardy's *The Open Sea* (Boston: Houghton Mifflin, 1965), and J. A. C. Nicol's *The Biology of Marine Animals* (New York: Pitman, 1960) provide excellent accounts of certain aspects of physiological ecology in invertebrates. A detailed synthesis of existing knowledge on one aspect of invertebrate nutrition is provided in the *Biology of Suspension Feeders* by C. Barker Jørgensen (Oxford: Pergamon, 1966). Surveying the coelom and locomotory mechanics in invertebrates, and discussing evolution with this particular bias, is the excellent *Dynamics in Metazoan Evolution* by R. B. Clark (Oxford: Clarendon Press, 1964). For more circumscribed groups, the volumes of *Physiology of Crustacea,* edited by T. H. Waterman (New York: Academic Press, 1960 and 1961); *Physiology of Insecta,* edited by M. Rockstein (New York: Academic Press, 1965–1974); *Physiology of Mollusca,* edited by K. M. Wilbur and C. M. Yonge (New York and London: Academic Press, 1964 and 1967); and the *Physiology of Echinodermata,* edited by R. A. Boolootian (New York: Interscience, 1966), contain several useful reviews and extensive bibliographies.

INDEX

A

Abalone, 360, 369, 422–423
Acanthocephala, 22, 162, 616
Acarina (mites and ticks), 111, 290, 296, 299–300, 600–601
Acartia, 258, 572
Acinetaria (see Suctoria)
Acmaea, 370, 374, 388–391
Acochlidiacea, 348
Acoela (flatworm order), 133–135, 138, 142, 149, 617
Acoela (molluscan, = Nudibranchia), 348, 360, 365, 380–383
Acoelomate, definition, 11–13, 157–158
Acontia, 77–78
Acorn-worms (see Hemichordata)
Acrania (= Cephalochordata), 3 4, 535, 536–544, 608
Acrothoracica, 243, 267–268, 284
Acteon, 379, 380
Actinopodea, 104, 109–111, 116, 117, 574–575, 596, 597
Actinotrocha larva, 495–496, 578
Actinulae (see Hydroid morphs; Larvae)
Actinulida, 45, 49, 57–58
Adapedonta, 348, 415–417

Adductor muscles, 406–407, 413–418, 420
Aeolids (nudibranchs), 380–381
Agametes, in Mesozoa, 99
Agassiz, Louis, 613
Agnatha (vertebrate), 536, 552, 608, 620
Agricultural pests
  insects, 312, 337–338
  nematodes, 163, 397
  slugs, 349–350, 397–398
Agriolimax, 350, 382
Ahermatypic corals, 82, 87
Ainiktozoon, 552–553, 565, 608
Alcyonaria (see Octocorallia)
Alcyonium, 72–74
Alimentary canal (and nutrition)
  in Annelida, 175, 185, 209–215, 223–224
  in Arthropoda, 238, 250, 263, 267, 282, 307–308, 322, 324, 326, 333, 337, 339
  in chordates, 536–538, 540–543, 544–548, 551, 555–558, 560, 561, 563–564
  in Cnidaria (gastrovascular system), 43–45, 48, 51, 59, 63–65, 67, 71–73, 77–78, 79–81, 86

627